# VLSI Design Methodology Development

Thomas Dillinger

For information about buying this title in bulk quantities, or for special sales opportunities (which may include electronic versions; custom cover designs; and content particular to your business, training goals, marketing focus, or branding interests), please contact our corporate sales department at corpsales@pearsoned.com or (800) 382-3419.

For government sales inquiries, please contact governmentsales@pearsoned.com.
For questions about sales outside the U.S., please contact intlcs@pearson.com.

Visit us on the Web: informit.com/aw

*Library of Congress Control Number:* 2019937605


ISBN-13: 978-0-13-573241-0
ISBN-10: 0-13-573241-7

1 2019

**Acquisitions Editor**
Malobika Chakraborty

**Managing Editor**
Sandra Schroeder

**Senior Project Editor**
Lori Lyons

**Copy Editor**
Catherine D. Wilson

**Production Manager**
Abirami Ulaganathan

**Indexer**
Erika Millen

**Proofreader**
Christopher Morris

**Cover Designer**
Chuti Prasertsith

**Compositor**
codeMantra

*To Pat, for his inspiration*
*and*
*To Martha, who loved to write*

# CONTENTS AT A GLANCE

## Topic V: Electrical Analysis



## Topic VI: Preparation for Manufacturing Release and Bring-Up



Register your copy of *VLSI Design Methodology Development* at informit.com for convenient access to downloads, updates, and corrections as they become available. To start the registration process, go to informit.com/register and log in or create an account. Enter the product ISBN 9780135732410 and click Submit. Once the process is complete, you will find any available bonus content under "Registered Products."

# CONTENTS

# PREFACE

This book describes the steps associated with the design and verification of Very Large Scale Integration (VLSI) integrated circuits, collectively denoted as the *design methodology*. The focus of the text is to describe the key features and requirements of each step in the VLSI methodology. The execution of each step utilizes electronic design automation (EDA) software tools, which are invoked by a script that manages the design configuration, assembles the input data, allocates the IT job resources, and interprets the output results. The script is commonly referred to as the *flow* for the specific methodology step. This book covers both the underlying EDA tool algorithms applied and the characteristics of the related flow. Specific attention is given to the criteria used to assess the status of a design project as it progresses toward the release to fabrication.

The audience for the text is senior-level undergraduates and first-year graduate students studying microelectronics. Professional engineers will also likely find topics of interest to expand the breadth of their expertise. In many cases, the discussion of a specific step extends beyond the design engineering considerations to include the perspective of a project manager, a design automation engineer, a fabrication technology support engineer, and, to be sure, a member of the project methodology team.

It has been my experience that graduating engineers pursuing microelectronic hardware design would benefit from broad exposure to all facets of a VLSI design project and an understanding of the interdependencies between the various engineering teams. The goal of this book is to provide a comprehensive discussion of a VLSI design methodology at a level of technical detail appropriate for a two-semester, project-oriented course of study.

The book is targeted toward a discussion style of presentation rather than formal lectures. The text often highlights the trade-offs that are evaluated when selecting a specific approach for a design methodology step. An interactive discussion among the students provides an opportunity comparable to the engineering environment as part of a design team.

There are no chapter problems provided in this book. However, many universities participate in EDA vendor programs that provide access to individual software tools. This text would work extremely well in combination with such a program. After reviewing a step as a constituent of the overall design methodology, students would be able to exercise the corresponding EDA tool. Projects of larger scope could be incorporated to align with individual student interests—including flow scripting, pursuing power/performance/area evaluations, designing (cell-based) circuits, and developing methodology policies and the software utilities to verify those design standards. Projects would typically culminate in a final presentation to the class.

The text is divided into six major topics. Topic I, "Overview of VLSI Design Methodology," is rather lengthy, intended to provide background on microelectronic hardware design. Students with prior exposure to these topics could quickly review this material. The subsequent topics include Topic II, "Modeling," Topic III, "Design Validation," Topic IV, "Design Implementation," Topic V, "Electrical Analysis," and Topic VI, "Preparation for Manufacturing Release and Bring-Up." The chapters in each topic describe individual flow steps. There is admittedly some overlap in the chapter discussions. For example, the task of embedding an engineering change order (ECO) in a design database nearing release to fabrication is mentioned in multiple chapters and described in detail in Chapter 17, "ECOs." This repetition reflects the importance of the ECO methodology for a design project. Another example is the pervasive impact of lithographic multipatterning in advanced fabrication process nodes. The decomposition of the design data for a mask layer into (individually resolvable) subsets needs methodology support throughout design implementation, analysis, and physical verification flows. The influence of multipatterning is therefore described in multiple chapters.

The references provided with each chapter are rather sparse and in no way reflect the exceptional research that has enabled the complexity of current VLSI designs. The references listed are often among the landmark papers in their specific disciplines. A search for the technical papers that have recently cited these references will enable the reader to develop a more comprehensive bibliography. The website http://www.vlsidesignmethodology.com provides links to errata and additional technical publications of interest.

Several technical areas deserve greater depth than the length of this book allows. Readers are therefore encouraged to pursue the "Further Research" sections provided at the end of each chapter.

Many colleagues have provided great insights to assist with the development of this text. The collaboration over the years with Tom Lin, Mark Firstenberg, Tim Horel, and Bob Deuchars has been pivotal. The technical review recommendations from Professor Azadeh Davoodi at the University of Wisconsin–Madison have been extremely beneficial. The support from Bob Masleid, Tammy Silver, William Ruby, Charles Dancak, and Dan Nenni is greatly appreciated. Bob Lashley deserves special mention, as his expertise and inspiration have been invaluable. Finally, thanks to my family for their encouragement, especially my wife, Suzi.

**Tom Dillingor**
Livermore, California

# About the Author

**Thomas Dillinger** has more than 30 years of experience in the microelectronics industry, including semiconductor circuit design, fabrication process research, and EDA tool development. He has been responsible for the design methodology development for ASIC, SoC, and complex microprocessor chips for IBM, Sun Microsystems/Oracle, and AMD. He is the author of the book *VLSI Engineering* and has written for SemiWiki.

TOPIC I

# Overview of VLSI Design Methodology

"There is no single right way to design a chip, but there sure are many wrong ways."

—*Anonymous*

This book describes the breadth of design, validation, and analysis steps associated with the development of an integrated circuit chip. Collectively, these steps are typically described as the overall *methodology* to be applied by the engineering teams throughout a chip design project. Some steps can be pursued concurrently, whereas others require successful completion of preceding steps as a project milestone before proceeding.

The design methodology encompasses more than just the steps to be followed. It also includes the policies in effect for chip hierarchy data management and the description of the model build configuration to represent the full chip logical and physical definitions. Each instance in the overall design configuration will be associated with a revision control system, with related version identifiers.

A key element of the design methodology is the application of electronic design automation (EDA) software *tools* used for each step, plus the software employed for design data management and revision control. Each step that invokes an EDA tool commonly executes a *flow script*, which ensures that the appropriate input files are available and requests the corresponding IT resources for job submission.

The design methodology includes specification of the various logical and physical data representations of the design, specifically the persistent file storage format(s) used for model input and output prior to and after tool execution. The design methodology is heavily dependent upon additional software scripts and utilities that transform the output data of one tool to be used as input to a subsequent tool. There are de facto and IEEE-sponsored file format standards that have been promoted by the EDA industry to simplify these tool interfaces.

An increasing trend is for an EDA vendor to provide multiple tools that share a *common data model* schema to represent the logical and physical design database. The EDA vendor typically offers a software *application programming interface (API)*, consisting of a set of functions to query and update the model. The EDA vendor is also focused on optimizing the in-memory model data structures, especially for tools that execute in either a multi-threaded or distributed mode on large datasets.

An integrated EDA *platform* incorporating a data model with associated tools, an API, and a user interface for interactive steps offers considerable productivity improvements. For example, consider the methodology flow that analyzes I*R voltage drop on the supply and ground power distribution network, using estimates for the (static and dynamic) circuit currents throughout the chip (see Chapter 14, "Power Rail Voltage Drop Analysis"). The results of that analysis could be displayed in the interactive layout editor tool as a graphical overlay on the physical design, highlighting the layout coordinates where the voltage drop exceeds design margins. The layout engineer is thus able to more quickly identify where physical design updates are required. The EDA platform promotes tighter integration between tools and enables faster design optimization.

An important facet of the methodology is the management of the output result file(s) from exercising a flow. The suitability to move to the next step and the readiness to submit the final chip design for fabrication depend upon a review of the status of these results, especially any error and/or warning messages.

Methodologies are not static by any means. The general chip development project phases—design capture, functional validation, physical design, electrical analysis, and preparation for fabrication—have not changed substantially over the years. The aspects of the methodology that have evolved significantly are the specific tools and data model features used for each step, in response to new fabrication requirements and/or to improve flow throughput and designer productivity. For example, advanced fabrication processes require the photolithographic patterning of the most aggressive dimensions on the silicon wafer using multiple mask exposures per layer (see Section 1.2 in Chapter 1, "Introduction"). Thus, it is necessary to decompose the full set of layout shapes data for that layer into separate subsets that could be independently resolved with the photolithography exposure wavelength, as depicted in Figure I.1.

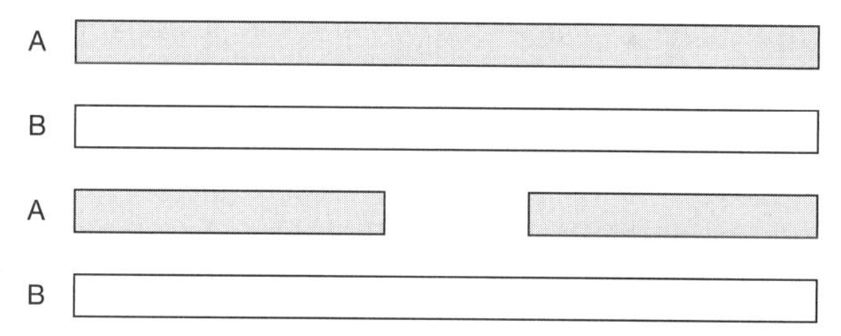


**Figure I.1** Illustration of the multipatterning decomposition requirement for shapes data on a mask layer into A and B subsets. Fabrication of the design layer is implemented with multiple lithographic patterning and etch steps. For example, metal layer M1 could be fabricated using an "LELE" sequence, with two mask layer subsets.

New tools for multipatterning decomposition needed to be developed. New flows were scripted and added to the methodology to support the mask subset assignment to satisfy the optical resolution constraints. The increasing adoption of multiple power states during chip operation is another example of a requirement to be addressed by the methodology. Different power *domains* within the chip may change between active and sleep states at different times. There was a growing need for new tools to efficiently represent and validate the logical design of the power state sequences, especially across domain interfaces; implement and verify the unique (gated) power distribution to each of the domains; and analyze the electrical response to a power state transition. As the realization of power states broadly impacts tools across logical, physical, and electrical flows, the IEEE also engaged in an initiative to establish a new standard, defining the power format description semantics for all EDA vendors to use (see Section 7.6 in Chapter 7, "Logic Synthesis"). Methodologies continue to adapt to new design and fabrication requirements.

To address designer productivity and flow throughput concerns when faced with chip designs of increasing complexity, many tools are also being re-architected. A tool may support a shared-memory multi-threaded mode (e.g., to execute on a multi-core server) or a fully distributed mode (e.g., where the model data and/or tool algorithms are partitioned across multiple servers and execute under the coordination of a master process on a host server). A methodology team is often faced with decisions related to how to incorporate a completely new flow step or insert a new tool for an existing step when preparing for a chip design project.

Another area where the design methodology is changing relates to the increasing importance of *chip-package co-design*, a collaborative effort among chip and package technology engineering teams. The teams collectively develop the connectivity of power/ground distribution and high-performance signals through the package to the chip and analyze input/output signal behavior between the chip and the overall system.

The most important characteristic of a successful design methodology is *planning*. The resources and cost expenses allocated to a chip design project are fundamentally linked to the chosen methodology. In addition to addressing new requirements imposed by the fabrication process or supporting new design features, the methodology and project plan need to be in sync relative to the estimated time required to complete each flow step. That project schedule strongly depends on the following:

- Engineering resources available
- EDA tool license quantities available
- IT resources available
- Opportunities for the *reuse* of existing (logical and physical) functionality from previous designs (often described as reuse of intellectual property [IP])
- The complexity associated with mapping existing IP functionality into new implementations to meet new specifications

These measures are linked to the hierarchical representation of the chip's logical and physical models. The complete chip design is composed of an integration of hierarchical blocks, both a logical partitioning and a physical partitioning (see Chapter 3, "Hierarchical Design Decomposition"). The validation and analysis of the full chip is typically performed in a bottom-up manner. Individual blocks are functionally validated and analyzed separately (and, ideally, concurrently if sufficient resources are available). Upon successful exit of block-level flows and promotion of the block model, full-chip integration is pursued. The chip logic model is exercised against an extensive set of simulation testcases. Electrical analysis is also pursued, potentially using block *abstracts* to reduce the data volume of the full chip model.

The design methodology needs to provide the project team with resource estimates for tool/flow runtime and compute resources required to assist with the trade-offs on hierarchical partition size and complexity. Small blocks will individually be completed more quickly but may be more logistically difficult to manage across an engineering team, and the volume of electrical abstract data to analyze at full chip is greater. Larger blocks may be more difficult to functionally validate before promotion and/or may stress the IT compute server resources available.

In addition, a significant interdependence between the methodology and design hierarchy relates to the block's clocking, power delivery, and manufacturing test strategy. Current chip designs often have numerous distinct clock domains, with synchronization circuitry at domain interfaces. As mentioned earlier, there are likely separate power supply domains. In addition to support for different power states, there will be optimum active power and performance characteristics per domain, with appropriate signal voltage level-shifting circuits at interfaces. To expedite flow setup and runtime execution, the methodology may impose guidelines (or restrictions) on how these domains relate to the block hierarchy:

- Where distinct clocks are sourced in the chip model
- Where signal interfaces at *clock domain crossings* should reside in the hierarchy (see Figure I.2)
- How power domains and domain isolation are modeled
- Where manufacturing test control signals are inserted into the model

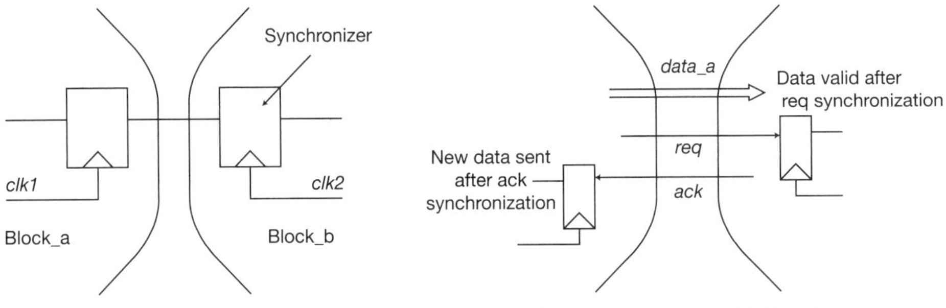


**Figure I.2** Examples of clock domain crossing signal interfaces between block boundaries. Specific synchronizer flop cells are used at clock domain crossings, typically residing at hierarchical block inputs.

The simplest block-level methodology would restrict the partitioning to a single clock and a single power domain; for example, no generated clocks within the block and all required synchronization circuitry at block inputs. Yet, the overall chip design hierarchy may not easily fit this restriction, necessitating a more robust methodology that can verify intra-block circuit timing paths sourced and captured by multiple clocks.

There are two key considerations, when finalizing the methodology and design project plan:

- What methodology guidelines are imposed for correspondence points between the logical and physical design hierarchies?
- How is the inter-block *glue logic* to be managed by the methodology?

## I.1 Methodology Guidelines for Logical and Physical Design Hierarchy Correspondence

Two separate hierarchies are maintained throughout the design project: a logical hierarchy developed for functional validation and a physical hierarchy of circuits and layout data that implement the logic, as illustrated in Figure I.3. These two models evolve (somewhat) independently.

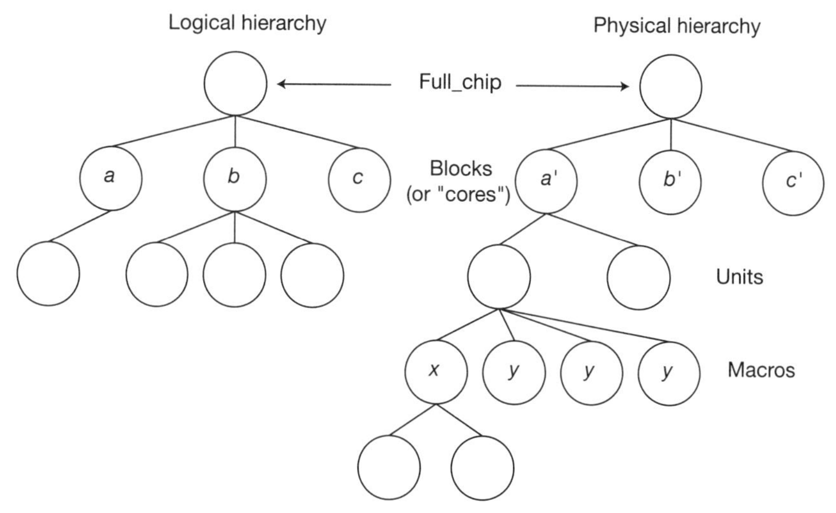


**Figure I.3** Logical and physical hierarchical models for an SoC design. Nodes *a* and *a′* are examples of "correspondence points" between the two hierarchies. Physical design reuse is a key productivity goal.

The logical model relates to the optimum functional blocks for sub-chip validation. For full-chip logic simulation, testcase development seeks to exercise block interfaces and interrogate the values of internal register and memory array values. The stability of these control/observe signal definitions in the logic hierarchy is paramount to expediting testcase preparation and maintenance. A relatively shallow logical hierarchy is common, to keep the number of sub-chip validation tasks in check.

Conversely, the physical hierarchy is typically much deeper. Clock and power domains strongly influence the physical definition. In addition, there is a goal to leverage any opportunities for circuit and layout design reuse due to the intensive engineering resources needed for a new *custom* implementation. As a result, the physical design typically consists of a hierarchy of *cells* instanced in *macros*, *units*, and *blocks*. (These are common terms for the levels of the physical hierarchy; a microprocessor- or microcontroller-level block is also typically denoted as a *core*.)

The project team must define the correspondence points between the two hierarchies to ensure that the final chip design is fully consistent. The choice of correspondence points requires an intricate trade-off. Deferring logical-to-physical equivalence checking to the full chip model introduces a significant risk late in the project schedule. Requiring correspondence deep in the hierarchies may be more expeditious but may also impose undue influence on how the logical/physical signal interfaces are defined and maintained. Ideally, the physical implementation should have minimum impact on the definition of logical functionality and vice versa. The typical project decision is to enforce and verify correspondence at the block (or core) level.

As an aside, an increasingly important design flow utilizes *synthesis* of a logic block to an equivalent model of interconnected basic logic functions from a *cell library* (see Chapter 7). Each library function has a physical implementation that has been previously qualified. These physical layouts are then automatically placed and interconnect routes generated from the synthesized *netlist*. This automated flow effectively generates correct logical-to-physical correspondence for the block.

## I.2 Managing Inter-Block Glue Logic

Inevitably, logic functionality and physical circuitry reside between the hierarchical blocks. The most common example is the need to insert signal repowering buffers or sequential circuit repeaters between the output of one block and the input to another (see Figure I.4).

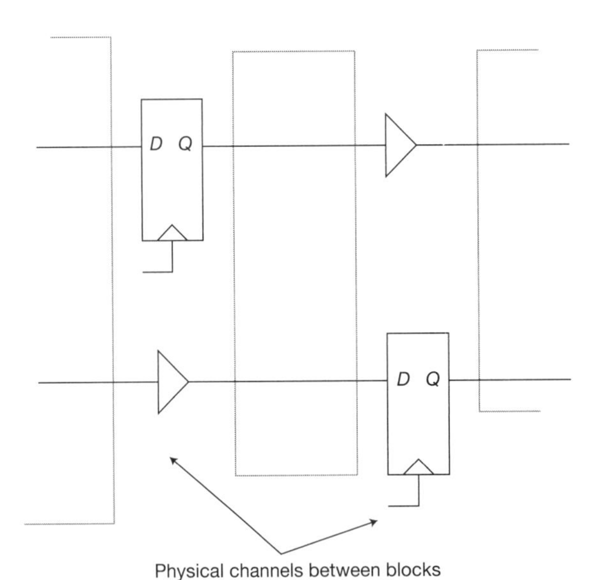


**Figure I.4** Functionality present between blocks is typically denoted as "glue logic." The figure illustrates both signal repowering (buffering) and insertion of a sequential "repeater" for a global signal between blocks.

These circuits prevent poor electrical characteristics on a signal transition due to the effects of a long signal wire or, in the case of a sequential repeater, to maintain synchronous timing between blocks. The cycle latency of a repeated signal is part of the chip functional model and needs to be validated. The physical circuits for the glue logic require allocated chip area between blocks, corresponding signal *wiring tracks*, and connectivity to power/ground (see Section 8.1 in Chapter 8, "Placement"). Another example of glue logic would be any manufacturing test multiplexing logic to toggle between functional operation and specific test modes for individual blocks.

The full chip methodology is likely to impose restrictions on how these glue functions are represented. It may be suitable (albeit perhaps unwieldy) to simply add these functions to the top of the chip hierarchy as a large collection of flat cell instances; however, these cells will be physically scattered, necessitating electrical analysis with these cells merged with all block abstracts. Alternatively, cells located in a physical *channel* between blocks could be grouped into a single hierarchical instance in the full chip model and analyzed as just another block. However, this second approach directly ties the physical implementation to the logical model of the chip functionality. As the physical design evolves, the inter-block signals buffered and/or repeated through a specific channel will likely change. With this alternative approach, the input/output pins and internal signal names within these glue blocks will be in flux, disrupting the logical validation testcases that may refer to these (hierarchically qualified) signals.

The most aggressive approach for managing global glue logic cells would be to allow them to be placed and connected within the physical boundary of a design block, perhaps to provide a more optimum buffer/repeater cell location along the length of the global connection. This "channel-less" design methodology is depicted in Figure I.5.

This would require the most intricate methodology for maintaining logical and physical models. Blocks would need to be analyzed, including the electrical and thermal influence of the embedded global circuits, introducing interdependence between block and full-chip design closure in the methodology.

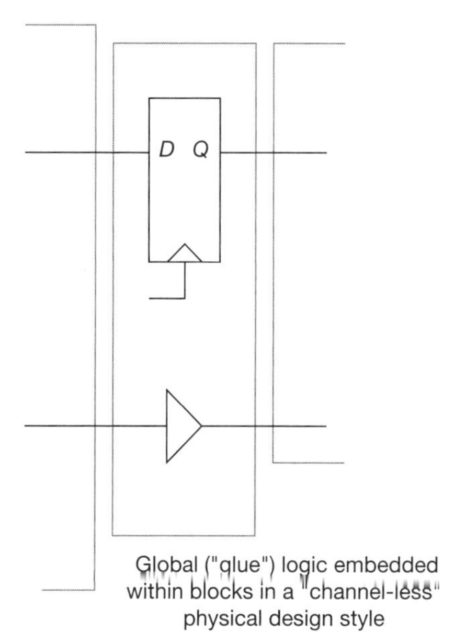


**Figure I.5** Physical placement of glue logic within blocks in a "channel-less" design methodology.

As mentioned at the beginning of this overview discussion, no single set of methodology flows and data management policies is best suited to all chip design projects. Cost, resource, and schedule constraints will necessitate trade-off decisions on what project milestones are established. Checkpoints will be defined that gate the integration of individual logical and physical blocks to the full-chip model, with criteria for the sufficiency of block-level functional validation and electrical analysis. This text describes the characteristics of individual flow steps, focusing on the key features and options that are pertinent to the methodology choices to be made.

One additional cautionary note is worth highlighting: As a new chip product proposal is being evaluated for funding and assignment of engineering resources, so must the methodology be incorporated into the review, including related EDA software and IT cost estimates. It is tempting to begin the design activity to meet project schedules, deferring any open methodology issues until nearer the project milestone associated with the issue. However, such a strategy simply shifts the schedule pressure to the methodology team, which must then ensure delivery of EDA tools and flows with the requisite functionality (and testing) on a project-critical deadline. Commencing a design with major open flow requirements will almost assuredly result in project delays and unforeseen resource impacts.

A successful chip project employs a well-defined, documented methodology that has been finalized prior to (or at least during) the project kickoff phase. Ideally, a trial design could be accelerated through block-level flows. This *trailblazing* design would be representative of the clocking, power delivery, and test micro-architecture prevalent throughout the full chip so that any lacking tool/flow functionality could be identified and promptly addressed. An early collaboration among the methodology team, the chip design lead engineers, the chip project manager(s), the EDA tool/flow support team, and the IT resource management staff on an appropriate methodology will result in more predictable project execution and milestone completion.

CHAPTER 1

# Introduction

## 1.1 Definitions

### 1.1.1 Very-Large-Scale Integration (VLSI)

The amount of functionality that can be integrated on a single chip in current fabrication processes is associated with the term very-large-scale integration (VLSI). Preceding chip design generations were known as small-scale integration (SSI), medium-scale integration (MSI), and large-scale integration (LSI). Actually, *VLSI* has been used to refer to a succession of several manufacturing process technologies, each providing improved transistor and interconnect dimension scaling. A proposal was made to refer to current designs as ultra-large-scale integration, but the term ULSI has never really gained traction.

The complexity of a new VLSI design project is used to estimate the engineering resources required. However, it is not particularly straightforward to provide a single metric for describing the complexity of a VLSI design and its associated manufacturing technology. The previous fabrication technology generations were predominately dedicated to either digital or analog circuitry. For digital designs, the number of *equivalent NAND logic gates* was typically applied as the measure of design complexity, and the logic density, measured in *gates/mm**2*, was the corresponding technology characteristic. The succession of VLSI manufacturing processes has enabled a much richer diversity of digital, analog, and memory functions to be integrated, making a logic gate estimate less applicable to the design and making gates/mm**2 less representative of the process. The embedded memory technologies available from the fabrication process have a major impact on the physical complexity of the chip design, with options for volatile (e.g., SRAM) and non-volatile (e.g., flash) array circuits. For designs with significant on-chip memory requirements, the specific memory cell implementation is a strong factor in estimating the final chip layout area (and, thus, cost). The number of metal interconnect layers available in the VLSI process—especially the electrical resistance and capacitance of each layer—also has a tremendous influence on the engineering resources needed to achieve design targets.

In summary, the complexity of a VLSI design is not easy to define when estimating project difficulty. Comparisons to the resources needed for previous designs are often insightful if a comparable methodology was used. As will be discussed shortly, the ability to reuse existing logical and physical intellectual property is a major consideration when preparing a project plan.

### 1.1.2 Power, Performance, and Area (PPA)

The high-level objectives for any new design are typically the power, performance, and full-chip area, typically referred to as the "PPA targets."

The chip power dissipation target is an integral specification for any product market, from the battery life of a mobile application to the electrical delivery (and cooling) for racks of equipment in a data center. The methodology includes a power calculation flow; actually, multiple flows are used to calculate a power dissipation value. Initially, signal switching activity from functional simulation testcases provides a (relative) measure of active power dissipation, using estimates for signal loading. Subsequently, a more detailed measure is calculated using the physical circuit implementation with resistive losses. Multiple modes of chip operation may require power dissipation data as well. For example, a standby/sleep mode requires a calculation of (inactive) transistor leakage currents rather than the active power associated with switching transients.

The performance targets of a design may involve many separate specifications, and analyzing each of them requires methodology flow steps. The most common performance target is the clock frequency (or, inversely, the clock period) applied to the functional logic timing paths between flops receiving a common clock domain signal. Logic path evaluation must complete in the allocated time period. In addition, a myriad of other (internal and external) design performance specifications are covered by the timing analysis flow, including the following:

- (Maximum) clock distribution arrival skew between different flops
- (Minimum) test logic clock frequency for the application of test patterns
- Chip output signal timing constraints, relative to either an internally sourced clock sent with the data or an applied external clock reference

The area of the chip design is a factor in determining the final product cost. The larger the area, the fewer die sites available on the fabricated silicon wafer. A larger die likely requires a more expensive chip package, as well. The manufacturing production yield is strongly dependent upon the die area due to the probabilistic nature of an (irreparable) defect present on a die site. The accuracy of the chip area estimate prepared as part of the initial project proposal is crucial to achieving production cost targets. The methodology does not have a direct influence on this estimate, perhaps, but it may still play an important role. As mentioned earlier, a trial block design exercise may be useful to test the flow steps and provide data for representative block area estimates. It will also provide insights into the number and type of interconnect layers needed to complete block signal routes and achieve performance targets. The specific *metallization stack* selected for the block and global signal routes will impact fabrication costs, with each metal layer providing particular signal delay characteristics and signal route capacity (measured in wiring tracks-per-micron on the preferred route direction for the layer).

### 1.1.3 Application-Specific Integrated Circuits (ASICs)

Electronic products were traditionally designed with commodity part numbers, using SSI, MSI, LSI, microprocessor, memory, and analog (including radio-frequency) packages on printed circuit boards (PCBs). Unique fabrication process technologies were introduced to add commodity programmable logic and programmable (non-volatile) memory parts to provide greater product differentiation. With the transition to VLSI chip designs and processes, a number of factors led to the development of entirely new design methodologies:

- The markets for electronic products grew tremendously and became much more diverse. Mobile products required intense focus on reducing power dissipation. Product enclosures pursued unique form factors, minimizing the overall volume. Both trends necessitated maximizing the integration of digital, memory, and analog functions. (Packaging technologies were also driven toward much higher pin counts and areal pin densities for PCB attach.)
- Product differentiation became increasingly important to capture market share from competitors.
- The advances in EDA tool algorithms were being transferred from academia and corporate research labs to the broader microelectronics industry, providing vastly improved productivity for designers to represent and validate functionality and implement and analyze the physical layout. Hardware description language semantics were introduced to allow functionality to be described at a much higher level of abstraction than the Boolean logic gate schematics used for MSI/LSI designs. For physical design, algorithms for automated circuit placement and signal routing were introduced. The circuits to be placed were selected from a previously released *cell library*, whose individual layouts had been verified to their logical equivalent and whose circuit delays had been characterized. The complexity of these library elements typically covered a wide range of SSI and MSI logic functions, small memory arrays and register files, input receiver and output driver pad circuits, and (potentially) simple analog blocks (e.g., phased-lock loop, ADC/DAC data converters).

These factors led to the adoption of a new set of products known as application-specific integrated circuits (ASICs).

A number of companies began to offer a full suite of ASIC services, such as developing cell libraries, releasing EDA tool software suites, assisting with PPA estimation, performing cell placement and routing, generating test patterns, releasing product data to manufacturing, executing package assembly and final test, and completing product qualification. Design teams were to follow the ASIC company's documented methodology. The handoff from design to services consisted of the final logical description, including a netlist of interconnected cells plus performance targets. The detailed netlist was synthesized from the hardware description language functional model, either manually or with the aid of logic synthesis algorithms. In return, the ASIC company provided an upfront services and production cost quote for completing the physical implementation and release to manufacturing, simplifying the budgetary planning overall.

There were still many integrated circuits (ICs) developed using a full custom methodology, where all functional blocks consisted of unique logic circuits and manually drawn physical layouts, without utilizing automated placement and routing of library cells. This custom methodology was used only for high-volume parts, where the intensive engineering resources required could be amortized over the sales volume. (As ASIC designs typically were between commodity LSI parts and custom VLSI designs in terms of complexity, with functionality implemented using an existing cell library, these designs were also referred to as *semi-custom*.)

### 1.1.4 System-on-a-Chip (SoC)

As VLSI process technologies continued to evolve, offering increasingly scaled transistor and interconnect dimensions, there were several shifts in the ASIC market:

- The complexity of physical library content required to meet application requirements grew—for example, larger and more diverse memory types (caches, tertiary arrays); programmable fuse arrays; high-speed external serial interfaces with serializer/deserializer (SerDes) physical units; and, especially, larger functional blocks (microcontrollers and processor cores for industry standard architectures).
- Design teams were seeking multiple, competitive sources of these new library requirements. Rather than relying solely on the library available from the ASIC services provider, design teams sought sourcing of IP from other suppliers. This initiative was either out of financial interests (specifically, lower licensing costs) or technical necessity (e.g., if the ASIC provider did not offer a suitable PPA solution for the IP required).

- Design teams sought to be more directly involved in the selection of the silicon fabrication and package assembly/test suppliers. Much as with the negotiations with IP suppliers for library features, design teams were willing to invest additional internal resources to investigate and select production sources. The goal was to more closely manage operational costs and balance supply/demand order forecasts rather than to work through the ASIC services company interface.
- EDA software tools began to be marketed independently from the ASIC services company, allowing design teams to develop their own internal methodologies. The breadth of responsibilities for the internal EDA tool support team grew. The design engineers providing (part-time) EDA software support for the ASIC design methodology were consolidated into a single "CAD department." This new and larger organization was established with the mission to provide comprehensive EDA software support for the internal methodology to all design teams. Comparable tools from different EDA vendors were subjected to benchmark evaluations to ensure suitability with the proposed methodology, confirm PPA results on representative blocks, and measure the tool runtime and IT resources required. These benchmarks also helped the CAD team assess the task of developing flow scripts and utilities around the tool to integrate into the methodology. Software licensing costs from the EDA tool vendor were also a major factor in the final competitive benchmark recommendations.

These shifts in the ASIC market resulted in the growth of several new semiconductor businesses and organizations:

- **Semiconductor foundries**—These companies offer silicon fabrication support to customers submitting designs that have been verified using their *process design kit* (*PDK*) of manufacturing layout design rules and transistor/interconnect electrical models.
- **Outsourced assembly and test vendors (OSATs)**—These specialty companies provide a broad set of services, including good die separation from the silicon wafer, die-to-package assembly, and final product test/qualification.
- **EDA tool vendors**—The EDA vendors vary in terms of their software tool offerings. Some focus on point tool applications for a specific meth-

odology flow step. The larger firms have worked to integrate multiple tools into an integrated platform, encompassing multiple flow steps.

- **Industry standards organizations**—To facilitate the use of EDA tools from multiple suppliers in a design methodology, a set of industry-standard data formats were developed for tool input/output at various common step interfaces. For high-impact flow steps, the Institute of Electrical and Electronics Engineers (IEEE) has taken ownership of these evolving standards. For example, the syntax and semantics of the hardware description languages Verilog (and its successor, SystemVerilog) and the VHSIC Hardware Description Language (VHDL) are IEEE standards, subject to periodic updates. The format for representing the electrical characteristics of signal interconnects extracted from a physical circuit layout is the Standard Parasitic Exchange Format (SPEF), another IEEE standard. Other data format standards are maintained by industry consortia, with approval committees comprised of member company representatives. The Open Artwork System Interchange Standard (OASIS) definition for the file-based representation of physical layout data is maintained by the Semiconductor Equipment and Materials International (SEMI) organization.
- **De facto standards**—There are some de facto data representation standards that are not formally approved and maintained. Their use became so widespread, often because of market-leading tools, that most EDA vendors support the adoption of the format to prevent their own products from being at a disadvantage to integrate into a flow step. The Simulation Program with Integrated Circuit Emphasis (SPICE) circuit simulation program reads in a netlist of interconnected transistor models and electrical primitives. After the transient circuit simulation completes, the signal waveform results are written. Both the SPICE netlist and output waveform file formats are de facto standards. The predecessor format to OASIS for layout description, the Graphic Database System (GDS)—along with its long-used successor, GDS-II—is another de facto standard. The Fast Signal Database (FSDB) format is a de facto standard for representing functional logic simulation results.

When evaluating a tool to integrate in the methodology flow, specific attention must be given to the consistency of the input and output file formats between steps. Data translation utilities added into a flow introduce the risk of data integrity loss if the utility does not recognize—or, worse, misinterprets—a data file record. This is especially important for de facto data formats. (It is also crucial to examine an EDA vendor's tool errata documentation for any semantics included in a standard specification that are not fully supported to determine the potential impact to the design team.)

- **Intellectual property suppliers**—The ASIC shift is perhaps best illustrated by the importance of companies that have focused on the development of complex intellectual property licensed to design companies for integration. Design teams (and/or their ASIC services provider) may not have the expertise or financial and schedule resources to develop all the diverse functionality required for an upcoming project. A separate IP vendor, often working in close collaboration with the foundries during initial development of a new process technology, could fabricate and qualify a large IP block. The emergence of standard (external) bus interface definitions (e.g., DDRx, PCIe, USB, InfiniBand) has accelerated this IP adoption method. Design teams need not be experts in the functional and electrical details of these protocols but can leverage the availability of existing IP for reuse. (Note that this discussion refers to IP functions that include a physical implementation, in a specific foundry's process technology; there are also functional model-only IP offerings, described shortly.)

Several EDA vendors have also recently expanded their product offerings to include IP libraries. There is a natural synergy between advanced EDA tool development and IP design for a new fabrication process. The EDA team collaborates with the foundry during process development to identify new tool features that are required. The IP team also has an early collaboration with the foundry to prepare designs for fabrication and testing so that the IP is available for leading customers when the process qualification is complete. As part of the EDA vendor engineering staff, the IP team also helps test and qualify the new software (with software licenses for free, to be sure). This synergistic relationship between tool and IP development is proving to be financially successful for the EDA vendor and also to improve the quality of new EDA software version releases.

For similar reasons, the foundries are also increasingly offering IP libraries. The foundries have internal process development teams focused on specific circuit requirements (e.g., device characterization and reliability, memory bit cell technology, fuse programmability). As with the EDA vendors, the foundries are also investing in enhancing internal circuit design expertise. It is a natural extension of these process development areas for the foundry to offer customers IP libraries consisting of diverse sets of functionality, including a base cell library, I/O pad circuits, memory (and fuse) arrays, and so on. The foundry may offer the IP library at an aggressive licensing price as a means of attracting more customer "design wins" to secure the corresponding manufacturing volume and revenue.

In summary, VLSI designs have evolved in complexity from the earlier ASIC approach, both in terms of the diversity of integrated IP and the unique methodologies applied by the design team. A description that better describes the current class of VLSI designs is *system-on-a-chip* (*SoC*).

## 1.2 Intellectual Property (IP) Models

SoC chip designs incorporate a range of physical IP components of varying functional and PPA complexities.

### 1.2.1 Standard Cells

Basic logic functions are collected into a library of physical layout cells with corresponding electrical models. Design blocks are implemented by the placement and routing of these cells to match the description of a logical netlist. The layout of these standard cells adheres to a *template* applied throughout the cell library. Specifically, the template defines:

- The position and width of (lower-level) metal layer power and ground rails
- The number of lower-level metal wiring tracks available for signal connections within the cell implementation
- Valid locations for pins to connect to the cells with signal routes

The cell layouts are thus constrained in one dimension (e.g., the vertical dimension in Figure 1.1) and typically span a variable width in the other dimension to complete the layout connections. In Figure 1.1, note that the standard cell template employs a shared power and ground rail design. The orientation of alternating rows of cells would be flipped in the vertical direction to share rail connections with the adjacent row. The width of the rails in the template would be designed to provide a low-resistive voltage drop for the current provided to both adjacent cell rows.

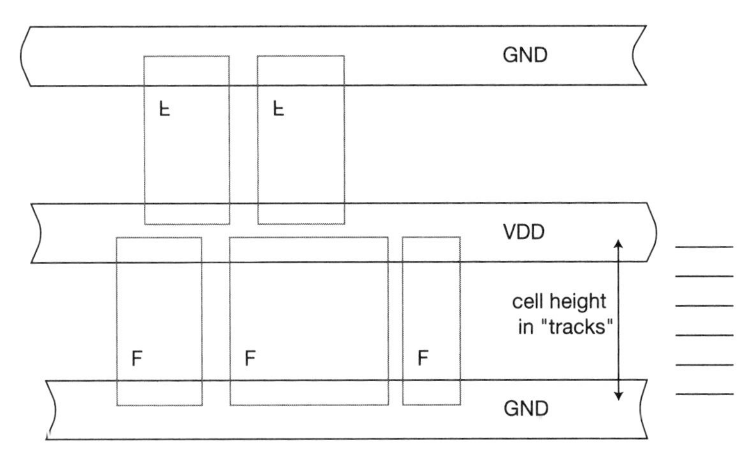


**Figure 1.1** Standard cell layout template example, illustrating shared power and ground rails. Logic cell layouts are one (perhaps two) template rows tall and of variable width. The template cell height spans an integral number of horizontal wiring tracks.

The standard cell library content is thus limited to the logic functions that can be successfully connected within the wiring track limit. Typically, flip-flop circuits are the most difficult to complete; their layouts often define the allocation of intra-cell tracks in the template.

Given the wiring track constraint, some cell layouts may add connections on an upper metal layer, which is typically used for inter-cell routes. In these cases, the placement and routing methodology needs to accept a partial *route blockage map*, honoring the unavailability of these tracks for signal routes due to the cell layouts. Some libraries may also include "two-high" cells, under the assumption that the placement flow can accommodate both one-high and two-high dimensions; this enables a richer set of library logic cells to be available.

#### *Cell Drive Strengths*

To assist with PPA optimization during physical implementation, each logic function in the standard cell library is likely to be offered in multiple *drive strengths*. For example, a function built with minimum-sized transistors may be denoted as 1X. Increasing transistor sizes may enable 2X, 4X, and so on alternatives, as illustrated in Figure 1.2. The template height constraint limits the transistor dimensions in the physical layout; larger drive strength variants need multiple transistor fingers connected in parallel, spanning a greater cell width. (A *transistor finger* is an individual device with common drain, source, gate, and substrate connections to the other fingers connected in parallel. The total device width is the sum of the individual finger widths. Section 10.2 discusses layout proximity effects, where the surrounding layout topology impacts the device performance. In this case, the device current model for the individual fingers varies, necessitating expansion of the total device width into the individual fingers for detailed circuit analysis.)

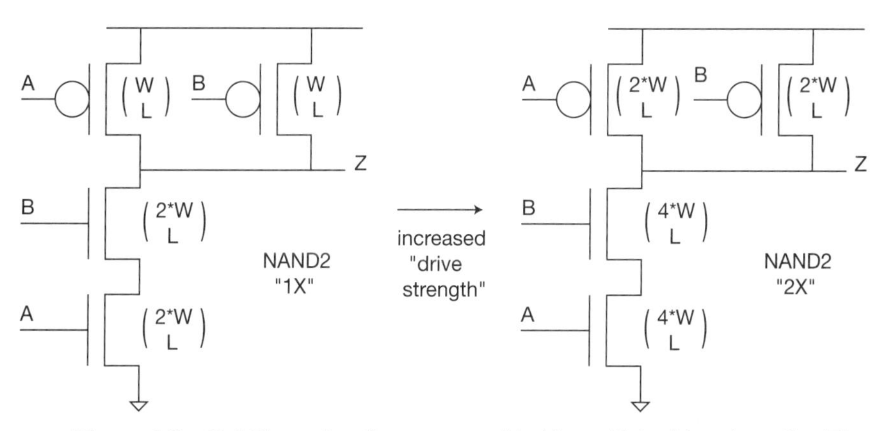


**Figure 1.2** Cell library functions are provided in multiple drive strengths. The wider devices used for higher drive strengths may require a layout using parallel fingers. The pin input capacitance is the sum of all device input capacitances, each of which is proportional to the product of device width and length.

The cell layouts must all observe a minimum *half-rule space* within the template boundary to allow adjacent cell placements to satisfy the process lithography requirements. Although the output impedance of a higher-drive-strength cell is reduced, providing improved signal transition delays, the capacitive loading presented to the circuits sourcing the input pins is increased, adversely impacting their performance and power dissipation. Performance optimization requires an incremental path-oriented algorithm for cell drive strength selection to evaluate the relative trade-offs between improved drive strength and increased loading for cells and interconnects in the path.

### Cell Threshold Voltage Variants

Another cell library option is to offer the logic functions with circuit variants using different transistor threshold voltages available in the fabrication process technology. For example, a single drive strength of a cell could be available using standard $V_t$ (SVT), high $V_t$ (HVT), and low $V_t$ (LVT) transistors. The lithographic layout rules from the foundry typically facilitate placing SVT, LVT, and HVT transistors in close proximity; that is, SVT, LVT, and HVT cells could be placed adjacent to each other. (In the most advanced process nodes, there are new lithographic rules impacting the adjacency of different device threshold types, necessitating additional spacing between cell threshold variants.)

The circuit transition delay for an LVT cell is improved over its SVT equivalent, without the cell area and input capacitive loading increase associated with a higher-drive-strength version. The trade-off is that the LVT cell static leakage current power dissipation is greatly increased. Whereas a 2X drive strength transistor size effectively doubles the leakage current over 1X, the leakage current for an SVT-to-LVT cell exchange increases exponentially. (If a circuit timing delay path is far less than the allocated clock period, the positive *timing slack* could be applied to the substitution of HVT cells for SVT equivalents, reducing leakage power substantially.) A visual representation of this trade-off is provided by the $I_{on}$-versus-$I_{off}$ curve from the foundry shown in Figure 1.3, which illustrates the process target for transistor on-current drive strength (in saturation mode) versus the off-leakage current, for a reference transistor size.

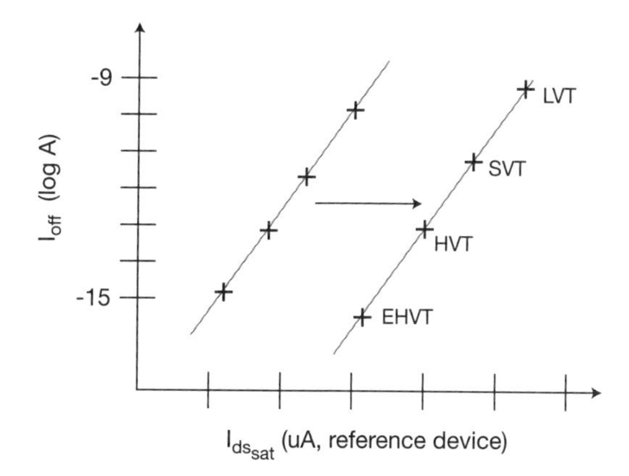


**Figure 1.3** Example of an $I_{on}$-versus-$I_{off}$ curve from the foundry, illustrating the process target for device currents with different $V_t$ threshold voltages (in saturation mode, for a reference device size). The curve allows a (rough) comparison between processes. The "device current gain at constant leakage power" is depicted for two process targets (with scaled reference device sizes).

Note that the graph axes in Figure 1.3 are log-linear, highlighting the exponential dependence of $I_{off}$ on transistor $V_t$. The line connecting the LVT, SVT, and HVT points is artificial; it simply illustrates the linear $I_{on}$ and exponential $I_{off}$ dependence on $V_t$. There are no transistor design offerings along the curve; they are strictly at the different $V_t$ fabrication options. Nevertheless, the curve is extremely informative, in case the foundry and its customers opt to evaluate an investment in another transistor $V_t$ offering for specific PPA requirements (e.g., extra high $V_t$ [EHVT]). Such a curve is also often used to compare successive process generations. The horizontal distance between the curves for two processes is a benchmark for *performance improvement at constant leakage power* (for the reference size transistor in the two processes). The vertical distance estimates *the leakage power improvement at constant performance*. It should be highlighted that these are transistor-specific data points; to realize these gains between processes, there are also scaling assumptions on interconnect lengths, interconnect R*C parasitics, and power supply values that need to be evaluated. The selection of the optimum combination of drive strengths and $V_t$ offerings for millions of cells in a netlist for PPA optimization is perhaps the most complex set of EDA tool algorithms.

### 1.2.2 General-Purpose I/Os (GPIOs)

The base cell library also includes input pad receivers and output pad drivers (not integrated into a specific physical IP interface block). These general-purpose input/output (GPIO) cells have numerous variants with different characteristics, such as:

- $V_{in}$ and $V_{out}$ logic threshold levels ($V_{out}$ measured with a specific DC load current)
- Output driver impedance, for matching to the package impedance to minimize transmission line reflections
- Bidirectional driver/receiver circuits (with high-impedance driver enable control), in addition to unidirectional drivers
- Receivers with hysteresis feedback to improve noise immunity
- Electrostatic discharge (ESD) protection (see Section 16.3)

The I/O cells have their own layout template, which is much larger than the standard cell template. These circuits incorporate very-high-output drive strengths with many parallel transistor fingers and much larger power/ground rails to deliver greater currents at low voltage drop. In addition, the voltage levels connected to these GPIO circuits are typically distinct from the internal IP voltages. For example, external interface voltages may be in the range of 1.2 to 1.8V, while the internal circuitry for current process nodes operates below 1V. The GPIO circuits may also use a different set of *thick gate oxide dielectric* field-effect transistors to support exposure to higher voltages, with additional layout constraints. Although some methodologies support the placement of GPIO cells scattered throughout the die area, the majority of methodologies allocate the GPIO templates (and related I/O power/ground rails) strictly to the die perimeter.

### 1.2.3 Macro-cells

The elementary logic circuits in the standard cell library are the components for the physical implementation of larger functional designs. However, additional productivity in generating the logic netlist could be achieved if more complex logic cells were available. A larger set of PPA optimizations would likely be enabled as well. For example, an n-bit register consisting of n flops sharing a common clock is a typical complex logical design element in a library. An n-bit multiplexer sharing common select signal(s) is also common. An n-bit adder is another, with a performance dependence on the specific adder implementation. In all these cases, there is a goal to minimize the delay of a critical timing path by reducing the loading on a high fan-out input connection or optimizing internal (multi-stage) logic circuit delays.

Two types of offerings have been added to cell libraries for these more complex *macro-cells*: a custom physical implementation and a logic-only representation. As mentioned earlier, physical design methodologies are often required to support larger cell dimensions than a one-high template row standard cell layout. Macro layouts can be added to the library if the methodology supports more varied physical cell dimensions. Within the layout, special focus would be given to ensuring low skew of high fan-out inputs and short delays for critical paths within the macro. (There is also the decision of whether the macro layout would be given the flexibility to discard the template power/ground rail definitions from the standard cell library and employ a unique rail design, with continuity only present at the macro edges for abutted cells.) However, offering these additional custom physical layout components would add to the task of library timing model characterization and electrical analysis.

An alternative would be to add a level of hierarchy to the cell library; in this case, a *logical macro* would be defined, and it would consist of a specific netlist of base standard cells to implement the function. Figure 1.4 depicts an example of an n-bit multiplexor, consisting of a hierarchy of n single-bit mux cells. Buffer cells could also be judiciously added to the netlist for high fan-out internal signals and, potentially, macro outputs.

The unique feature of the logical macro would be the inclusion of *relative placement* directives with the library model. The placement flow would encounter the logical macro in the netlist, expand to its primitive cell netlist, and apply the relative placement description to treat the constituent cells as a single component that would be moved in unison during placement optimization.

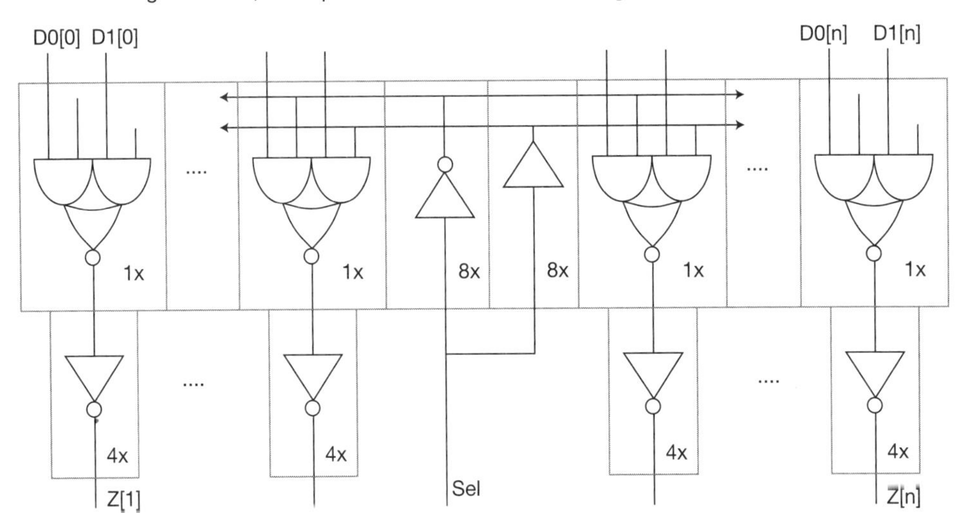


**Figure 1.4** A logical macro defines a complex function composed of existing standard cells. An n-bit 2:1 multiplexor is depicted. Critical signals within the macro definition would receive specific buffering. Relative cell placements optimize the routing track demand and interconnect loading within the macro.

There are several advantages of using logical macros, including the following:

- No additional characterization and electrical analysis of the power and timing delays of a complex function are required. The final netlist of expanded (characterized) cell primitives is used.
- Subsequent cell-based optimizations are available. Once the logical macro is part of the physical implementation, cell-based power and performance optimizations are applied, such as the substitution of HVT cells to save leakage power. The drive strength of macro output cells could be optimized for the actual loading in the specific context of the larger physical block layout. (The cell dimensions could vary during optimization; the placement algorithm would resolve any cell overlaps while maintaining the relative positioning.)
- The relative cell placement should (ideally) guide the router to short connections for critical signals. With alignment of cell pin locations resulting from the relative placement, simple and direct route segments are likely. The risk of a circuitous connection on a critical macro signal can

be further reduced by utilizing router features that accept a signal priority and/or preferred route layer setting for specific signals; these router directives could also be added to the logical macro library model.
- Layout engineering resources required for library development are reduced. It is typically much easier to develop and verify the logical macro (and relative placement and router directives) than to prepare a custom circuit schematic and layout for the function.

The primary disadvantage of the logical macro approach is that the resulting implementation typically has less optimal PPA characteristics than a custom design. Whereas there is a clear granularity in logic cell drive strengths in the library (e.g., 1X, 2X, 4X), a custom schematic design could utilize greater variability, tuning the drive strength optimally for the loading within the custom macro layout. Also, due to the discrete size of the template in terms of wiring tracks, individual cell layouts may not fully utilize the area of a single-row 1xN (or double-row 2xN) template. A custom layout would likely be more area efficient.

These are complex trade-offs that need to be made by the library development team. The decisions on these trade-offs also have a strong interdependence with the features of the physical flows in the design methodology.

### 1.2.4 Hard, Soft, and Firm IP Cores

The evolution of SoC design integration has been enabled by the availability of complex IP functions that design teams can quickly integrate. The most prevalent examples are large memory arrays; analog blocks for high-speed external signal interfaces; and (especially) microprocessor, microcontroller, and digital signal processing cores. More recent complex IP examples include cryptographic units, graphics/image processing units (GPUs), and on-chip bus arbitration logic for managing communication between cores, potentially using advanced network communication protocols. (The IP core is also commonly referred to as a specific application "unit," as in GPU or CPU.)

These IP designs are typically not developed by the design team but are licensed from external suppliers, each of which offers expertise in a specific microprocessor architecture or high-speed interface design standard.

Complex IP may be provided in different formats that have been given a unique terminology. The following sections discuss the various terms.

### Hard IP

A *hard IP* offering is a large, custom physical layout implemented and qualified in a specific foundry technology. The IP vendor has developed the implementation to its own PPA targets, which hopefully align with customer requirements. The IP layout typically fully utilizes internal signal interconnect routes on several metal layers—more than for the standard cell library and macros. The IP vendor has also made an assumption on the appropriate metallization stack choices for the entire SoC, up to the top metal layer used in the hard IP layout. The physical integration of this layout in an SoC design involves providing global power distribution and signal routing to the layout pins (on the corresponding pin metal layers and route wire widths) that satisfy the vendor's specifications.

The vendor provides a functional model(s) for the IP to use in SoC validation. For microprocessor or microcontroller core IP, a software development kit (SDK) is included for the generation of system firmware. Functional simulation *testbench suites* may also be provided to the SoC team to assist with validation of the processor bus arbitration logic with other units. The IP vendor will also have implemented a specific manufacturing test strategy for the core. A special test mode may be defined to allow the core to be functionally isolated from the remainder of the SoC, to enable the vendor's test patterns to be applied and responses observed. Indeed, an industry-standard *core wrap* architecture has been developed to support isolating a core in an SoC for production testing.[1,2]

The hard IP release provides the vendor with the greatest degree of protection of proprietary information. (Although license agreements include strict policies on unauthorized disclosure of the vendor's model data, IP leaks are nevertheless a major concern for the vendor.) A special feature that has been incorporated in EDA vendor functional simulation tools is the capability to link a binary executable when building the full SoC model. The IP vendor may thus provide the hard IP simulation model as a compiled executable rather than as a source logical description. A unique programming interface standard has been defined by EDA tool and IP vendors. This code is added to the model to allow the IP vendor to limit access within the simulation environment to selected micro-architectural signal values for query and debug.[3,4]

The main disadvantages of the hard IP release are the direct ties to a specific foundry process and metallization stack (which must be consistent with *all* the IP on the SoC, of course) and the lack of configurability of micro-architectural features. The other forms of IP release offer greater flexibility in IP core features and production sourcing, with a corresponding reduction in IP vendor information protection.

Memory arrays and register files for SoC integration are a special case of hard IP. The diversity in required array sizes is great, as are specific features that may be desired, such as partial write (read/modify/write). Register files vary widely in size, as do the number of uniquely addressable (and simultaneously accessible) read/write ports. It would not be feasible to release a specific hard IP design for each unique application, nor would it be attractive to SoC design teams to be required to adapt their architectures to a very limited set of array and register file offerings. However, these functions require an aggressive layout style, as their PPA is typically critical to the overall SoC design goals.

IP vendors have addressed these customer requirements by offering *array generators*. Customers provide array configuration parameters as inputs to the generator. The output is equivalent to a hard IP release—that is, a physical layout (ideally, close to the density of a full custom layout); a functional simulation model; timing models for clock/address/data read-write access operations; and a production test model. The generator infrastructure allows the IP vendor to more easily support multiple foundries for customers. The foundry may offer different, proprietary high-performance or high-density array bit cell designs for use in the generators, enabling a wider range of PPA support.

There are some minor considerations when incorporating an array or a register file generator into the design methodology:

- The range of sizes supported by the generator must fit the SoC architecture.
- The timing model calculator within the generator must fit the operating temperature and supply voltage ranges of the target SoC application.
- The test model from the array generator must fit the overall SoC test architecture for applied patterns and response observability. Specifically, large register files present a trade-off in terms of whether the test model represents an addressable array or expands to individual register bits, with the corresponding test architecture for each bit as a flip-flop.

### *Soft IP*

A *soft IP* release is strictly a functional model, without a corresponding physical implementation. The model is typically provided using the semantics of a hardware description language (HDL). The common HDLs, SystemVerilog and VHDL, both readily support passing configuration parameters down through the logical model hierarchy, offering considerable flexibility in micro-architecture exploration before finalizing the functional design.

The physical implementation is realized through the synthesis of the HDL model to a target standard cell plus macro library (and the mapping of arrays to a generator). EDA tool vendors provide logic synthesis tools to compile the HDL model and subsequently implement the function as a netlist of cell library components, optimizing to PPA constraints.

The use of soft IP provides an easy path to evaluating multiple foundries, using different libraries when re-exercising logic synthesis and physical design flows. Similarly, it allows for rapid experimentation with different PPA targets. The primary design disadvantage is the larger area and reduced performance compared to a hard IP implementation.

### *Firm IP*

A less common format is for the IP vendor to provide a cell-level netlist rather than an HDL model. The netlist may be in terms of instances of a cell library targeting a specific foundry or consisting of generic Boolean logic primitives. The latter would allow for multiple foundry evaluation by a straightforward, but perhaps suboptimal, synthesis *mapping* of Boolean primitives to different cell libraries.

This *firm IP* release resembles the logical macro methodology. The cell-based implementation would enable drive strength and power optimizations during physical design, from the "manually synthesized" firm IP netlist. Some limited configurability may be offered by the vendor providing the firm IP, adapting the output cell netlist to customer input parameters. With the ongoing improvements in logic synthesis algorithms and the increased experience with synthesis in SoC design teams, the release of firm IP formats by vendors has declined in favor of releasing soft IP HDL models.

The soft IP and firm IP approaches provide greater design flexibility in micro-architecture and foundry sourcing. In return, the IP vendor relinquishes some degree of information protection.

### 1.2.5 "Backfill Cells"

A special class of cells are included in the library and are crucial to all block implementations. These cells are added to complete the physical layout without a corresponding instance in the netlist.

The cell template dimension is typically chosen to enable intra-cell signal connections to be completed on the most complex circuits, on the allocated metal layers for cell layouts. The access points to the cells for inter-cell routing are associated with *pin shapes* in the cell layout, placed to align with wiring tracks on upper-level metals. The metallization stack utilizes thicker metals for improved electrical characteristics on upper layers, with corresponding increases in the (minimum) wire width and space; thus, the wiring track density is reduced for these layers. In addition, the cell library may include high fan-in logic primitives for efficiency in the total logic gate path length to realize a timing-critical function. For example, wide AND*OR cells are useful in many datapath-centric designs (e.g., a 2-2-2-2 AND*OR with each AND gate receiving a data and select input). Placement of these cell layouts with high fan-in is likely to introduce a high local pin density. The reduced number of available tracks on the inter-cell routing layers combined with a high local pin density may result in *routing congestion* when attempting to complete the block physical design. The routing tool may be unable to find suitable (horizontal and vertical) segments for all the netlist connections, resulting in incomplete signals (known as *overflows*).

The physical block layout area allocated to route the complete netlist may thus need to be enlarged. Cell area utilization substantially less than 100% is used to gain confidence in achieving a fully routed solution. The library vendor provides utilization guidelines, based on previous experimentation; for example, a recommendation of ~80% utilization is common. (Placement algorithms evaluate local pin density during cell positioning and space cells apart to reduce route congestion.) The net result is a significant vacant area within the block layout, both within and around the perimeter of the block boundary.

The library contains *backfill cells* (of varying widths) to insert into vacant locations in the block. These completion cells can be quite varied in content and purpose:

- **Dummy cells for photolithographic uniformity**—The backfill cell may add *dummy transistors* to maintain a more uniform grid of transistors between functional cells to reduce the variability in lithographic patterning of the critical dimension transistor gate length.
- **Local decoupling capacitance**—Expanding upon the addition of non-functional transistors in vacant cells, a relatively simple connection of devices provides a capacitance between the VDD supply and GND rails (see Figure 1.5). These decap cells can be easily inserted in the unoccupied cell locations after routing and prior to electrical analysis.[5]

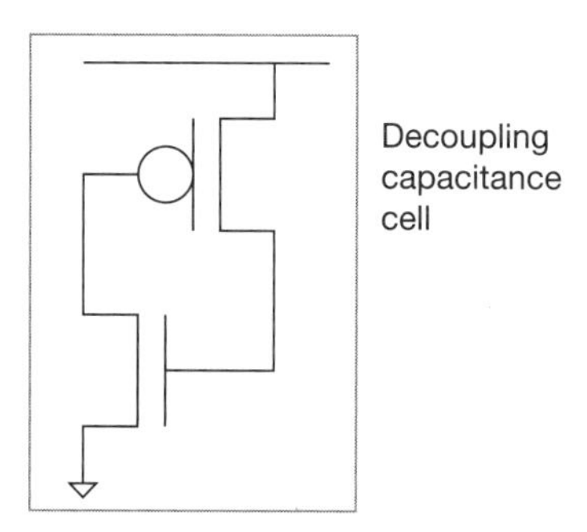


**Figure 1.5** Schematic for the "cross-coupled devices" decoupling capacitance cell, inserted in vacant cell locations. Both devices are operating in linear mode, with maximum Cgate-channel.

The switching activity of logic cells draws supply current. Local decoupling can efficiently provide charge to this transient, minimizing the resistive I*R voltage drop along the rails and without the delayed inductive response through package pins. Indeed, a key facet of the power distribution network (PDN) analysis flow step is to calculate the total dynamic switching transients, the local (inherent circuit and explicit decoupling) capacitance, and the PDN electrical response characteristics to ensure rail voltage transients are within design margins (see Section 14.3).

### *Gate Array Logic ECO Cells*

After an SoC design is submitted for initial fabrication, it may become evident that a logic design modification is required. A change in product requirements may arise while awaiting the *first-pass* prototypes from the foundry and OSAT. Or, more likely, a bug in the functional design may have been discovered during the ongoing system model validation, after release of the design to the foundry. In either case, there is a requirement to apply an engineering change order (ECO) to the (first-pass) netlist.

The simplest approach would be to simply remove any erroneous standard cells (and their routes) from the physical layout, insert the standard cells to fix the bug, and reroute the updated netlist connections. Indeed, physical design tools commonly have an incremental mode, which accepts a unique netlist syntax for cell instance adds/deletes. The goal of this ECO methodology is to leave the existing physical design as unperturbed as possible to minimize new issues arising during electrical analysis. However, adding and deleting standard cells for a second-pass design implies that all lithography layers for the SoC will likely have been edited, starting with transistor fabrication. The cost of a full set of new lithographic masks will be considerable, and the schedule impact of the full fabrication cycle time will be maximum before updated packaged parts are available.

An alternative approach is to backfill some percentage of the vacant block area with uncommitted transistors. A specific ECO logic function would be realized by adding local metal segments between transistors, vias, and pin locations. An array of transistors could be personalized to a logic function, using only metal and via physical layout edits.

The collection of individual logic functions implemented by the metal *personalization* of uncommitted transistors is known as the *gate array* library. An IP library provider may include gate array cells with the standard cell release. Typically, the gate array offering will be a small subset of the standard library, with limited drive strengths and transistor $V_t$ variants available. The ECO design with gate array logic inserted (and, likely, signal buffers added) will not be optimized to PPA design targets. There is certainly a risk that the gate array backfill cells will not be able to implement a set of logic ECO changes, due to limited availability, location, or the impact on overall performance that no longer meets design specifications.

If a gate array library is available, the ECO methodology needs to ensure that all flow steps are consistent with incremental design at the cell netlist level (see Chapter 17, "ECOs"):

- Placement and routing tools must accept the incremental netlist syntax and restrict additions to unpersonalized gate array locations.
- The full-chip SoC data model used for test pattern generation must reflect the incremental updates.
- The extraction of signal interconnect parasitics for electrical analysis must (ideally) also be able to recognize the incremental inter-cell route segments and efficiently update the parasitic network.
- A complete logical-to-physical equivalence ECO flow step is required; a netlist update applied directly to the physical layout also needs to be reflected in the HDL source model for functional validation. Designers make functional changes to the HDL model related to the ECO netlist adds and deletes and recompile the HDL for functional simulation. The ECO methodology flow needs to verify the logical equivalence of the updated cell netlist to the revised HDL.
- The data version management policies adopted by the methodology need to apply a nomenclature that maintains the evolution of the first-pass design and any succession of ECO netlists. When first-pass prototypes are received from the foundry, hardware system bring-up and debug will be referenced to the original release data while ongoing updates for a subsequent fabrication pass are being developed.

### *Standard Cell Logic ECO Backfill*

A couple of methodology considerations regarding backfill in preparation for subsequent ECOs are worth specific mention. If a gate array cell library is not available, an alternative methodology would be to add a small quantity of standard cells throughout the vacant block area. The inputs to the standard cell would initially be routed to a power or ground rail to tie all inputs to a fixed logic value. Unlike the flexibility of the gate array logic function provided by the personalization of uncommitted transistors, a spare standard cell implies a specific logic function available at that location. The number and diversity of spare standard cells added to the block is a judgment choice.

There is a risk that an ECO netlist change may not be efficiently realizable with metal-only lithography updates if the spare logic standard cells are limited in location and function.

Another methodology decision pertains to whether spare flip-flop cells would be inserted to support an ECO that adds a sequential state signal. The gate array backfill library offering may not include flip-flop cells; the connections between transistors to implement a flip-flop are the most difficult within the standard cell template and may not be viable for the uncommitted transistor array. Alternatively, flip-flops from the standard cell library could be selected for insertion and judiciously placed in vacant areas.

The more difficult decision pertains to what consideration should be given to the clock input(s) to the spare flops. During physical design, an effort is made to provide clock signal buffering and load balancing to minimize the arrival skew at the clock signal fan-out endpoints—namely flip-flop inputs and clock gating cells. If spare flops are provided, the ECO routing updates to add the clock connection to the flop will perturb the existing balanced skew; previously valid timing paths (unrelated to the logic ECO modification) may now fail due to increased clock arrival skew. To avoid the issue, the spare flop clock inputs could be part of the original physical optimization, so their subsequent use would not adversely impact the clock timing. However, in this case, spare flops that remain unused result in wasted power dissipation due to the additional clock loading. And, for logical-to-physical equivalence verification, these visible spare flop cells would need to be added to the logical model for the block despite having no functional contribution. A potential requirement to include spare flops in the test model for the original design also needs to be assessed. If the test architecture includes unique test mode clocking, the spare flop may need to be included in the functional description.

The methodology approach selected to optimize the cost/schedule impact of ECO updates requires several trade-off assessments with regard to the specific spare cell library chosen, the number and placement of spare cell logic functions, the number of mask layout layers impacted to implement the ECO, and, for the insertion of spare flops, the impact on the (functional and test) clock distribution.

### 1.2.6 Cell Views and Abstracts

The cell library release includes numerous *views* for each element for different methodology flow steps:

- **Functional model for netlist-level validation**—Example flows are logic simulation and logical-to-physical equivalency.
- **Test model**—A description of the prevalent circuit-level manufacturing *faults* is needed to investigate with manufacturing test patterns.
- **Timing model**—The timing model includes cell pin-to-pin delay arcs, with delays represented as a function of supply voltage, temperature, input pin signal slew, output pin loading, and (potentially) conditional arcs based upon the logic values at other input pins; sequential functions have additional timing data, such as clock-to-data setup and hold constraints. Recent timing models also include detailed data on the output pin transient current waveforms for more accurate signal interconnect propagation.
- **Physical model**—A representation of the physical cell layout, used for placement and routing, includes cell size, route blockages, and pin shapes (including either "connect to" or "full cover" route constraints for the large area pins).
- **Power model**—The power model includes the static leakage power and internal power dissipation for arc transitions, with comparable coverage of voltage, temperature, and signal transitions as the timing model.
- **Electrical and thermal models**—A collection of data derived from the physical layout is used for other flows (e.g., pin capacitance, input pin noise pulse rejection, output pin impedance, power and ground rail currents for signal transitions, self-heating thermal energy).

These views are developed by library generation methodologies to provide model abstracts that are most efficient for the cell-based flows. Library model generation is also commonly referred to as *characterization* (see Section 10.2). EDA vendors have utilized unique abstract data best suited to their tools. The vendors have collaborated with IP providers to prepare these specific formats as part of library generation methods. Due to the critical importance of these models for all tools and SoC methodologies and the costs associated with library generation, the IC industry is increasingly pushing for (de facto) standards for model view abstract data. Several EDA vendors have responded to this trend by releasing their formats into the open source community.[6] Nevertheless, methodology development requires coordination between specific tool requirements and the cell abstracts released by the IP library supplier.

The automated logic synthesis of a cell-based netlist from an HDL functional description employs optimization algorithms that use data from multiple views for initial logic mapping and subsequent PPA optimizations. There are assumptions inherent to these tool optimizations about the relative ordering of the drive strength and Vt cell variants of each logic function in the library release. For example, when seeking to reduce path delay during timing optimization, the algorithms seek to deploy the "next higher" performance entry for the mapped logic function. The methodology team and IP library supplier should review the criteria used to determine the ordering for cell variants of a logic function to ensure that these criteria are consistent with the SoC design goals. If a design is extremely constrained by static leakage power, the relative choice of LVT cell variants could be reduced (or excluded altogether, with a separate "don't use" directive to the synthesis flow).

### 1.2.7 Model Constraints and Properties

In addition to the abstract data that describe the cell to the EDA tools, there is a view for some cells that reflect any restrictions on the cell usage in the context of the full netlist. The cell design may have incorporated specific assumptions about its usage, which need to be provided as *constraints* for subsequent validation. For example, a flip-flop cell with separate clock inputs for functional data and test data capture requires the clocks to be mutually exclusive as the circuit behavior may be indeterminate ("undefined" to a validation tool) if both clocks are active simultaneously. The model constraints view would include this restriction, to be validated against the netlist model.

The number and complexity of IP usage constraints also grows with the size of the IP design. An IP core or memory array release will likely include a set of functional *properties*, which describe the interface requirements. EDA vendors have developed unique property specification language semantics to efficiently describe the required behavior. The property would commonly be compiled with the functional simulation model. If the property is violated during functional simulation, the simulator tool will be directed to report an error. EDA vendors have also released (static) property prover tools to use in lieu of functional simulation flows. If the prover determines that the logic connected to the IP core or array could potentially violate the interface behavior described in the property in any possible scenario, the tool will flag an error (and derive a functional counter-example for further debug).

The increasing number of discrete power supply domains on current SoC designs has resulted in a significant number of additional cell constraints. An example would be a "level shifter" cell, intended as the logic interface between two different voltage domains on the SoC. The level shifter cell usage is restricted to receiving a lower-voltage signal logic 1 level at the input to a higher-supply-voltage domain.

The addition of power-gated "sleep" functionality to IP cores adds to the usage constraints. As mentioned in the introduction to this chapter, the complexity of power domain management on-chip has motivated the EDA industry to define a unique *power format description language* to capture the voltage/power states of cores and their interface signals.[7] Tools specifically developed to confirm the power format constraints against the SoC model need to be incorporated into the validation methodology.

### 1.2.8 Process Design Kit (PDK)

A foundry releases a process design kit (PDK) to IP developers and SoC customers that provides documentation, models, techfiles, and runsets to enable designs to be released for fabrication. Transistor models are used for detailed circuit simulation. The process definition for physical design is provided in *techfiles*, for use with physical layout, placement/routing, and electrical analysis flows (e.g., lithography mask layer nomenclature, electrical characteristics of interconnect and dielectric layers for parasitic extraction from a layout). A *runset* refers to a sequence of operations exercised against the physical design. For example, runsets for physical design checking include mask layout geometry operations to identify devices and signal continuity between layers, perform measures on devices, and execute checks to ensure that the layout satisfies photolithography rules and manufacturing yield guidelines. In addition, the foundry may provide software scripts and utilities to enhance layout productivity (e.g., device layout generation from a schematic instance, the insertion of *dummy fill* shapes for improved lithography uniformity).

The formats for the physical and electrical techfiles are specific to the EDA vendor tool. Multiple formats are actively used to represent the process cross-section and resistance/capacitance layer data for parasitic extraction.

(There is an effort under way to promote a de facto standard interoperable PDK format [iPDK] and adopt a common representation for this process data.[8]) Similarly, there are different runset formats for layout operations, measures, and checks. EDA vendors regard their runset language semantics and geometric algorithm optimizations as highly proprietary and as being product differentiators. As a result, a foundry's PDK and design enablement teams release design kits that are qualified for multiple *reference EDA tools*. The SoC methodology team needs to work closely with the foundry to ensure that flows utilize the EDA vendor tools for which foundry support is provided.

The foundry may engage early SoC customers for a new process still in development and provide preproduction PDK data prior to the version 1.0 process qualification release; for example, PDK v0.1, v0.5, v0.9, and so on could be made available to key customers. Certainly, IP providers may also want to engage with the foundry using these early process descriptions to have silicon test chip hardware measured and qualified in time for SoC designs using the v1.0 PDK. For these early adopter customers, in addition to addressing the risks associated with designing to preliminary process models, the schedule availability of preproduction PDK techfiles and runsets for a specific EDA tool needs to be reviewed by the methodology team.

As mentioned earlier, a foundry is typically also a provider of cell libraries to IP vendors and SoC customers (e.g., memory array bit cells, ESD structures to integrate with I/O circuits, process control monitoring and metrology structures to add within the die area [used during fabrication], perhaps even a base standard cell logic IP library). These foundry libraries are distinct from the PDK releases and are covered by separate license agreements.

The v1.0 production PDK is also likely to be superseded by subsequent releases (e.g., v1.1, v2.0). After achieving production status with v1.0, the foundry will pursue continuing process improvement (CPI) experiments to optimize performance, power, or, especially, manufacturing yields. New materials and/or process equipment may be introduced to enhance electrical characteristics and/or reduce statistical variations in the fabricated devices, interconnects, and dielectrics. The transition of these ongoing process improvements to the production fabrication lines will result in subsequent PDK releases.

An SoC design underway using a specific PDK release may need to assess the impact of moving to a new PDK version in terms of project schedule, resource, and anticipated manufacturing costs. Discussions with the foundry will indicate whether an existing PDK release will continue to be supported (and for how long) or whether the transition to a new PDK is required. The methodology team is an integral part of this review, to evaluate the impact of a new PDK on existing tools/flows—specifically whether any new EDA tool features are also required to be supported. If the new PDK is accepted, the methodology and CAD teams will coordinate the transition to this PDK release. The release update involves qualifying EDA tools (especially any new features), releasing the new PDK, and recording PDK version information as part of the flow step output to verify consistency across the design project.

## 1.3 Tapeout and NRE Fabrication Cost

The culmination of an SoC design project is the release of the physical design and test pattern data to the foundry for fabrication, a project milestone known as *tapeout* (see Chapter 20, "Preparation for Tapeout"). Fabrication data used to be written on magnetic tape and sent to the foundry. Today, although encrypted data is sent electronically to the customer dropbox maintained by the foundry, the *tapeout* name continues to be used.

Tapeout is a major milestone for any project. SoC design data has been previously *frozen*. A snapshot configuration of design data, flows, and PDK releases is recorded so that all full-chip electrical and physical verification steps can be performed on the database targeted for tapeout release. Often, special IT compute resources are allocated prior to tapeout to ensure that the requisite servers with suitable memory capacity are readily available for these steps. Full-chip flows may utilize parallel and/or distributed algorithms for improved throughput; appropriate servers from the data center may need to be reserved.

A number of tapeout-specific methodology checks are performed to collect the results of the full-chip flows exercised on the tapeout database. Any error/warning messages from electrical analysis and physical verification flows that are highlighted by the tapeout checks need to be reviewed by key members of the design engineering and methodology teams to assess whether the PPA impact can be waived. These final full-chip checks (with the documented review team's recommendations) are collectively denoted as the "signoff flow."

Any errors/warnings from physical verification flows using PDK runset data also need to be reviewed with the foundry customer support team to determine if they can be waived (e.g., a minor yield impact) or whether a design modification prior to tapeout is indeed mandatory.

Tapeout is also a key project milestone for financial considerations. The SoC project manager communicates the tapeout date to the foundry, with sufficient advance notification to reserve a slot in the mask generation and wafer fabrication pipeline. This tapeout slot is associated with payment to the foundry for lithographic masks and a quantity of wafers suitable to provide an adequate number of parts for product bring-up and stress-testing qualification. This payment to the foundry is often denoted as a *non-recurring expense* (*NRE*) when preparing the initial project cost estimates. The NRE amount is dominated by the foundry quote for a set of masks rather than the subsequent wafer fabrication costs. The OSAT is also notified of the "expected wafers out" date, based on the target tapeout date plus wafer fabrication turnaround time. The NRE payment to the OSAT is another project cost estimate line item.

A yield estimate provided by the foundry based on manufacturing defect densities (not a circuit design yield detractor based on power/performance targets) is used to determine the number of wafers in the prototype tapeout fabrication lots. Multiple lots may be started in succession to provide enough parts and/or to provide a reserve of wafers for subsequent ECO design submission. Assuming that a second-pass ECO design submission could indeed be implemented with physical layout changes limited to only metal and via layers using a cell backfill approach, a number of first-pass wafers could be held after initial processing prior to metallization. The second NRE would be limited to only the new masks, and the reduced wafer cost to re-commence fabrication of the wafers that have been held in reserve during first-pass fabrication.

The foundry is likely to offer engineering prototype lot fabrication on a unique manufacturing line, different from the volume production facility. Foundries promote the feature that multiple lines for the same process use a "copy exact" approach to allow the customer to confidently assume that prototype evaluations are directly applicable to production parts.

The prototype-focused line allows for some experimentation that would not easily be supported in high-volume production. Specifically, the design team may request *split lot* processing, in which a set of intentional variations are introduced. A relatively small number of prototype wafers would normally not provide a statistically significant sample that would be representative of volume production parts. The foundry may offer a split lot option with intentional process variations to provide a wider range of slow-to-fast parts for the design team to better assess any circuit sensitivities.

The prototype line may also offer expedited processing, which would not typically be available in production. The fabrication line may support a limited number of customer designs that receive priority scheduling at each process module station. There is a considerable NRE surcharge to the customer for a "hot lot" turnaround schedule.

The IP vendor has a different set of tapeout criteria. The IP will be much smaller than for a full SoC and may only need *wafer probe-level testing* for characterization rather than requiring the full package assembly and final test services from an OSAT. The foundry may offer an option for a *multi-project wafer (MPW)* fabrication slot. A number of different IP-level designs could be merged together into a single set of mask data. Each IP submission would be allocated an area on the MPW die site and would use an array of pads added to the IP layout matching the test probe fixture(s) available at the foundry. The MPW provides a cost-effective option for the IP vendor, as the NRE costs are apportioned among the IP contributors. The disadvantage is that the foundry is likely to offer only limited MPW slots (also known as "shuttles"), scheduled infrequently. In addition, the foundry may have limited test engineering resources available to exercise the IP test specification and collect characterization data. As a result, the IP vendor may need to seek additional external test engineering services. The IP vendor needs to collaborate closely with the foundry support team on the MPW area allocation and test strategy and then work aggressively to meet the committed MPW shuttle date.

## 1.4 Fabrication Technology

### 1.4.1 Definitions

#### *VLSI Process Nodes and Scaling*

Although each silicon foundry offers a unique set of fabrication processes and features, a foundry generally follows a lithography roadmap, as outlined by the (evolving) International Technology Roadmap for Semiconductors (ITRS). As a result, VLSI technologies are commonly associated with a *process node* from the roadmap. The succession of nodes in recent history has been 0.5um, 0.35um, 0.25um, 180nm, 130nm, 90nm, 65nm, 45/40nm, 32/28nm, 22/20nm, 16/14nm, 10nm, 7nm, and 5nm. The duration between the availability of these processes (in high-volume production) has typically been on the order of two years, following a trend forecasted by Gordon Moore over 50 years ago, in what has become known as *Moore's law*.[9]

Note that there is a consistent *scaling factor* of 0.7X between successive process nodes. The goal of each new technology offering has been to provide a physical layout dimensional scaling of 0.7X linear, or equivalently 0.5X areal, effectively doubling the transistor density (per mm**2) with each new node. The PPA benefits of scaling were captured in a landmark technical paper by Robert Dennard; project teams have applied "Dennard's rules" when planning product specifications for the next process node.[10,11]

Indeed, for older process nodes, where the light wavelength used for photolithographic exposure was significantly smaller than the minimum dimension on the mask plate, scaling of (essentially) all layout design rules was the fabrication goal. The focus for process scaling was to improve material deposition and etch process steps, and to reduce defect densities; photolithographic scaling was readily achievable. IP development in these older process nodes was rather straightforward. Existing IP physical layout data were easily scaled by 0.7X, and electrical analysis was performed using the new process node techfiles. The Dennard scaling of interconnects resulted in an increase in wire current density, necessitating a materials transition in the late 1990s from aluminum to copper as the principal metallurgy due to electromigration issues (see Chapter 15, "Electromigration Reliability Analysis").

More recent process nodes utilize wavelengths longer than the mask plate dimensions. The drawn layout must undergo sophisticated algorithmic modifications prior to mask manufacture, applying optical diffraction and interference principles to the light transmission path from mask to wafer. As illustrated in Figure I.1, the most recent process nodes and exposure wavelengths necessitate decomposition of the layout data for a dense design layer into multiple masks prior to optical optimizations. In addition, the scaling of interconnect wires and via openings in dielectrics has required greater diversity in metals (and the use of multiple metals in the wire cross-sections) for both electromigration and resistivity control. As a result, although the ITRS node designation continues to use a 0.7X factor, the actual scaling multiplier between process nodes for individual layer layout design width, spacing, and overlap rules varies significantly, from 0.7X to 1.0X (i.e., no scaling). Some layers will continue to use fabrication process modules and materials unchanged from the previous node, with no dimensional scaling; this is especially true for the upper metal interconnect layers.

The engineering effort to prepare an IP design in a new node has thus increased, as the physical layout needs to be re-implemented. The quantization of allowed transistor dimensions for new VLSI technologies also necessitates a design re-optimization. (FinFET devices are discussed later in this section.)

### *Shrink Nodes and Half Nodes*

In addition to the ITRS scaling roadmap on the Moore's law cadence, some foundries offered an intermediate offering, a layout "shrink" of 0.9X applied to design data. The production process steps for the base process node needed to have sufficient latitude and variation margins to use the same materials and fabrication equipment for lateral dimensions scaled by a 0.9X factor.

This "half-node" offering was achieved by adjusting the optical lens reduction in the photolithography exposure equipment (nominally, a 5X reduction). As a result, new mask plates were not required. This option was extremely attractive to design teams. An improvement to the PPA specifications was provided at minimum NRE expense. Typically, the cost of transitioning an existing design to a half-node consisted of fabrication of a small quantity of wafers to complete an updated qualification of the new silicon before ramping production. This allowed an existing part in volume production to be offered with a "mid-life kicker" in performance and/or reduced customer cost (or improved profit margins) from the reduced die size. The half-node process introduction occurred on a production schedule between the major node transitions.

More recently, half-node introduction has not been viable. There is insufficient process latitude at current nodes to apply a 0.9X scale across all existing fabrication modules without new optical mask data generation and additional process engineering. There has been some confusion introduced in the nomenclature used by foundries when describing their fabrication process capabilities, especially since lithographic scaling is much more constrained. Some foundries chose to focus solely on half-node process transitions without offering an ITRS base node (e.g., 40nm, 28nm, 20nm, 14nm, 10nm). Some foundries describe their process using a general ITRS process node designation but diverge from detailed lithography dimension targets in the ITRS specification. Process comparisons currently require much more detailed technical evaluations, using transistor and wiring density estimates, materials selection and dimensional cross-section data, and $I_{on}$-versus-$I_{off}$ device measures (refer to Figure 1.3).

### *"Second Sourcing"*

At older VLSI nodes, it was not uncommon for customers to expect a foundry to accept tapeout data developed for a different manufacturer. Customers were seeking a "second source" of silicon wafers, both for competitive cost comparison and as a continuous supply chain if a specific foundry experienced an unforeseen interruption in production. Foundries were expected to have sufficient process latitude to accommodate a design completed to layout rules from another manufacturer. At current design nodes, layout design rule compatibility between foundries is no longer the norm. Although multiple foundries offer a 7nm process, for example, there is an increasing diversity on layout design rules and power/performance targets. The SoC design project decision to select a specific foundry involves an early and thorough assessment of the PPA, available IP, and production cost; it is no longer straightforward to redirect a project to a different foundry at the same node. Concerns about continuity of production supply are typically covered by the multiple production lines available for each process at a single foundry, ideally manufacturing at geographically separate fabrication facilities, or "fabs."

### *Process Variants at a Single Node*

The application markets for current SoCs are increasingly diverse. Consumer, mobile, automotive, military/aerospace, and medical equipment products span the gamut of performance, power, and reliability specifications. Foundries may offer several process variants at the same lithographic node, specifically addressing these different markets. For example, a low-leakage mobile process would set transistor $I_{on}$-versus-$I_{off}$ targets to minimize leakage for SVT and HVT devices, whereas a process variant targeting high-performance computing customers may establish different targets for SVT and LVT devices.

In addition, there may be different IP library offerings for these market segments. For example, a standard cell library for aggressive cost, lower-performance applications may use a smaller number of wiring tracks in the template definition. A high-performance library would benefit from larger devices in the circuit design (e.g., for 1X and 2X drive strengths) and thus might incorporate a taller standard cell template definition.

The diversity of the product application requirements across these markets is driving the need for additional process and library options at each advanced process node. The engineering resource investment required from the foundry and IP providers is certainly greater, to capture a broader set of customer design wins.

### 1.4.2 Front-End-of-Line (FEOL) and Back-End-of-Line (BEOL) Process Options

The fabrication process steps are typically divided into two subsets, denoted as the front-end-of-line (FEOL) and back-end-of-line (BEOL). The front-end steps correspond to the fabrication of all devices. The back-end steps fabricate the metal interconnects, dielectrics, and vias, through the metallization stack up to the pad metallurgy. A fraction of prototype wafers may be held after FEOL steps are completed, while the remainder complete all BEOL steps. The held wafers could subsequently be used for a second-pass, metal-only ECO submission. Each foundry process includes optional FEOL and BEOL process modules (at additional cost), which are selected by the design team to satisfy the set of structures used by the SoC and its constituent IP.

The BEOL metallization stack defines the number of metal/via layers, as depicted in Figure 1.6.

Typically, several metal wire thicknesses are available from the foundry, each with unique electrical characteristics. Thicker metals have larger minimum width and spacing dimensions (and, more recently, are embedded in insulators with lower dielectric constants). Thus, the metallization stack is "tapered." The foundry provides guidelines on the required and optional number of layers of each thickness type and the appropriate transitions between types. The design team needs to collaborate with the foundry to choose an optimal stack corresponding to the PPA requirements (e.g., 3Mx_3My_2Mz_1Mr for an eight-layer metal design with a top pad metal redistribution layer). The PDK techfile from the foundry is adapted to support this cross-section.

The circuits used for I/O cells are likely to use different transistors to support the higher applied voltages. The FEOL process definition includes the "thick gate oxide" device process module.

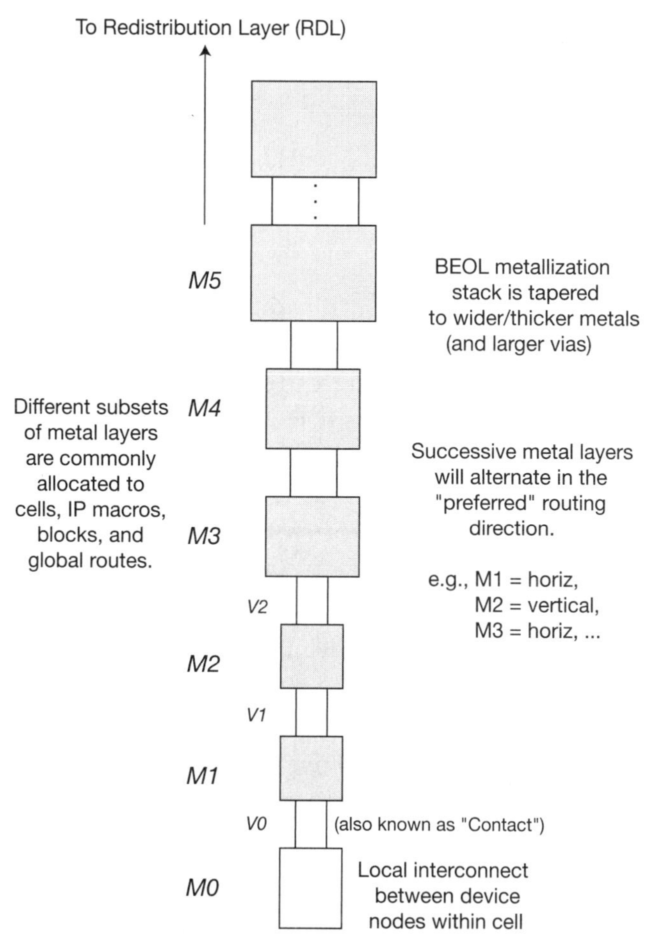


**Figure 1.6** Illustration of a BEOL metallization stack, consisting of a specific sequence of metal layers (with corresponding via definitions between layers).

All field-effect devices receive a substrate voltage through contacts to the bulk (for nFETs) and n-well regions (for pFETs), using a p-type wafer substrate. A "triple-well" option in the FEOL process allows n-channel field-effect transistors (nFETs) to receive a unique substrate voltage, which is electrically isolated from the voltage applied to the volume bulk substrate of the die.

Analog IP designs may incorporate a variety of optional FEOL process modules:

- p-n junction diodes
- Resistors, with both (area-efficient) high-resistivity layers and high-precision resistive layers
- Bipolar junction transistors (SoC processes are based on field-effect transistors.)
- Voltage-dependent capacitors (“varactors”)

Memory array IP layouts are extremely dense and thus are especially sensitive to manufacturing process defects. For SoC designs that utilize large amounts of memory, there is a significant yield risk of a killer defect located in the arrays. The IP vendor may include redundant rows and/or columns in the array implementation, which can be switched in to replace an array location found to be defective in a manufacturing test (see Section 19.3). To enable this yield-enhancing feature, the foundry commonly needs to provide a process option that includes FEOL electrically programmable fuses. (In addition, programming fuses at manufacturing test enables a specific set of personalization information to be embedded in each die.)

The process may offer a unique BEOL metal-insulator-metal (MIM) structure, with a thin dielectric and special intermediate metal plate located between interconnect layers, for an area-efficient decoupling capacitor to connect to the power distribution network.

On-chip inductors are typically implemented using annular coil layouts using the top two BEOL (thick, low-resistive-loss) metal layers, below the pad redistribution metal.

### 1.4.3 Fabrication Design Rules

The foundry PDK includes documentation that describes the physical layout rules—that is, the *design rule manual* (*DRM*). The PDK also includes the EDA tool runset, a sequence of geometric operations that check the layout data against the DRM rule dimensions. All design rule measures are described in reference to a base layout grid (e.g., 0.025nm). All circuit layout dimensions are integral multiples of this grid increment.

The complexity of layout rules has increased dramatically in more recent process nodes due to the requirements to resolve wafer-level exposures using a wavelength greater than the mask dimensions. The design rules for older VLSI nodes primarily consisted of the following:

- Minimum width and spacing for line segments (transistors and interconnects)
- Minimum area for all shapes (especially applicable to device contacts and inter-level vias)
- Minimum overlap/enclosure (e.g., metal over vias)

The sum of the minimum width and spacing is commonly denoted as the *pitch* for the mask layer.

There were typically no (or very few) design rules that imposed a maximum width/space/overlap, and all dimensions greater than the minimum were allowed to be any integral multiple of the base grid dimension. (The main exception to these general rules pertained to the layout of chip pads on the top metal layer. Pads utilized fixed geometries for the size and top dielectric passivation layer opening for die-to-package pin attach metallurgy.)

As process nodes progressed, manufacturing steps were added to provide improved wafer surface planarization after metal deposition/patterning. These *chemical-mechanical polishing* (*CMP*) steps led to the introduction of density-based rules (i.e., minimum density, maximum density, and density gradient limits for the shapes on a CMP metal layer, measured over a layout window of interest). EDA tool vendors needed to add algorithms and runset operations to support these additional design rule requirements.

As SoC designs integrated more memory, the array circuit density became an increasingly influential factor in the chip area. The foundry engineering team pursued specific bit cell designs that were typically more aggressive than the general DRM descriptions for transistor dimensions, contact spacing, and local metal segments. (These special cells in the tapeout database receive unique mask data optimizations.) The memory IP supplier works with the foundry PDK runset team to ensure that these special cells are checked against their array-specific rules and excluded from the general checks applicable to the remainder of the design. The most common methodology approach is to add a "cover shape" over the special cells on a non-manufacturing layer designation specific to array checking. Layout data intersecting the cover is checked against a separate set of array design rules from layout data outside the cover.

As the disparity between mask plate dimensions and exposing wavelength increased, design rules became significantly more complex, including the following:

- **Metal spacing based on the *parallel run length* (*PRL*) to adjacent wires**—The greater the parallel run length, the greater the required spacing.
- **Forbidden pitches**—Invalid spacing ranges between (minimum width) wires are due to light diffraction from mask edges interfering to create phantom shapes.
- **Required non-functional "dummy" shapes**—These dummy shapes are placed next to isolated design shapes for better (local) mask data uniformity.

At current process nodes, both the photolithographic resolution and material etch process modules exhibit much greater local sensitivity. A regular pattern improves the fidelity of the fabricated result to the drawn shapes. Dummy data may have a distinct set of width/spacing design rules, as they are non-functional and, thus, non-critical. The layout model needs two separate data designations for a single layer (i.e., functional and dummy) so that specific rule checks can be applied accordingly. During tapeout release, these two designations are merged to a single set of mask data for the layer.

The methodology for physical layout assembly needs to provide support for the addition of dummy data for CMP density and isolated-line checks. The points in the physical design hierarchy where non-functional data is added need to be chosen judiciously. Dummy data added at deep levels of the physical hierarchy may be unnecessary if design rules would be satisfied by integrating the IP layout into higher levels. Conversely, deferring the addition of dummy data until the full-chip SoC layout nears completion may result in design rule errors that are difficult to resolve.

Regardless of where the dummy data is added in the physical design hierarchy, the design rule checking flow needs to include a representative *context cell* around the cell/block being checked to avoid false errors around the perimeter of the design data, as depicted in Figure 1.7. The context cell includes shapes data for each layer at a "half-design rule space" from the context cell boundary. This ensures that a corresponding half-design rule space is present within the cell/block layout itself so that no design rule errors arise when cells are abutted. The context cell may also include dummy data patterns to provide a larger window around the block perimeter for the layout density checks. And, to provide a representative environment for layout dependent effects upon device behavior, the context cell is also used for parasitic extraction of the block layout prior to electrical analysis.

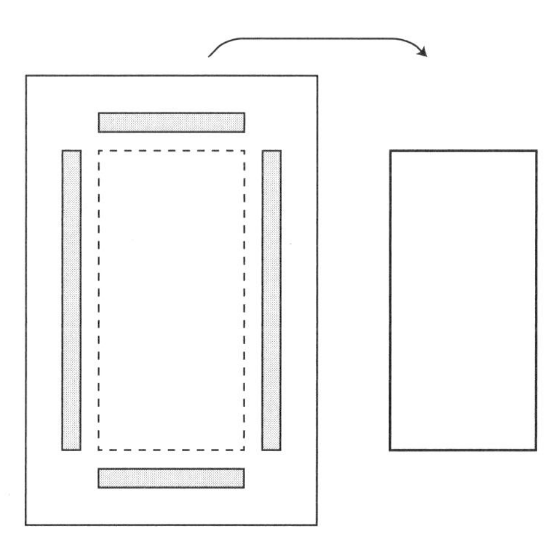


**Figure 1.7** Layout design rule verification requires addition of a surrounding layout context cell. The context cell is intended to provide a "representative" environment around the layout to satisfy local spacing checks and minimize the risk of a larger-scale density error when integrating the cell into higher nodes of the physical hierarchy.

### *Preferred Line Segment Orientations*

At current process nodes, the exposing illumination intensity is very non-uniform over the mask field. The illumination pattern (along with the optical corrections applied to the design data when generating the mask) assumes a preferred orientation to the layer data. Design rules are optimized for shapes in the preferred direction, with larger widths and spaces required for any non-preferred "wrong-way" segments. Some layers may not allow *any* wrong-way segments at all but only rectangles in the preferred direction. EDA tool algorithms thus need to include support for direction specifications in the run-set commands.

#### *Multipatterning, Layer Data Decomposition, and Cut Masks*

The design rule change with the greatest impact in recent process nodes is the requirement to divide the design data on a single layer into multiple mask plates for fabrication. For example, using 193nm wavelength exposing illumination, with the added resolution benefit of an immersive refracting liquid between lens and wafer, the finest resolvable pitch resolution is ~80nm. A line width plus space less than this dimension would not be printable, necessitating mask data decomposition.

At the most advanced nodes, foundries are working closely with semiconductor equipment and materials manufacturers to transition photolithography for the most critical mask layers to shorter exposing wavelengths, using "extreme ultraviolet" (EUV) sources. This transition involves much more than introducing a new light source and requires significant engineering development focused on new reflective and transmissive mask materials, new mask inspection techniques, new mask data correction algorithms for the source-to-mask-to-wafer optical path, and new EUV-sensitive wafer photoresist coatings, to name but a few. Although shorter exposing wavelengths will help simplify some of the lithographic design rules, multipatterning decomposition will still be required to enable future node scaling.

The scaling of interconnects led to a (lowest-level) metal layer pitch less than 80nm at the ITRS 22nm node. New algorithms were required across the full physical design and checking methodology to decompose layer data into separate masks. Figure 1.8 expands on the data in Figure I.1, illustrating a two-color A/B decomposition and a *cyclic layout* configuration (at minimum spacing) which is not divisible and is thus a design error.

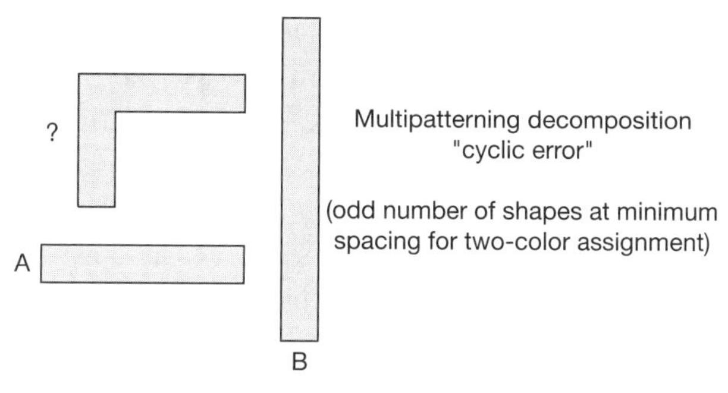


**Figure 1.8** The decomposition of mask-layer data into two multipatterning subsets is depicted. A cyclic layout topology is highlighted, which is a layout design rule error.

The number of DRM rules expanded with multipatterning to represent the (min/max) width, (min/max) spacing, (min/max) density, and (min/max) density gradient requirements for each of the layer data decomposition subsets.

Note that the dimensional range of a decomposition algorithm extends well beyond the full size of a shape, unlike the rather limited dimensions associated with all other design rule checks. The color-assignment algorithm needs to verify that no cyclic dependency exists for every shape and all its neighbors. EDA vendors needed to expand their design rule checking runset commands beyond the traditional geometric operations to identify and measure even and odd cycles among (potentially very large) sets of neighboring shapes. If a cyclic dependency is found, changes to wire lengths and/or wire spacing are required. (Although it would be algorithmically feasible to dissect a single shape and assign the dissections to different colors to avoid a cyclic error, this would introduce additional fabrication and electrical modeling issues due to the overlay tolerances between the masks. It would also preclude the foundry from introducing a mask bias specific to the distinct subsets, as highlighted below.)

The impact of multipatterning on the design methodology is pervasive. All physical design checking tools need to be prefaced by a decomposition algorithm before applying shape operations. An interconnect routing tool also needs a decomposition feature to avoid introducing cyclic errors during detailed routing when assigning segments to specific tracks. If the A/B color design rules are symmetric and there is no fabrication "bias" resulting in different electrical characteristics between A and B segments, the cyclic rule checking algorithms need not be exactly the same as that subsequently used by the foundry during mask data processing. If the design rules and/or electrical parameters distinguish between the decomposition subsets, the methodology needs to ensure that a color assignment from the layout engineer or EDA tool is made during physical design and remains as a shape property throughout all verification and analysis flows and, subsequently, in the tapeout release.

Similar to the considerations with the addition of fill shapes data, the methodology for assigning and/or checking decomposition needs to evaluate where in the physical design hierarchy the color property is applied. An individual layout cell may pass decomposition checks but could fail at the higher physical integration levels when adjacent shapes are added. The methodology needs to assess how to ensure that physical assembly does not introduce new decomposition errors. The distance between the physical shapes in a cell and the cell abstract could be increased; layouts could thus be abutted without introducing a decomposition error—at a minor impact in area.

Alternatively, wiring tracks could be pre-assigned corresponding colors. Valid placement locations for a cell or larger IP layout would align the colors assigned to the pin shapes with the wiring track definition. This alternative is emerging as the default for advanced process nodes—that is, the tracks on a multipatterning layer are assigned a specific color, and the shapes aligned to that track (both within cells and for routed wires) are by default assigned that color. There are some subtle nuances to this method, as routed designs may include a mix of minimum and non-minimum width wires on a layer (see Chapter 9, "Routing").

The shapes in Figure 1.9 illustrate the significance of line-end-to-line-end spacing in terms of the difficulty in resolving the segment ends with high fidelity. To support additional process node scaling, an alternative lithographic and fabrication approach is being applied. Rather than attempt to lithographically resolve two segments with aggressive line-end spacing on a single mask, distinct wires are merged into a single shape, which are then physically separated by a "line cut" masking layer.

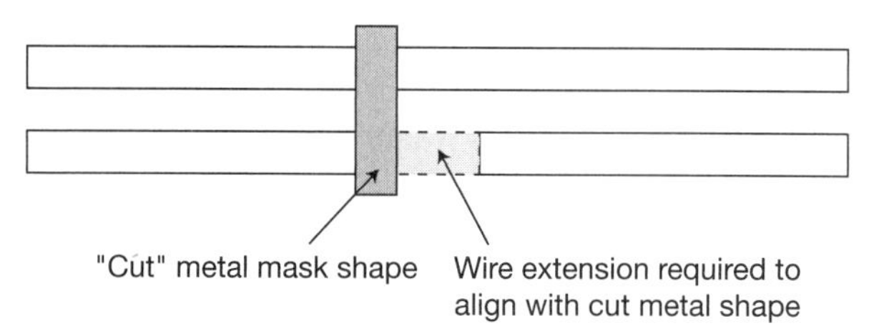


**Figure 1.9** The resolution of line-end spacing at advanced process nodes requires merging (and potentially extending) line segments in support of a separate "cut mask" lithography step. Note that the (minimum area) cut mask shape will span multiple tracks. Line ends for wires may need to be extended past their original endpoint to align with the cut mask.

The methodology team needs to collaborate with the foundry support team on the specific mask layer and checking operations in the PDK techfiles and runsets to merge two logically and physically distinct design nets into a single wire plus a cut mask shape. The routing flow needs to ensure that the final wiring track segments will satisfy a subsequent merge-and-cut operation.

As VLSI process nodes continue to scale, the "two-color" multipatterning approach has continued to evolve. One option is to decompose the shapes data into an increasing number of subset colors (continuing with the 193i illumination wavelength). Another option is to accept stringent limits on the flexibility in width and space rules and use process material deposition thickness as a defining dimension. Figure 1.10 illustrates a *self-aligned, double-patterned* (*SADP*) fabrication flow. The number of mask layer exposures is reduced with SADP. In return, the post-etch width of the underlying material is fixed by the dimensions of the "sidewall spacer," a dielectric material isotropically deposited and anisotropically etched.

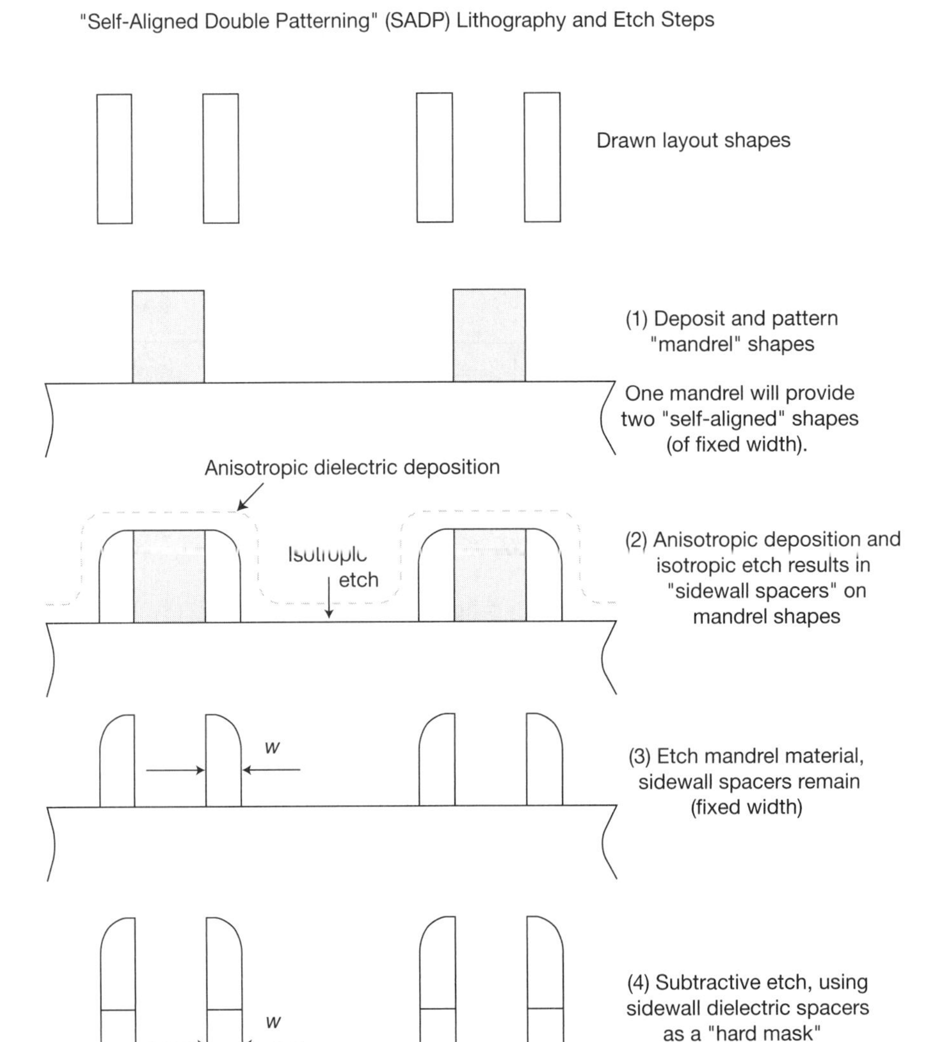


NOTE: Subtractive etch steps shown — e.g., patterning of silicon "fins" for FinFET devices. SADP lithography and etch for a deposition process (additive, "damascene") differs slightly.

**Figure 1.10** A self-aligned, double-patterned process flow utilizes the deposition of a sidewall spacer on a patterned shape as a subsequent masking layer. The final (fixed) resolution of the etched material is defined by the width of the spacer.

If the final etched material is a dielectric, the subsequent deposition of a metal will result in a fixed space between adjacent wires. For the case of a FinFET transistor (discussed in the next section), the etched material is the silicon "fin" used for the device substrate; in this case, the thickness of the silicon fin is fixed by the SADP process. Successive applications of this technique offer a self-aligned, quad-patterned (SAQP) flow for continued dimensional scaling; the original SADP-etched material serves as the mandrel for a subsequent SADP process sequence.

### 1.4.4 Bulk CMOS Technology

The origins of IC logic technology began with bipolar transistor circuit configurations (e.g., emitter-coupled logic [ECL]), transistor-transistor logic [TTL]). The emphasis on VLSI scaling resulted in the emerging process maturity of metal-oxide-semiconductor field-effect transistors (MOSFET) and specifically logic circuits consisting of complementary n-channel (nFET) and p-channel (pFET) device topologies, or CMOS. (There was also a short-lived logic technology offering merging CMOS logic devices with bipolar transistor output drivers, but BiCMOS did not continue through subsequent process node scaling.) This section briefly reviews the prevalent CMOS process technology options, with an emphasis on how their different characteristics impact the design methodology.

Advanced fabrication technology research is being pursued to define transistor topologies that will continue to enable scaling for the 5nm ITRS node and below—for example, (horizontal or vertical) nanowires with the device input gate "all around" the nanowire semiconductor. Regardless of the likely emergence of new transistor materials and topologies to support the demand for PPA improvements, the methodology considerations for those technologies will have much in common with the three CMOS processes highlighted in this section.

The traditional CMOS process technology utilizes a silicon wafer substrate as the foundation for fabrication of n-channel and p-channel FETs. Specifically, Figure 1.11 illustrates a cross-section of the two device types. A starting p-type bulk wafer consists of an epitaxial crystalline layer grown on a substrate; the epi layer has a smaller concentration of introduced p-type impurities, and the substrate has a much higher concentration. The pFET requires an n-type background material, which is realized by the introduction of n-type impurities, creating an *n-well* in the p-epi layer. Electrical isolation between device nodes is improved by the addition of a recessed oxide, or "shallow trench" oxide isolation (STI), dielectric introduced into the substrate.

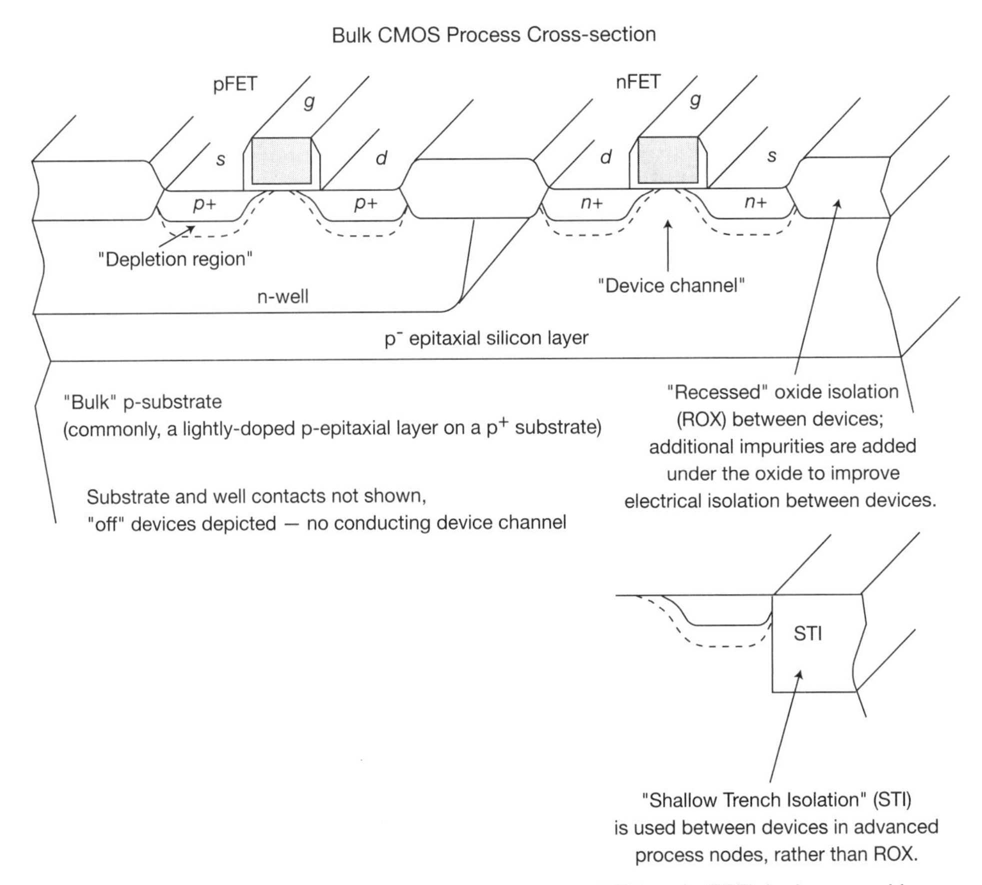


**Figure 1.11** Silicon wafer cross-section for nFET and pFET devices used in complementary metal-oxide-semiconductor (CMOS) logic circuits.

The nFET and pFET devices consist of four terminals: gate, drain, source, and substrate. The input gate terminal is isolated from the *FET channel* by a thin gate oxide dielectric. The conductivity between drain and source is modulated by the voltage difference between gate and source ($V_{gs}$), with an indirect influence of the source-to-substrate voltage difference, $V_{sx}$. (The substrate contacts to the p-epi for an nFET and the n-well node for the pFET are not shown in the cross-section figure but will be added liberally to circuit layouts.)

The electrical charge carriers in the surface channel region with $V_{gs} = 0V$ isolate the drain and source nodes, denoted as the "accumulation" region of transistor operation. As the input voltage $|V_{gs}|$ increases—$V_{gs}$ positive for nFETs, $V_{gs}$ negative for pFETs—the charge carrier concentration in the channel changes from the majority carriers of the substrate to the minority carriers of the source/drain nodes. A depletion region is initially formed in the channel as $|V_{gs}|$ increases, where the concentration of free charge carriers is reduced. The electric field across the depletion region provides the drift transport for any injected carriers. As $|V_{gs}|$ increases further, a concentration of minority free carriers is present at the surface channel. The conductivity of the channel between drain and source increases through "weak inversion"—with $I_{ds}$ at leakage current levels—to "strong inversion." The input gate voltage where the minority carrier concentration in the channel is (roughly) equal to the majority carrier substrate concentration in the substrate is defined as the transistor threshold voltage, $V_t$:

$$V_{tn} = (V_g - V_s) > 0V; \; V_{tp} = (V_g - V_s) < 0V$$

As a CMOS logic input signal transitions between logic voltage levels, the drain-to-source conductivity of the nFET and pFET devices receiving that logic gate input transitions in complementary fashion, with one device reaching inversion while the other reverts to accumulation. During this input transition, there is a "cross-over" period when both devices are conducting simultaneously, as depicted in Figure 1.12. This cross-over interval results in significant power dissipation between VDD and ground and detracts from the device current to the interconnect loading to propagate the logic gate output value.

A key methodology step is to ensure that the input logic signal *transition time* does not exceed a suitable limit to minimize the cross-over power dissipation; otherwise, buffering of the circuitry driving the input would be necessary.

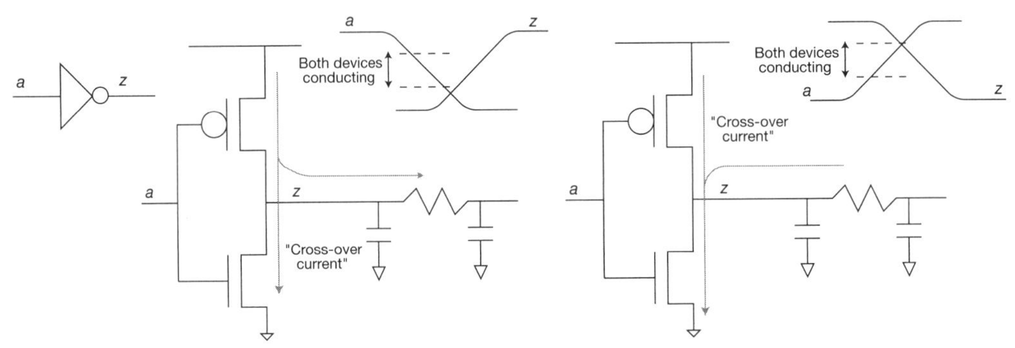


**Figure 1.12** An input transition to a CMOS logic gate results in complementary behavior of the nFET and pFET devices. A cross-over current flows within the circuit during the interval when both devices are conducting.

With the evolution of VLSI technologies, the device channel length has been the benchmark for the process node definition. The "drawn" channel length in layouts has typically been the same as the node name (e.g., L_drawn = 10nm). The actual "electrical channel length" differs from the drawn length as a result of the process bias—a combination of lithography, deposition/etch, and impurity introduction steps that result in a wafer-level dimension that differs from the drawn dimension. The transistor simulation models in the foundry PDK incorporate this bias so that the circuit netlist with drawn values provides accurate results.

The scaling of device dimensions in advanced process nodes has also necessitated reducing the applied supply voltage to keep the electric fields between device gate and channel through the thin oxide below the reliability concerns of dielectric breakdown. Similarly, the electric fields between source/drain and substrate need to observe limits associated with p–n junction reverse bias avalanche breakdown. As supply voltages have scaled, from 5V at the 2-micron node to sub-1.0V for current processes, the magnitudes of the threshold voltages have also necessarily been reduced to maintain circuit performance. As a result, devices with a $V_{gs}$ = 0V input may indeed be in the weak inversion mode. The *subthreshold leakage currents* in a CMOS logic circuit (i.e., $I_{ds}$_leakage for $V_{gs} < V_t$) are a major concern for the (static) power dissipation of SoCs intended for mobile applications. The electric field between drain-and-source of an "off" device in weak inversion (i.e., $|V_d - V_s| = VDD$, $V_{gs} = 0V$) is also a reliability concern, potentially resulting in *device punchthrough* current. As VDD supply voltage scaling has slowed with successive nodes, fabrication processes have transitioned to an effective gate length significantly greater than the drawn length.

A fabrication option is commonly provided that locally varies the majority surface carrier concentration in the channel and thus provides different transistor $V_t$ values. (The previous section highlights library standard cells using LVT, SVT, and HVT devices.) A low-$V_t$ device, with fewer (net) majority carriers in the channel switches sooner upon an input transition and thus provides a faster overall logic propagation delay. The trade-off is that the low-$V_t$ device has increased subthreshold leakage current when the device is off. In the weak inversion mode, the $I_{ds}$ subthreshold current is exponentially dependent upon the $V_{gs} - V_t$ voltage difference. As illustrated earlier in the log-linear plot in Figure 1.3, a low-$V_t$ device has an exponential increase in leakage current compared to a nominal "standard" $V_t$ device. The PPA optimization steps in the methodology flow evaluate performance versus (static) power dissipation when considering a logic cell swap between circuit implementations with different $V_t$ devices.

### Body Effect and $V_t$ Dependence on $V_{sx}$

The device channel operating mode depends on the source-to-substrate voltage difference. As the magnitude of $|V_{sx}|$ increases, the magnitude of $|V_t|$ also increases. To first order, this is a square root dependence; that is, $|V_t|$ is proportional to ( $|V_{sx}|$**0.5 ). This *body effect* has a significant impact on circuit behavior and the overall design methodology.

One benefit of the body effect is that it offers an opportunity to readily provide a sleep state to an IP core. For example, application of a pFET n-well voltage above the VDD supply rail results in a higher $|V_{tp}|$, as illustrated in Figure 1.13. The leakage currents in a static logic network (e.g., with inactive clocks) are significantly reduced with this *body bias*.

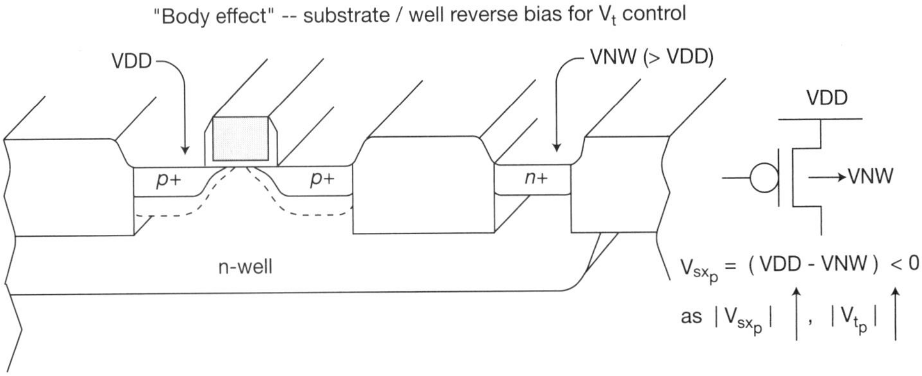

"Body effect" -- substrate / well reverse bias for $V_t$ control

**Figure 1.13** An n-well voltage above VDD applies a body bias to pFET devices, increasing the $|V_t|$, and significantly reducing the static leakage current.

Commonly, the n-well bias connection uses the VDD rail in the cell template. If a separate n-well voltage rail distinct from the logic VDD supply is provided, and a power management controller is available to generate a unique supply above VDD, the body effect circuit topology is a relatively straightforward means to achieve a reduced-leakage sleep state. The complementary nature of CMOS processes implies that a similar body bias implementation is available by applying a negative voltage to the p-type bulk connection for nFET devices; the nFET leakage currents are reduced with $(V_s - V_x) > 0V$.

Note that there is a significant capacitance $C_{sx}$ that would need to be charged/discharged to use body bias to transition to/from the low-leakage condition. Also, there is a strict limit on the magnitude of the source-to-substrate voltage bias that can be applied due to limits on the electric field present in the surrounding depletion region. Additional fabrication steps can be taken to increase the body bias limit, with tailoring of majority carrier concentration near the source/drain nodes and below the surface channel, including a "lightly doped drain" (LDD) profile near the channel surface, halo implants at the source/drain device nodes, and buried layers of higher-majority-impurity concentration in the bulk substrate. Nevertheless, the effectiveness of body bias in scaled VLSI process nodes for bulk CMOS is reduced due to the applied voltage and electric field limits.

Another characteristic of the body effect is the impact on circuit performance for FET devices connected in series. Figure 1.14 illustrates the circuit topology for the series nFET devices from a three-input NAND gate. (The pFETs connected to the logic gate inputs in the complementary parallel topology are not shown in the figure.) Series devices need to be significantly wider—typically, implemented as parallel fingers—to reduce the nominal "on resistance," as depicted earlier in Figure 1.2. A logic input transition that results in a current through a series nFET stack is illustrated in Figure 1.14.

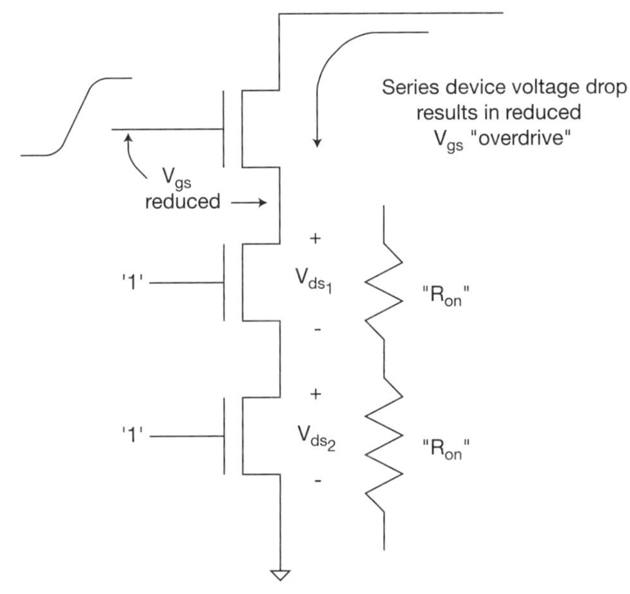


**Figure 1.14** A logic transition current through a series stack of devices generates a $V_{sx}$ voltage difference at intermediate nodes, increasing the device $|V_t|$ and reducing the stack conductivity.

The transient current results in a $V_{ds}$ voltage across the resistance of the individual devices in strong inversion; the voltage drop at the intermediate nodes in the stack provides an inherent body effect. During this logic transient, the switching device conductance decreases, adversely impacting the logic transition delay.

In the bulk CMOS substrate device cross-section in Figure 1.15, there are additional characteristics of note.

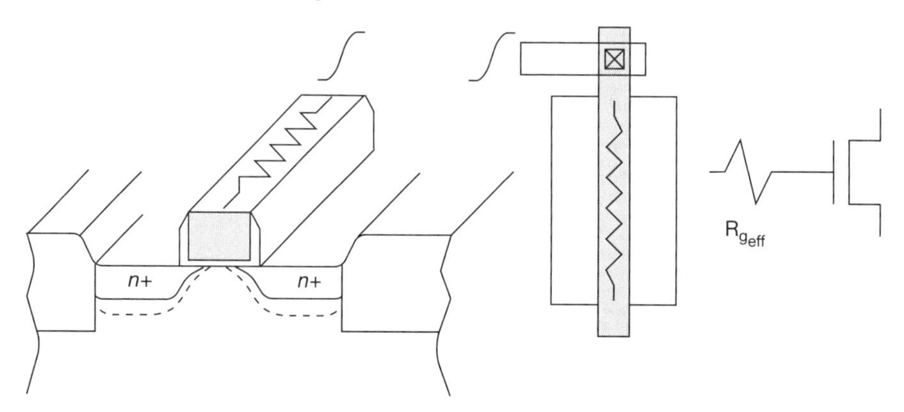


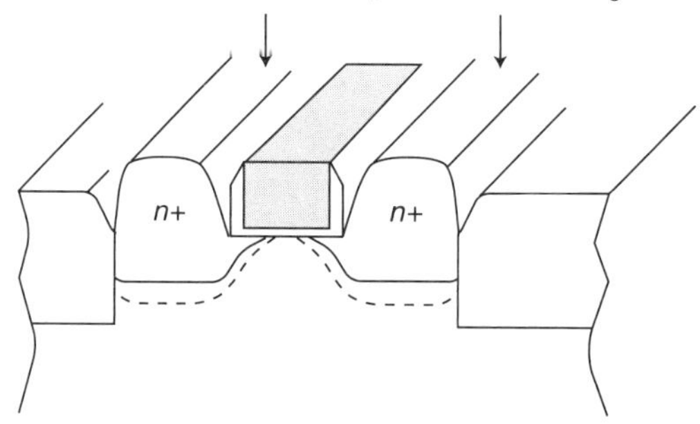


**Figure 1.15** Electrically, the device gate input connection is a distributed R*C network, as the input transition charges the channel capacitance through the resistive gate. Circuit simulation models for the device reflect this behavior using an "effective" $R_g$ element. Additional process steps are introduced to add to the silicon volume of the source/drain nodes to reduce the series resistances $R_s$ and $R_d$ in the device model. The "raised" nodes incorporate a high impurity concentration (n-type for nFETs, p-type for pFETs) during the silicon growth step.

### *Low-Resistivity Gate Material*

The FET gate is more accurately modeled as a distributed $R*C_{channel}$ network. A transient on the device input propagates down the resistive gate, charging the local channel capacitance. Various process options are available from the foundry, with different gate material resistivity (e.g., poly-crystalline silicon with a high impurity concentration, polySi with a metal silicide top layer, refractive metals). At advanced process nodes, to reduce the impact of the increase in gate resistance with gate length scaling, a metal gate material is used. The device models include an "equivalent" parasitic $R_{gate}$ so that simulation with the model device capacitances more accurately represents the input transient.

The contact potential between the gate material and the gate oxide dielectric also helps define the threshold voltage of the device. In addition, the foundry offers local device $V_t$ options through the introduction of a small concentration of impurities at the channel surface, using a *threshold implant* mask.

### *Raised Source/Drain Nodes*

Complex process steps are incorporated to increase the silicon volume of the source/drain nodes, concurrently introducing minority carrier impurities. The device model includes parasitic resistances $R_s$ and $R_d$ in series with the device channel. The high-impurity concentration "raised" nodes reduce this parasitic resistance. The top of the source/drain nodes typically includes a metal-silicide material layer to further reduce the source/drain resistivity. The method for introducing the raised source/drain nodes may also transfer compressive (pFET) or tensile (nFET) stress into the silicon crystal lattice of the device channel. These lattice forces are used to improve the inversion minority carrier mobility, improving the $I_{ds}$ device current for a given $V_{ds}$ voltage difference.

### *Spacer Dielectric Between Gate and Raised Source/Drain Nodes*

In addition to the $R_s$, $R_d$, and $R_g$ elements, another key device characteristic is the parasitic effects of $C_{gs}$ and $C_{gd}$. The transient operation of the device through accumulation and inversion involves charging the internal gate-dielectric-channel capacitance, $C_{ox}$; in the device model, the voltage-dependent $C_{ox}$ is allocated between the elements $C_{gs}$, $C_{gd}$, and $C_{gx}$. The fixed parasitic capacitances in the device structure are added in parallel to the internal $C_{gs}$ and $C_{gd}$ elements (see Figure 1.16).

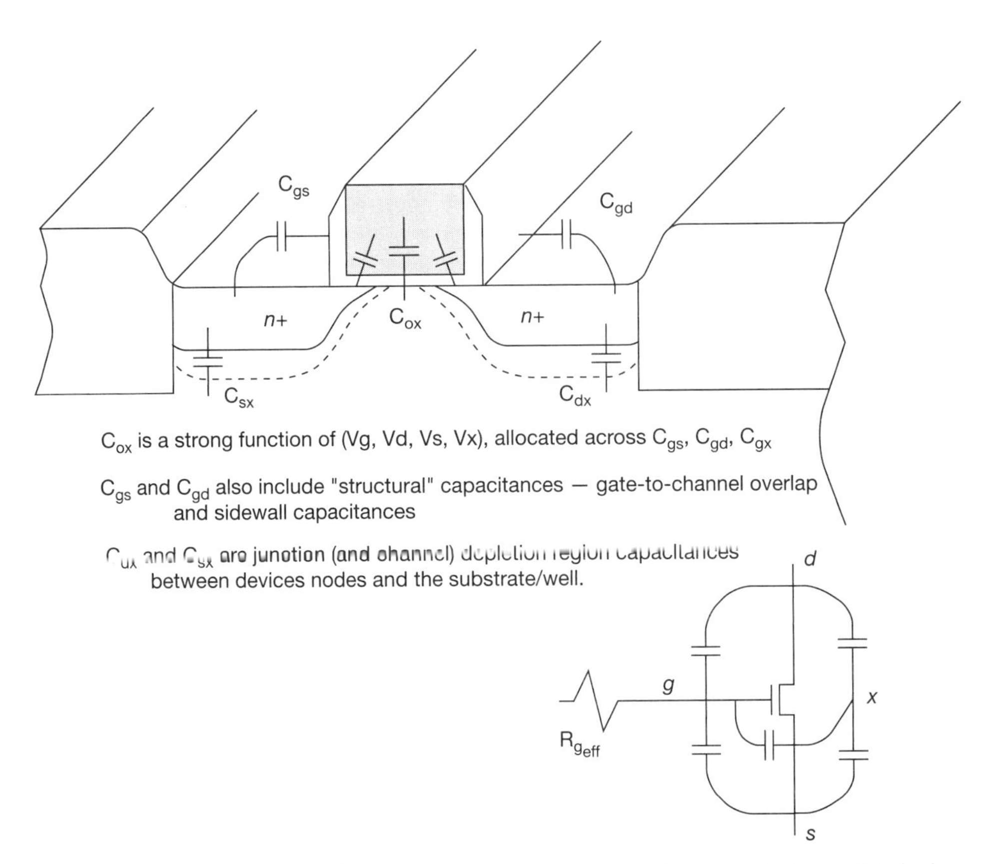


**Figure 1.16** Capacitances $C_{gs}$, $C_{gd}$, $C_{gx}$, $C_{sx}$, and $C_{dx}$ are included in the device model to reflect the gate-to-channel input capacitances, the source/drain junction capacitances to the substrate, and the configuration of the dielectric materials between the device nodes. For simplicity, raised source/drain nodes are not shown in the figure, but their topology contributes significantly to $C_{gd}$ and $C_{gs}$.

The dielectric surrounding the gate defines the parasitic capacitances added to $C_{gs}$ and $C_{gd}$. The materials choice and thickness of the "spacer" is a critical process optimization. A larger spacer reduces the $C_{gs}$ and $C_{gd}$ parasitics but adversely impacts the layout pitch for the gate-to-device contact-to-gate dimension. A smaller spacer improves the *contacted gate pitch* but adversely impacts the capacitive parasitics.

In Figure 1.16, also note that there is a (fixed) $C_{gs}$ and $C_{gd}$ parasitic contribution due to the gate-to-source and gate-to-drain overlap at the channel surface. Although it is required to ensure a continuous, conducting channel in inversion, significant process engineering development resources are applied to reduce the extent of this overlap.

From a design methodology perspective, the main implications of selecting a bulk CMOS fabrication process are as follows:

- All circuit simulation flows need to ensure that the parasitic elements—$R_g$, $R_s$, $R_d$, $C_{gs}$, $C_{gd}$, and $C_{gx}$—are enabled and accurately reflected in the device model, using data from the foundry process PDK.
- The primitive cells in the IP library with a high fan-in (i.e., a large number of devices in series) are logically powerful. However, these cells have the disadvantage of requiring large-width devices to reduce the $R_{on}$ device resistance. There is also the performance impact of the transient body effect. The SoC methodology and design teams need to evaluate the PPA benefits of these cells relative to design goals and determine if any should be excluded from the available cell library. (To some extent, the IP cell library provider has already made this decision in terms of the diversity of its logic offering—for example, "no logic gates with series stack height greater than three.")
- If the SoC design team intends to utilize the (limited) back bias device characteristic for leakage control, this introduces additional voltage settings for electrical analysis flows. The addition of a distinct back bias rail from the circuit VDD (or ground) rail impacts the definition of the cell template and available routing tracks. Power distribution network analysis needs to include the additional back bias rail (e.g., the I*R voltage drop on the rail due to leakage currents, the transient rail currents to/from sleep state transitions). A special layout physical design rule check applies for the back bias rail, as the voltage difference between this rail and internal circuit nodes exceeds VDD. The methodology team needs to review with the foundry how the "high voltage differential" spacing checks are to be applied in the PDK runset operations and ensure that the back bias rail connections are designated with the corresponding property.

There are two additional very important reliability considerations for selecting a bulk CMOS process:

- **Latchup**—Adjacent nFET and pFET devices in a bulk CMOS process introduce a p-n-p-n junction topology, typically associated with a silicon-controlled rectifier (SCR). If a transient (capacitive-coupled) event were to inject sufficient carriers into the "triggering input," and if the current gain of this composite structure were greater than one, the SCR action would result in a high-parasitic, sustained current; this *latchup* condition must be avoided. Foundry layout design rules provide guidelines on the density of well/substrate contacts and provide requirements for separate injected charge collection "guard rings" surrounding circuits subject to capacitive triggering. Both precautions reduce the likelihood of a latchup event. A latchup analysis methodology flow and a set of circuit and layout checking criteria is needed for a bulk CMOS design (see Section 16.2).

- **"Soft errors"**—The exposure of an SoC design to cosmic radiation and/or close proximity to radioactive material decay results in the trajectory of a high-energy particle toward the bulk substrate. As the particle traverses through a high electric field depletion region near a device source/drain, lattice collisions generate free carriers that are swept to a device node by the field. The resulting collected charge on the node introduces a voltage differential. If the active current at the node is small, this differential could result in erroneous (detectable) circuit behavior. As the error is not due to a permanent defect, such as a dielectric breakdown, it is denoted as a "soft error."

The most prevalent example of circuits subject to soft errors in bulk CMOS is a node dynamically storing charge, such as an array location in an embedded dynamic RAM IP block on an SoC. The lack of an active current from the dynamic node results in a high charge collection/retention efficiency. SRAM bit cells and flip-flop circuits also rely on very-low-current feedback loops to retain state; these circuits are also susceptible to soft error event upset.

The foundry adds features to the bulk CMOS process to reduce the sensitive free carrier generation and collection volume when a particle strike occurs. For example, devices are commonly fabricated using a bulk substrate with a thin, low-impurity concentration epitaxial crystalline surface layer, as depicted in Figure 1.11. Free carriers generated in the high-impurity concentration bulk have a high recombination rate rather than drift to a sensitive circuit node.

Even with these process features, critical design and methodology steps are needed to reduce the soft error rate. The layout design area of any dynamic or weak feedback node needs to be minimized to reduce the collection volume. The methodology flow needs to include an analysis of the probability of a single-event upset (SEU) based on the sensitive layout collection volume, anticipated particle flux and energy distribution, and carrier generation probability (see Section 16.4). The resulting SEU probability calculation is used by the design team to determine the need for additional parity generation and/or error detection and correction (EDC) logic on sensitive memory array word or stored register values.

The IP vendor has to address a trade-off when designing a function with SEU-sensitive circuits, as the end customer SoC applications could be quite varied. The addition of EDC functionality reduces the SEU susceptibility, at the cost of additional area and performance, which may detract from the marketability for very cost-sensitive end products. Medical and mil-aero applications have the most stringent SEU correction requirements. Indeed, IP qualification for a mil-aero application is likely to require more than the typical stress testing of the IP from a foundry's shuttle lots. Special electronic testing facilities offer a high radiation flux exposure chamber similar to a high-altitude environment to more accurately evaluate whether SEU correction methods are sufficient; this additional qualification adds considerably to the IP development cost.

### 1.4.5 Fully Depleted SOI (FD-SOI) Technology

Silicon-on-insulator (SOI) process technologies have been available in production for several VLSI process generations. As illustrated in Figure 1.17, a thin silicon layer with "near intrinsic" (very low) impurity concentration is used to fabricate the devices, similar to a bulk CMOS process. In this case, a dielectric layer exists between the devices and the substrate. The presence of a dielectric introduces several differences to a bulk CMOS process:

- The bulk CMOS process parasitic reverse-bias leakage currents for the p-n junctions between source/drain and bulk/well nodes are eliminated.
- The parasitic $C_{sx}$ and $C_{dx}$ capacitances are now a silicon-dielectric-bulk structure rather than the p-n junction capacitance.
- The body effect is now controlled by the dielectric capacitance between the substrate and device channel.

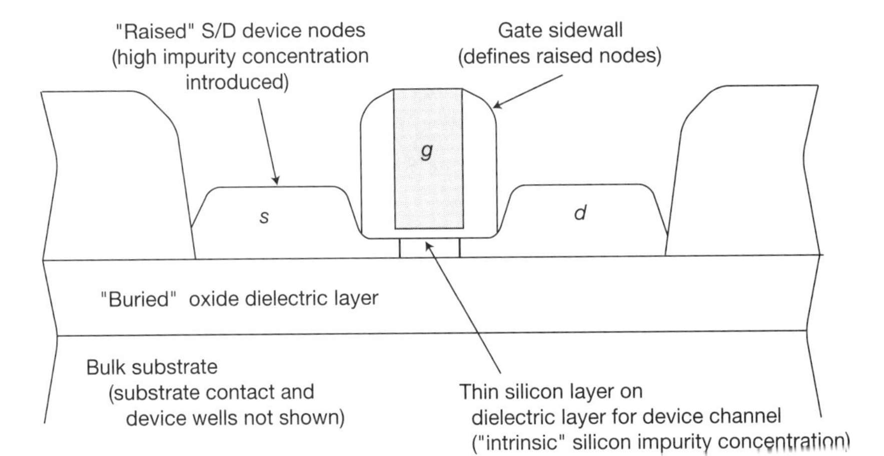


**Figure 1.17** In this cross-section of a fully depleted silicon-on-insulator device, note the dielectric layer between the device channel and the wafer substrate. The substrate/well connections are etched through the thin silicon and dielectric top layers. (The well profile below the dielectric layer is not shown in the figure.)

As VLSI technologies have scaled, the thickness of the SOI process silicon device layer has also been reduced. At current process nodes, the depth of the source/drain impurity regions extends throughout the silicon layer. The silicon layer and device gate oxide thicknesses are selected such that the gate input has a strong electrostatic control over the channel. For example, the silicon layer thickness is roughly one-third of the device electrical length (distance between drain and source nodes). As a result, the device current falls quickly as the gate input turns off, which reduces CMOS cross-over current and improves logic circuit performance. This is expressed as the device model parameter known as the *sub-threshold slope*—mathematically, $S = dV_{gs}/dI_{ds}$, ($V_{gs} \sim V_t$). Alternatively, this device measure is expressed as the difference in $V_{gs}$ that results in a 10X decade reduction of $I_{ds}$ in the sub-threshold region—for example, S = 70mV. The gate oxide-channel electrostatics are such that the channel volume is fully depleted of majority free carriers when the device is "off"; this specific SOI process option is denoted as *FD-SOI*.

The FD-SOI process has some unique characteristics:

- The elimination of source/drain junction leakage currents and the improved device sub-threshold slope result in lower overall leakage power than a comparable bulk process node.

- Latchup is effectively suppressed, as p-n-p-n junctions are absent.
- The dielectric layer between device channel and bulk substrate offers the opportunity to apply either a negative or positive body bias to the device.

  As with a bulk process, a negative bias increases $|V_t|$, reducing leakage currents further. The magnitude of the applied negative body bias can be larger than in a bulk CMOS process, as it is limited by the electric field across the dielectric rather than the electric field between the source/drain and bulk junctions.

  The SOI dielectric also offers the possibility of a (limited) forward body bias relative to the channel, reducing $|V_t|$ and providing an active device current increase. (A forward body bias in a bulk CMOS process is discouraged due to the forward bias on all p-n junctions.) A performance boost could be realized from forward body bias, either temporarily or from an adaptive controller that generates a bias to minimize performance variations over process, voltage, and temperature ranges.
- Multiple $V_t$ device variants are implemented differently in FD-SOI than for a bulk process, as the controlled introduction of additional impurities into the very thin SOI channel is more difficult. Instead, the contact potential between the materials can be varied (i.e., the gate-oxide, oxide-channel, channel-insulator, insulator-well interfaces). For example, the (metal) gate composition can be altered to adjust the gate-to-oxide contact potential. Alternatively, the impurity concentration (and type) in the local well below the insulator can be modified to adjust the insulator-to-well potential. Nonetheless, it is more complex (and expensive) to introduce device $V_t$ options in FD-SOI than for a bulk process.
- The volume of high-impurity concentration source/drain nodes is constrained by the thickness of the silicon device layer; $R_s$ and $R_d$ are significant model parasitics in FD-SOI. As with a bulk process, raised source/drain node process engineering is pursued. Nevertheless, there remains a series resistance contribution to $R_s$ and $R_d$ from the thin FD-SOI silicon layer.
- The physical layouts of bulk CMOS circuits are relatively easy to migrate to an SOI process (at the same node). As the device channel

cross-sections between bulk CMOS and SOI are similar, the device and contact layouts are likewise comparable.

To transition a bulk CMOS layout to an FD-SOI equivalent, specific layout migration support is needed for bulk CMOS $V_t$ device variants and bulk/well contacts.[12] (As mentioned earlier, in FD-SOI, the well contacts require etching through the top silicon and insulating layers.)

- Despite the relative ease of device layout migration, the FD-SOI process does not easily support the addition of some circuit elements used in bulk CMOS analog IP (e.g., bipolar npn transistors, p-n junction diodes).

With FD-SOI, unlike with bulk CMOS, there is no inherent (vertical) p-n junction with precise impurity concentration profiles. Analog IP design requires adaptation to the available circuit elements. (This also applies to other IP circuits using p-n junctions, such as ESD protect structures and temperature sensors.) For example, a lateral p-to-intrinsic-to-n junction may be available from the foundry by introducing different impurity types for the source and drain nodes, separated by the intrinsic device channel. Alternatively, an FD-SOI nFET device with the gate input tied to the drain as anode with the source as cathode provides a similar two-terminal structure.

The methodology for an FD-SOI design is similar to that for a bulk CMOS design. A foundry may offer both processes at the same node, and an SoC design team may want to pursue a competitive PPA analysis of critical IP blocks. Layout migration may be a new flow to add to port existing bulk layouts to FD-SOI for re-extraction and performance characterization.

The increased range of (negative and positive) body bias in FD-SOI offers a variety of operating mode conditions and requires a review of the multi-corner, multi-mode (MCMM) analysis matrix (see Section 10.2). As with bulk CMOS designs, the applied FD-SOI body bias requires support for high-voltage signal spacing checks.

### 1.4.6 FinFET Technology

The FD-SOI device technology is based on a topology in which the gate input voltage has a strong electrostatic influence on the thin, fully depleted silicon layer. An alternative topology that also provides this electrostatic impact is realized by the FinFET device, whose cross-section is illustrated in Figure 1.18.

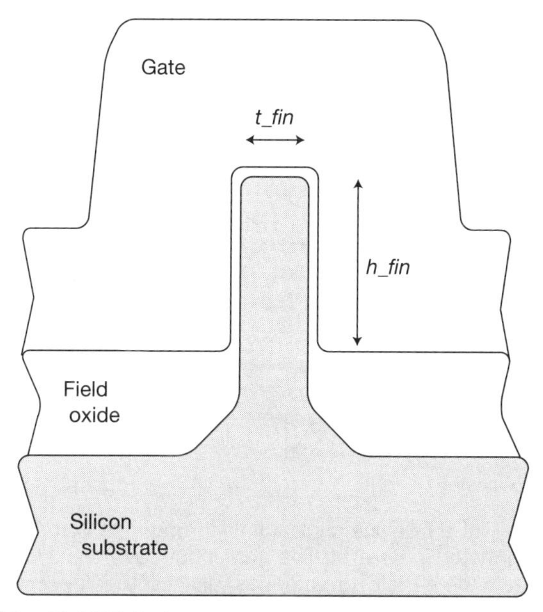


**Figure 1.18** FinFET device cross-section. The device gate length (and direction of $I_{ds}$ current flow) is perpendicular to the page.

The device channel volume is defined by a vertical pedestal, or *fin*, of silicon, that is defined by a sequence of process steps:

- Silicon wafer etch using SADP to define a uniform fin thickness, based on the width of the spacer. (Note that the silicon etch is deeper than the final fin height.)
- Dielectric deposition
- Wafer polish
- Precisely controlled oxide etch-back to expose the fin

The device gate oxide is grown on the fin, followed by gate patterning. The final device consists of a common gate covering all three exposed facets of the fin. The device current between drain and source nodes flows laterally through the fin, as depicted in Figure 1.19.

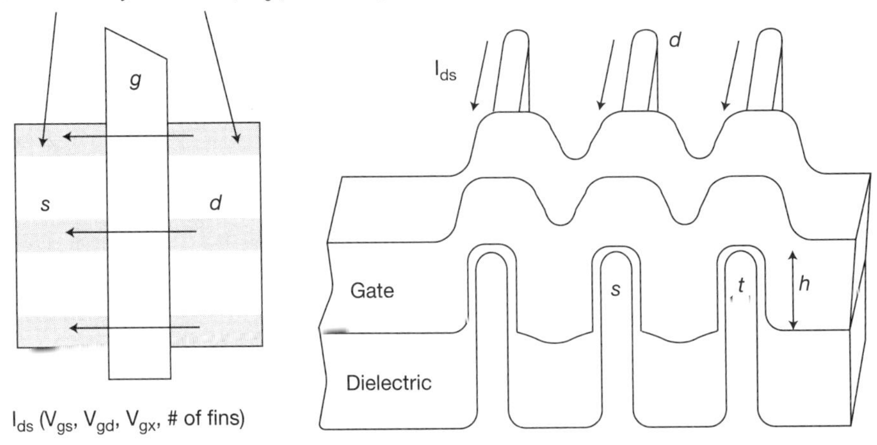


**Figure 1.19** FinFET device current flow through the vertical fin. The figure depicts three fins connected in parallel. The gate input traverses all the fins. A local metal shape connects the parallel drain nodes, and another local metal shape connects the parallel source nodes (not shown in the figure).

Note that the effective width of the device channel is defined by the fin geometry: Current flows through a silicon surface channel that spans the two vertical sides and the top of the fin. Thus, the effective device width of a single fin is $W = ((2 * h\_fin) + t\_fin)$. The fact that the FinFET width is not a continuous circuit design/layout parameter but rather is quantized in multiples of $(2h + t)$ has significant impact on the methodology flow (as discussed shortly). For process simplicity, the fin height and thickness are (nominally) uniform across the wafer. In addition, no combination of fins and planar FET devices are typically offered. Also, for advanced lithography VLSI process nodes, there is typically very little flexibility allowed in drawn gate length.

The fin thickness is sufficiently small, such that the fin volume for an "off" device is fully depleted of majority carriers. The subthreshold slope of the FinFET is attractive, much as with FD-SOI. There is a bulk drain-to-source leakage current path in the base of the fin, at the interface between the gate and thick dielectric. Additional process steps are pursued to alter the impurity concentration in the base of the fin to reduce this leakage current and reduce the risk of drain-to-source punchthrough current at a high $V_{ds}$ voltage. Although FinFET devices could be fabricated using an SOI wafer substrate, where the silicon layer thickness above the insulator would define the fin height, all FinFET high-volume fabrication is currently using bulk silicon substrates.

### *FinFET Characteristics*

There are unique characteristics of device variation in FinFET fabrication as compared to planar (bulk or FD-SOI) devices:

- **Fin profile**—The silicon pedestal etch and subsequent oxide etch-back process steps to expose the silicon fin introduce a source of variation in the fin height and thickness across the wafer. Actually, the corners of the fin are not rectangular but intentionally rounded to avoid high gate-to-channel (horizontal and vertical) electric fields at the top of the fin (see Figure 1.20). The initial FinFET device models assumed a rectangular profile. The PDK model parameters (and statistical variations) required fitting the model to fabricated silicon characterization data. More recently, extended device models that accept a more general (non-rectangular) fin profile have been developed.[13,14]

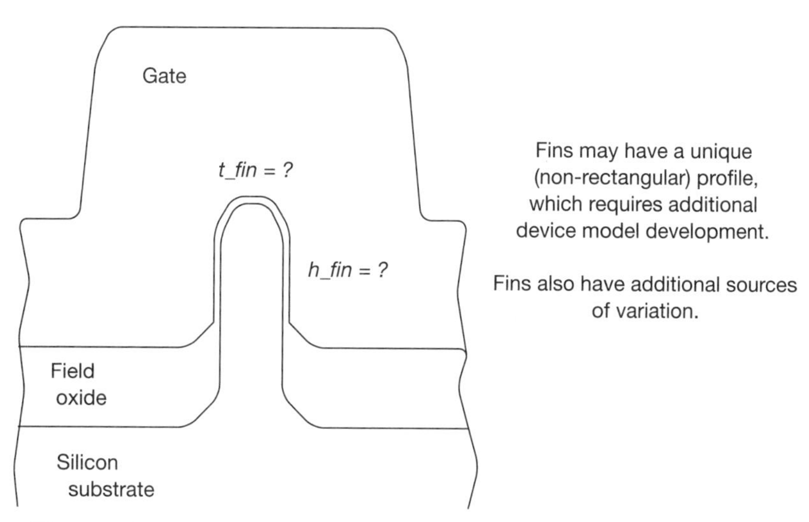


**Figure 1.20** FinFET corner profile. The fin cross-section is typically closer to being a trapezoid than to being a rectangle.

- **$V_t$ variation (SVT device)**—The silicon fin utilizes a very low (near intrinsic) background impurity concentration, resulting in low threshold voltage variation in the "standard" $V_t$ device.
- **Fin surface roughness**—The FinFET-specific process steps result in a risk of vertical surface roughness and subsequent variations in local oxide thickness and device current.
- **Gate line edge roughness (LER)**—Although planar devices are also subject to LER gate length variations along the width of the device, the FinFET gate must traverse the vertical topography of the fin, which is a difficult patterning process step to control.
- **Gate oxide thickness at the FinFET base**—At the base of the fin, the gate material interface transitions from the thick dielectric over substrate to the thin device oxide layer. Ideally, this transition would be abrupt, restricting channel current to the vertical fin surface. As mentioned previously, there is (active and sub-threshold) device current along the base transition that is more difficult to model.

### *FinFET Advantages*

The FinFET technology offers some significant advantages over the planar device processes:

- **Layout density**—The vertical fin provides an effective device width of $(2h + t)$ in a drawn dimension of width $t$. To realize a larger device width, multiple fins are connected in parallel, as shown in Figure 1.21. A single gate input spans all fins, and local metal interconnects short together the individual source and drain nodes. Figure 1.21 illustrates three fins in parallel, with a total width of $W = 3 * (2h + t)$, in a drawn width of $(3t + (2 * space\_fin))$. Assuming an appropriate fin height relative to the fin space, the total device current density is higher than for a planar FET layout. As mentioned previously, the total FinFET device width is not a continuous dimension but rather is a multiple of the individual fin effective width.

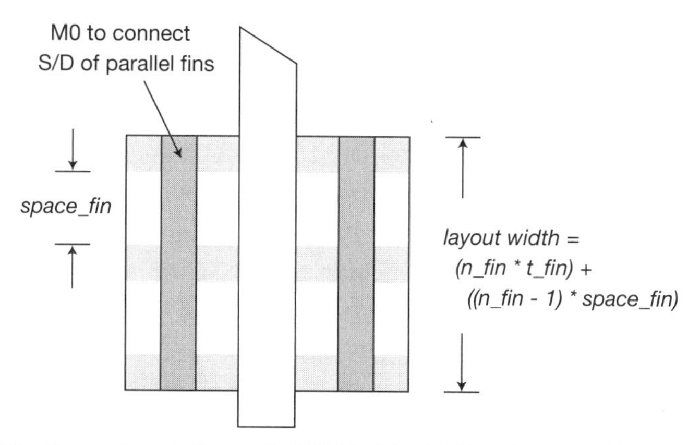


Effective device width = n_fin * ((2 * h_fin) + t_fin) > layout_width
FinFET device layout current density is improved over a planar device layout.

**Figure 1.21** Multiple fins connected in parallel. The device width is an integral multiple of ($2h + t$). The current density is improved for FinFETs compared to planar (bulk or FD-SOI) devices, as the effective width of multiple fins is greater than the layout width.

Note that the fabrication process design rules impose strict requirements on the fin positioning. The space between fins is fixed. Fins for separate devices throughout the die are "on grid"; the fabrication process for the silicon pedestals does not readily support staggered fin alignment in adjacent cells. The design rules may impose limits on the number of parallel fins per device. Even if there are no stringent lithographic limits, the parasitic gate resistance ($R_g$) and parasitic capacitances ($C_{gs}$ and $C_{gd}$) adversely impact an input transient for a large fin count. The maximum number of fins may also be constrained by the local metal interconnects: The increased current density through these metals introduces a reliability concern due to electromigration (see Chapter 15). As with planar device technologies, I/O pad driver circuits still require many device fingers in parallel to provide the large transient currents rather than using very large fin counts per device.

- **Very weak body effect**— The FinFET topology, the strong electrostatic control of the gate over the active fin volume, and the punchthrough stop impurity introduction below the fin imply that the substrate voltage has little influence on the FinFET device's behavior. Unlike the bulk CMOS and FD-SOI planar devices, the body effect for FinFETs is very weak. Although the subthreshold leakage and subthreshold slope are attractive features, the application of the body effect to further optimize static leakage power in a sleep mode is not as effective for FinFET devices.

The lack of a body effect also implies that the series device stack performance penalty is reduced. The combination of the improved layout current density and low series stack penalty suggests that a FinFET IP library may incorporate base logic cells with higher fan-in (series stack height) than for a planar process.

- **Multiple $V_t$ device variants available**—The local introduction of additional impurities into the fin or an alternative metal gate composition with a different oxide contact potential enables multiple FinFET device $V_t$ options at a relatively low additional process cost.

#### *FinFET Constraints*

For IP circuit designers, FinFET technology introduces new constraints:

- **Parasitic $C_{gs}$, $C_{gd}$, and $C_{gx}$**—The topography of the gate traversing multiple parallel fins adds parasitic capacitances, as highlighted in Figure 1.22. The improved layout current density is mitigated somewhat by the additional device input capacitance.

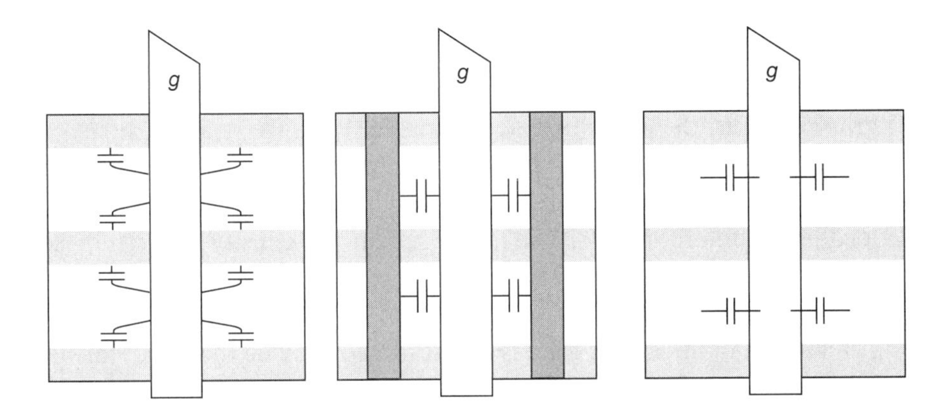


Additional $C_{gs}$, $C_{gd}$ capacitances — to fin sidewalls and M0 Additional $C_{gx}$ capacitances below gate

Parasitic capacitances for the gate traversal between fins

**Figure 1.22** Illustration of the additional parasitic capacitances $C_{gs}$, $C_{gd}$, and $C_{gx}$ for the gate input traversing over the dielectric between parallel fins. There are additional contributions from the gate to fin sidewall capacitance ($C_{gs}$ and $C_{gd}$) and the gate over thick oxide between fins to the substrate ($C_{gx}$).

The optimum FinFET circuit sizing to reduce total logic path delay is an intricate function of device drive strength, internal capacitances, and the external loading from signal interconnects and fanout gate inputs. For small interconnect loading, increasing the logic gate drive strength along a path quickly reaches a "gate-limited" condition, where increased currents are countered with comparable increases in internal and gate fanout capacitive loads.

- **Quantized device effective width**—The primary design impact of the FinFET technology is the limit on device widths, in multiples of the fin dimensions ($2h + t$). Analog circuits no longer have the ability to arbitrarily tune to specific bias currents. Fortunately, many analog designs rely on device matching to a greater extent than the absolute magnitude of the bias currents. The low variation in the FinFET standard $V_t$ (with intrinsically doped fins) and the weak body effect offer good differential matching characteristics.

  Flip-flop circuits often rely on weak feedback of the output value to an internal node to retain state. The feedback circuit needs to provide necessary node leakage currents yet must be easily overdriven when writing a new flop value. The smallest single fin device width (and limited gate length options) may result in a stronger feedback circuit than would be needed to sustain internal circuit node voltages and, thus, higher internal switching power dissipation for these flip-flop circuit types.

  The FinFET width constraints also have a strong influence on SRAM bit cell design. For a planar process, the sizes of the cross-coupled inverters and word line access devices in the bit cell are optimized for PPA, for a target number of bit cells located in a column between a differential pair of bit lines. The SRAM bit cell offerings in a FinFET process technology are limited by the quantized fin width. A high-density bit cell could be designated by the foundry as using one fin throughout for access, pullup, and pulldown transistors (e.g., a 1-1-1 cell). A high-performance option could be implemented with (asymmetric) device widths, such as the 2-2-1 design depicted in Figure 1.23. The memory array IP provider integrating the foundry FinFET bit cell is constrained by these sizes.

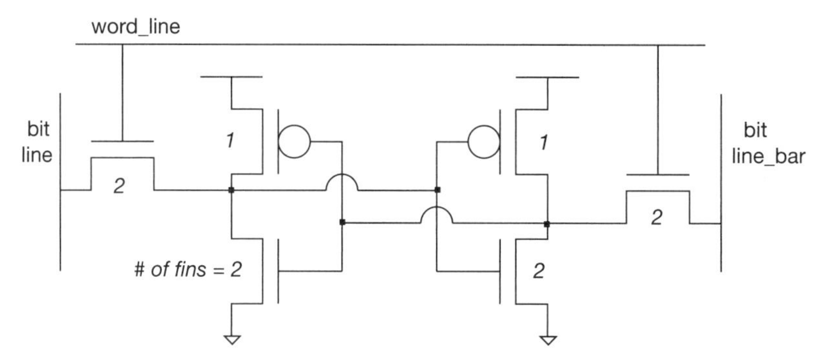


**Figure 1.23** FinFET 6-T SRAM bit cell design example, illustrating the device widths as an integral numbers of fins (all with minimum gate length).

These FinFET bit cell characteristics may introduce a unique SRAM (sub-block) architecture if the high-density or high-performance array circuit design results in a different number of bit cells per column than would be defined for a planar process.

- **Electromigration and self-heating**—The higher FinFET layout current density implies that the local metal wires and contacts will also be carrying a greater current density. The risk of electromigration failure in these segments is increased. The fin profile also results in a constrained thermal path. The ($I_{ds}$ * $V_{ds}$) power dissipation in the device channel generates heat; the energy flow follows the path of low *thermal resistance*. In a bulk CMOS process, the silicon substrate provides a low thermal resistance under the channel. In an FD-SOI process, the thin insulator layer is less thermally conductive, and an increased fraction of the heat flow traverses through the device nodes. For a FinFET process, the relative dimensions of the fin (with the surrounding dielectrics between fins) results in an inefficient thermal path to the silicon substrate. A large percentage of the thermal energy flows through the device nodes and interconnects. This energy results in a local self-heating temperature rise. The activation of metal atom transport associated with electromigration in a current-carrying segment increases exponentially with temperature. Electromigration analysis needs to include the corresponding temperature rise from active device self-heating in the neighborhood of the interconnect segment.

### *FinFET Design Methodology*

The methodology for IP design in FinFET technology also must address several new constraints:

- All cell placements must result in the underlying device fins being on-grid.
- All device schematic widths must be a multiple of $Nfin * (2h + t)$; all device layout shape widths must be $(Nfin * t) + ((Nfin - 1) * s)$.

  For simplicity, the schematic symbol notation for FinFET devices could replace the ($W$, $L$, *N_fingers*) parameters for bulk devices with (*Nfin*, $L$, *N_fingers*); indeed, the PDK device simulation models are likely to be defined using *Nfin*. To reduce visual complexity, the fin process lithography layer need not be explicitly drawn in circuit layouts; a single device area shape on-grid and of vertical dimension equal to $(Nfin * t) + ((Nfin - 1) * s)$ could equivalently represent the parallel fins. Layout parasitic extraction and design rule checking flows would internally convert to the required fin geometry.
- Inactive gates drawn on the edge of the active area are required for litho and process uniformity.

As with the planar process technologies, it is necessary to expand the silicon volume of the drain and source nodes to reduce $R_d$ and $R_s$. A spacer dielectric is again deposited on the sidewalls of the gate. The FinFET geometry is unique, as the additional silicon grows in multiple (crystalline) directions off the exposed source/drain fin surface. To improve process uniformity, the step for expanding the fin nodes benefits from a symmetric topography of gate dielectric spacers on both sides of the node. As a result, it is necessary to add a dummy gate at the edge of an active area. The spacers on both sides of the source/drain fin define the extent of the raised node. Figure 1.24 illustrates the addition of a dummy gate at the edge of an active area. The figure also illustrates dummy gates inserted adjacent to the active device areas for improved photolithographic uniformity.

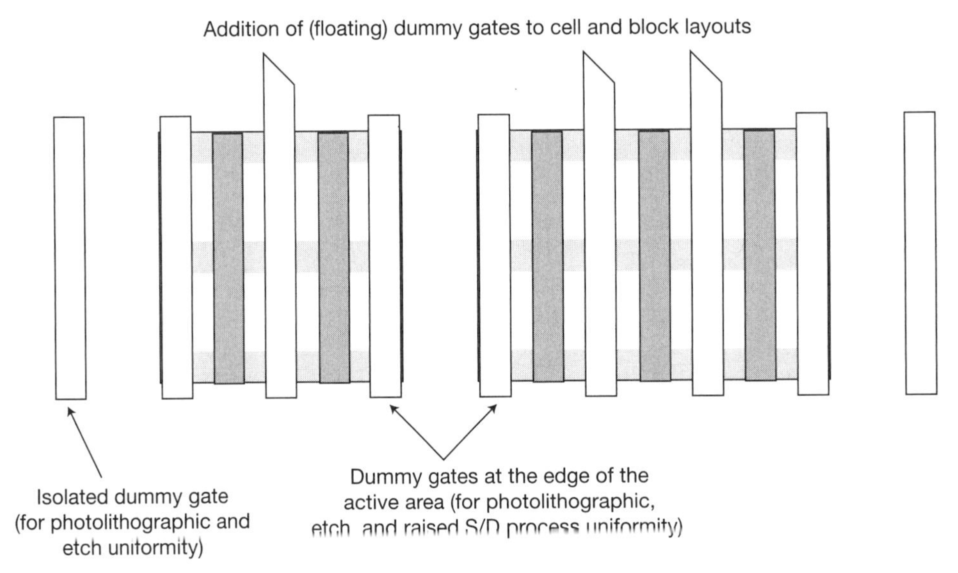


**Figure 1.24** FinFET device layouts require "dummy gates" at the edges of active areas for lithographic and fabrication uniformity; the gates on the active device edge are electrically floating. An alternative design style would merge the dummy gates at the edges of separate active areas, tied off to electrically isolate adjacent nodes.

The dummy gate on an active area edge requires unique methodology consideration. Device recognition algorithms need to exclude this non-functional device from layout-to-schematic (LVS) correspondence checking. Layout parasitic extraction needs to include the parasitics associated with this abbreviated device (e.g., a "three-terminal" FinFET instance in the extracted circuit simulation instance with floating gate and no source node).

To minimize the overhead of these dummy gates, layout designers may seek to consolidate active areas, merging adjacent FinFETs. In this case, the single dummy gate between devices must be tied to an "off" voltage. The layout analysis methodology must recognize the presence of the off device isolation and apply the correct interpretation for various flows (i.e., a non-functional device for layout-versus-schematic correspondence checking yet included in layout parasitic extraction for the additional parasitics and leakage currents between adjacent circuit nodes).

Extending this approach further, it would be feasible to use off devices between adjacent library cells. The cell template would need to readily support connecting gate inputs to the "off" logic value power rail for both nFET and pFET devices. Special "context cells" would need to be abutted when analyzing an individual cell to estimate the final environment. Special "row end" non-functional cells would need to be added to the placed layout. A complication of the use of off devices between cells arises when abutting cells with different numbers of fins in the active areas, as depicted in Figure 1.25.

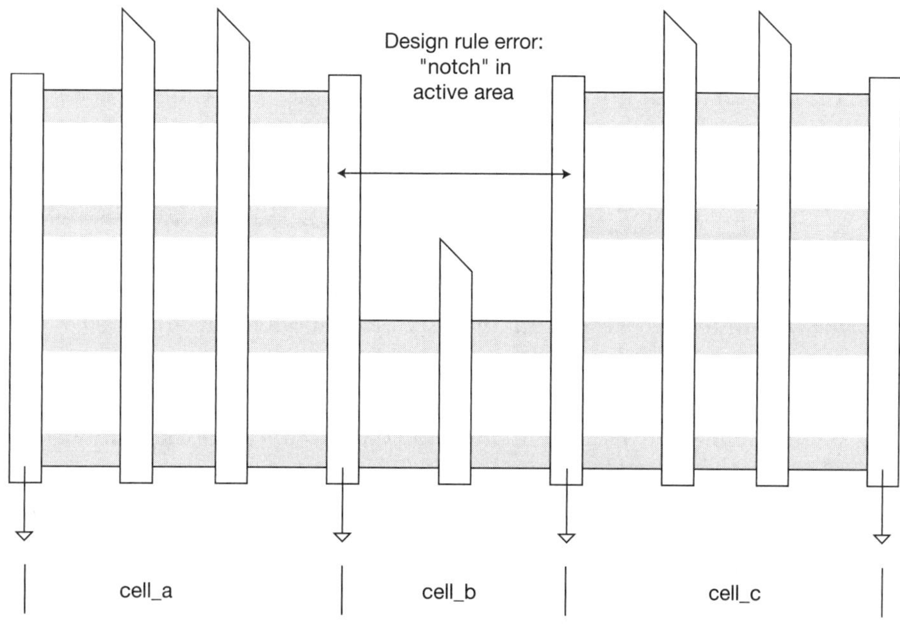


**Figure 1.25** Illustration of a merged dummy gate layout style between adjacent library cells. A design rule error is introduced by the resulting notch in the merged active area.

Adjacent cells may result in a "notch" in the active area that violates process lithography design rules. The methodology flows to support (limited) abutment of the active area in different cells require additional consideration for cell characterization and IP block physical layout completion.

Much as with FD-SOI, the FinFET process does not easily provide for a (high-quality, vertical) p-n junction or bipolar transistor. Analog IP design requires adaptation to the available circuit elements.

### 1.4.7 Operating Corners and Modes

Logic path timing closure, power dissipation calculations, and electrical analysis all need to be confirmed over the full range of fabrication process variation, temperature extremes, and operating voltages. Specifically, there are local and global variations to factor into the analysis. The supply voltage applied at the chip pads minus the power distribution network voltage drop margin is used to assign the local circuit voltage. A chip thermal map helps determine the local rise in temperature above the ambient environment. In addition, there are tighter process variations locally than would be measured globally across the die or wafer or fabrication lot; in this case, the "tracking" of local circuits helps reduce the range of electrical and dimensional parameter variations to consider.

The evaluation of timing, power, and electrical characteristics across the full extent of process, voltage, and temperature (PVT) ranges is impractical. Rather, a set of *PVT corners* is defined for analysis. For example, an "sslh" corner would imply that circuit analysis was conducted using a "slow" nFET, "slow" pFET, low applied VDD at the chip pads (minus the PDN voltage drop margin), and high temperature. This corner would typically result in the slowest circuit delays for path timing and the data setup-to-clock timing checks at flip-flop inputs. Conversely, the "ffhl" corner-based timing delays would be used for data hold-to-clock path timing checks.

Product marketing may include performance sorting, or "binning," of parts based upon their measured performance; higher-performance parts from the statistical fabrication distribution might command a premium price. To assist with the estimation of the fab yield by bin, it is common to add a nominal, or "typical," transistor model corner for analysis. The "ttlh" corner reflects a nominal fabrication process, with the voltage and temperature extremes to which the part will be subjected in the final product application.

Although the highest operating temperature has typically been associated with the slowest circuit performance, due to both device current characteristics and interconnect resistance effects, that assumption is no longer neessarily valid. The device current is a complex function of two temperature-dependent parameters: the threshold voltage ($V_t$) and the inverted channel carrier mobility. The device $|V_t|$ and carrier mobility both decrease with increased temperature; the decrease in $|V_t|$ provides greater device current, while the decrease in mobility reduces device current. At older process nodes, the mobility reduction was dominant, and device current decreased at higher temperature. At newer nodes, the scaling of the nominal device $V_t$ (commensurate with the reduced VDD supply voltage) results in a more complex interaction between threshold voltage and mobility. As a result, the high temperature corner may indeed result in greater device current, a phenomenon known as *temperature inversion*. However, interconnect resistance still increases with temperature, further complicating the electrical behavior.

The process variation ranges are applicable not only to the transistor characteristics but also to the tolerances on the fabrication of interconnects and vias. The use of chemical-mechanical polishing (CMP) for wafer planarization after damascene metal deposition results in wire thickness variations. The use of multipatterned decomposition and multiple lithography/etch process steps for interconnect shapes on a single layer will result in tolerances in the distance between adjacent wires due to mask-to-mask alignment variation. As a result of CMP-based and MP-based process steps, there will be ranges of interconnect resistance and wire sidewall-to-sidewall coupling capacitance. The definition of an analysis corner thus needs to be expanded to include variations in the extracted interconnect parasitics. For example, the additional parasitic calculations could include min/max C_total, min/max R*C_total, and min/max C_coupling. As with the temperature inversion behavior, it is not necessarily definitive that a specific corner is always the *worst case* or the *best case* for path timing checks and electrical analysis.

In short, multiple PVT corners need to be evaluated across all methodology flows. EDA tool vendors have adapted their algorithms and data structures to support concurrent, multi-corner analysis on the network of devices and interconnect parasitics.

In addition, the SoC design likely has multiple logical operating *modes*, each with corresponding power/performance requirements. For example, there may be design states that include:

- **Performance boost**—Technically, a new PVT corner, with a higher applied supply voltage to support higher clock frequencies
- **Sleep mode**—Inactive clocks and perhaps a reduced supply voltage with logic state and memory array values "retained"
- **Deep sleep**—Supply or ground rails isolated from logic networks using series "off" sleep devices, with the lowest leakage currents, as depicted in Figure 1.26

- **Test modes**—For example, serial scan shifting of test pattern stimulus and response data, at a reduced test clock frequency

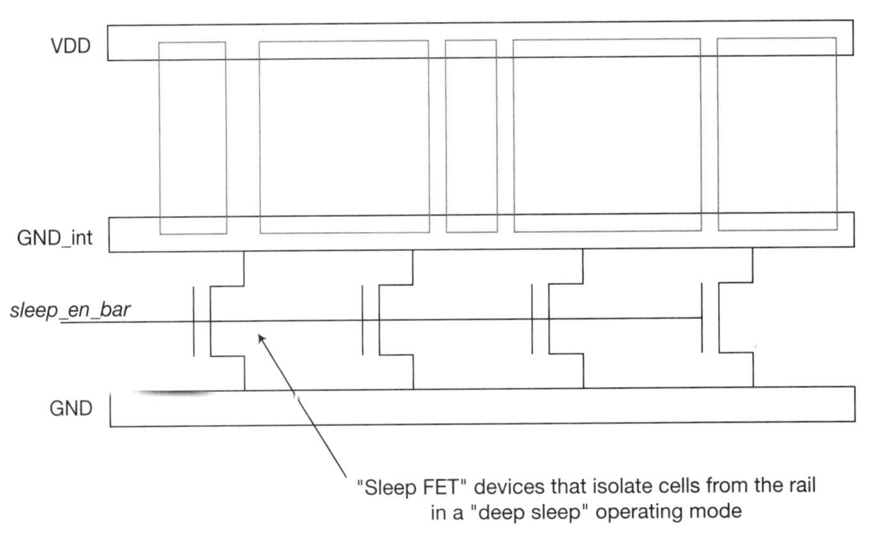


**Figure 1.26** Circuit topology for a deep sleep operating mode, with the addition of sleepFET devices between an internal rail and a global rail. nFET sleep devices and a split GND distribution are shown.

It is therefore necessary to create a "cross-product" of operating modes and corners to analyze. EDA tools have been expanded to accept definitions of different modes to sensitize the related logic paths; a concurrent multi-corner, multi-mode (MCMM) analysis is then performed.

The SoC methodology and design teams need to review the mode-corner matrix to determine the necessary conditions for each analysis flow step. Some combinations can be excluded for a particular flow. For example, test mode operation would not be applicable to a performance boost voltage corner. (The various sleep modes require unique analysis flows, to be discussed shortly.) Nevertheless, the number of MCMM combinations can be large; the computational workload to run MCMM analysis on the full-chip SoC model would be prohibitive. The methodology needs to evaluate IP blocks efficiently over the MCMM matrix and provide appropriate abstract models for full-chip analysis. For IP received from an external supplier, these abstracts are a key deliverable, spanning the corresponding modes and corners for the final SoC application.

There is an additional methodology consideration related to special analysis requirements for mode state transitions. For example, an IP core transition from a deep sleep state to an active state results in unique circuit current distributions. During the deep sleep duration, network nodes are disconnected from one of the rails; thus, these floating nodes drift over time to an unknown voltage level due to various leakage currents. At the transition back to an active state with the network reconnected to the rails, there are significant rail currents to charge/discharge the internal network nodes. Commonly, the resistance of the logic network to global rail connection is gradually reduced during this mode transition, through the turning on of successive sleep devices in parallel over multiple "wake-up" clock cycles. The rail in-rush current needs to be analyzed over the allocated number of clock cycles to ensure that rail peak current values are within appropriate limits. (SoC functional validation also needs to ensure that the clock cycles are logically allocated to this transition and that logic functionality is not assumed before the wake-up transition is complete.) Once active, electrical analysis is required for the effective "on" resistance of the parallel sleep devices to ensure that the active network current does not adversely impact I*R voltage drop margins. The methodology needs to ensure that flows cover the unique requirements of mode transitions in addition to the MCMM tool capabilities.

### 1.4.8 Process Variation–Aware Design

The previous section describes how the expected range of process and environment variation is represented by analysis corners. When the set of process parameters is selected for a corner, there is an implicit assumption on the statistical distribution of the corresponding circuit delay. For example, if the circuit delay is measured repeatedly by sampling across all process parameter variations, a composite normal (Gaussian) distribution would result. The target best-case (BC) or worst-case (WC) process variation is a specific number of standard deviations from the nominal, mean performance – e.g., a "three-sigma" delay. Path timing closure and electrical analysis using the corner parameters will effectively represent an n-sigma point on the overall performance/power distribution, as depicted in Figure 1.27.

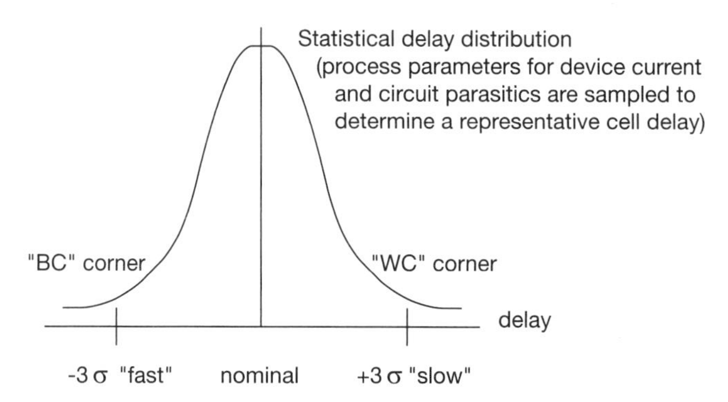


**Figure 1.27** The process parameter set selected for a (worst-case or best-case) PVT corner represents a composite of many individual parameter variations to provide an effective n-sigma circuit delay.

Note that at advanced process nodes with aggressive VDD supply voltage scaling, the statistical delay distribution results for a characterized logic circuit across process variations is decidedly non-Gaussian; in this case, a different method to define process corner definitions is required.

For more accurate variation analysis, an extended approach is warranted. Superimposed on the overall global process distribution is a local distribution. Circuits that rely upon matching characteristics for device pairs need to perform offset analysis, using assumptions on local device variation. Similarly, interconnect R*C parasitics have on-chip variation (OCV), which is an important consideration for calculating clock skew and signal data path-versus-clock arrival timing tests at sequential circuit inputs. A modified global corner parameter definition is used as a starting point; a local distribution is added, and its standard deviation increases with the distance between which circuits are placed on the die. The endpoints of the local distribution provide a new parameter set that can be applied to individual instances in analysis flows to model OCV extremes. Alternatively, multiple analysis simulations can be performed by repetitive (Monte Carlo) statistical sampling from the local distribution. The results data from multiple simulation iterations provide an analysis output distribution for further yield review.

The complexity of timing path networks is very high: Rather than use a statistical sampling method for entire paths, a unique approach to modeling OCV for timing analysis is commonly used instead, applying *derating multipliers* to the delay arcs in each path (see Section 11.4).

### *Variation-Aware Array Design*

Large memory array IP presents a unique consideration for variation analysis. The sheer number of devices in the array suggests that a much higher-sigma set of analysis parameters should be used. Three-sigma circuit-limited yield (CLY) design closure on a large array would imply a potentially significant number of marginal "weak bit" cells and, thus, a larger allocation of spare array rows/columns to maintain sufficient fabrication yield. The goal to operate large arrays on a VDD_min supply voltage domain to reduce leakage power only amplifies the requirement to ensure that very few fabricated weak bit cells are present. Rather than use a 3-sigma CLY corner definition, array performance verification over process variation requires a statistical confidence analysis to a sigma value commensurate with the acceptable weak bit probability.

Array circuit simulations employ a high-sigma method to ensure valid read/write operation over the process variation distribution. Demonstration of array functionality to 5-sigma or 6-sigma yield involves sampling the parameter distributions, simulating the array operation, and measuring the results. The most direct circuit sampling approach employs a random Monte Carlo selection of parameter values; however, the number of simulations required to compile sufficient results data to demonstrate high-sigma yield would be prohibitive. EDA vendors have developed unique products for high-sigma circuit simulation. At a minimum, this simulation requires a full bit array column and sense amplifier network. Designers specify the measurement criteria that define a successful operation—for example, a written array value reaching x% of its final circuit node voltage by the end of the operation cycle or a sufficient sense amplifier bit line differential at the end of an array read. The EDA tool algorithm selects a sequence of parameter samples that efficiently explore the extremes of the results distribution for the measurement criteria. An n-sigma yield (with statistical confidence limits) can be calculated with far fewer simulations than a brute-force Monte Carlo sampling distribution would require.

High-sigma simulation with advanced sampling for array weak bit evaluation can be extended to other IP circuit types. For example, the capture of asynchronous data by a clocked flip-flop requires detailed characterization to analyze the probabilistic risk of the synchronizer entering a metastability fail condition. This synchronizer flip-flop characterization also benefits from a high-sigma simulation approach. (Note that the confidence level for a high-sigma sampled simulation is only as good as the accuracy of the parasitic extracted network surrounding the devices. Chapter 10, "Layout Parasitic Extraction and Electrical Modeling," discusses some of the extraction accuracy, capacity, and runtime trade-offs.)

#### *High-Sigma Simulation Fails*

The methodology for high-sigma analysis is an adjunct to the conventional CLY corner-based IP characterization methods. IP library developers need to make this EDA tool investment to accompany their existing simulation flow. Of specific interest are the sampled parameters for any simulation testcases that fail the measurement criteria. An IP developer compiles a results summary for failing simulations to review with the foundry. The foundry's continuous improvement process engineering team benefits from awareness of the critical process parameters, for which critical IP has a high CLY sensitivity. The foundry team can then focus on steps to reduce the variation of these key parameters.

Note that circuit-limited yield is a metric that is a function of IP design and fabrication parameter variation. CLY is distinct from the *defect density-limited* yield, which is a function of the manufacturing defects (per square millimeter of a specific size or larger) and the *critical area* of the SoC physical design sensitive to the presence of the defect. The overall fabrication yield is a product of these two factors; ideally, by design, the impact of the defect-limited yield is much larger than that of the circuit-limited yield.

### 1.4.9 Process Retargeting and Process Migration

The discussion on VLSI technologies in Section 1.4.1 indicates that direct lithographic scaling of physical layout to a new node or to an alternative foundry process at the same node was relatively straightforward through (roughly) the 130nm generation. EDA tool vendors offered layout migration and "layout compaction" tools to facilitate the retargeting of a design and automatic fixing of any (minor) layout design rule errors for the new process implementation. The diversity among different foundry PDK rules at more advanced nodes has increased, and the design rules between nodes at a specific foundry no longer broadly follow a predominant scaling factor. The introduction of multipatterning mask layers and the transition from planar to FinFET devices further complicate the migration of physical layouts. As a result, the recent usage of compaction tools has diminished.

To facilitate the efficient retargeting of layout designs, alternative approaches have emerged. For IP circuit design migration, EDA tool vendors provide parameterized layout cells and relative cell positioning, as described in the following sections.

### *Parameterized Layout Cells (pCells)*

The circuit schematic includes transistor instances that correspond to *parameterized layout cells*, commonly known as *pCells*. Invoking the pCell software function generates a transistor layout that corresponds directly to the schematic parameters (e.g., W (or Nfins), L, N_fingers). The pCell function may also add the source/drain/gate contacts and local metal interconnects; this is a key advantage of invoking pCells rather than attempting layout migration, as the rules for these connections vary widely between process nodes and foundries. The "generate-from-schematic" feature in the EDA tool platform exercises the pCells and seeds the new physical cell layout with the individual transistors.

The relative positioning of the seeded pCell layouts can follow the positioning of the schematic drawing itself or utilize a preferred pattern selected from an existing pattern library. (For analog IP with critical device matching requirements, individual fingers of separate devices can use an interdigitated common centroid pattern to minimize process variation sensitivity.) The generated layout data retains the schematic connectivity between devices. The generated parameterized layout cells can be repositioned, and common schematic nodes can be merged in the layout, based on the underlying connectivity model. The parameterized cell function includes the layout updates to share a source/drain node with an adjacent device. Alignment of nFET and pFET devices (horizontally) enables the common gate input to be extended (vertically) to connect both. After pCell positioning and merging, a device-level router can be invoked to complete the IP layout to the schematic connectivity. (Additional parameterized code is invoked to add the necessary dummy shapes around the devices.) Although more complex than automated layout compaction, custom layout assist tools provide considerable productivity to generate and complete a correct layout from a (migrated) schematic.

#### *Relative Cell Positioning*

To facilitate efficient IP core-level layout retargeting, EDA tool vendors offer an algorithm to maintain the timing optimization characteristics of the core. Starting with an existing cell-level netlist and a physical implementation, a relative layout cell position description can be derived. This description can then be applied during placement with the retargeted cell library to retain an optimized topology. (If the retargeted library does not have a corresponding logic function for all the instances in the original netlist, a cell-to-cell remapping tool can be used.) A set of net routing priority constraints ensures that critical routes and the relative cell placement retain the focus on the timing-optimized paths from the starting design. After layout retargeting, the path timing and electrical analysis flows are exercised using the new process technology PDK.

### 1.4.10 Chip-Package Co-Design

The focus of this text is on the methodology and flow steps for VLSI SoC designs. However, a comparable methodology and development engineering investment is made for package substrate design and analysis.

The package design rules and package composite layer stackup technology files define the physical layout and electrical analysis characteristics, similar to the PDK data from the silicon foundry. EDA tools specific to package physical design import the stackup and electrical materials data, the chip pad image, and the corresponding package pin data. A combination of automated route and interactive layout features completes the chip-package connectivity. The package netlist is likely to also include additional (surface-mount technology [SMT]) components to be integrated with the SoC die (e.g., SMT decoupling capacitors). The package layout tool also applies designer-provided rule checks to maintain matching topologies for critical busses and differential signal pairs. Shielding constraints can also be applied to ensure signal routes are located adjacent to power/ground wires and/or between supply planes in the package cross-section, to minimize switching noise on the signal. Specific package layer via definitions are often used to maintain an electrical impedance for (high-speed) signal transmission to minimize discontinuity reflections.

Due to the unique (and more interactive) requirements of package design, the EDA tools used are distinct from the physical design tools for SoCs. As a result, there is an important "chip-package co-design" interface required to/from the SoC methodology flows.

### *Early Chip-Package Floorplanning*

With VLSI scaling, the number of SoC I/O signals and supply voltage power domains has increased tremendously. Compared to the initial LSI packages using wire bonds from the chip pads to a lead frame prior to encapsulation, current VLSI SoCs predominantly use a flip-chip face-down die orientation, with a raised metal "bump" for chip pad-to-package substrate lead attach on a very aggressive pad pitch. This has enabled both the SoC signal I/O count to grow and the allocation of bump pads internal to the die to connect package supply planes directly to the power distribution networks on the top redistribution metal layers of the SoC metallization stack.

To ensure optimum package pin assignment, the initial chip-package co-design activity is to *floorplan* the chip signal pad locations and the corresponding package pins to evaluate the package signal trace lengths and trace congestion (see Figure 1.28).

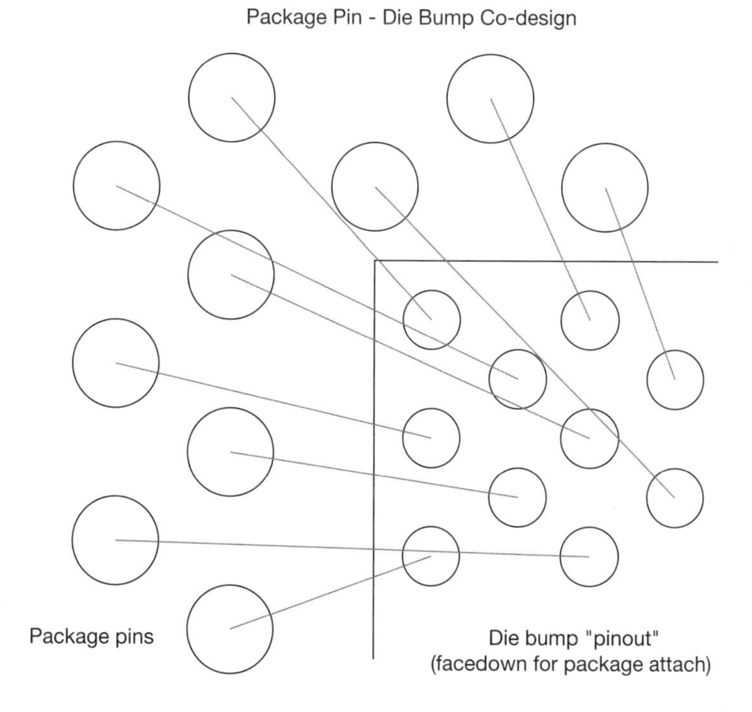


**Figure 1.28** Illustration of SoC die pad-to-package pin "flightlines," providing visual feedback during package floorplanning to guide physical location assignments for the die and package.

At this point, no detailed package routes are available; a visual flightline view is typically sufficient to identify opportunities for improved SoC bump pad assignments and, ultimately, corresponding SoC IP placements on the chip floorplan.

As the package design flow is separate, an efficient exchange of bump pad coordinate and signal assignment to/from the package design toolset is an additional required SoC methodology flow step. For a flip-chip bump SoC attach technology, note that the SoC coordinates used by the package reflect the face-down die orientation. (The author is aware of a catastrophic impact to a chip design project where the "face-up" bump coordinates were used for the chip-package model exchange, and this went undetected until after assembly of packaged prototype parts.)

### *Chip-Package Model Analysis*

Additional model exchanges between package and SoC domains are required for electrical and thermal analysis, as well.

For thermal analysis, a chip "power map" abstract is promoted to the package model. The (local) dissipated SoC power generates thermal energy that flows through both the die substrate and the die surface metal interconnects. A typical cross-section of a flip-chip implementation is shown in Figure 1.29, with an additional thermal interface material (TIM) layer compressed between die and package.

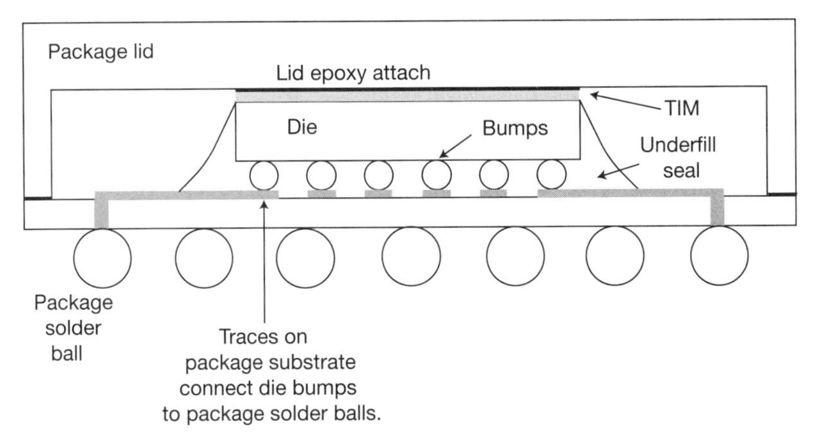


**Figure 1.29** Die-package cross-section, illustrating the addition of a TIM layer for improved thermal conductance. Different thermal expansion coefficients result in mechanical stress across the interfaces between the underfill and die attach materials.

The thermal resistance of this model is used to determine the heat flux to the ambient. The product designer integrating the SoC subsequently uses the results of the package thermal analysis to confirm the suitability of heat removal in the product enclosure.

The heat flow through the die surface (and resulting thermal model solution) introduces another analysis requirement. The different interconnect and dielectric materials present in the redistribution metals, bumps, underfill, and package substrate have different thermal expansion coefficients. A mechanical stress is present across these material interfaces, which requires "stress fatigue" analysis to ensure reliability across an appropriate number of power on/off thermal cycles.

The chip power model abstract also needs to contain sufficient detail to represent the time-averaged and transient current through the power and ground bumps. The SoC simulation methodology captures this data from on-chip power distribution network I*R voltage drop analysis and adds to the power abstract. A package analysis flow uses the bump currents from the SoC abstract to simulate and verify that the supply/ground voltage transients are within the allocated design margins that were used when characterizing IP circuit delays. The package simulation model should include the extracted parasitic inductance of the package power network, as well as explicit decoupling capacitors.

To date, there is no industry standard for the chip power abstract model used by package analysis tools. The SoC methodology team needs to collaborate with the package design team to determine what data format is required by their flows and derive the power abstract from SoC-level PDN analysis methods.

#### *Chip Electrical Analysis with Package Models*

High-speed I/O interfaces require frequency-dependent circuit simulations to confirm that the signal losses between drivers and receivers are sufficiently managed to maintain voltage levels that are accurately interpreted at the receiver. For serial interfaces with source-synchronous clocking, the high-frequency losses and adjacent signal crosstalk result in increased inter-symbol interference (ISI) between unit time intervals, impacting the accuracy of data recovery. The package team needs to provide extracted parasitic networks for (sets of) chip interface signals to enable the SoC I/O designer to confirm proper electrical driver/receiver behavior. The package signal models are often provided as a multi-port S-parameter matrix to represent the frequency-dependent insertion, reflection, and crosstalk losses of the package. The SoC designer also includes a reference load model for printed circuit board losses connected to the package model to complete the network description from driver to receiver—part of an overall chip-package-system (CPS) simulation methodology.

The other main chip-package co-design interaction for the SoC designer is the flow for electrical analysis of simultaneous switching outputs (SSO), as depicted in Figure 1.30 (see also Section 14.3).

Electrical Analysis Required for SSO Noise, with Detailed Die and Package VDD_IO and GND_IO Parasitic Model

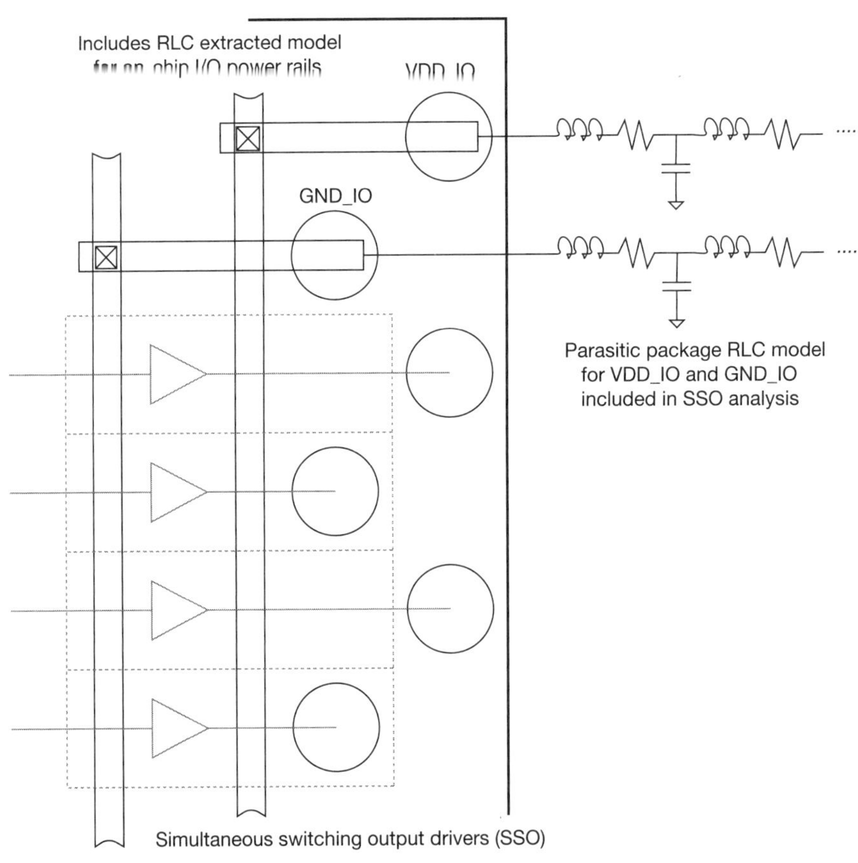


**Figure 1.30** Simulation model to evaluate simultaneous switching outputs (SSO) signal noise propagation, including both the local VDD_IO and GND_IO distribution and package electrical models.

A set of chip pad drivers that may switch in unison in a narrow time window will result in a transient current spike through the local PDN. The SoC methodology flow for SSO analysis requires excising of parasitics from the package PDN model, addition of a reference load, and attachment of the SSO pad driver circuits and related on-chip VDD_IO and GND_IO networks for circuit simulation. Potential SSO issues to investigate are:

- **The magnitude of the transient I*R and inductive noise on the VDD_IO and GND_IO rails**—The delay of the drivers is adversely impacted by excessive noise on the rails. As a result of noise on the VDD_IO and GND_IO supply rails, receivers are subject to changes in the input voltages that are adequately interpreted as the different logic levels (e.g., VIH, VIL). Quiet drivers on the same VDD_IO and GND_IO supply network propagate the PDN noise through to their pad outputs, as well.
- **"Overshoot" and "undershoot" on rails significantly beyond VDD_IO and GND_IO**—Inductive "ringing" *di/dt* voltage transients beyond the rail values forward bias p-n junctions connected to the rails, injecting current into the substrate/wells. To reduce the risk of latchup, pad drivers commonly include additional injected charge collection "guard rings" to suppress any p-n-p-n triggering current.
- **Excessive fast/slow output signal slew rates**—The I/O driver specification to the system designer needs to include min/max signal slew rates into the reference load; these slews are to be valid across the range of PVT corner simulations, which is a difficult design objective. I/O driver cells may include a set of PVT "compensation" inputs, which adaptively control the number of transistor fingers switched on in parallel (see Figure 1.31).

If compensated drivers are to be used, the methodology flow needs to verify the correct connectivity of the drivers to the PVT sense macro outputs and include the full (compensated) network in SSO simulations.

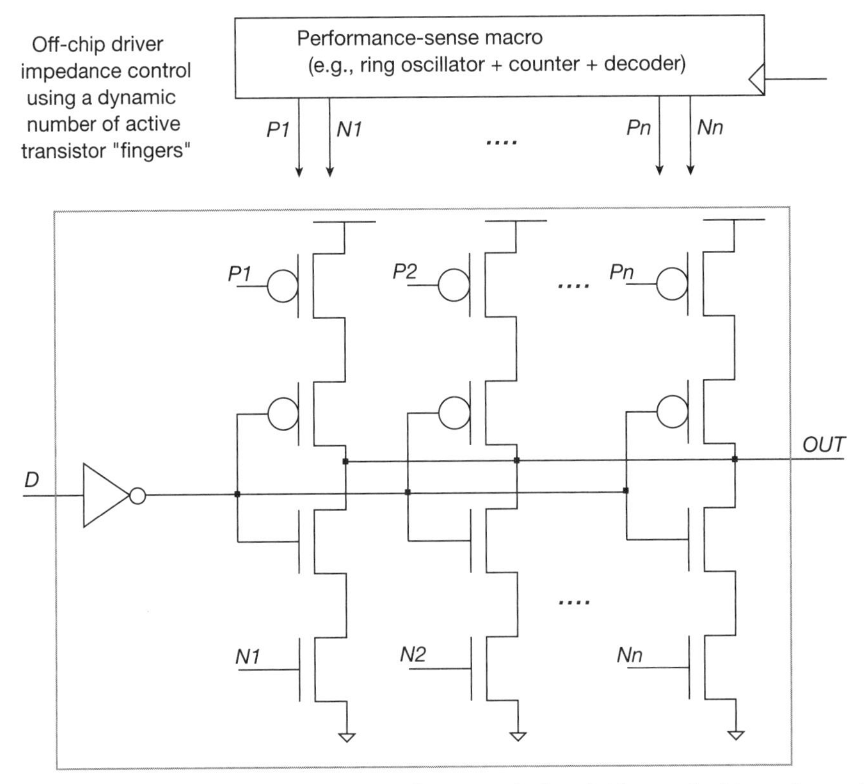


**Figure 1.31** Schematic for adaptive control of pad driver output impedance. The number of active transistor fingers in the driver is dynamically varied, based on input from a PVT sensing circuit.

Chip-package co-design optimization requires that the SoC methodology export pad geometry data and electrical power abstracts to EDA tools outside the conventional SoC flows. Similarly, the methodology needs to import (excised) parasitic models from the package extraction tools for a variety of circuit simulation tasks. Due to the unique package design tools and (non-standard) model formats that may need to be exchanged, special consideration should be given to additional model validity checking in these flows. An example would be to verify that an S-parameter model for package parasitics used for I/O signal simulation is indeed fully passive. As networks for circuit simulation are merged using chip and package parasitic elements, additional checks for dangling nodes are appropriate. The complexity of chip-package flow development is further expanded by the need to ultimately perform chip-package-system analysis using actual system model data in lieu of supply voltage drop margins and reference load models.

The discussion in this section on package analysis assumes a relatively straightforward single-chip module, with a multi-layer package substrate for signal and power plane distribution. There are an ever-increasing variety of multi-chip module (MCM) package technologies available, providing higher signal density and improved performance for inter-chip interfaces contained within the MCM. These advanced packages offer the capabilities of placing multiple dies adjacent to each other with connectivity through substrate layers or multiple dies stacked vertically using additional connectivity through an *interposer* separating the dies. The same chip-package co-design considerations discussed in this section extend to these MCM technologies (i.e., collaboration on floorplanning of chip pads and package substrate/interposer routes, utilization of chip power abstracts for package electrical and mechanical analysis, merging of package parasitics for I/O interface signal quality and SSO simulations).

### 1.4.11 Chip Thermal Management and Designing for a Power-Performance Envelope

For many high-performance SoC designs, there is an opportunity to boost the performance of the design if it can be ascertained that the (local) operating temperature is far from the extremes used for PVT corner circuit delay characterization. Conversely, it may be necessary to throttle performance if the operating temperature approaches the corner value. An SoC architecture may choose to employ a *dynamic voltage, frequency scaling* (*DVFS*) approach, in which a power management control module can vary the IP core supply voltage (and clock source frequency) corresponding to operating environment feedback.

Initial DVFS implementations added an external programmable power management integrated circuit (PMIC) to set supply voltages, typically communicating with the SoC through an I2C bus interface. More recently, an increasing number of SoC designs are using an integrated voltage regulator as part of the on-chip power management strategy. The integrated regulator offers faster response time (lower latency) for power state transitions, improved regulator input noise filtering/rejection, and extendibility to multiple core voltage domains, all with higher power efficiency.

#### *Temperature Sensing*

The on-chip power management design utilizes a temperature sensor IP macro from the circuit library. A variety of *thermsense* circuit implementations are employed, and their outputs are a function of temperature. Examples of thermsense macros include:

- **Bandgap voltage reference generator**—The voltage difference between active junctions operating at different currents is (linearly) temperature dependent. The bandgap would be connected to an analog-to-digital converter (ADC) to provide a digital code for the measured temperature that can be readily routed to a power controller.
- **Sub-threshold current sensor**—This sensor uses a similar topology as the p-n junction bandgap, using the difference in $V_{gs}$ between two devices of different size operating in the sub-threshold current region.[15]
- **Ring-oscillator frequency variation**—A ring oscillator feeds a reference counter to directly encode the temperature digitally.

The architecture for on-chip power management includes the thermsense macros, a power state controller, and programmable voltage regulators and clock generators, as illustrated in Figure 1.32.

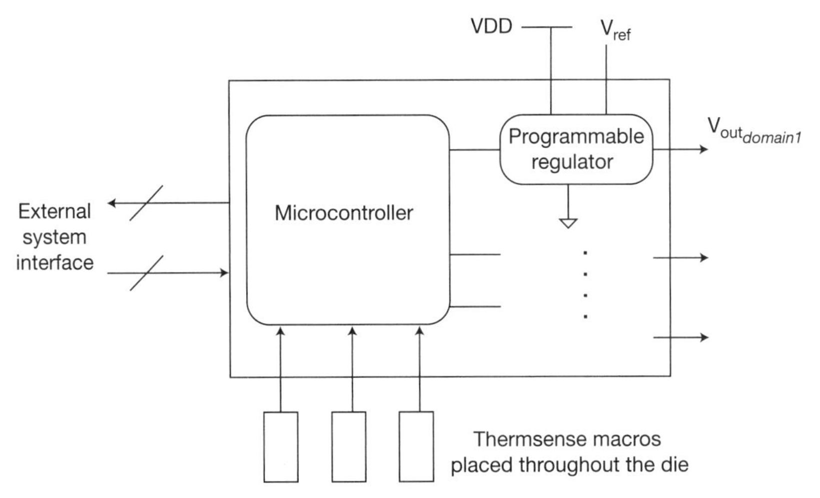


**Figure 1.32** A representative block diagram for the on-chip power management architecture.

The microcontroller may also interface with the operating system to communicate environmental conditions and receive instructions for external control of boost/throttle/sleep states.

The methodology for adding power/performance management on-chip starts with a review of the IP library offerings for the voltage regulator and thermsense functions. The specifications for this IP need to be reviewed against the SoC requirements for DVFS operation:

- Temperature range and temperature reporting tolerance (over process variations)
- Reference voltage input(s) required to the regulator
- Regulator programmability
- Regulator output voltage characteristics, as a function of current load (and dynamic load variation), and additional recommended capacitive filtering
- Latency for boost/throttle transitions
- Robustness of regulator output voltage behavior during a DVFS transition

An SoC incorporating temperature-based boost/throttle support introduces new methodology flow constraints:

- Circuit delay/power dissipation characterization needs to be extended to the boost voltage.
- Electrical analysis corners need to likewise reflect the extended boost operating range.

Specifically, noise coupling between nets is impacted by the voltage transition signal slew, which is faster as a result of applying the higher boost voltage. Electromigration current density is also impacted.

Additional macro usage checks should be employed to confirm that the thermsense macro is correctly inserted and connected to the power management module inputs. In other words, the thermsense macros and voltage regulators become an integral part of the functional validation of the power state microcontroller. The thermsense macros do not have a digital model, so they are "stubbed out" during functional model compile and elaboration. Their inputs and outputs become response and stimulus points in the testcases exercising the power controller. For SoC physical implementation, the placement of the temperature sensors requires consideration of the anticipated *hot spot* locations. The routing of the thermsense outputs and power management controls requires an allocation of (global) wiring track resources.

#### *Chip Thermal Management*

The SoC power management architecture also introduces new production test considerations. For efficiency, a selected subset of boost/throttle conditions could be examined for delay-based test patterns by sending specific codes from the controller to the voltage regulators and clock generators, bypassing the thermsense outputs. The capability to set IP cores to a lower power dissipation (inactive) state could be employed during production test as well. Rather than requiring full active power delivery through the test probe fixture, a production test could use a sequence of power state transitions and core-specific patterns. This would be particularly attractive during *burn-in stress test* screening, in which multiple packaged parts are socket-mounted in parallel on a burn-in board. Although the pattern test time would be longer, more parts could be stressed in parallel within the power supply delivery and thermal constraints of the burn-in chamber.

The logical, physical, and test requirements for the thermsense and regulator models suggest that they are directly included in the core hierarchy; it would be more difficult to represent them as separate global instances in the SoC hierarchy.

In addition to the functional validation of the interface between the power controller and thermsense and regulator macros, there is a complex set of power state transition testcases. Cell usage checks are required to confirm correct instantiation of level-shifter circuits and state retention flops. Logic inserted to block undefined signal propagation at the interfaces between cores at sleep and active states also needs to be verified. To facilitate the efficient application of these cell usage checks and the (automated) generation of functional validation testcases, EDA vendors have enabled SoC methodology flows to incorporate a concise *power intent specification* file (see Section 7.6).

## 1.5 Power and Clock Domains On-chip

The previous section introduces some of the design considerations for incorporating multiple SoC power supply domains. This brief section highlights the interface requirements between power domains. The interface between clock domains also introduces specific clock usage and checking requirements.

### 1.5.1 Power Domain Constraints

Methodology flows are added to check valid and required cell usage at the interfaces between distinct power domains on the SoC. As mentioned previously, a power format file is a separate functional description that is used to identify the specific interfaces in the SoC hierarchy, where EDA tools apply the necessary library cell checks. Separately, EDA vendors have developed tools to assist with identification of clock domain crossing paths and implement the appropriate cell checks.

A level-shifter cell accepts an input logic signal at a logic '1' voltage level below the VDD supply voltage of the receiving domain, as illustrated in Figure 1.33. It is required at interfaces when the sourcing domain operates at a lower supply voltage.

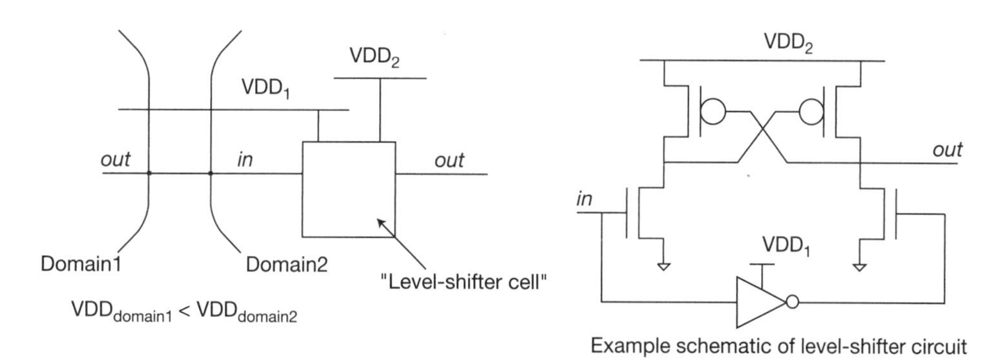


**Figure 1.33** A level-shifter cell adapts the logic '1' voltage level from a sourcing block to the (higher) supply voltage of the receiving power domain. Note that both supply voltages are required to the level-shifter cell, necessitating a more detailed power distribution. Typically, the level-shifter cell is located within the higher VDD domain.

If a core enters a full-sleep power state—either by the disabling of series *sleepFET* devices or by being controlled by a voltage regulator—the output signal values of the domain reach an unknown voltage and thus become logically undefined (see Figure 1.34).

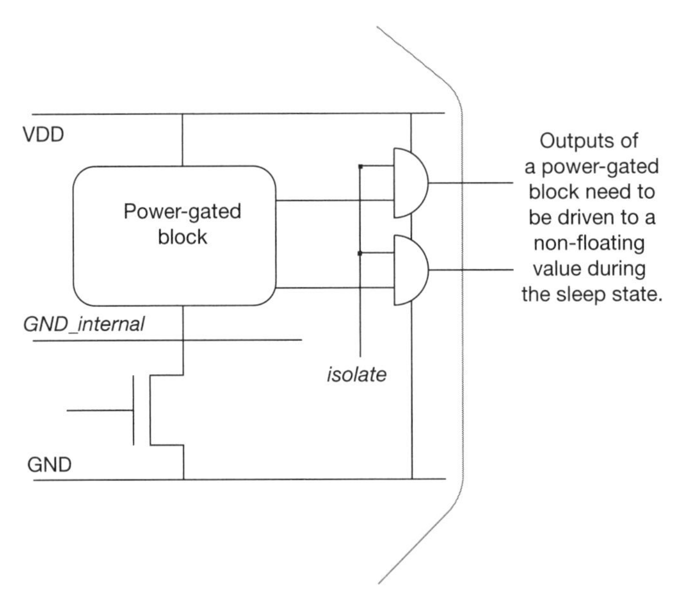


**Figure 1.34** Isolation cells are added at the outputs of a power domain to provide fixed logic signal values to fan-outs when the domain is power gated. Note that the isolation cells connect to the non-gated rails within the domain to always remain powered.

It is necessary to block these undefined logical (and signal voltage) levels from propagating into active domains. Usage checks are needed to confirm correct interface behavior.

If a core enters a full sleep condition, by default all internal logic state values are also lost. In order to minimize the time required to recover from the sleep condition, special state value retention flop library cells can be employed. These cells necessitate using a cell template that includes access to the active supply rail during the sleep state, as shown in Figure 1.35.

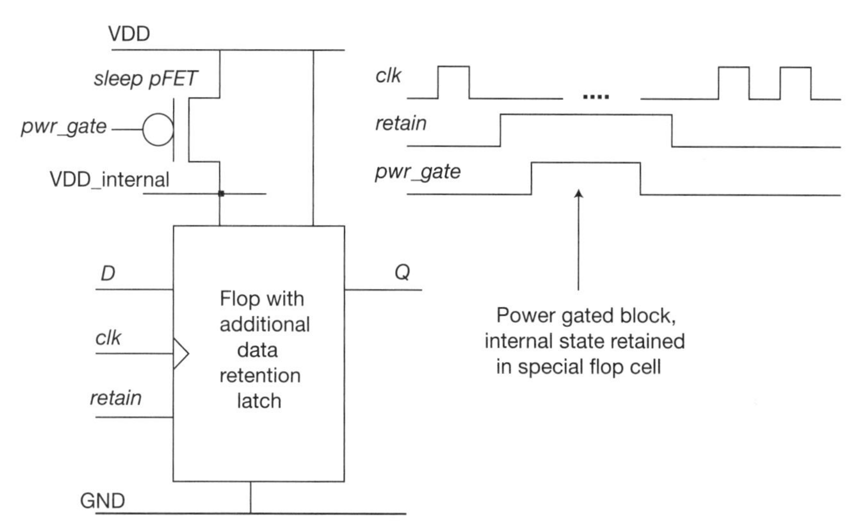


**Figure 1.35** State-retention flop cell to retain internal state values when a block is power gated and restore state when the block returns to active operation.

The functional validation team needs to confirm the sequencing of the "retain" signal to save the flop state prior to the sleep transition and restore the value as part of the transition to the active power mode.

### 1.5.2 Clock Gating

An efficient method of (active) power reduction within a domain is to implement *clock gating*. If it can be logically determined that the logic state values of a network are valid for one or more future clock cycles, the clock input transitions to registers can be suppressed, saving the power associated with the clock pin loading and any logic switching activity not influencing the network state. Figure 1.36 shows a rudimentary example in which a free-running clock to a register is converted to a gated clock, in the case where the register values would be unchanged.

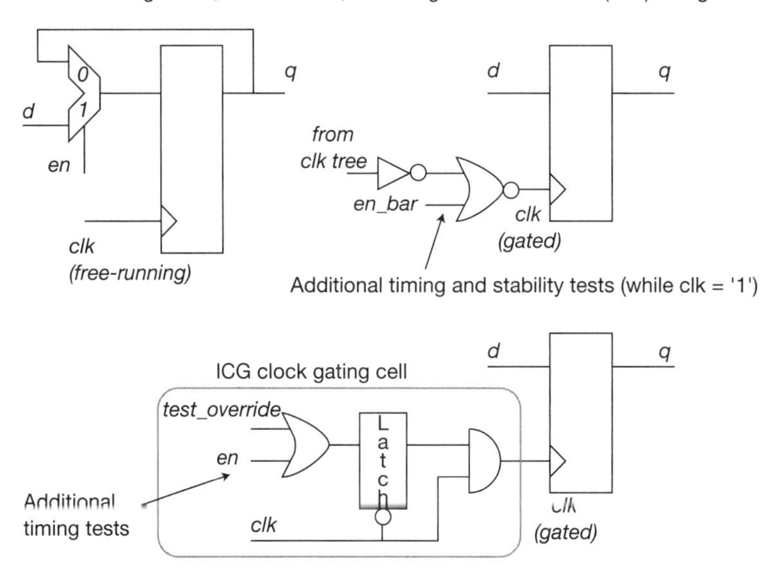


**Figure 1.36** Clock gating logic implementations. To simplify enable input timing constraints and ensure that no spurious clock pulses propagate, an integrated latch is typically used, denoted as an *ICG* cell.

Note that the clock gating library cell has special design requirements. To maintain the fidelity of the clock pulses and minimize skew, the clock gate needs delay characteristics comparable to those of the clock buffer being replaced. Also note that there are strict constraints on the clock gate enable timing. There must not be any spurious (or short, truncated) clock pulses due to enable signal transitions while the clock phase is active. A common clock gate implementation inserts a latch between the enable signal and the clock gate, where the latch blocks enable transitions on the active clock phase. There is a setup time constraint at the clock gate for the enable signal, as well. The specific clock gating cells in the IP library include timing constraints so that the required path timing tests at their (combinational) gate pins are enforced during the timing analysis flow.

The combinational clock gating topology reduces wasted power associated with a static network state. Greater savings are achievable with sequential gating, as illustrated in Figure 1.37.

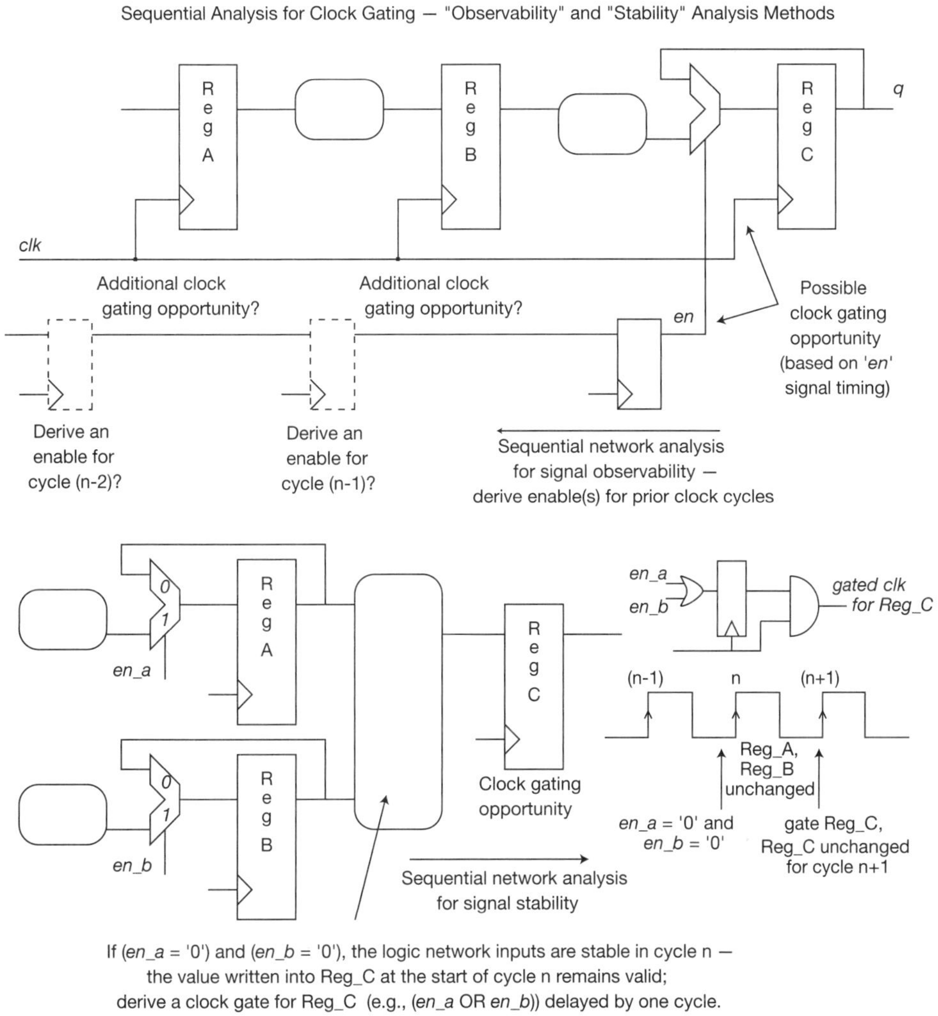

**Figure 1.37** Illustration of sequential clock gating logic. Both sequential signal observability and sequential signal stability examples are depicted.

EDA vendors have developed tools to evaluate hardware description language designs and identify potential combinational/sequential gated clock enable expressions and the corresponding logic design transformations. If these clock gating tools are employed, the SoC methodology team needs to confirm that the functional equivalency flow is capable of (independently) proving that the new gating model is logically identical to the original, before accepting the transformation.

### 1.5.3 Clock Domains

The clock domains on an SoC are interdependent with the design hierarchy. For a hierarchical interface between timing models in the same clock domain, a set of timing constraints relative to the common clock needs to be maintained. Figure 1.38 depicts an elementary example.

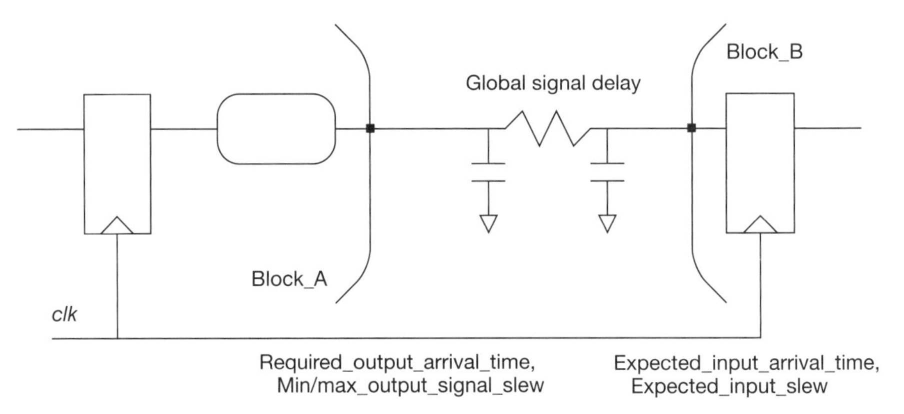


**Figure 1.38** Examples of timing constraints to be provided for block-level analysis, relative to a common clock reference; timing paths are launched and captured at state elements. Constraints are required for each PVT corner used for timing analysis.

To enable path timing analysis and optimization between the two models in the figure, designers need to specify constraints such as required_arrival_output_time_A, estimated C_load_A, min/max_signal_slew_A, expected_input_arrival_time_B, expected_C_in_B, and expected_input_slew_B. These timing constraints are relative to the common clock reference and are required for each timing corner. The path timing analysis methodology flow adds compensation for the potential clock arrival skew at the launch and capture state elements (see Section 11.3).

The data management tools in the SoC methodology are responsible for associating the constraints file with each model revision. In addition, the methodology should include checking utilities that aid designers, including the following:

- The required output arrival times are always less than the expected input time on the net; in equation form, (RAT_A < EAT_B).

- The Cload_A constraint is consistent with the sum of the fan-out input pin capacitance.
- The output pin slew and input pin slew constraints are consistent (i.e., slew_A < slew_B).
- Perhaps most importantly, from individual block-level path timing analysis results, the constraint checking utility could identify cases in which the positive and negative *slacks* for the output and input pin arrival times at the blocks need to be re-apportioned (or "re-budgeted") across the interface.

Clock domain interface timing constraints enable designers to work independently on (partial) path optimization within their respective blocks. It is important to ensure that opportunities to accelerate closure on failing paths between blocks are identified and timing constraints adjusted accordingly.

In cases where the interface between blocks represents paths that are associated with unrelated clocks, a synchronization flop is required to capture the signal to the new clock reference, as shown in Figure 1.39.

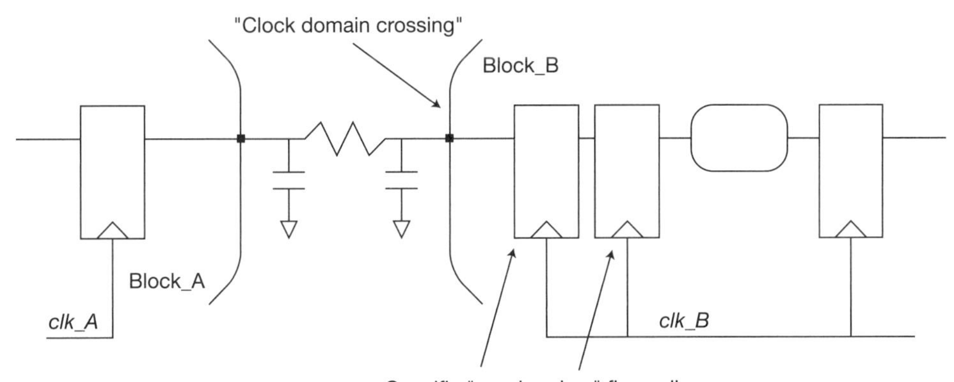


Synchronizer flop circuits are designed to minimize internal R*C time constants. Often, multiple synchronizer flops are placed in series, to further reduce the probability of a metastable signal propagation failure.

**Figure 1.39** A unique synchronization flop cell is required at clock domain crossing interface. The circuit design of this cell specifically focuses on reducing metastability output transition behavior.

The launch-to-capture path between two blocks utilizes unrelated clocks if there is no timing relationship between the two clocks and there is no corresponding expected arrival time at the receiving block pin. A specific IP library flop cell is provided for asynchronous interfaces. The circuit characteristics of this cell differ from more conventional flops. Synchronization requires reducing the probability that the flop will enter and linger in a *metastability state* due to the indeterminate data arrival-to-clock transition interval. Internal circuit node capacitances and feedback transistor drive strengths are optimized in the synchronizer to minimize the node transit time to a stable state. The metastability probability can be further reduced by placing multiple synchronization flops in series at the asynchronous input pin.

The SoC methodology flow needs to include two features specific to asynchronous clock domain interfaces:

- Metastability failure rate calculations, which are a function of the input data rate, capture clock frequency, input signal slew, internal time constant of the synchronizer circuits, and flop output capacitive load
- Checks that the synchronizer flop is present at the interface or multiple flops in series, based on the failure rate calculation

EDA vendors offer tools that identify asynchronous clock domain crossings in a hardware description language model, tagging signals with their sourcing clock and propagating through logic expressions to the capture clock. (Note that conventional hardware description language simulation is not designed to reflect the very low probability of asynchronous interface metastability issues.) The methodology flow utilizes the relationship between a signal and the sourcing clock for the synchronizer checks.

This discussion on clock domain interface design and constraint management has made a simplifying assumption of a single clock reference for sourcing signals and a single (perhaps asynchronous) clock reference for capturing logic path results. SoC designs commonly employ more complex clocking features, with the possibility of using divided functional clocks and/or multiplexed clocks within and between domains. The chip reset clocking and wake-from-sleep state also introduce special timing constraints. Another typical example is the addition of production test clocks within a domain to implement the logic for serial shift-register scan of test stimulus/response data (see Section 7.3) or the clocking of dedicated built-in self-test (BIST) logic networks (see Section 19.3).

The SoC methodology and design teams conduct a project kickoff review of the clocks used throughout, the MCMM modes for timing analysis (IP blocks and full chip), and the timing constraint hierarchical management. The outcome of this review defines the flow inputs required and the results data to be tracked by the methodology manager application (see Section 2.3).

## 1.6 Physical Design Planning

### 1.6.1 Floorplanning

The initial methodology step in SoC physical implementation is the positioning of IP blocks on the die image, commonly denoted as *floorplanning*. This flow often utilizes a combination of automated placement algorithms and (iterative and interactive) manual refinement, using specific features of EDA vendor layout editing tools.

Floorplanning involves many optimization decisions during IP block coordinate assignment within the die:

- Hard IP will have a fixed extent and aspect ratio. The netlist for soft IP cores is not available for early floorplanning; cell count and corresponding area estimates are needed. Further, the aspect ratio of the area allocated to the soft IP is somewhat flexible. Note that a high aspect ratio for the estimated area may be useful to fit between other blocks but is likely to result in significant routing congestion within the core in the narrow dimension. Also, the soft IP area estimate should reflect an appropriate cell utilization percentage for internal routability and the addition of decoupling capacitance.
- The horizontal and vertical dimensions of the IP core blocks need to reflect a multiple of the global wiring track grid. The origins of the blocks need to snap-to-grid for FinFET technologies.
- The global routing requirements between IP are a crucial floorplanning consideration. Initially, the visual feedback in the floorplan layout editing tool highlights the flightlines between IP cores, based on the hierarchical top-level connectivity between blocks. (The origin and destination of the flightlines use the center of the block, prior to detailed pin assignment.) The density and length of the flightlines can indicate potential global route congestion issues, including the need for additional global signal buffering and/or sequential repeater flops. The floorplan

tool may calculate additional data that are useful to IP placement optimization (e.g., a total global wire length estimate, a comparison of the number of flightlines crossing sections of a [coarse] grid to the number of wiring tracks associated with each grid segment). Any global wiring track blockages associated with hard IP placement need to be included in the flightline calculation.

- The floorplanning tool allows the addition of a predefined global track allocation to critical busses; in this case, IP flightlines connect to these global tracks.
- The floorplanning of SoC I/O pad arrays is a key element of chip-package co-design (see Section 1.4.8). The flightlines from internal IP to the pads are a critical part of the IP placement optimization.

#### *Pin Assignment*

Pin assignment is a key optimization phase as the floorplan begins to stabilize. For hard IP, the pin locations for each core input/output signal are fixed; nevertheless, there is likely an opportunity to flip/mirror the hard IP layout view to minimize global signal length and optimize performance. For example, many SoCs incorporate multiple instances of the same processor core, which communicate across a multi-point internal bus architecture; to optimize the core-to-bus interface performance, a hard IP instance is flipped so that pin locations are shortest to the global tracks allocated to the bus. For soft IP, there is more flexibility in pin assignment during floorplanning to address several constraints:

- Pins need to be assigned to a specific metal layer, corresponding to the expected union between the global routes and the internal IP physical implementation.
- Pins need to be located on a global wiring track. In addition, if the pin metal layer corresponds to a lithography layer that will be part of multi-patterning decomposition, the pin may also need an assigned color.
- The width of the pin may need to be larger than the minimum wire width of the metal layer. If the signal pin is part of a critical timing path, the global route (and internal IP metal to the pin) may use a *non-default rule* (*NDR*) as a routing constraint, as illustrated in Figure 1.40.

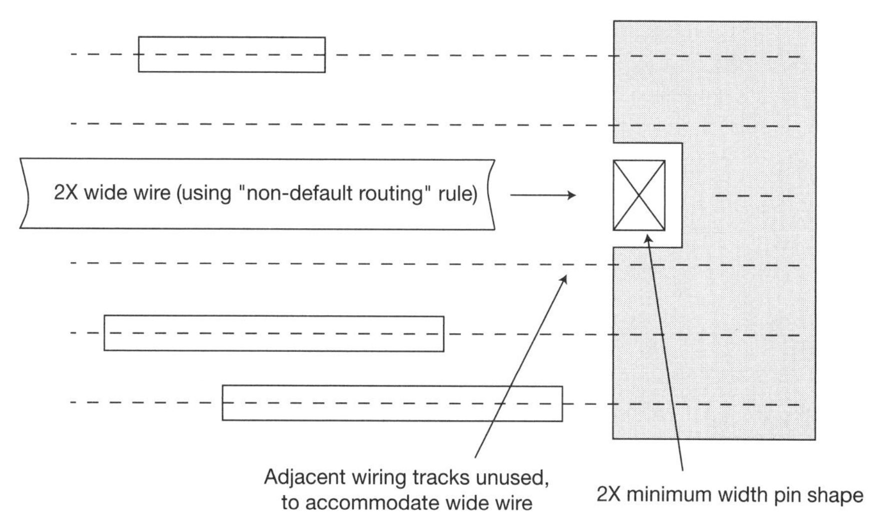


**Figure 1.40** The pin shape for an IP block defines the width and metal layer for the global route. Based on the internal IP circuitry, wider (non-default) global wire connections may be required.

In a library cell layout, it is common to provide a physical pin shape that spans multiple wiring tracks to offer the router greater flexibility in completing local routes. The physical abstract of the cell has a *pin property* which indicates to the router that any valid track connection is acceptable. For an IP block model in a floorplan, the pin property (and global signal routing directive) typically requires full pin coverage. For example, the IP designer must make assumptions on the driver strength required to connect to global busses. The pin output drive strength determines the corresponding global signal wire width and, thus, the size of the pin shape in the IP block abstract used during floorplanning.

A routing rule assigned to a global signal may require wide interconnect segments for optimal performance. However, a fan-out of that signal may terminate on a narrow width IP input pin necessitating that the global route be tapered, as depicted in Figure 1.41. The methodology team needs to review the foundry PDK design rules for wire segment width tapering and ensure that the router will observe the NDR global route constraint up to the tapering distance.

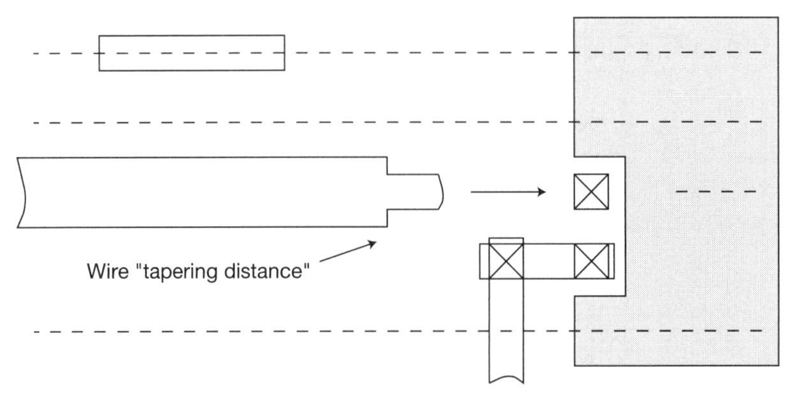


**Figure 1.41** A global route may utilize a non-default wire width for performance, tapered at the IP pin shape; the metal layer design rules for tapered shapes need to be observed.

In addition to pin width selection, pin location assignment is also a complex optimization during soft IP block floorplanning. The goals of pin location assignment are varied:

- Minimize the global wiring track demand.
- Optimize performance for critical timing interfaces between IP blocks.
- Avoid internal wiring congestion during subsequent IP block physical implementation.
- Maintain the accuracy of the timing constraints and the IP timing abstract models used for path timing analysis.

The last goal introduces a physical design constraint. The block input pin data should be able to accurately reflect the capacitive loading within the block. As a result, the input pin should be physically close to the corresponding fan-out cells. It would be difficult to represent a long wire between pin and cells in the IP timing abstract for the calculation of the *effective capacitance* load on the driving circuit (see Section 11.1). Figure 1.42 illustrates the complexity of estimating the load at the block pin for the IP timing abstract for a long internal R*C wire.

For a hard IP layout, this physical constraint implies that pin locations internal to the IP boundary may be present. For soft IP, although internal area pins need to be supported by the methodology, it is much more common to assign pins to the edge of the IP physical boundary in the floorplan. The expectation is that the subsequent soft IP physical implementation will place cells logically connected to a pin with a strong affinity to the floorplan pin location.

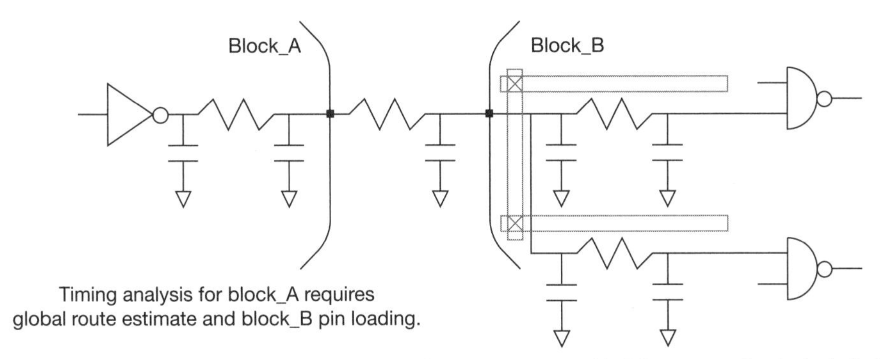


**Figure 1.42** A distributed RC parasitic from a block pin to internal fan-out is difficult to model as an effective capacitive load at the pin for the block abstract; pin-to-fan-out connections should be short.

An enhanced floorplanning (and soft IP block routing) tool feature would allow a pin to be included in a *pin group*, with the group assigned to a range of wiring tracks on the boundary edge, as illustrated in Figure 1.43. The pin group would allow some optimization flexibility during the detailed block routing flow yet still provide sufficiently accurate data for early IP floorplanning. The detailed pin assignment after routing needs to be reflected in a floorplan version update to complete the global routes.

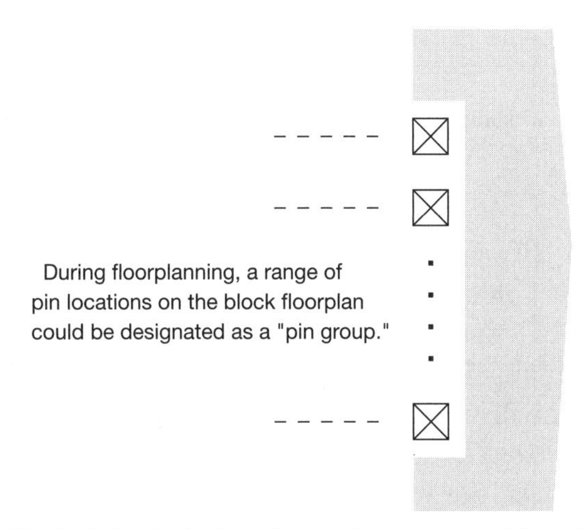


**Figure 1.43** Soft IP block pins allocated to a range of floorplan wiring tracks during floorplanning, as part of a pin group.

The requirement for accuracy in the soft IP timing model also typically precludes the creation of multiple physical pins for a single logical output signal. Figure 1.44 illustrates the timing model inaccuracy associated with multiple block IP physical pins for a logical output.

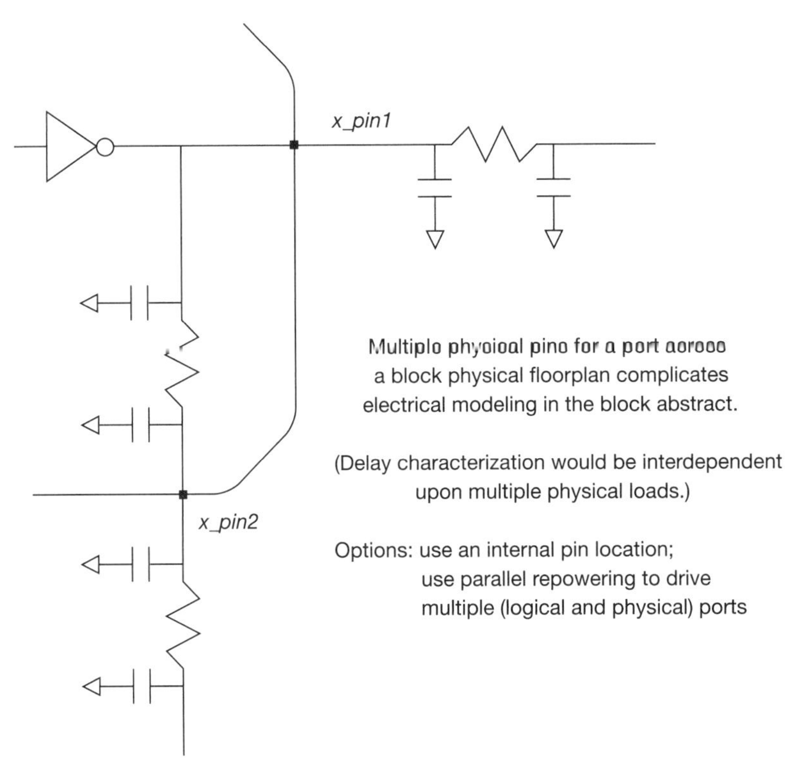


**Figure 1.44** Multiple block physical pins for a logical output are difficult to model accurately in a block timing abstract.

The initial "firm" SoC floorplan release allows IP designers to proceed with their physical block design and allows the physical SoC integration team to work on global signal buffer and repeater requirements. The floorplan needs to be maintained and updated, as SoC logic validation and production test development will inevitably result in changes to the IP block input/output signal interface list. The floorplan is the link between logical and physical hierarchical models; the version release management policy for logic models impacts the floorplan data, as well. The EDA floorplan tool needs to input the (revised) logical model, compare against the current physical design, and report/track any pin discrepancies. The SoC project management team must continually assess the stability of logic model revisions and decide when the connectivity of the logic model and of the physical floorplan need to be aligned and a new floorplan version released.

### 1.6.2 Power Grid and Wiring Track Planning

The quality of the SoC floorplan relies on a calculation of the available global wiring tracks, compared to estimates of the wiring segment demand across the chip. The global power grid definition subtracts from the available signal tracks and needs to be incorporated into the floorplan.

The global power grid is actually a set of separate grids for individual power domains. The floorplan designer needs to ensure that the power domain assumptions for each block are reflected in the placement of corresponding grids. The connectivity of the power grids to chip pads is also part of the floorplanning task (i.e., routing from pads/bumps—perhaps internal to the die area—through top-level redistribution metal to connect to power grid access points).

The term *power grid* is used somewhat loosely. If the foundry process supports the stacking of vias between successive metal layers, the actual PDN metal utilization may be rather sparse on (some of) the intermediate layers. The power grid points may consist of a sequence of via arrays and metal coverage shapes between upper and lower layers, as illustrated in Figure 1.45.

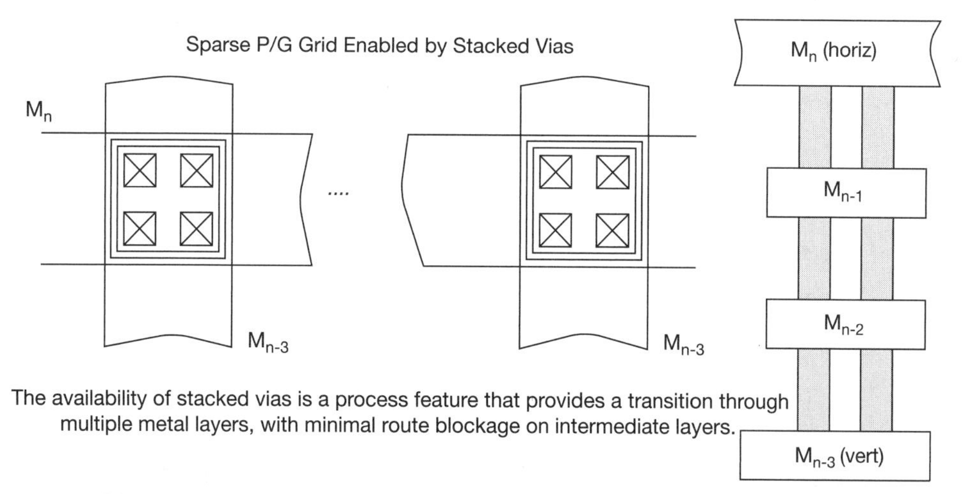


**Figure 1.45** Illustration of a sparse global power grid, leveraging the availability of stacked via arrays between metal layers.

The via arrays provide a lower-resistance current path to the thicker top-level metals, at the expense of blocking multiple adjacent tracks for the covering metal. Nevertheless, compared to a denser grid of intermediate-layer metal segments, the sparse grid with stacked via arrays is likely more effective overall, both electrically and for signal wiring track layer availability. On the other hand, some signals may require that the router provide shielding (e.g., routing a signal in a track adjacent to a power/ground rail) to minimize capacitive coupling noise from neighboring wire switching transients. A denser PDN grid of metal segments throughout the metallization stack offers more opportunity for the router to assign these critical signals adjacent to the grid to serve as a shield. There are complex trade-offs to be considered when developing the PDN grid topology for a power domain.

A key parameter in the design of the grid is the estimated power dissipation of the corresponding domain. The number of chip pads and pad redistribution and the density of grid points are adapted to satisfy the power dissipation of the domain. For example, if the switching activity is low, via grid points could potentially be sparser; the electrical analysis of the local I*R voltage drop would provide closure on the suitability of the grid (see Chapter 14, "Power Rail Voltage Drop Analysis").

The connectivity of the floorplan power grids to the SoC blocks (and global circuits) differs for hard IP and soft IP blocks:

- A hard IP block includes an internal grid within the layout. The floorplan grid needs to cover the power pin shapes added to the IP physical model abstract. An analog hard IP block will likely require special focus on the power grid density for its power domain to minimize (static and dynamic) voltage drop.
- The physical implementation of soft IP blocks is not available during initial floorplanning. Rather, the cell library template and local track definition include the local power grid; the floorplan grid aligns with the access points on the cell library template grid.
- The floorplan power grid design may include a population of additional specific decoupling capacitance elements. The fabrication process typically provides the option of adding a thin intermediate metal layer, used to implement a metal-insulator-metal (MIM) structure. The MIM provides an area-efficient capacitance implemented within the upper-level metals, as illustrated in Figure 1.46. The power grid template now

requires connection to the parallel plates of the MIM structure. (The intermediate metal layer is not a routing layer.) Note that the MIM decoupling element is electrically present between the local decap library cells interspersed with logic circuits and the SMT capacitors added to the package substrate.

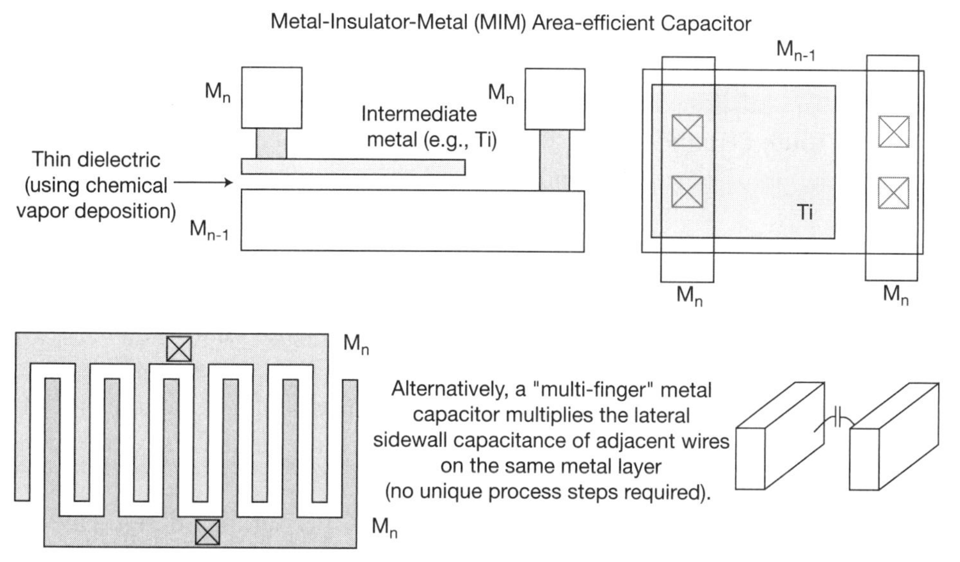


**Figure 1.46.** Cross-section of a metal-insulator-metal (MIM) capacitor structure residing among upper metals. The intermediate metal layer is provided by the foundry as a process option. The intermediate metal plate of the MIM is shaded in the figure. A thin dielectric is deposited over layer $M_{n-1}$ prior to patterning of the intermediate metal. This parallel plate capacitor is contacted on metal layer $M_n$ in the power grid template. An alternative decoupling capacitor structure is also shown: interleaved multi-finger metal "combs" on metal layer $M_n$, where the sidewall capacitance between the fingers provides the decoupling.

The floorplan power grid for any cores implementing a DVFS or sleep state feature requires additional design considerations. If an integrated voltage regulator is employed on the SoC, the regulator power connections for Vin and Vout need special power grid design. If an IP block utilizes a sleepFET topology to turn off the core, a global power grid for VDD and VSS needs to align with the block VDD and VSS grid, plus the allocation of local rails consistent with the cell template for VDD_core (pFET sleep) or VSS_core (nFET sleep) distribution.

As the SoC floorplan evolves, the corresponding power domain grids need to track the changes. The floorplan methodology flow needs to confirm the correct grid overlay for each IP block, which satisfies any specific IP features, supports the power dissipation estimates, and enables accurate global wiring track resource calculations.

### 1.6.3 Global Route Planning

During initial floorplanning, the flightlines between IP cores highlight the global route demand. These flightlines ultimately need to be converted to a combination of horizontal and vertical segments, also known as a *Manhattan routing* style.

Note that early VLSI process nodes supported more general routing segment directions for global signals (e.g., allowing 45-degree segments to switch tracks on a single wiring layer). Commonly, non-orthogonal segments are now permitted only on the top redistribution layer. (There was a short-lived methodology effort with some EDA support to define global routing tracks with entire layers dedicated to a 45-degree orientation, with related vias to orthogonal metal tracks on layers above and below.)

As VLSI scaling has continued, the PDK lithography design rules have reduced the routing flexibility to preferred directions for alternating metal layers, with more conservative widths/spaces for any orthogonal *wrong-way segments* (routing track *jogs*) on the layer. When estimating the wire length between pins, the Manhattan distance is applied, with the assumption that the final detailed global route will not experience an excessive *scenic path*. For a multiple fan-out global signal, a Steiner tree topology estimate is used, as illustrated in Figure 1.47.

For performance-critical nets, the wire track demand calculation promotes segments to the thicker upper global layers in the metallization stack.

The floorplan design may incorporate dedicated wiring tracks allocated to a signal bus, which directly provides a Steiner topology for wiring track estimation. The prerouted bus segments may be greater than minimum width for the assigned layer and/or may incorporate adjacent shields.

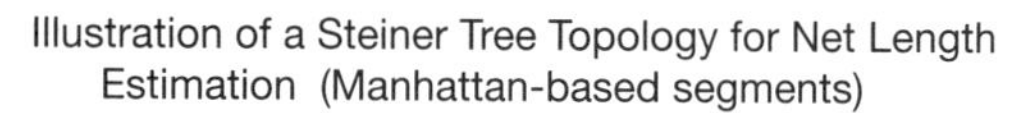


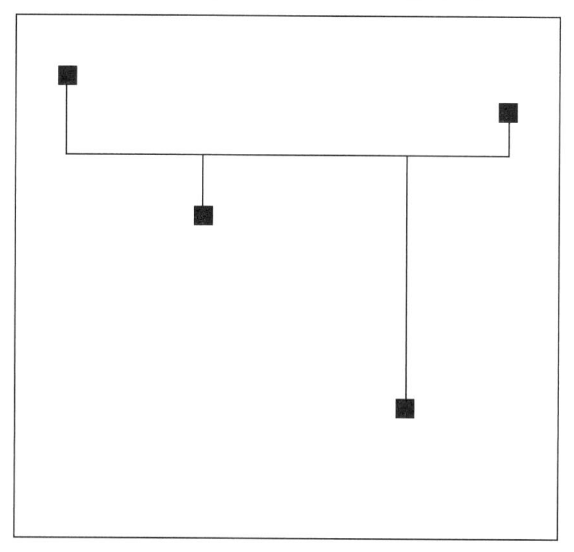

**Figure 1.47** Interconnect estimates for Manhattan routing wiring layers typically utilize a Steiner tree topology.

The IP cell library needs to include a specific set of buffers and sequential repeaters for global signals. The (Manhattan) wire length for global signals may necessitate inserting a buffer or repeater to improve path performance and avoid excessive signal slew rates. A preliminary simulation analysis of the global cells in the IP library with different loads should provide guidelines on the optimum segment length between inserted buffer or repeater cells, given the cell drive strength and global metal routing layer(s). Note that a floorplan designer has multiple options for the topology of a global signal, with the insertion of inverting and/or non-inverting cells.

If the global signal path (across a common clock domain) cannot satisfy timing constraints with the insertion of buffers, it is necessary to insert a sequential repeater flop cell. If a global sequential flop and/or an inverting buffer cell is inserted, the SoC top logic model needs to include the corresponding change in signal latency and/or polarity for logic simulation. (Of specific importance is the task to ensure that any separate high-level, C-language performance model of the SoC also reflects the additional interface latency due to the insertion of global sequential flops.) The completion of the initial chip integration project milestone requires updating the SoC logic model to reflect any global functionality introduced during floorplanning.

The space between IP block boundaries in the floorplan is called a *channel*. The inserted global cells assigned to a channel are grouped, and a small image of valid cell placement locations and power rails needs to be generated. The buffers and sequential repeaters are typically high-drive-strength cells; local decoupling capacitor cells should be included in the channel, as well. Some spacing between placed cells in the channel image may be beneficial, as the vias between the global signal upper metal layers and the cell input and output pins introduce wiring track blockages on intermediate layers.

When designing floorplan channels, the foundry requirements for insertion of layout cells used for process monitoring also need to be considered. For example, cells provided by the foundry to be inserted throughout the chip layout provide lithographic alignment measurement verniers. The floorplan also needs to allocate area for the insertion of identifying alphanumeric data, from the mask-layer nomenclature required by the foundry to the company part number and revision information. Individual die are lithographically stepped across the wafer, including a specific distance between die required for separation, denoted as the *scribe channel*. The floorplan includes the *edge seal* design around the top physical cell of the chip, a specific set of foundry layout design rules for all mask layers to provide a clean scribe break with minimal risk of layer delamination or die edge contamination.

### 1.6.4 Global and Local Clock Distribution and Skew Management

The SoC floorplanning design phase includes the placement of clock generation IP and the global distribution of clocks to their respective domains. A phased-lock loop (PLL) IP block is often used as the clock source; this circuit provides an internal clock that is a multiple of a reference clock input. (The supply power grid to this analog IP domain requires special design to minimize noise injection.) The clock distribution from this source to the SoC IP associated with the clock domain requires significant buffering and load balancing to minimize the arrival skew at each local IP clock pin input.

There are two prevalent global clock distribution styles: a grid and an H-tree. A global clock grid overlaps an IP core in a manner similar to a power grid. The global grid needs to cover all the IP clock input pins. This is a unique case, in which multiple physical input pins on an IP block correspond to the same logical input. For logic path timing analysis within the block, the model effectively collapses the multiple physical pins into a single source; an assumption is made about the clock arrival skews across the block. For flop setup timing analysis checks, this skew assumption is subtracted from the clock period; for "short path" flop hold timing, the assumed skew is added to the path delay between launch and capture endpoints. A separate methodology flow performs a high-accuracy parasitic extraction and PVT circuit simulation of the clock distribution to confirm the validity of the skew assumption (see Section 11.3). As with power grid distribution, the global clock grid is designed to provide a low-resistance path from the main clock driver to any point on the grid. Several advanced SoC flows have been developed to further tune the grid physical implementation once the detailed clock loading is known, such as tapering or truncating ineffective grid segments to reduce total clock dissipation power.

An H-tree is a network of symmetric segments at each level of a multistage tree, with buffers at the ends of the segments sourcing the next level, as shown in Figure 1.48.

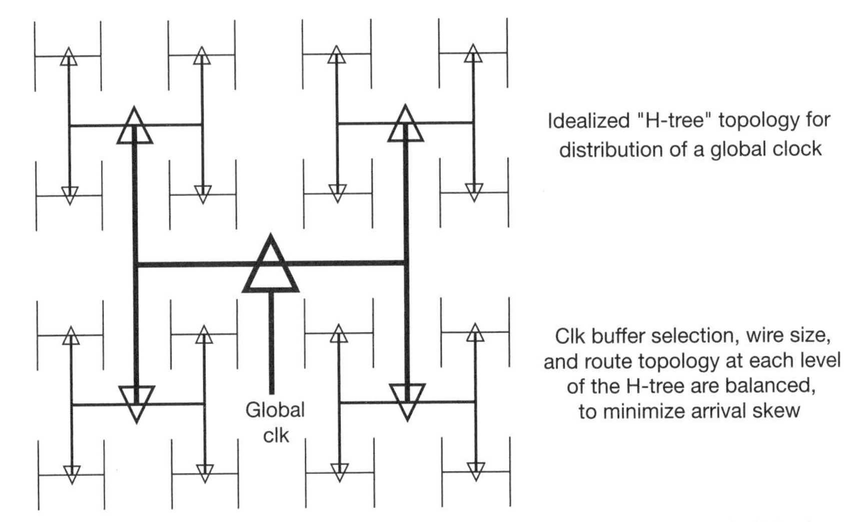


**Figure 1.48** Illustration of an H-tree clock distribution, with physical design optimizations to maintain low-skew arrivals and balanced cell delays at each level of the tree.

The H-tree relies on maintaining identical cell and R*C interconnect delays to the buffers at the next level. Note in the figure that the number of buffers increases at each level of the tree. The floorplanning of the H-tree is more complex than for a clock grid, as there are relatively strict requirements for the placement of the buffer cells and tree routes. Interfacing the buffers at the last level of the global H-tree distribution to the local clock inputs of the IP blocks also requires careful load balancing. On the other hand, the H-tree is likely to dissipate less total clock power than the global clock grid. In addition, it offers opportunities for clock gating earlier in the distribution if a buffer is replaced by a clock driver with a logic enable. A hybrid design implementation of a global clock H-tree and grid is also commonly used.

An additional design consideration for global clock distribution is the required shielding of clock signals by dedicating adjacent wiring tracks to a non-switching rail (perhaps at a relaxed spacing to the clock wire to reduce the total capacitance as well).

## 1.7 Summary

This introductory chapter has covered a broad set of material as background for more detailed discussions of specific methodology flows in subsequent chapters. A major point of emphasis in this chapter is the requisite methodology planning that is part of the initial SoC project definition. A methodology team must coordinate a wide range of SoC project characteristics, from high-level hierarchical data management and version control policies to detailed optimization decisions on foundry technology selection. The methodology team must also collaborate with a diverse set of engineering teams, both those that are directly involved in the SoC design (e.g., architects, logical and physical designers, test engineers, project managers) and teams that support the SoC design. Specifically, the methodology team maintains a close working relationship with three support teams: (a) the engineering team working with the foundry and OSAT for technology support, (b) the engineering team working to qualify IP from external providers, and (c) the CAD team providing flow development, EDA tool support, and methodology testing. Indeed, the methodology team is the focal point for any SoC project; the ultimate success of the project in meeting its technical specifications and its development resource/cost goals is fundamentally dependent upon the success of the methodology defined for the engineering teams to pursue.

## References

[1] IEEE Standard 1500: "Embedded Core Test," http://standards.ieee.org/findstds/standard/1500-2005.html.

[2] Marinissen, E., and Zorian, Y., "IEEE Std 1500 Enables Modular SoC Testing," *IEEE Design & Test of Computers*, Volume 26, Issue 1, Jan–Feb. 2009, pp. 8–17.

[3] ANSI/IEEE 1499: "IEEE Standard Interface for Hardware Description Models of Electronic Components," http://standards.ieee.org/findstds/standard/1499-1998.html.

[4] Dunlop, D., and McKinley, K., "OMI: A Standard Model Interface for IP Delivery," IEEE International Verilog HDL Conference, 1997.

[5] Su, H., Sapatnekar, S., and Nassif, S., "Optimal Decoupling Capacitor Sizing and Placement for Standard-Cell Layout Designs," *IEEE Transactions on Computer-Aided Design of Integrated Circuits and Systems*, Volume 22, Issue 4, April 2003, pp. 428–436.

[6] Liberty modeling format; open source licensing is available from Synopsys: https://www.synopsys.com/community/interoperability-programs/tap-in.html. (The Liberty Technical Advisory Board participates in the

format definition and evolution, under the auspices of the IEEE: https://ieee-isto.org/member_programs/liberty-technical-advisory-board/.)

[7] ANSI/IEEE 1801-2015: "IEEE Standard for Design and Verification of Low-Power, Energy-Aware Electronic Systems," http://standards.ieee.org/findstds/standard/1801-2015.html.

[8] An open iPDK standard is maintained by the Interoperable PDK Libraries (IPL) Alliance, https://www.iplnow.com.

[9] Moore, Gordon, "Cramming More Components onto Integrated Circuits," *IEEE Solid-State Circuits Society Newsletter*, Volume 11, Issue 3, September 2006, pp. 33–35. (Reprinted from the original article in *Electronics*, Volume 38, Issue 8, April 19, 1965.)

[10] Dennard, R., et al., "Design of Ion-Implanted MOSFET's with Very Small Physical Dimensions," *IEEE Journal of Solid State Circuits*, Volume SC-9, Issue 5, October 1974, pp. 256–268.

[11] Bohr, M., "A 30-Year Retrospective on Dennard's MOSFET Scaling Paper," *IEEE Solid-State Circuits Society Newsletter*, Volume 12, Issue 1, Winter 2007, pp. 11–13.

[12] SOI Industry Consortium, "Considerations for Bulk CMOS to FD-SOI Design Porting" whitepaper, , September 2011, http://semimd.com/wp-content/uploads/2011/11/Considerations-Bulk-to-FD-Release-0-1-a.pdf.

[13] Duarte, Juan-Pablo, et al., "Unified FinFET Compact Model: Modeling Trapezoidal Triple-Gate FinFETs," 18th IEEE International Conference on Simulation of Semiconductor Processes and Devices (SISPAD), September 2013, pp. 135–138.

[14] Khandelwal, S., et al., "New Industry Standard FinFET Compact Model for Future Technology Nodes," Symposium on VLSI Technology, 2015, pp. 6-4–6-5.

[15] Ituero, P., Ayala, J., and Lopez-Vallego, M., "Leakage-Based On-Chip Thermal Sensor for CMOS Technology," IEEE International Symposium on Circuits and Systems, ISCAS 2007, pp. 3327–3330.

## Further Research

This text includes suggestions for additional research and project activities, rather than chapter problems. The majority of these activities involve brief investigations into facets of VLSI design methodology development introduced in the text. A few recommendations delve more deeply into emerging technologies that will have a broad impact on microelectronic design.

### Dennard Scaling

For a traditional scaling factor s ($s < 1.0$) between successive process nodes, list the target for each (bulk CMOS) device dimension and impurity concentration. Assume a "constant electric field" as the process development constraint. For example, tox → (s * tox), which necessitates VDD → (s * VDD).

Describe how the device current and device input capacitance are scaled. Assuming a path delay dominated by device loading rather than interconnect loading, describe the path performance improvement with scaling (assuming that the path delay ~ ((C*VDD)/I)).

Describe the scaled circuit density and power dissipation density.

Note that with a reduction in supply voltage scaling, the traditional scaling definitions above have become less applicable.

### VLSI Fabrication Options

#### *CMP*

Describe the chemical-mechanical polishing process step in terms of the slurry composition, the chemical reactions occurring in the slurry, the mechanical force applied to the wafer, and the methods used for determining the polishing duration.

#### *Low κ Dielectrics*

Describe the low κ materials used for inter-level metal dielectrics, including material deposition and etch steps. Describe the process steps for the recently introduced "air gap" cross-section in the inter-level metal dielectrics.

## Self-Aligned Lithography Options

### *SAQP*

Figure 1.10 illustrates the fabrication process for self-aligned, double-patterning (SADP). Recent process development has introduced self-aligned, quad-patterning (SAQP). Describe the process steps to implement SAQP.

### *Self-Aligned Contacts*

Another innovation required to continue process scaling is to pursue aggressive layout dimensions for metal contacts to device nodes. Describe the recently introduced "self-aligned contact" process steps.

### *Analog Layout Design*

Analog IP specifications often require close matching of device characteristics and parasitics, such as for differential signal input pairs. To minimize the sensitivity to mask overlay tolerances, interdigitated device fingers are used in analog layout design. Describe "analog common centroid" layout styles.

### *Metastability*

Figure 1.39 illustrates the use of synchronizer flop cells at clock domain crossing interfaces. As the data input to the synchronizer is unrelated to the clock, data setup and hold time constraints cannot be guaranteed.

*Metastability* refers to the probabilistic risk that an asynchronous data input transition in a narrow window around the capture clock transition will result in an indeterminate output voltage for an extended "settling time," resulting in subsequent path delay fails at the next clock cycle.

Describe how the metastability failure rate is modeled as a function of the input data rate and slew, capture clock frequency, synchronizer cell characterization, output capacitive load, and allowed settling time interval. Describe circuit design and layout techniques for the internal synchronizer flip-flop nodes to reduce the metastability risk.

A common design technique is to use multiple synchronizer cells in series at the asynchronous interface. Describe how the metastability failure risk is reduced with this configuration.

CHAPTER 2

# VLSI Design Methodology

## 2.1 IP Design Methodology

The design of an IP core entails the development and release of a set of models for subsequent SoC methodology flow integration. These models are described in the following sections.

### 2.1.1 A Functional Model for Logic Validation

The functional IP model is compiled into the SoC simulation environment. The model source is commonly provided as part of the IP license, typically in a hardware description language (HDL) format. For hard IP, for added security of the intellectual property, a compiled binary is licensed, with a set of EDA simulation tool application program interface functions to initialize, exercise, and query the behavior. An HDL model can be developed at different levels of functional abstraction—that is, a logic gate–level netlist, a register transfer–level module, or a model utilizing higher-level semantics.

For IP with analog content, the IP model may include extensions to the HDL constructs used, as industry-standard HDLs provide support for mixed-signal behavior. The functional simulator used for an SoC project needs to accept any model abstraction level; for models with mixed-signal content, the simulator maintains both a logical event-driven and analog continuous-time network evaluation engine with a synchronization interface between the two model subsets.

The IP functional model needs to include additional support to ensure proper usage in the SoC design. Specifically, HDLs include *assertion statements* that do not represent additional functionality but rather observe signal behavior in the model to check for events that invalidate IP design assumptions. A simple example would be a combinational assertion that is connected to a set of model inputs that are to be mutually exclusive—say either "zero-or-one-hot" or "zero-or-one-cold." The severity of the assertion identifies how the simulation tool responds when the assertion condition is invalidated; for example, a *fatal* assertion would halt execution, with the time, assertion message, and network state available for direct debugging.

Functional models may also include additional statements to record key events and/or sequences during a simulation test. The complexity of these *event monitors* ranges from a simple capture of important signal values to more complex statement sequences with internal counters of specific operations executed during simulation. The SoC functional validation plan needs to confirm that the testbench suite goals for thorough model coverage are consistent with the event activity and monitor data available from the IP model.

Note that an increasing percentage of the SoC functional validation workload is being allocated to accelerated simulation hardware platforms (see Section 5.3), with a diminishing focus on software tool simulation. The accelerator platforms have unique requirements for the model semantics that can be compiled, partitioned, and mapped to the hardware. If the SoC validation plan includes acceleration, the IP functional model needs to offer corresponding support for the hardware platform.

### 2.1.2 A Performance Model (Optional, Perhaps)

Functional simulation models include the full detail of the IP behavior (e.g., necessary for SoC validation but too slow for SoC architecture performance optimization). Consider the case where processor cores, caches, and bus interface models are being designed to achieve a target performance for program code examples. A high-level, instruction set architecture (ISA) execution model of the core is required to gain sufficient program execution throughput to identify any performance bottlenecks.

An IP core provider may be required to provide a (C-language–based) performance model, in which accurate ISA evaluation and internal cycle latency are the key features.

### 2.1.3 A Logic Synthesis Model

For a hard (or firm) IP core, there is no need to synthesize an HDL model to a library cell implementation. For a soft IP design, a functional model that can be used by a logic synthesis flow is required. Typically, the synthesis models are based directly on the functional simulation model. Any non-functional code in the simulation model, such as assertions and monitors, would include directives (also known as *pragmas*) to inform the synthesis flow to bypass this section of HDL code, as illustrated in Figure 2.1 (for Verilog and VHDL).

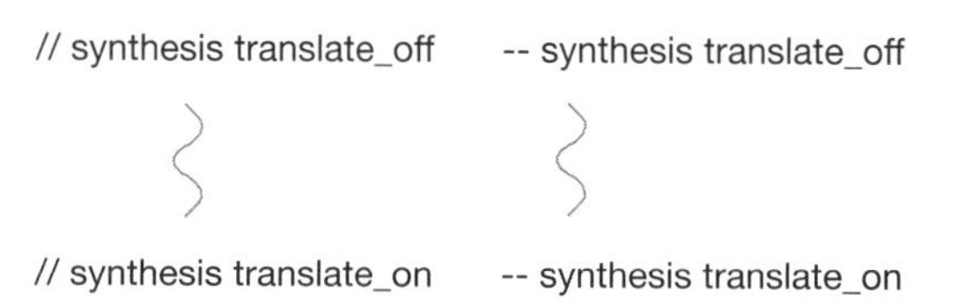


**Figure 2.1** Two examples are depicted to direct the logic synthesis flow to skip a section of HDL code when generating a cell-level netlist model. The HDL "comment" syntax is used with a specific keyword string.

The HDL model for the soft IP is commonly provided at the register-transfer level (RTL) abstraction. Dataflow registers and control flow state flip-flops are fully defined in the model, as is the specific clocking behavior for data operations and state machine sequencing. Combinational logic operations are written to describe the register and state machine inputs each clock cycle. The RTL model approach provides a suitable trade-off between simulation throughput and the expected specificity of IP logic performance. For IP cores where simulation testbench throughput is a priority, the HDL functional model may include more abstract constructs. For example, cores providing digital signal processing on serial data streams require higher simulation throughput and would be more concisely and easily written in terms of do/while and for loop statements. In this case, a more general sequential synthesis flow is required (see Section 7.1). The synthesizable model is also typically used to map the IP functionality to a simulation acceleration hardware platform.

In addition to delivery of the IP model for logic synthesis, the constraints used by synthesis algorithms are required (e.g., input clock definitions, input/output pin timing requirements (relative to the clock), timing modes for MCMM path timing analysis). If the IP is designed to support power states, a power format file description is provided, as well; the synthesis algorithms use this description to ensure that the proper cells are selected for level shifting, sleep state retention, output isolation, and sleep gating.

The synthesis model may include details about test-specific connectivity within the IP or may defer to *test insertion* algorithms within the synthesis flow. A key methodology decision pertains to whether the initial HDL validation environment is to include simulation of testability features or whether test architecture validation is to be deferred to later functional simulation steps. If the former, the HDL model needs to include related built-in self-test (BIST) and scan-shift test logic; no test insertion is required. (The final serial scan-shift flop ordering may be updated during IP physical implementation, as described in Section 8.3.) If testability validation is part of a post-synthesis flow (e.g., using a simulation accelerator platform), the IP model only needs to include sufficient detail for the test insertion synthesis step.

### 2.1.4 A Test Model, Test Patterns, BIST, and Wrap Test Architectures

For hard IP, several test approaches are available. A detailed test model could be provided to merge with other SoC models for pattern generation. Alternatively, logic could be embedded within the IP to exercise test patterns directly in support of BIST operation, with both internal pattern generation and response signature compaction features (see Sections 19.3 and 19.4). Alternatively, the hard IP may support a *wrap test* operation. In this architecture, the IP internal functionality is surrounded by a serial scan-shift register. In wrap test mode, the primary input and output connections to the hard IP are deselected, and the internal scan-shift register is multiplexed to provide the input values and capture the output responses. The test patterns are applied using a scan-shift input sequence, and the response values are then captured and shifted out for observation. The IP is logically isolated from the rest of the SoC during the wrap test. Or, in the most basic approach, test patterns are provided by the hard IP vendor, and the SoC design is responsible for the multiplexing functionality outside the IP to make the pins controllable and observable for pattern application and response capture. Figure 2.2 illustrates this *embedded macro test* approach.

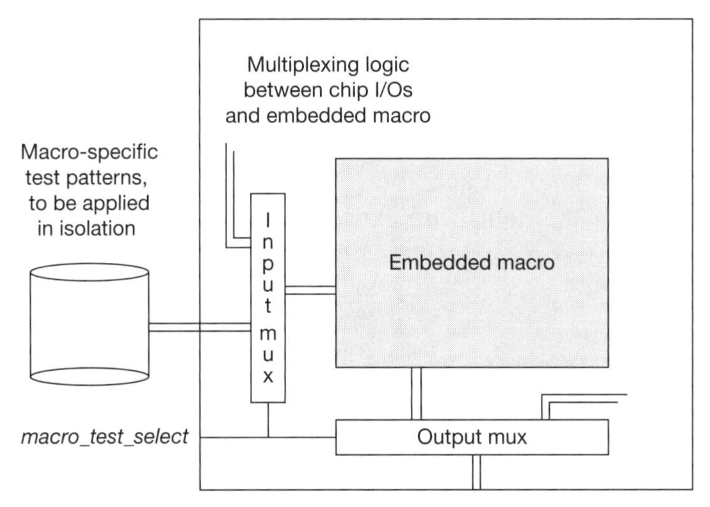


**Figure 2.2** An embedded macro test architecture provides pattern controllability and observability connections to the IP directly from SoC I/O pads through multiplexing logic. (Also see Section 19.3.)

Analog hard IP includes a unique set of test specifications and the interface description to stimulate and capture the corresponding signal measurements. For hard IP associated with high-speed chip I/O serial interfaces, a unique test methodology includes a *loopback* function, where the transmit (Tx) stream output is connected to the serial receiver (Rx) lane, as depicted in Figure 2.3.

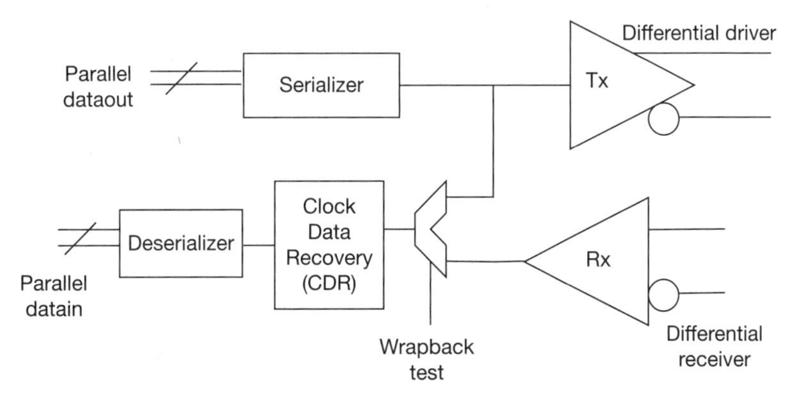


**Figure 2.3** Simplified block diagram of an I/O interface IP design that incorporates a loopback test feature, a common approach for high-speed SerDes interfaces.

The test model for large memory array IP presents some additional considerations if the diagnostic results of an array test are to be used for programming fuses for array repair (see Section 19.3).

For soft (or firm) IP, the netlist model from logic synthesis serves as the basis for the test model. Each library cell includes a specific fault model. The cell instances in the netlist are expanded to provide the potential faults in the network for test pattern generation algorithms to provide production test vectors.

### 2.1.5 A Physical Model

The physical model for a hard IP block consists of a detailed layout and an abstract for use in floorplanning and global routing flows.

For IP security, it is feasible for the model delivery to omit the detailed layout and provide only the abstract. During mask data preparation after tapeout, the foundry—or another *trusted partner*—would insert the layout data and re-confirm the physical verification steps.

Note that device models are dependent on nearby layout structures (e.g., adjacent devices and impurity implants for wells and nFET/pFET source/drain nodes). When developing the timing model and exercising electrical analysis, a hard IP provider assumes a representative layout context around the IP layout. The methodology team needs to review these context assumptions to ascertain whether the *layout-dependent effects* (*LDEs*) are accurate for the target SoC design. The methodology team may need to consider re-extracting and characterizing the hard IP layout (if available) with a different layout context. Alternatively, the IP vendor license may include a request for an updated set of models with a specific context from the SoC design team at an additional cost.

The physical abstract is a text file that conveys the information needed to integrate the hard IP into the physical implementation flows. This text file is typically derived from a combination of designer input and the IP layout data, where the layout includes shapes on specific non-manufacturing layer (and layer purpose) definitions. An *abstract generation* methodology flow step (commonly denoted as *abgen*) at the completion of the IP design produces the abstract file by merging the designer configuration input with the coordinate information of these specific layer shapes. The abstract describes the IP layout in terms of numerous data records, including those described in the following sections.

### IP Boundary

The IP boundary is derived from data on a specific non-manufacturing boundary drawing layer. The hard IP abstract boundary is used during floorplanning to place the IP layout cell. A single shape on the boundary layer is usually sufficient for the abstract to encompass all the layout data on the mask layers associated with devices and local signal routes. The extent of the boundary includes all dummy data added for lithographic uniformity and half-rule spacing for abutting other abstracts. Most floorplanning tools accept rectilinear boundaries, although some require a rectangular representation.

The boundary created for soft (and firm) IP is a design optimization objective during floorplanning. It represents the placement-and-routing boundary (also known as the *prBoundary*) for subsequent IP implementation. Typically, the floorplan boundary for soft IP is rectangular, as cell placement and routing algorithms may not readily adopt route planning and detailed track assignment that accept a rectilinear area boundary.

### Routing Blockages

If the hard IP layout includes any data on layers allocated to global signal routes, the abstract includes additional non-manufacturing *blockage layer* shapes associated with the metal layer(s) and routing tracks used by the hard IP. If there are critical nodes in the layout for which the capacitive coupling noise of a global signal routed over the IP would be problematic, an additional blockage shape may be added. (A global power rail over the node may be acceptable; the IP design methodology could provide separate blockages for global signals versus rails if the floorplanning and routing flows can recognize the distinction.) The abstract generation flow translates any IP layout shapes on global metal layers into route blockage information in the abstract file.

### Flip/Mirror Placement Orientation Options

Hard IP floorplanning may benefit from the capability to place the block after mirroring the original orientation across the vertical and/or horizontal axes. A 90-degree rotation orientation has also been an option for IP that is associated with SoC I/O pads, to indicate that the IP can be placed along any of the four sides of the die. However, more recent process nodes require that *all* device gates share a common (vertical) orientation, precluding the ability to rotate IP and/or library cells. As a result, two separate physical layouts are required for I/O cells on current SoC designs: one for N/S sides and one for E/W sides. The designer input to abgen describes the allowed mirroring and any die edge placement restrictions for the IP abstract.

#### *Pin Data*

The abstract includes all the physical pin information necessary to complete the floorplan and guide the global track routing algorithm, such as the pin name, pin directionality (e.g., input, output, bidirectional), pin metal layer, pin dimensions, and connect constraint. This information is derived by abgen from the IP layout (top) cell, using additional shapes data on pin-specific layer definitions. Note that power and ground pin shapes are also present in the IP physical abstract. In this case, additional pin information provided by the IP designer is required to associate the power or ground pin with the corresponding voltage definition, as illustrated in the abstract snippet example in Figure 2.4.

```
cell (sample ) {
   pg_pin ( P1 ) {
     voltage_name: VDD1 ;   pg_type:  primary_power }
   pg_pin ( G1 ) {
     voltage_name:  GND1 ;  pg_type:  primary_ground }
   ...
```

**Figure 2.4** IP physical abstract text snippet for power and ground pin definitions.

The pin is visually represented in the layout by a pin shape, using a non-manufacturing layer purpose (e.g., m4_pin). There is also a (non-manufacturing) alphanumeric label included in the layout with each pin name; the origin of the label is placed on the pin shape. Abgen derives the data for the pins section of the abstract file from these layout shapes. Floorplanning tools display the pin shapes and label information from the abstract.

The dimension of the pin shape in the IP layout is an important design consideration. A configuration constraint used during abstract generation indicates whether a (minimum) valid connection to the pin shape is sufficient or whether the entire pin shape must be covered by a (non-minimum) global route. The circuit design of the hard IP may provide an *area pin* for two potential global implementation options: to offer multiple potential routing track access points or to ensure that wide metal is used to cover the pin for electrical requirements. Figure 2.5 shows both global route topologies.

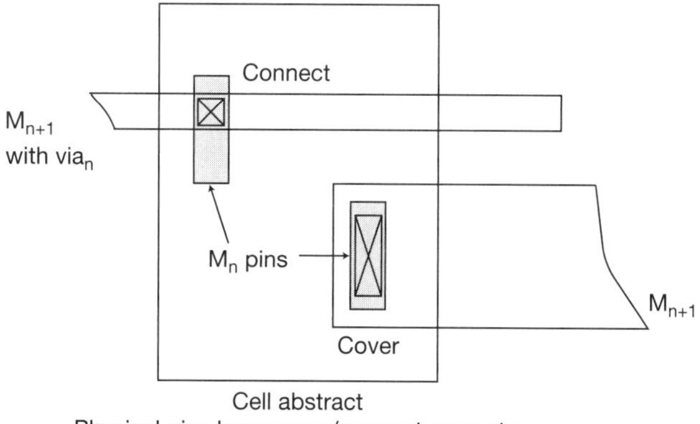


**Figure 2.5** Area pins in a cell abstract with connect and cover properties.

The designer input to abgen for each pin is reflected in the abstract pin property to guide subsequent global track allocation and routing.

For soft (and firm) IP, pin data is typically assigned during floorplanning: The output of the floorplanning flow is an IP abstract for detailed physical implementation.

Cell abstract generation is also a key flow for library development, and it uses the same format as the abstract for a hard IP core design. The cell layout has a different abutting approach than a hard IP core for any circuit structures that are to be continuous through adjacent cells (e.g., the device fabrication well). As a result, a different set of layout design checks between the boundary shape and the internal layout data is needed for cells than for hard IP cores. For cells, area pins are preferred, if possible, to provide the detailed routing algorithm with multiple access points. There is a minor methodology issue associated with area pins pertaining to parasitic extraction: An assumption is made that the area pin is a single electrical node and all access points are electrically equivalent. For an area pin spanning two wiring tracks, this assumption is reasonably accurate. For larger area pins, a single node in the electrical parasitic model may not be accurate for a route connection at an extreme of the pin shape.

### *Schematic/Circuit Netlist (Pre- or Post-extraction, Optional)*

The hard IP (and library cell) delivery may include a device-level netlist model. If the IP provider utilizes a trusted partner rather than licensing the layout view, a netlist is typically not provided.

Access to the netlist model allows circuit simulation at PVT corners for the SoC design that are outside the characterization suite used to derive the IP timing model. Alternatively, the IP vendor license may include a request for analysis at more corners at an additional cost.

Figure 2.6 depicts the constituent views included in the IP model delivery.

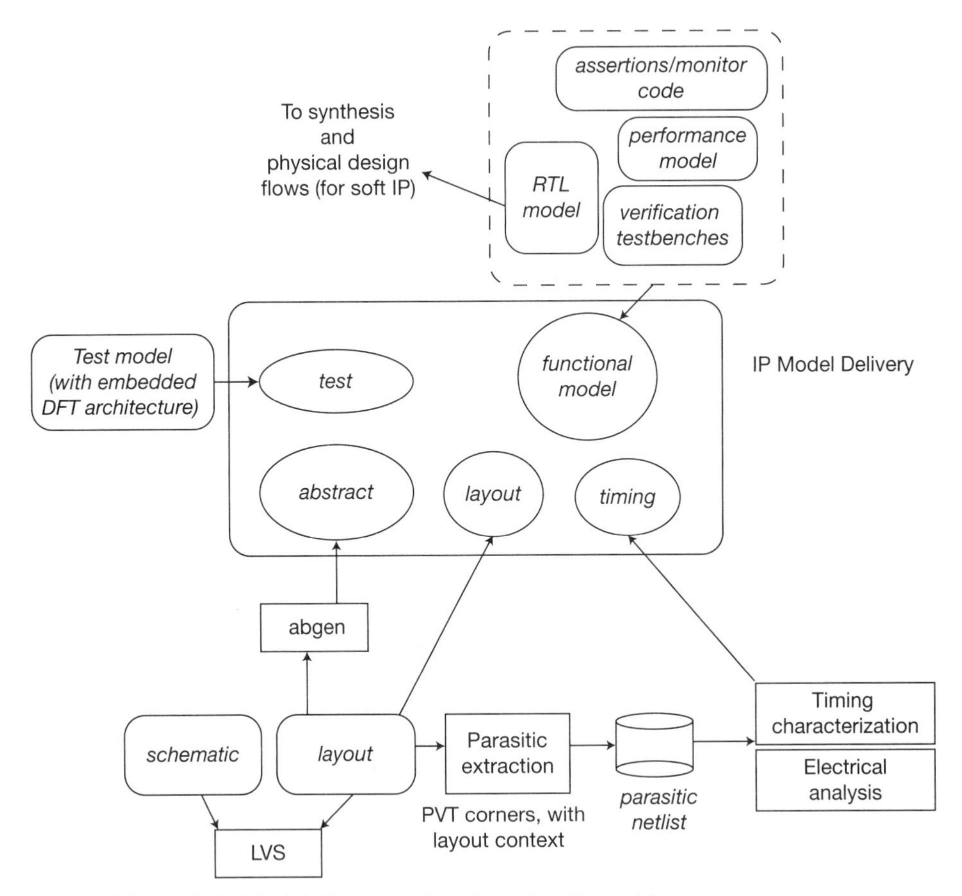


**Figure 2.6** Model views required from the IP provider.

## 2.2 SoC Physical Design Methodology

### 2.2.1 Floorplanning

The SoC physical design begins with floorplanning, which consists of the following steps:

1. Chip area design
2. Hard IP placement/orientation
3. Soft IP area optimization and pin assignment
4. (Global) power domain distribution design
5. (Global) clock domain design (skew estimates/margins determined)
6. Global bus route planning (wire widths, wire spacing, wire shielding, preliminary track assignment)
7. Global test architecture planning
8. Global signal buffer/repeater planning
9. I/O pad assignment (signals, power/ground)

### 2.2.2 Physical Integration

Physical integration of IP blocks, global cells, and global routes is an iterative process that commences soon after the project milestone for the initial floorplan release. As the implementation of each soft IP block evolves, the physical integration team incorporates successive releases of these blocks into the floorplan. The soft IP block updates reflect functional bug fixes and, likely, new and/or updated pin definitions (e.g., input loading, required input arrival times, expected output times, estimated output loading). Logic added to the soft IP RTL model for bug fixes may stress the originally allocated floorplan block area. Local signal routing closure may highlight the need for a different block aspect ratio to adjust the available horizontal or vertical wiring tracks. Addressing either increased logic utilization or routing congestion issues within a block may necessitate significant floorplan changes.

The increasing detail available from power estimation flows may highlight changes required to the global power distribution grid. The SoC functional validation tests provide representative switching activity data, which are used to estimate the active power dissipation for all IP blocks. The resulting power map may offer opportunities to reduce the density of the global grid for IP with low activity or may dictate a more robust grid for IP with high activity. In either case, the impact to the global signal route topology estimates versus available wiring tracks is reevaluated, and existing global signal track assignments are potentially subjected to a "rip-up and reroute" step.

A key methodology feature supporting physical integration is the configuration management tool used to define the block IP release versions that comprise an integration *snapshot*. For early physical integrations, the functional model is quite fluid. Functional debug/fix iterations occur frequently and must be addressed in a timely manner. The SoC soft IP RTL revision branches update pin definitions and global signal connectivity at a rate that is (initially) likely to exceed the bandwidth of the physical integration resources. The SoC project management team needs to coordinate when the functional model configuration will be tagged as suitable for a physical integration snapshot. The SoC functional top-level hierarchy becomes the new global signal connectivity model, and at this point it is out of sync with the existing physical integration netlist. Resolution of the logical-to-physical discrepancies is a key outcome of the next physical integration, including area and pin updates for individual soft IP blocks.

The release promotion of a soft IP block physical view for SoC integration is subject to a set of methodology quality checks. The active IP block design occurs on a "branch" version from the current physical integration configuration. Promotion of this branch to the "main line" integration development requires successful completion of the appropriate methodology checks. Commonly, a *scoreboard* of IP checking and analysis flow results is maintained for each active IP block version. The set of IP quality checks appropriate to promote a revision for the next physical integration becomes more stringent as the SoC project progresses. Initially, a minimum quality level may be acceptable to allow the integration team to adapt to any significant area and pin updates; a successful pass through the abstract generation flow for the IP block could be a sufficient check for release to early integration iterations. The methodology for subsequent soft IP releases for physical integration incorporates additional checks, including:

- RTL logical-to-cell netlist functional equivalency
- Test architecture consistency checks and IP-level testability measures
- Detailed signal route completion (confirming no congestion issues)
- Full physical block completion
- Final IP model completion

A physical integration update may include an SoC configuration with a revised top-level netlist, with IP abstracts that may not yet reflect all the global signal changes. The integration team goal would be to identify the magnitude of the top-level changes and assess the impact to the floorplan and existing routes. The integration of the global netlist and abstracts is subjected to a methodology flow that performs a *connectivity check*. The output of this flow is a report of the model discrepancies to review against the existing floorplan:

- Un-driven IP block input pins (*sourceless*)
- IP output pins without any fan-outs (*sinkless*)
- "Dotted" IP output pins (typically disallowed for SoC global signals)

High-impedance drivers on global nets introduce intricate performance, testability, and electrical reliability issues. As a result, dotted output pins are commonly denoted as invalid.

The combination of (a subset of) new IP physical abstracts and top-level SoC netlist for each integration imposes a challenge to maintain the chip area, floorplan channels, I/O pad assignments, and routing track plan that will converge in the final integration project phase.

### 2.2.3 ECO Mode and Chip Finishing

The physical integration iterations use netlists and abstracts of increasing quality (and logical-to-physical consistency). After completion of the final integration step, the global signal and clock routes, the global power distribution, and the channel buffer/repeater cells are all in place. The SoC project functional validation team should have only a minimal set of active bugs under investigation. After exiting the final integration, the project management team will transition to *engineering change order* (*ECO*) mode. In ECO mode, three types of model update requests may be presented to the ECO methodology flows: functional model updates, netlist updates for timing and power optimization, and chip finishing updates.

#### *Functional Model Updates*

Logic changes to fix functional bugs uncovered late in system validation require judicious attention to minimizing the disruption to the existing physical design, whether to a soft IP block or to a global signal. Normally, functional bug fix updates are initially made to the RTL for validation. In ECO mode, however, functional logic updates are commonly made directly to the cell netlist, applied as a set of library cell and signal adds and deletes. The goal is to avoid the disruption of re-synthesizing a cell netlist from an updated RTL model and to instead work through the logic changes needed to fix the bug with the existing detailed netlist. Note that the correlation between RTL signals and the synthesized netlist signals is limited; debugging functional bugs at the cell level is not particularly straightforward. The physical integration team works through the cell change netlist, using the interactive editing features of the layout tool.

Concurrently, the RTL model is updated with the functional bug fix and recompiled for validation. As the RTL and netlist models have been edited independently in ECO mode, the methodology flow for RTL-to-netlist logical equivalency is invoked (see Figure 2.7).

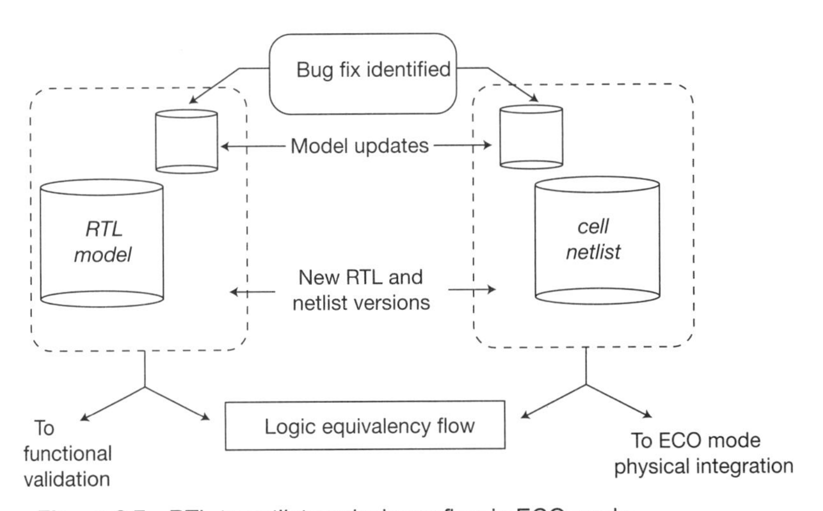


**Figure 2.7** RTL-to-netlist equivalency flow in ECO mode.

Note that EDA vendors have pursued development of logic synthesis tools with an ECO synthesis feature.[1] The intent is to allow designers to implement the bug fixes directly in the RTL rather than at the cell netlist level. To date, however, the typical methodology flow applies physical edits based on a manually generated ECO netlist, with proven logic equivalency to an RTL revision.

### *Netlist Updates for Timing and Power Optimization*

In ECO mode, the design team is focused on closing any remaining path timing issues and optimizing power dissipation (especially leakage power). Non-functional cell netlist updates are applied to (a series of) physical model revisions. To address performance path issues, a lower-$V_t$ variant of a cell may be selected, a higher-drive-strength cell may replace an existing cell of the same logic function, and/or an alternative flop cell with different setup time and hold time constraints may be selected. Note that a higher-drive-strength cell is likely to occupy greater area; after substitution, there may be cell overlaps that have to be resolved. The ECO mode of the physical layout tool can identify the overlaps, allowing the layout designer to shift cells (and their route connections). Hopefully, this *annealing* of cell placements will result in minor changes in route lengths and minimal new path timing issues.

There are many other netlist-level timing optimization options available, related to series-parallel repowering topologies of high fan-out signals, as shown in Figure 2.8.

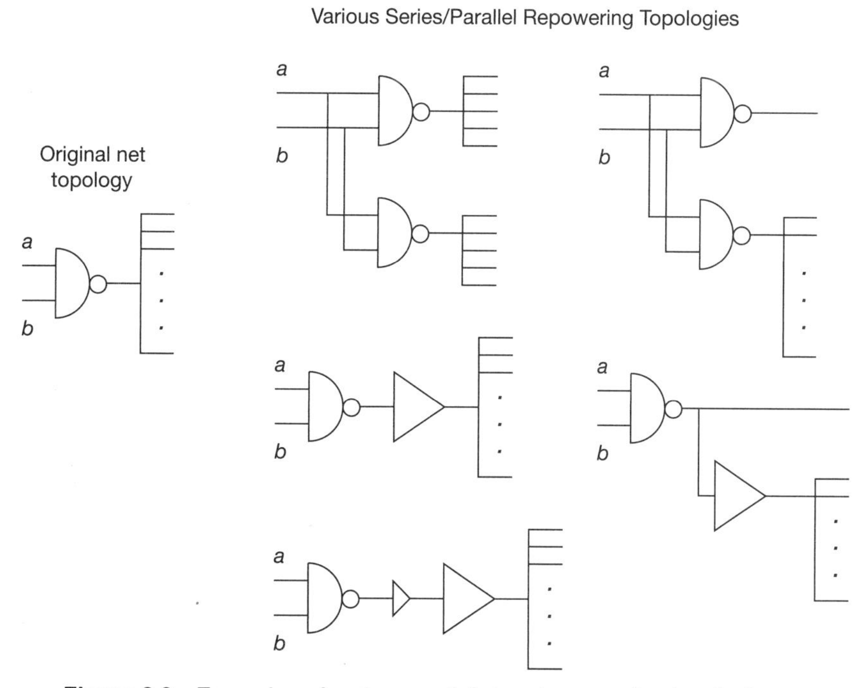


**Figure 2.8** Examples of series-parallel signal repowering topologies.

Ideally, the more complex gate and route topologies illustrated in Figure 2.8 would have been optimized previously, during the physically aware logic synthesis flow (see Section 7.2), as these repowering updates are much more disruptive to the physical SoC model in ECO mode.

Another critical timing optimization that is often completed in ECO mode is the insertion of delay cells to resolve short-path hold time constraints (see Figure 2.9). The shortest timing path delay between two sequential flops needs to be sufficient to ensure that the capture flop does not erroneously record the "next cycle" data from the launch flop at the current cycle.

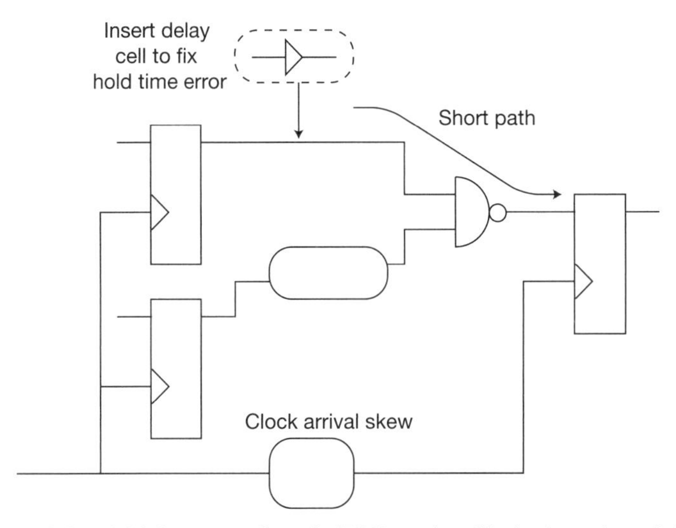


**Figure 2.9** Hold time error fixes in ECO mode with the insertion of delay cells.

It may be necessary to add delay cells to ensure that the capture flop data input is stable past the capture clock arrival (with skew) plus the flop hold time constraint. As logic networks are likely to have both long timing paths and short paths, a judicious assessment is required to determine where to insert the delay buffer(s) so as not to adversely impact existing setup time checking results. EDA path timing analysis tools offer recommendations for the network nodes to insert delay cells, based on the hold time test results. Also, the hold time constraint must be satisfied at *all* PVT corners, especially the fast corner(s). This can be problematic, as gate delays scale differently across PVT ranges than the clock wire and fan-out tree load imbalances that may be contributing to the clock arrival skew between flops; the delay cells added to fix short-path issues at a fast corner result in significantly different path timing updates at other corners.

The cell library limits the maximum delay cell to satisfy constraints on maximum signal transition slew rate at the slow PVT corner. As a result, it may be necessary to insert multiple delay cells in series to provide sufficient path timing adjustment to fix hold time issues. As with the substitution of a higher-drive-strength cell to address path setup time issues, the delay cells added for hold time fails in ECO mode may result in area overlaps that need to be resolved.

Although some short-path delay padding may be invoked prior to ECO mode (based on flop-to-flop *logic path depth*), the hold timing analysis results from the detailed physical design are commonly used. This ensures that the most comprehensive set (and only this set) of issues are addressed after clock arrival skews are accurately determined.

A unique consideration for SoC projects is that a (first-pass) silicon prototype design with known setup time issues (at some corners) may still be useful for system bring-up, operating at a reduced clock frequency. However, a hold time fail results in non-functional behavior. If a project is approaching its tapeout date, the SoC project management team may choose to prioritize hold time fixes and release the design for fabrication once those fixes have been implemented. The schedule pressure to enter system bring-up with prototype parts may necessitate deferring remaining setup fixes until a subsequent tapeout release. This second-pass tapeout addresses both functional bugs found during system test (or ongoing SoC validation) and the deferred setup timing path fixes (see Section 20.3). The tapeout sign-off methodology flow needs to record any setup time issues from path timing analysis that have been reviewed and deferred by SoC project management.

For leakage power optimization in ECO mode, the most common update is to replace a cell with a higher-$V_t$ variant if the additional gate delay does not introduce new setup time issues. The common metric to identify candidate cells for power optimization is the timing slack on the gate output (see Section 11.4). If the gate output *arrival time* is less than its *required time*, the timing slack at the cell is positive and could potentially accommodate a slower cell variant with lower leakage current.

In ECO mode, there is also an option to reduce the drive strength of a cell with positive timing slack, reducing input load capacitance and leakage currents to save power. However, the drive strength may have been previously selected by synthesis algorithms to maintain a suitable output signal slew rate. In this case, a reduction in cell drive strength may result in an increase in active power due to the cross-over current of active pullup and pulldown devices in the fan-out cells receiving a longer input signal slew transition. Cell updates for power optimization require an evaluation of the impacts to both active and static leakage dissipation.

All these timing and power optimization netlist updates are non-functional changes to the SoC design. Nevertheless, after *any* netlist-level revision, successful completion of the RTL-to-netlist logic equivalency flow is required before the updates can be promoted to a new version level.

### Chip Finishing

In ECO mode, a set of *chip finishing* methodology steps are invoked. These steps relate to the addition of library cells and shapes to the layout database to complete the physical design, improve manufacturability, and prepare for design edits in a subsequent tapeout. These steps include:

- Insertion of well/substrate contact cells
- Insertion of decoupling capacitance cells
- Insertion of spare logic cells
- Insertion of active and gate fill patterns in open areas (for photolithographic exposure and etch rate uniformity)
- Insertion of metal fill patterns (for uniformity of metal density to minimize CMP *dishing*)
- Insertion of *slots* in wide metals
- Insertion of (reverse-bias) diodes to metal *antennas*

The cell utilization area percentage in a soft IP block should allow addition of sufficient well/substrate contacts and decaps (e.g., 80%–85% functional utilization). Similar to metal fill in open areas, wide metals require slotting to minimize CMP non-planarity. Note that slotting affects the effective resistivity and I*R power rail voltage drop, as illustrated in Figure 2.10. The metal shapes in a net collect electric charge during fabrication. A net topology of sufficient area not connected to a device junction until upper metals are fabricated may need to have a reverse-biased diode explicitly added to remove the collected charge and protect device gates from damaging electric fields (see Figure 2.11).

- Insertion of foundry-provided process monitoring layout cells
- Insertion of alphanumeric layout data, including layer information, corporate logo and copyright, part number, tapeout revision

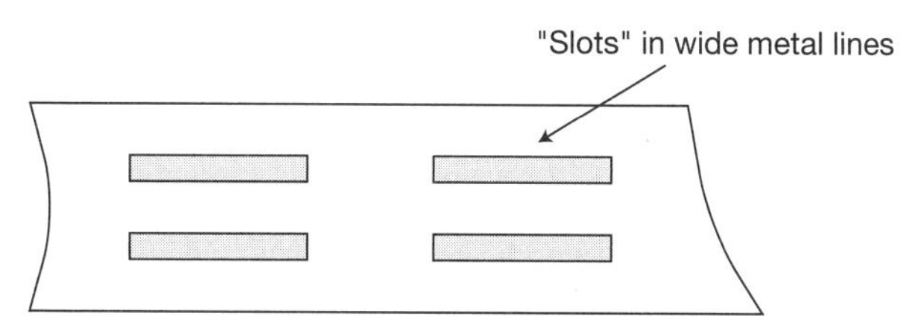


**Figure 2.10** Wide metal lines require dielectric "slots" to reduce dishing during CMP.

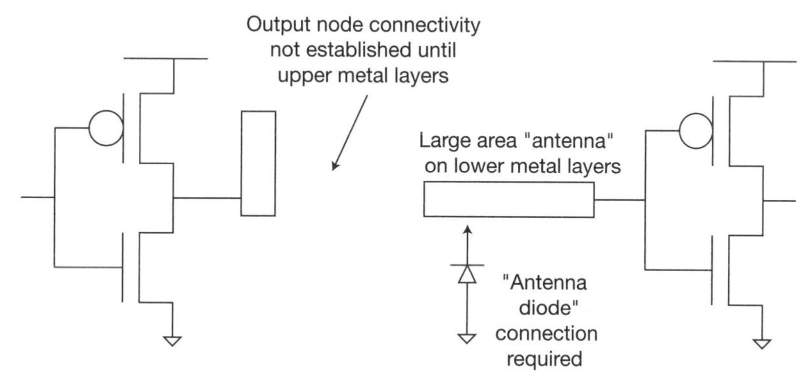


**Figure 2.11** A net of sufficient metal area connected to device gate inputs but not to a device junction until upper metals are fabricated may require insertion of an "antenna diode."

### 2.2.4 Frozen Physical Database for Tapeout and Use of Sign-off Flows

ECO mode relates to netlist-level cell changes applied to the SoC physical data, and functional equivalence needs to be proven against a (potentially revised) logical model. Chip finishing flows add non-functional layout cells necessary for fabrication. After exiting this project phase, the complete physical (and equivalent) logical model configuration is "frozen" in preparation for tapeout release.

The SoC methodology shifts to exercising *sign-off* flows against IP blocks and the full SoC integration database. Sign-off flows for IP blocks typically utilize the same underlying EDA tool algorithms used earlier for analysis and checking. However, in sign-off, the block-level pin constraints used for block implementation are replaced by data derived from the SoC implementation.

The sign-off step includes a combination of physical verification and electrical analysis flows. During block physical design, analysis and verification flows may have reported errors; however, release version promotion to a new SoC configuration may still have been allowed. In sign-off, the flow reports are parsed for all error (and warning) messages that might represent potential issues that require project management review and approval.

### *Physical Verification Flows for Sign-off*

#### *Layout-Versus-Schematic (LVS)/Cell Netlist Verification*

The physical layout is compared to the SoC netlist to demonstrate equivalency. In the most detailed comparison, the full layout hierarchy is expanded into each custom IP cell, transistor recognition and net tracing are exercised, and the physical topology is compared to the SoC schematic netlist. Commonly, the layout hierarchy expansion algorithm ceases when it reaches an instance corresponding to a cell "stop list"; the assumption is that the IP library release has already proven layout-versus-schematic equivalency within the library cell. With a stop list, LVS verification traces signal nets through metals/vias between cell pins and compares the SoC layout-based connectivity to the cell-based netlist. Note that the LVS comparison algorithm requires knowledge of symmetric cell pins (or transistors, for full custom verification), whose connectivity is functionally equivalent. Figure 2.12 illustrates an example in which the physical SoC signal connectivity to a cell is logically equivalent to the netlist and thus may be deemed to be LVS correct if the LVS flow settings allow pin swaps.

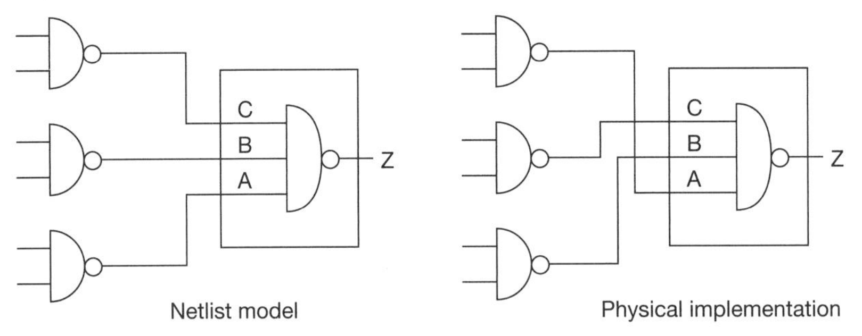


**Figure 2.12** LVS connectivity checking may be enabled to accept logically equivalent pin swaps between schematic/netlist and layout models.

Power net analysis is also a key part of LVS verification. The SoC netlist model exported from the physical design flow is denoted as a *PG netlist*, with additional pins and connections on each cell instance to include all the power and ground nets. (Recall that the cell abstract also includes power and ground pin definitions, as illustrated in Figure 2.4.) Layout tracing includes the power, ground, well, and substrate connectivity for LVS sign-off comparison to the PG netlist.

The proof that the functional RTL model is fully equivalent to the physical layout for tapeout is transitive. The functional equivalency of the RTL (plus power format file directives) to the PG netlist is proven as part of exiting ECO mode. The sign-off LVS flow proves that the layout matches the PG netlist. No LVS flow waivers are allowed when conducting the sign-off flow report review.

The methodology for the sign-off LVS flow includes allocating sufficient compute resources to complete the full-chip verification in a schedule-critical project phase. EDA vendors have addressed LVS tool throughput by implementing support for parallel (shared memory) and distributed comparison algorithms. An additional performance enhancement is realized by leveraging multiple identical IP cores in the PG netlist and layout hierarchy. Preprocessing of the layout and netlist models identifies opportunities to demonstrate LVS correctness for a corresponding instance in the two models, and that proof can subsequently be applied directly to other identical correspondences in the two model hierarchies.

The methodology team needs to accurately forecast the SoC model size and LVS runtime goals with the available compute servers to determine the required number of tool licenses. In addition to the hierarchical correspondence preprocessing feature, the EDA vendor may offer multiprocessing support consisting of a "master" plus "slave" license product offering or may perhaps allocate a ratio of execution threads to product licenses. For these long-running, large resource jobs, the EDA vendor may also provide a tool *checkpoint/restart* feature in case of an unexpected compute error. The methodology team prepares sign-off flows to manage compute resources and tool licenses to provide the requisite throughput; this is likely to also involve establishing policies with the IT team for reserving critical compute servers, setting job queue dispatch priority, and (potentially) suspending existing jobs unrelated to sign-off to acquire enough tool licenses.

### *Design Rule Checking (DRC) Verification*

The frozen layout database is verified against the fabrication process design rule checking runset, which is provided in the PDK release from the foundry. As with LVS tools, EDA vendors have implemented throughput enhancements incorporating hierarchical preprocessing and multiprocessor execution. In addition, the methodology team needs to offer a sign-off flow that maximizes IT throughput on the full SoC layout dataset.

The potential for the foundry (or trusted partner) to subsequently insert layout cells after tapeout for any *black box IP* in the design implies that there will be overlay shapes in the layout covering the black box for which checking is to be skipped in this area. The sign-off flow needs to confirm that any non-manufacturing shapes data that may influence DRC results are correctly placed on the corresponding layout cell abstracts (and for only those cells).

### *Recommended Design Rules and Design-for-Yield*

For VLSI process nodes that first introduced mask data smaller in dimension than the exposing light wavelength, foundries introduced *recommended design rules* (*RDRs*). Layouts were presented to the RDR checking runset as well. SoC design teams were encouraged to review the RDR error results and address as many fixes as the schedule and design resources would permit. An example of an RDR would be the metal line end extension past a via opening (see Figure 2.13). A longer extension would improve the robustness of the via coverage to mask overlay variation and line end “pullback” fabrication. However, implementing an RDR without impacting routing density may not be feasible in all cases.

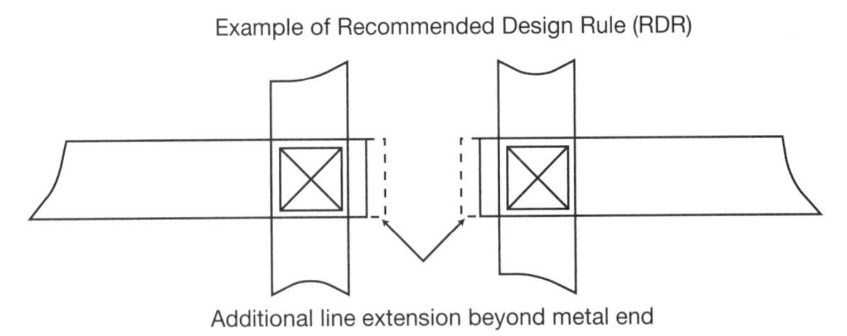


**Figure 2.13** For older process nodes, the foundry commonly provided recommended design rules to enhance fabrication yield, such as increasing the metal line extension past a via.

With successive process nodes, the inclusion of RDRs has diminished. The lithographic requirements have become stricter, in terms of allowable shape widths, spaces, and enclosures. In place of RDRs and the DRC reports from the corresponding checks, foundries and EDA vendors have collaborated on a set of layout analysis and shape manipulation utilities denoted as *design-for-yield* (*DFY*) optimizations that are to be applied directly to the physical layout. The most prevalent DFY modification is the addition of multiple redundant vias for an interconnect wherever possible (without violating PDK design rules). Figure 2.14 shows several potential topologies, including *dog-bone* and *icicle* metal extensions to facilitate adding another via. At the most advanced process nodes, these metal shape extensions in support of redundant via insertion are likely no longer resolvable, necessitating different via strategies, such as *via pillars*.

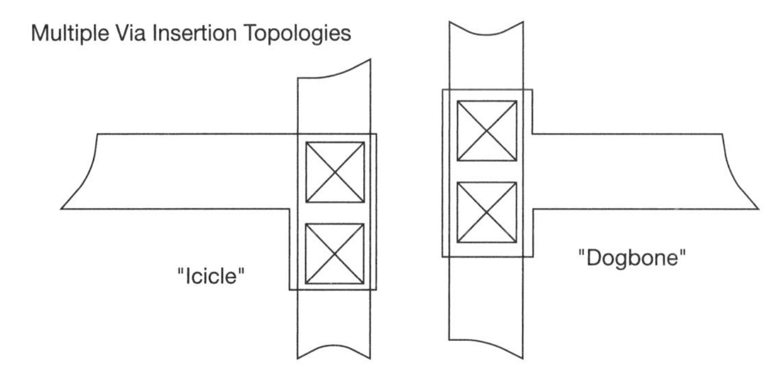


**Figure 2.14** DFY layout optimization examples related to multiple via insertion.

Yield loss due to incomplete via etch/fill is reduced with redundant via insertion, and electromigration reliability is improved. The importance of DFY for current VLSI process nodes is so critical that the optimizations have been integrated into earlier methodology flows, specifically detailed track assignment and routing steps. These considerations can no longer wait until the design freeze tapeout phase is approaching.

*Design-for-Manufacturability Verification Flows*

A separate class of additional layout DRC-related checks is part of the full-chip tapeout preparation phase; it is a collection of methodology flows typically denoted as *design-for-manufacturability* (*DFM*) verification. Whereas DFY optimizations are intended to address yield-detracting process defect densities, and thus are (to some degree) optional, the DFM flows include mandatory checks, to ensure the following:

- Mask data generation will be successful.
- Lithographic patterning will have a sufficient exposure window.
- BEOL chemical-mechanical polishing (CMP) will provide adequate surface planarity.

The two prevalent DFM flows that cover these requirements are lithography process checking (LPC) and CMP analysis.

**Lithography Process Checking (LPC)** The final data pattern after mask synthesis at the foundry reflects an *inverse lithography* computation. The final (5X enlarged) mask plate data is a result of the inverse analysis of the target wafer-level pattern exposure, interacting with the wavelength and illumination pattern of the light source, compensating for diffraction and light interference through the entire optical path from source to photoresist-coated wafer. Design rules are intended to ensure that the inverse litho computation at the foundry's *mask house* is successful. Despite the increased DRC complexity at current process nodes, there is still a risk that an exception is present and might prevent mask generation from converging on both the mask data and source illumination pattern. The LPC methodology flow exercises a more exhaustive set of algorithms against the final layout prior to tapeout release to the foundry. LPC consists of two primary approaches:

- **Pattern-based checking**—The foundry may provide a set of layout patterns that have proven to be problematic during process qualification. The custom designs used in process development wafer lots and early IP shuttle lots provide representative layout examples for mask data generation. Although the goal would be to capture any issues as regular design rules, there may be an accumulated set of (obscure, custom) layout patterns that should be identified and excluded from the full-chip design. These patterns consist of a small number of shapes, typically on one to three layers. EDA vendors have included pattern matching features to their DRC/LPC tools. The matching algorithms are quite sophisticated, and the widths, spaces, enclosures, and parallel run lengths of the pattern shapes may include a range of dimensions.

- **Model-based checking**—LPC tools from the EDA vendors also include the capability to execute complex algorithms, similar to the detailed inverse lithography computation engine used at the mask house. Manufacturability is analyzed over a range of focus and exposure variations. The computationally intensive model evaluation is mitigated somewhat by the relatively small layout area windows and mask layers over which LPC verification is performed and the ability to distribute the execution of the model analysis for the layout windows to separate processors.

The expectation is that LPC will not uncover any issues and that the DRC checks are indeed sufficient. As a result, LPC is usually deferred until full-chip layout freeze. The computational cost of running LPC can be very high, and it might be skipped during earlier SoC physical implementation project steps due to the low probability of errors. Nevertheless, fixing any errors reported by the LPC flow is mandatory. In addition, the errors should be reviewed with the foundry customer support team so that additional DRC rule development can be pursued.

**CMP Analysis** Chemical-mechanical polishing is a fabrication step used to planarize the wafer surface after metal deposition (which commonly uses an additive, or damascene, method). The wafer is placed face-down in a slurry consisting of both abrasive and chemically active compounds. Backside pressure is applied to the wafer, which, combined with (asymmetric) rotary motion, provides the polishing action. Areas with a low metal wire density are subject to local erosion of the surface topography.

Metal fill shapes are added to the layout to improve the planarization uniformity. Design rule checks are included to ensure a minimum (and maximum) utilization on each CMP metal layer, measured in a window stepped across the full-chip layout. Additional rules check that the density gradient between neighboring windows is limited. To reduce the profile "dishing" of wide metal wires, dielectric slotting is required, as depicted in Figure 2.10. The foundry makes a trade-off when establishing the CMP design rules and the stepping window; extremely tight min/max/gradient rules are difficult for design teams to achieve, whereas looser rules can adversely impact manufacturability. An additional tapeout sign-off methodology flow performs a CMP analysis of the full-chip layout data to simulate the fabricated surface profile and to ensure that depth-of-focus (DoF) for subsequent photolithography steps is within variation limits.

As with LPC, the expectation is that the CMP analysis flow will not identify any DoF issues, as layout blocks throughout the SoC hierarchy will have applied metal fill utilities during physical design to address metal density DRC errors. The CMP analysis flow can certainly be exercised during implementation phases, as it also offers insight into metal wire thickness variations that could impact interconnect delays and timing margins. However, it is often deferred until sign-off. Fixing any errors uncovered during CMP analysis is mandatory.

#### *EDA DFM Services*

Note that the full-chip DFM flows are typically deferred until the layout sign-off project phase. And, these steps are *very* computationally intensive. EDA vendors realize that many design customers may not be able to afford to license these tools (to execute on a large number of distributed processing cores) for a relatively short burst of usage. EDA vendors may collaborate with a foundry to offer DFM verification services, where the SoC physical data is provided to the EDA vendor to exercise the tool on the IT resources optimized for LPC and CMP analysis. This option may be a cost-effective approach to satisfying these sign-off requirements.

#### *DRC Errors/Waivers*

The sign-off DRC flow ideally does not report rule errors on the full-chip layout after the chip finishing steps are invoked. (Evaluations of full-chip DRC results during SoC physical integration phases can identify issues early and help test and qualify the chip finishing flows.) However, sign-off errors may indeed be present; consider the case of leading SoC designs under way concurrently with foundry process development, using preproduction PDK release data that are still evolving. A DRC runset release update during a design project may result in hard IP and/or SoC layouts that are no longer "DRC clean." The SoC project management team may need to respond to DRC errors.

The foundry establishes design rule measures that reflect trade-offs between PPA and high-volume manufacturability, using the fab equipment assigned to the process flow. Design rules are not strictly binary in that a rule check error does not necessarily imply zero yield fabrication. For a sign-off

DRC report with errors, the project management team may consider the following options:

- **Fix the layout DRC errors.** If the errors are readily fixed, this approach is certainly the most straightforward. The impact of the fixes on other facets of the sign-off flows needs to be assessed; rerunning electrical analysis and path timing verification flows at full chip may introduce project tapeout delays and additional expense.
- **Review the DRC errors with the foundry and request waivers.** If the errors cannot be fixed without adversely impacting the project schedule, the project management team may wish to review the DRC flow report with the foundry customer support team to determine if the errors can be waived. If the impact of the layout errors on the fabrication yield is estimated to be manageable, a collective decision with the foundry to proceed with the tapeout flow may be reached. This is especially true for a first-pass prototype tapeout (in a multi-pass project schedule), where the silicon bring-up schedule is more critical than the yield of the prototype fabrication wafer lots.

The sign-off methodology flow needs to record the jointly approved DRC waivers as part of the tapeout release documentation. Subsequently, a decision needs to be made on which DRC errors will be addressed and which will maintain their waiver status despite the potential (production) yield impact. The methodology team needs to provide a mechanism for recording waived DRC errors associated with a specific revision of a layout cell and excluding those errors from ongoing DRC flow results. Note that the waived error record should be self-contained within a layout cell; if a change in the context of the cell instance in the SoC physical hierarchy influences the associated design rule, the waiver approval may no longer apply.

#### *Electrical Rule Checking (ERC)*

There are a few fabrication layout and DFM rules for which physical verification sign-off requires more than the traditional geometrical checks. These rules require an electrical analysis of the connectivity through circuit topologies. EDA vendors offer electrical rule checking (ERC) tools to evaluate the related PDK runset checks. The two most prevalent sign-off ERC checks are electrostatic discharge (ESD) protection on chip pads and voltage-dependent design rule checks.

**ESD Checks** Electrostatic discharge mechanisms are a subset of the more encompassing issues of electrical overstress (EOS). Each SoC pad requires an ESD protect structure to be attached (see Section 16.3). An SoC design team may request an exception to the ESD rules from the foundry if the additional parasitic capacitance of the ESD protect layout cell is problematic; however, exceptions are rare.

The purpose of the ESD protect structure is to provide a separate discharge path from the pad circuitry in the case where the pad may be contacted by an external means and an electrostatic charge buildup has accumulated on that external source. This alternative discharge path is meant to protect the functional I/O pad circuitry from exposure to a transient that would likely result in a permanent failure, such as a gate oxide dielectric breakdown or an open connection between pad metal and the I/O circuitry. This is the *human-body model* (*HBM*) representation for electrostatic buildup, as might occur when either assembly equipment or engineers contact the package pins. There is another mechanism that requires the ESD protect structure: The *charged-device model* (*CDM*) reflects the assumption that the wafer substrate may accumulate charge, which would subsequently be discharged through electrical contact to a pad through a test probe. For either discharge model, ESD protect circuitry is required.

A wide variety of ESD circuit topologies are available, from relatively simple (overshoot/undershoot) diodes connected to the pad to more sophisticated techniques using device avalanche current mechanisms. The foundry may offer a recommended set of ESD layout cells to include with the I/O pad circuits.

The ERC sign-off flow needs to confirm the presence of the ESD protect cell connected to each SoC I/O and, significantly, needs to analyze the electrical network between the pad, the protect circuits, and the functional I/O circuitry to ensure that the discharge path will be effective.[2] The analysis checks include:

- Limits on the interconnect resistance between pad and ESD protect circuits
- Current density limit checks on the metals/vias that are part of the discharge path (e.g., using I-V curves of clamp diodes and avalanche devices and the CDM/HBM excitation models)

Note that whereas ESD is a short-term event, EOS more generally refers to any exposure to voltages and/or currents exceeding specifications and is lower in magnitude but potentially much longer in duration than an ESD discharge. Examples of EOS exposure include power/ground (inductive ringing) noise voltages on chip pins. A latchup triggering transient on a chip pad circuit would also be considered an EOS event. Failures caused by the energy dissipated during an EOS condition include contact metal spiking through a silicon diode junction, metal deformation, and defects introduced in die-to-package attach connections. The additional circuitry added for ESD protection also assists with diverting current from an EOS occurrence. However, the responsibility for EOS avoidance falls primarily onto the system designer, who must provide appropriate supply voltage filtering, low-impedance ground current return paths, and matched impedance on high-speed I/O signal transmission lines.

**Voltage-Dependent Design Rule Checks** The other main ERC category requiring electrical analysis to evaluate sign-off rule checks pertains to the subset of layout design rules for voltage differences between signals greater than VDD_internal. Most of the design rules in the PDK release apply to circuits operating at VDD_nominal +/- x% on-die margin. Even if the SoC implements DVFS power-performance optimizations, the maximum supply voltage is limited by the (die internal) process specification. For I/O pad circuits, however, a larger supply voltage (e.g., VDD_IO) is often required to provide signal voltages that meet an external interface definition. For example, at advanced VLSI process nodes, the SoC core IP operates at sub-1.0V, while external interfaces may require 1.2V to 1.8V.

The devices that are connected to a higher supply voltage require a different definition and layout geometry due to the thicker gate oxide and longer channel length required to accommodate the larger electric fields. In addition, the spacing between wires that have a higher potential difference needs to be increased. A specific set of PDK high-voltage design rules is provided by the foundry for circuits requiring thick oxide gate devices, in addition to design rules for their interconnect wires.

An ERC flow uses a (minimal) set of user-provided guidelines to refer to network nodes associated with higher supply voltages and electrically traces how these voltages can propagate throughout the layout. The shapes in the electrical network derived by the ERC tool are then evaluated against the high-voltage design rules. Note that internal signal wires adjacent to I/O cells in the global floorplan also need to be verified to the voltage-dependent rules.

In addition, some SoC designs require isolation between the ground distributions of different power domains to reduce the injected noise into sensitive analog IP from adjacent switching circuits, as illustrated in Figure 2.15. The ERC verification flow analyzes the SoC ground distribution to confirm that the required isolation is present. Any ERC sign-off errors reported by the flow need to be addressed—either fixed prior to tapeout or added to the approved waiver flow.

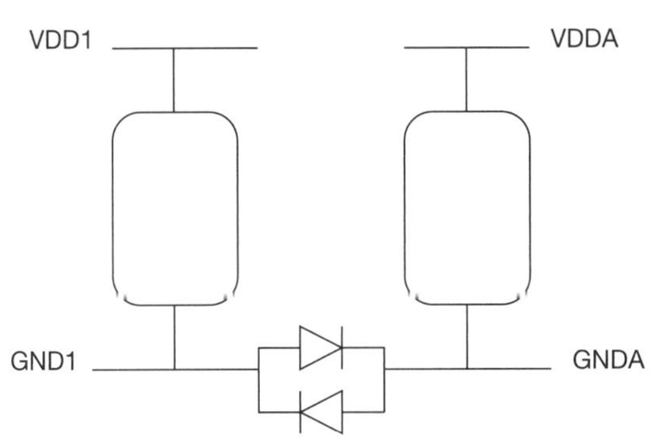


**Figure 2.15** ERC analysis evaluates the topology of power domains to confirm that the required noise isolation circuitry is provided.

### *Electrical Analysis Flows for Sign-off*

While the sign-off physical verification flows are being run on the frozen physical database, electrical analysis flows are also exercised. Several general characteristics of these flows require specific consideration:

- **Responsibility for electrical sign-off**—The results of the physical verification flows using PDK checking runsets are reviewed with the foundry as part of the tapeout procedure. The foundry assumes responsibility for wafer fabrication within specifications and, thus, an acceptable manufacturing yield. Conversely, the results of electrical analysis flows are not (typically) shared with the foundry; the SoC design team is responsible for ensuring functionality, performance, and lifetime reliability that meet product goals.
- **Full-chip model size reduction**—Electrical analysis at the IP block levels of the SoC hierarchy may provide an opportunity to reduce the scope

of the full-chip model used for sign-off analysis. Three methodology approaches play important roles in reducing the complexity of the full-chip electrical model and, thus, the number of post-analysis results to address:

- **Incorporating a (conservative) global overlay with the IP layout for parasitic extraction**—When analyzing the IP block layout, an overlay cell is merged with the block prior to parasitic extraction. The additional capacitive coupling of an estimated global routing interconnect environment makes the block analysis results more robust. The block methodology needs to address how to denote the "nets" associated with the global overlay cell so that all the extracted parasitics correctly annotate to the block schematic netlist. The overlay nets could be defined such that the additional parasitic capacitances are grounded or are regarded as potential coupling switching aggressors to the signals within the block.
- **Electrical abstracts generated by the block-level electrical analysis flows**—Analysis flows exercised at the block level may provide an abstraction model suitable for use with the full-chip SoC model. For example, an *interface-level model* (*ILM*) is an abstraction useful for path timing analysis at the SoC level, reducing the block detail to just the circuits and parasitics between the block and a path launch/capture test point, as depicted in Figure 2.16. Another example is a power abstract, which represents the supply/ground currents through block pins that connect to the global power distribution network. Note that suitable abstracts are required for all analysis corners and operating modes.

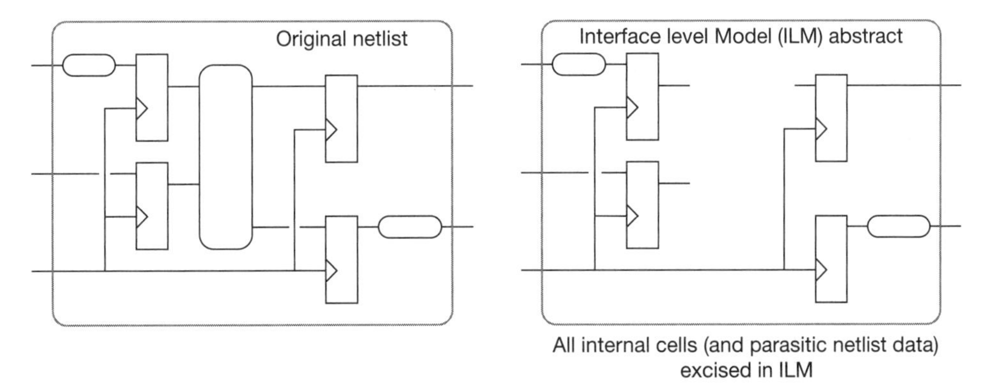


**Figure 2.16** Illustration of an interface-level model (ILM) electrical abstract for a block.

- **Parasitic extraction of global SoC nets utilizing *gray box* visibility of block layouts**—The full-chip parasitic model need not include detailed netlists from all block instances. However, the parasitic capacitances extracted for global nets are the most accurate when visibility to IP block internal nets is included. The nomenclature used by EDA extraction tool vendors is that the IP blocks would be identified as *gray box layout cells*, with capacitances included for annotation to the global netlist; in this case, the gray box capacitances would be grounded (see Figure 2.17).

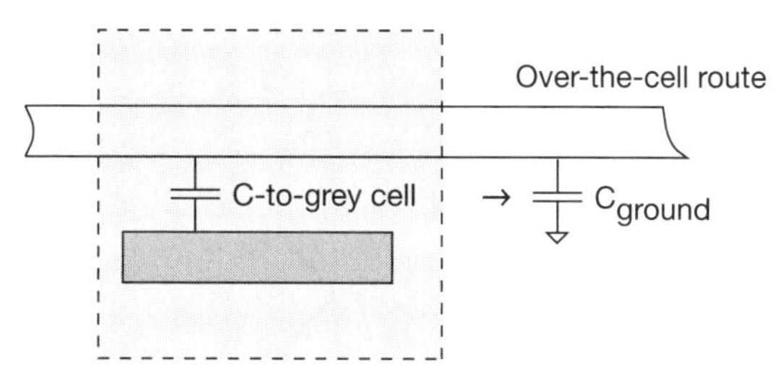


**Figure 2.17** Gray box visibility of block layouts for global route extraction. Capacitances to gray box block nets would be grounded.

- **Sign-off versus EDA platform flows for electrical analysis**—Increasingly, the methodology flows for physical design implementation include optimization algorithms that address electrical criteria. Path timing and power dissipation have been objectives of logic synthesis, cell placement, and routing flows for some time; recently, placement and routing have integrated additional analysis capabilities. For example, capacitive coupling from adjacent routes impacts interconnect delays and injects transient noise onto a quiet victim signal. Recent routing and timing optimization algorithms incorporate an analysis engine to calculate the signal integrity impacts of coupling noise, to be included as part of the optimization criteria. As another example, custom layout editing tools have integrated a parasitic extraction engine to provide interactive feedback to designers on nets that need to satisfy specific parasitic resistance and capacitance criteria. Calculation and propagation of device currents within the layout editing tool assist with identifying potential electromigration issues. The set of integrated tools and engines providing analysis results on a common data model is commonly referred to as a *design platform*.

The EDA tools platform strategy is intended to reduce the magnitude and severity of design issues when executing stand-alone analysis flows. The integrated platform engines are ideally the same as the stand-alone analysis algorithms. However, to manage the impact on the (interactive or batch) execution time, the engine may adopt a reduced implementation of the stand-alone flow—especially for layout parasitic extraction detail. The result is that there remains a methodology requirement to execute all electrical analysis flows in full-chip "sign-off mode" prior to tapeout. The electrical analysis flows to review as part of tapeout include:

- Path timing analysis for all MCMM corners and operating modes (see Chapter 11, "Timing Analysis")
- Signal noise analysis (see Chapter 12, "Noise Analysis")
- Power dissipation calculation, including the generation of a die thermal map (see Chapter 13, "Power Analysis")
- Power distribution network I*R voltage drop analysis (see Chapter 14, "Power Rail Voltage Drop Analysis")
- Power and signal electromigration analysis, incorporating the thermal profile of the die from active device switching self-heating (see Chapter 15, "Electromigration Reliability Analysis")
- Miscellaneous full-chip electrical analysis flows, such as *soft error rate* calculation and power state transition robustness verification (see Chapter 16, "Miscellaneous Electrical Analysis Requirements")

### *Test Analysis for Sign-off*

The frozen tapeout database serves as the basis for the development of production test patterns. The *design for testability* (*DFT*) architecture is an integral part of the IP block and full-chip SoC design. The DFT goal for the design project is to provide sufficient internal signal controllability and observability from the production test patterns to ensure high-percentage coverage of prevalent manufacturing faults. The final, post-ECO logic netlist is input to *automated test pattern generation* (*ATPG*) tools, with the cell library and IP vendors providing related fault models.

Specific test patterns may be released by hard IP providers with related wrap test (or macro test) DFT architecture support. Memory IP providers and EDA tool vendors have invested in memory built-in self-test (MBIST) architectures, with internal pattern generators inserted into the netlist during implementation. The decision to insert BIST functionality for general logic networks (LBIST) is also proving to be a key cost trade-off of additional die area versus reduced tester time. The sources of logic and memory test patterns are varied and need to be prioritized and integrated into the production test pattern set.

Analog IP has unique requirements, both in the DFT architecture to enable isolated testing and in the test specifications to be measured for each fabricated die. As a simple example, general-purpose I/O (GPIO) cells require additional current/voltage/impedance measurements to confirm Vin and Vout voltage interface levels at the specified load. More generally, analog IP test specifications include parametric measures at various operating conditions and may necessitate a project plan to use separate test equipment from the automated test equipment (ATE) used for the remainder of the SoC.

Recent process nodes have resulted in additional fault mechanisms. Whereas the initial test engineering focus had been on detecting circuit-related faults focused on the devices, contacts, and local connections, the scope of manufacturing faults related to global interconnects has become increasingly significant. *Bridging faults* between adjacent routes result in a low-impedance path between (logically unrelated) signals; these faults may arise from incomplete metal removal during the CMP polishing step. To define bridging fault candidates, the (final) physical layout database needs to be analyzed to identify high-likelihood fault locations and generate related (sequences of) patterns to exercise specific signal transitions. Resistive faults, due to low wire thickness and/or high via resistance, may result in correct logical behavior at lower tester speed but invalid path timing at target clock frequencies; these are known as *delay faults.* The output of the sign-off timing analysis flow identifies critical paths that would be candidates for delay fault patterns. Applying the bridging and delay fault test pattern flow on preliminary SoC integration phases gives a confidence level on the resulting fault coverage, but the final tapeout physical design is required.

The SoC project management team needs to address several trade-offs related to test pattern generation and tapeout sign-off. The overall project must have selected fault models and established test pattern fault coverage measures to meet *shipped project quality level* (*SPQL*) objectives; releasing a faulty product *test escape* to system build (and, potentially, to an end customer) has significant cost and contractual implications. The principal SoC project trade-off on production testing relates to schedule. After tapeout, wafer-level testing commences after the fabrication cycle time, which is potentially several months. The final production test pattern set is not required at tapeout and can evolve during wafer processing. However, the SPQL targets require satisfying fault coverage goals. Increasing fault coverage may require the addition of control/observe logic to the design (obviously prior to tapeout). The SoC project management team needs to establish criteria for test coverage at sign-off, with confidence that the post-tapeout test pattern development will meet targets with the existing logic and DFT architecture.

Note that the project management team also needs to define what patterns will be exercised (if any) during burn-in stress testing of packaged parts and what patterns will be applied during retest after burn-in to identify and cull any *infant fails* (see Section 21.1).

The SoC design methodology for sign-off is an intricate mix of physical verification, electrical analysis, and testability analysis flows. The flow results require extensive review of error and warning messages to determine what needs to be addressed and what may potentially be waived. Updates required to the frozen physical database are extremely disruptive; the time required to fix errors and subsequently rerun sign-off flows may impact the tapeout release date established with the foundry. Exercising full-chip flows on preliminary physical integrations can help identify potential analysis problems prior to the final version for tapeout. From the early integration efforts, the methodology team gains insight into the IT resources that need to be allocated to expedite the throughput of the final sign-off flows, many of which are submitted concurrently. In addition, incorporating representative overlay and context cells during individual IP block analysis reduces full-chip integration issues.

## 2.3 EDA Tool and Release Flow Management

Previous sections describe the importance of version management and configuration policies established to define the complete design hierarchy for intermediate and final SoC model builds.

A common tool used by SoC design project lead engineers is a *methodology manager*, which provides a summary view into the flow results associated with individual IP versions that are part of a full-chip SoC configuration (see Figure 2.18).

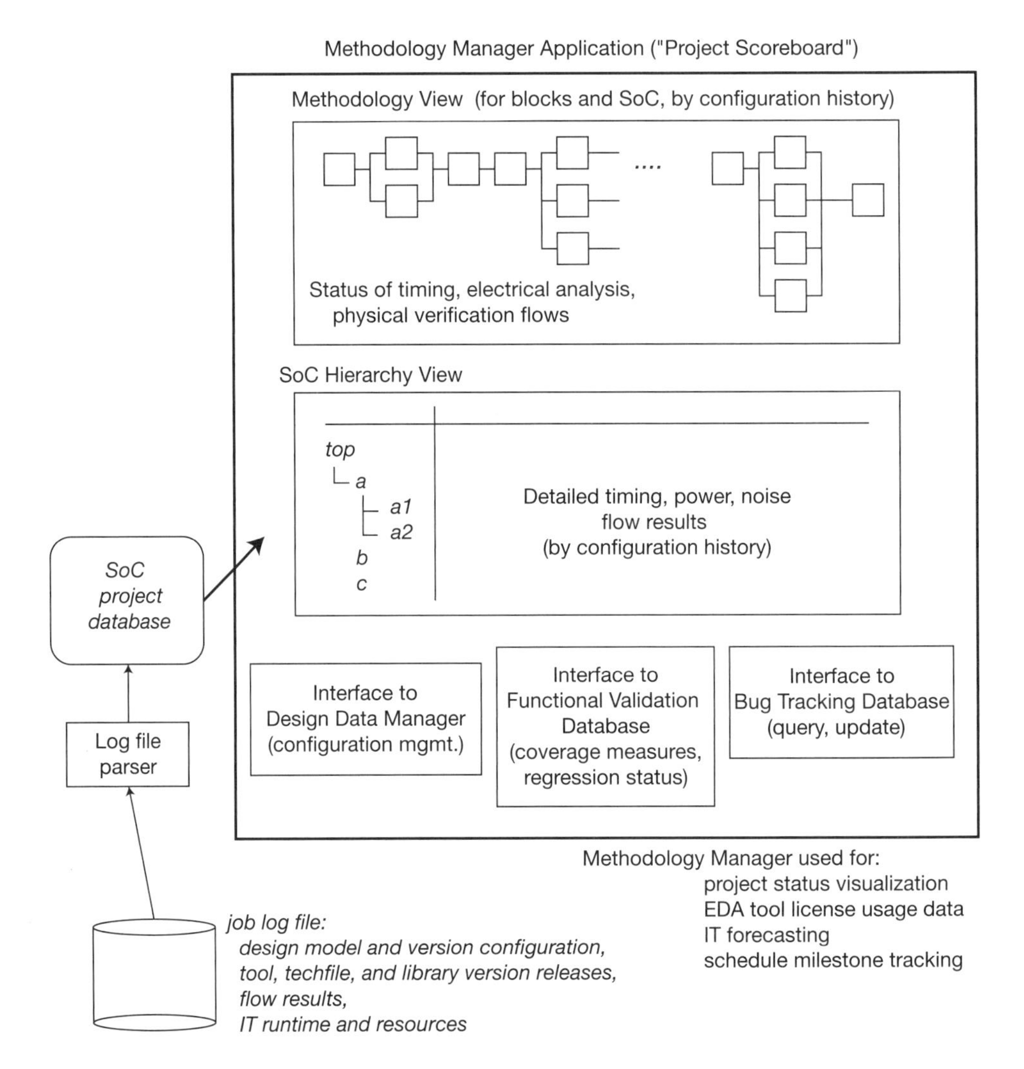


**Figure 2.18** A methodology manager application provides a summary view of the current project status and records flow results (and functional bug status) for the design hierarchy.

The flow results log file is parsed to provide a summary status of the errors and warnings from each flow. Project managers utilize this information to gauge the status of each IP block and to determine readiness for subsequent physical integration. An ongoing assessment is made to determine if specific IP blocks may need additional engineering resources to accelerate their progress.

(An engineering project management principle states that additional resources may actually impede progress in many cases.) The functional validation team also utilizes a similar *scoreboard*, linked to the methodology manager, to track the number and severity of functional and testcase bugs for each model build configuration.

In addition to tracking errors and warnings, the summary of the flow results needs to record EDA tool/flow version information and foundry PDK version data. During any SoC project, new tool versions will be released by the EDA vendor. The methodology team and the EDA vendor need to review new features, tool bug fixes, and ongoing software support commitments for existing tool releases; depending upon the outcome of such review, the methodology team may choose to adopt a new EDA tool version into IP and SoC flows. Adopting a new tool version as a flow default is commonly done by changing a project-wide environment variable to point to the updated tool executable install path. Individual *trailblazing* designers can test a proposed version by setting an environment variable user override when invoking the flow. As an SoC project approaches its schedule target date for database freeze, the project management team also freezes the set of environment variables; other SoC designs that are under way may still receive the EDA tool version update.

A key job of the methodology manager reporting tool is to accept a range of EDA tool and PDK versions as valid and invalidate results from an unapproved version. The methodology and SoC project management teams need to determine which versions are acceptable for an IP block to promote to a new SoC configuration model build and, ultimately, which tool versions are valid for sign-off. The use of IP abstracts for full-chip analysis provides considerable benefit to model size and runtime execution. However, the IP analysis and abstract generation flows may have been executed and saved on earlier tool versions. A resource cost is incurred if it is necessary to evaluate the applicability of the IP abstracts through subsequent EDA tool releases. If the IP originates from an external supplier, the resource cost includes the review of the tool and PDK version with the supplier and, potentially, the additional expense and time associated with the release of updated models and abstracts.

The implementation of the methodology manager depends on the identification of key summary information from the flow results. In addition to the job information about the flow execution, the flow results typically embed EDA vendor tool log files; parsing log file results requires that specific keywords and records be present. Unfortunately, the format of the EDA vendor tool log file is not necessarily consistent between tool releases. Adopting a new tool version may require updating the parsing code in the methodology manager, as well. (It would also be fruitful to review the methodology manager features with the EDA vendor, which might then focus on log file stability between tool versions.)

## 2.4 Design Methodology "Trailblazing" and Reference Flows

Each new fabrication process node brings unique challenges, from new device model features (and variation effects) to additional physical layout design constraints to the increased SoC physical and electrical model data volume. EDA tool vendors collaborate with the foundry PDK development team to add features for the new process node, whether for higher model accuracy, richer algorithm capabilities, or improved designer productivity. The methodology team is faced with adapting existing flows for a new node to address technical requirements and have the qualified flow releases available to meet SoC design project schedules. There are three major facets to preparing a new methodology release: reference flow reviews, tool evaluation, and trailblazing.

### 2.4.1 Reference Flows

EDA tool vendors partner with the foundries using early PDK data to verify new software features. Specifically, the integration of multiple products into an EDA design platform is tested by the vendor to confirm the scope of the data model and the interoperability of analysis and optimization algorithms. The completion of this testing culminates in the EDA vendor announcing the availability of a *reference flow* for the foundry process.

An EDA vendor may partner with IP providers to exercise the reference flow and to demonstrate the applicability of new features on representative designs. The IP providers accrue the benefit of accelerating their development for tapeout release to the foundry on (early PDK version) shuttles.

The SoC methodology team should review the reference flow with the EDA vendor to best understand the scope of the testing performed, as well as the associated compute environment and IT resources required. Of specific interest to the methodology team are additional tool commands and settings, new input constraint files, and output log file changes so that the impact to existing SoC design flows invoking the tool can be assessed. (In addition, enhancements to interactive tools incur a training phase as part of the designer's learning curve.) The methodology team can then extrapolate the reference flow release data to plan the overall flow availability for upcoming design projects.

The SoC design methodology may not fully utilize the toolset integrated into an EDA vendor platform but may instead use individual "point tools" from multiple vendors. The flows thus need to work with varied data files used by each tool that are read for input and written for output to persistent storage rather than using a common data model in memory. For new point tool integration into flows, the methodology team needs to review any EDA vendor changes to the tool interfaces to ensure correct and complete input/output data model updates. The goal of using (de facto) industry standard file format definitions is to minimize any model "handoff" issues between flows. However, these definitions (and the programming interfaces to access binary model data) are also evolving. The adoption and release of a new point tool and the corresponding flow invoking the tool is gated by the results of a tool evaluation.

Note that the criteria used for the platform versus point tool adoption decision are varied and complex (e.g., "best of breed" algorithm capability, tool maturity, designer experience and familiarity, licensing cost).

### 2.4.2 Tool Evaluation

A new point tool is initially subjected to an evaluation by the methodology team. Flow environment variables refer to the new tool executable version installation, and a set of selected design testcases are applied. The SoC design team and EDA vendor are likely to assist with the qualitative and quantitative evaluation criteria to determine the efficacy and value of the new tool. The foundry may also be asked to provide a preproduction PDK release for the evaluation. If the new tool release is not covered under an existing license agreement, the EDA vendor is asked to provide temporary software license keys for the duration of the tool evaluation.

Given the potential to introduce a new point tool into design flows for upcoming projects, it is common to consider competing products from multiple EDA vendors. The scope of a new tool evaluation is often expanded to include concurrent testing of different products; the allocated evaluation resource needs to be increased accordingly. The duration of the multi-tool evaluation becomes more difficult to predict, as the EDA vendors may have addressed new process requirements with features that have vastly different commands. The flow modifications required to support a complex evaluation also involve additional resources and duration.

If an evaluation concludes with a decision to pursue a software product that is not currently licensed, the schedule for trailblazing and ultimately the release to the design teams needs to include the time required for the license purchasing negotiation process.

### 2.4.3 Trailblazing

After flows are updated with new tools, the final milestone to releasing an updated methodology is the trailblazing process. A representative design is selected and exercised through a full set of physical implementation, analysis, and sign-off flows. The primary goal of trailblazing is to exercise data model interfaces between flows and verify the flow results data rollup into the methodology manager. The characteristics of the selected trailblazing design should ideally be diverse in order to include both hard and soft IP blocks with multiple clock and power domains. There are potential logistical issues with the trailblazing step that need to be addressed:

- **The library IP in a new process node may not yet be fully released.** Rather than wait for library IP to complete characterization and release, a simplified process-scaled set of abstracts and timing/electrical models from an older library may suffice.
- **Existing designs as potential trailblazing testcases may be lacking.** An upcoming design project in a new process node is likely to incorporate new features not present in existing designs (e.g., updates to clock distribution and skew management, new power distribution network design, new operating modes). Ideally, any new proposed design styles could be evaluated during trailblazing rather than requiring flow development concurrently with an SoC design project.
- **The resources required to complete the trailblazing step are significant.** Exercising new flows during trailblazing resembles the complexity of an SoC project itself, albeit with little need to correct or optimize design errors. The resources to perform the trailblazing step are likely to involve the support of the design engineering team. The availability (and cost) of the engineering staff is a trade-off against the value of trailblazing.

(Design engineering managers may regard trailblazing as "throw-away work.") As a result, the original trailblazing plan is dynamic, as resources allow. Some methodology flows may receive abbreviated testing if other similar flows provide sufficient data.

The final checkpoint of the trailblazing phase is review and approval of updates to user documentation to ensure that the design guidelines reflect the latest flow environment setup instructions and output results format.

The culmination of platform/tool evaluation and trailblazing is the release of a full methodology for upcoming SoC projects. A key catalyst to a major methodology version update is typically the latest tool (and PDK) features associated with the introduction of a new process node.

## 2.5 Design Data Management (DDM)

Section 2.3 discusses version and release management policies for EDA tools and flows as an integral part of the methodology manager used to track SoC design progress. Those policies are a subset of the broader design data management (DDM) applied to SoC project data. This brief section provides an overview of DDM. The methodology team needs to select how to adapt the general features of a DDM application to be consistent with SoC design flows and consistent with the project schedule milestones.

A DDM application is typically characterized as providing a *virtual filesystem interface* to designers. The application receives file open requests (for read/write/append) and applies a set of rules to direct the operating system to the appropriate file. The DDM may utilize its own specific *data repository* for the project data under its control. Each version of a "source" design data file is (initially) tagged with a label by the DDM; subsequent actions taken to identify a specific model build configuration may assign additional labels. These labels are part of the *metadata* that accompany the file, maintained by the DDM.

When a methodology flow is invoked for a design block, the DDM monitors the file-related operations and saves both the source and derived file information associated with the design and the flow. This record is captured as part of the flow results output. The derived model files may also be tagged to be used as part of a subsequent model build.

Note that the functional validation environment also follows the same DDM approach as SoC design data. Project source files also include simulation testcases (and testcase generators), bus interface transactors, monitors and assertions, and so on.

### 2.5.1 Configuration Specification

A configuration specification—often abbreviated as *configspec*—is a list of design files, each with a specific label, as depicted in Figure 2.19. A specific configspec is enabled by executing a DDM command prior to submitting the flow script. When a flow requests a design file, the DDM virtual filesystem applies the configspec to pinpoint the version of the file to provide for the requested operation.

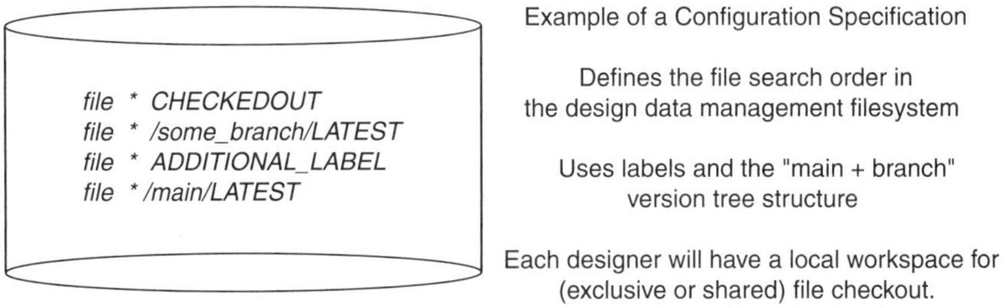

**Figure 2.19** Illustration of a DDM configspec, which represents a configuration of specific design file versions to submit to a flow.

As mentioned in Section 2.3, the EDA vendor tool executable, flow software, and foundry PDK data are also key parts of SoC project management. There are two potential approaches to associating these files with a project:

- Include the EDA tool and flow software development filesystem as part of the DDM repository and as part of each configspec
- Establish the EDA file version via a set of operating system environment shell variables

In the latter case, designers invoke a script to assign environment variables to point to a specific version of EDA flows and PDK techfiles (e.g., export EDA_FILESYSTEM = /…/). This script would be maintained by the methodology team and CAD flow qualification team. As part of tapeout auditing, flows would need to report the environment variables used, in addition to the design data configuration represented by the DDM configspec.

### 2.5.2 Checkout and Checkin

When a design source file is being updated, the user executes a DDM checkout request to enable editing. (The CHECKEDOUT tag is also applied.) When the editing and testing are completed, a checkin operation is performed to make the design file revisions visible to others on the project team.

An exclusive checkout could be requested; such a checkout applies a lock on the design file to preclude other checkout requests. Alternatively, a checkout could be non-exclusive, allowing multiple concurrent edits on the design file. As an example, consider the case where an RTL design block contains several functional validation errors to be debugged. For efficiency, different engineers could be assigned to examine different errors, necessitating concurrent checkouts. Rather than invoking a checkin command, the multiple checkouts would trigger the need for a "merge" operation. (Note that this makes sense only for design text files.)

### 2.5.3 Main, Latest, and Branch Configurations

The design evolution occurs along the "main" line of DDM configurations. There is a default LATEST tag applied to the most recent file checkin/merge, as illustrated in Figure 2.20.

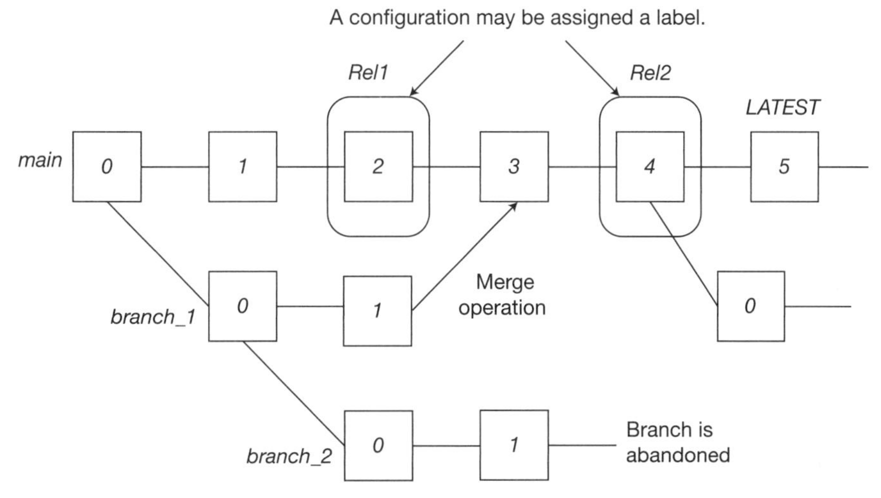


**Figure 2.20** Illustration of the DDM topology for design configuration management.

Concurrently with the evolution along the main line, design development and debugging may also occur along a branch. For example, a functional error uncovered during validation may need detailed investigation, requiring additional time and resources. Rather than work along the main line and potentially be impacted by ongoing releases to the latest version of design files, a specific configspec representing a design branch can be created to allow directed debugging to proceed in isolation from other design updates. Another common reason to create a branch is to enable an investigation into a micro-architectural change, separate from the current design specification. After the impact on performance and power is evaluated, the proposed change may be adopted and the branch changes reflected back into the main line, or the branch may simply be discarded.

A configuration branch is also typically created in anticipation of a design tapeout. Significant revision development may proceed on the branch to enable design ECOs to proceed quickly, without necessarily making frequent merges to the ongoing main line. The goal is to expedite tapeout preparations, with minimal interaction with other project activities. For example, for a first-pass prototype tapeout, functional validation is likely continuing on the main line. After tapeout (and the related configspec *snapshot*), the design merge to the main line is a key project management step. The tapeout branch is kept active for silicon debugging, and another crucial design merge to the main line occurs after bugs in prototype silicon are addressed.

### 2.5.4 DDM and Physical Design Data

It should be noted that most DDM tools used in SoC design have their roots in the configuration build managers for large software development projects. In many ways, the RTL and C files that comprise the functional and validation descriptions for an SoC are very similar to those for a software project; indeed, embedded software is an integral part of many SoC models. The concept of a (text) file merge from a configuration branch originates from concurrent software development policies.

However, the physical design data for an SoC project is composed of a potentially very large number of binary data files, many of which are also very large in size. The traditional software DDM tools may not be ideally suited to the characteristics of physical design data. The EDA vendors providing physical design tools may include a DDM offering integrated into their design platform. File checkout/checkin and hierarchical instance configurations may be enabled by the design library management features in the platform and optimized for a network filesystem.

The SoC methodology team is faced with a crucial implementation decision, with two distinct options:

- Impose a single DDM tool (and policy) across functional, validation, test, and physical environments.
  - Ensure that the performance of DDM operations is adequate for project text and binary files.
  - Ensure that DDM operations work seamlessly with EDA physical design platforms.
- Adopt two DDM approaches, one for functional/validation/test development and one for physical implementation.
  - Utilize the DDM features of the physical design platform.
  - Develop the methodology to sync models between the two design data managers at correspondence points in the logical and physical design hierarchies.
  - Export the necessary file representations from the physical domain to the validation and test flows (e.g., physical layout data to enable tools to identify manufacturing fault candidates for production test pattern development).

A key precursor to a successful SoC project is establishing proven policies for design data management—that is, the DDM tools and commands to be used, the handling of EDA tool and PDK techfile versions, and the steps (and approvals) for creating and merging configuration branches. If the DDM methodology is being developed concurrently with the SoC design, project schedule and resource forecasts will undoubtedly incur significant risks.

## 2.6 Power and Clock Domain Management

The power distribution from chip pads to local IP blocks is a global circuit design responsibility. Similarly, the generation and (low skew) distribution of clock signals to design blocks is a global consideration. Indeed, these design tasks are closely interrelated:

- Robust power delivery must be available to global clock buffering cells.
- Global clock signals are often routed adjacent to power rails to eliminate capacitive coupling switching noise.

- The switching of power domain blocks between active and sleep states needs to be coordinated with the clock(s) to the domain to ensure that active clocking is shut off during power transitions.

Due to this tight relationship, SoC projects typically assign joint global power and clock design responsibilities to the same engineering team.

The introduction of a power format design description standard has helped provide a specification format that can be pervasively used across SoC design and validation flows. The global power/clock design team utilizes this domain definition and state transition description in its planning and implementation.

The global power distribution design may need to adhere to additional constraints that are not represented in the functional descriptions in the power format file. In addition to unique power supply values associated with I/O interface circuits, analog IP, digital block domains, and on-chip memory arrays, there may also be production test limitations. The test probe station may not be able to contact all (internal) die bumps to provide power/ground connections or may not be able to provide sufficient current to the die under test if all blocks are active simultaneously. The global power design, from bumps to IP blocks, may need additional partitioning to accommodate multi-pass application of production test patterns.

Several SoC methodology decisions relative to global power and clocks are needed early in the project planning phase:

- The metal layers allocated to global distribution versus block-level IP (and any shared layers with both global distribution and block segments)
- The fabrication process metal layer thickness for global distribution (the *metallization stack*)
- The (non-minimum) width of power rails and the width of global clock signals
- The via strategy at the intersection between global and IP power/clock signals
- The percentage of total I*R voltage drop margin allocated to the global power distribution (including L*di/dt switching noise)
- The decoupling capacitance on the global power distribution (e.g., using MIM structures)
- The percentage of upper metal layer routing resources allocated to global power/clocks

The clock signal width decision involves multiple trade-offs to address both low skew and low power design goals: Wider clock wires are less affected by manufacturing tolerances and provide improved skew, at the cost of greater capacitance and switching power.

The SOC test architecture may also introduce global clock distribution requirements. As illustrated in Figure 2.21, the scan shift of applied/captured test pattern data may utilize either a global test enable (for mux_scan) or a specific global test clock requiring low skew distribution.

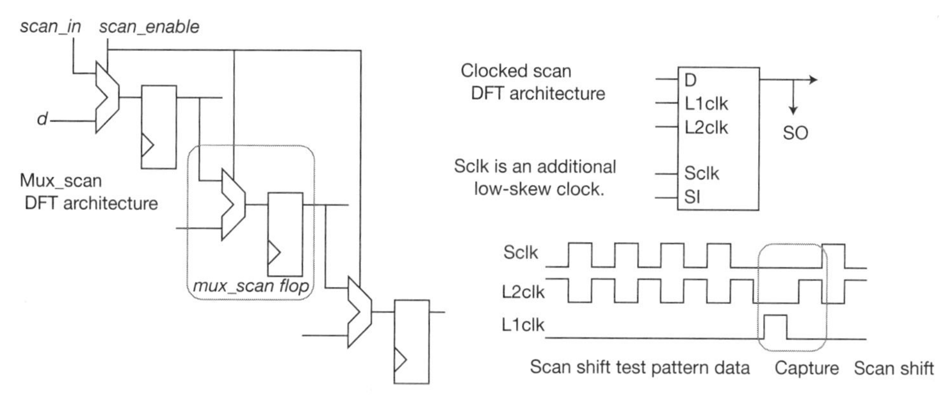


**Figure 2.21** Scan-shift DFT architecture utilizing a (low skew) global test clock.

The previous section highlights some of the considerations for establishing project design data version and configuration management policies. In addition to maintaining configurations for the SoC power format description and the global power/clock physical implementation, the project design data management team must ensure that any glue logic and buffer/repeater cells residing at the top level of the SoC hierarchy adhere to the same overall DDM policies.

## 2.7 Design for Testability (DFT)

SoC methodology decisions throughout the design, functional validation, and physical implementation flows are influenced by the test architecture and by the test pattern generation and application methods to be used.

At the highest abstraction functional modeling level, the SoC test features may not be included. The design specification and validation testbenches are focused on ensuring correct and complete SoC functionality. Designers may not want to bother with adding model detail that will not be exercised during (initial) validation phases.

At an appropriate point in the project schedule, the SoC test architecture and corresponding logic must be inserted into the design description. The test architecture description could involve the insertion of a scan-shift register implementation and/or a logical built-in self-test unit capable of generating patterns and capturing/compacting the resulting test signature. As illustrated in Figure 2.22, a common approach is to add scan-shift test features during the logic synthesis flow.

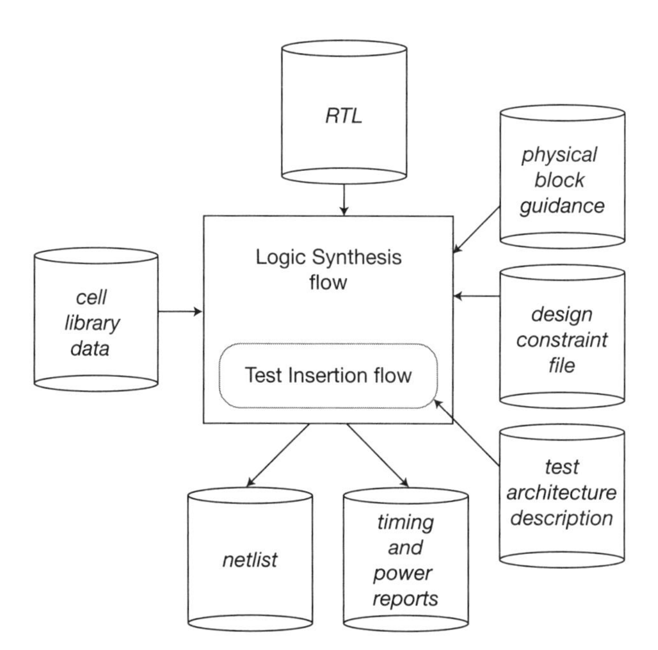


**Figure 2.22** DFT architecture insertion into a synthesized logic model; scan-shift register insertion is depicted. (BIST controller insertion would typically be applied at the RTL level.)

As mentioned in the previous section, the SoC test architecture typically requires the distribution of global test signals/clocks throughout. Validation of the global test design with local IP blocks needs to be deferred until after test insertion. Although the model with test features added is typically no longer at an abstract level and thus simulates more slowly, the simulation testbench associated with test validation is usually not very demanding in the number of testcases. Also, note that the insertion of test logic involves modifications to the ports of design blocks for the additional test signals; methodology flows that support functional equivalency checking between different model abstraction levels need to be aware of these port discrepancies (and the inserted internal logic).

Rather than utilize EDA tools for test insertion, the SoC test architecture could be included starting with the initial design phase, even at the highest levels of model abstraction. This decision would likely depend upon how and when the validation project phase needs to focus on two key features: SoC reset and real-time error capture/recovery.

The reset of an SoC (or system) can take a variety of forms. One approach is the application of specific global reset signals that direct logic to initial values for register/flop capture. Another reset approach is to use the scan-shift test chains to serially shift reset values. Logic inversions judiciously added to the test chains can enable reset to '1' or '0' values in registers at the end of the shift sequence, with a static input value to the chain. Similarly, upon detection of a system error, the SoC may transition to an internal state that addresses error data capture, typically utilizing the scan-shift features of the test architecture to export flop/register values for error diagnostic triage.

In these cases, the methodology team may decide to include test functionality in higher abstraction level models, to best reflect the design specification and support the corresponding validation testbenches.

### 2.7.1 Test Pattern Generation and Pattern Ordering

Chapter 19, "Design for Testability Analysis," provides additional details on the nature of physical manufacturing faults and the production test patterns intended to detect their presence.

From the perspective of methodology development, once the test, reset, and error detection SoC architecture is defined, the methodology team focuses on the efficacy and sufficiency of production test pattern generation. Test patterns may originate from various sources:

- Scan chain serial shift tests (verifying proper scan chain response)
- Built-in self-test logic (specifically MBIST)
- Automated test pattern generation algorithms (exercising logic networks between flops/registers, shifting stimulus patterns, and capturing responses through scan chains, potentially with on-die pattern expansion and/or compaction techniques)

- Timed test sequences to exercise specific logic paths
- Patterns to exercise physically adjacent nets
- Functional validation testbench sequences (applied at tester pattern rates)
- Embedded software routines at functional clock rates

The various sources of test patterns cover different subsets of the overall target fault space. Although a pattern set may focus on a specific fault type, applications of these patterns naturally cover other faults. As the production test set grows, the patterns are provided to a fault simulation flow to mark off additional fault coverage.

The methodology and test engineering teams are not only responsible for defining the fabrication fault types (in collaboration with the foundry customer support team) but also the strategy for pattern set ordering to be most efficient at detecting failing dies at wafer-level testing (and, subsequently, at package-level testing).

Several special-case DFT methodology considerations may also be applicable to an SoC design:

- **Static leakage tests**—For designs targeting mobile, battery-powered product applications, additional testing of the die static leakage currents is crucial. Functional testing may succeed over a wide range of fabricated device characteristics in active mode, with a potentially wide range of leakage currents. For static leakage testing, pattern sequences that place the die under test in various power states need to be generated. After pattern application, the static die leakage current is measured, and a pass/fail criterion is applied. Indeed, this is often the very first pattern set applied. A high-leakage condition is often an effective screening method for what are likely additional functional and/or timing issues, saving tester time.
- **Analog IP tests and I/O electrical measurement tests**—Test pattern application includes measurement of valid (static) electrical interface signal levels at I/O pads—for example, Vin_high and Vin_low levels at receivers, Vout_high and Vout_low at drivers (with the specified DC resistive load). The pattern sets for these electrical measurements are easily generated if the SoC test architecture includes *boundary scan* chain(s) throughout the I/O pads.

Complex analog IP integrated on the SoC involves specific test pattern strategies. The analog IP patterns may be merged with the functional patterns if the same production test equipment can be used. An analog IP block providing high-speed serializer/deserializer (SerDes) interface signaling could be tested in a wrapback configuration, where an on-chip test mode directs a serializer output back through a corresponding deserializer input. (This approach would not reflect the expected signal losses and inter-symbol interference from the actual package and printed circuit board topology or the capabilities of the deserializer to recover the data from the noisy signal.)

Analog IP presents unique methodology considerations relative to converting simulation testbenches to production test data and, specifically, evaluating the manufacturing defect fault coverage. Whereas digital circuitry has traditionally used a proven set of fault models (e.g., node stuck-at-1/0, interconnect transition delay faults, interconnect bridging faults), analog circuit test methods entail correlating manufacturing defects to circuit parameter sensitivity and, ultimately, to a detectable failure compared to the IP specification. Given the large space of potential defects, a *maximum likelihood* and corresponding *sensitivity analysis* are needed to provide sufficient test coverage.

The transient, DC, and AC circuit simulation resources required to establish the analog IP production test set need to be included as part of the SoC project IT plan. Digital logic tests typically offer limited capabilities for parametric testing of mixed-signal circuits. If wafer-level testing of the analog IP is to utilize separate parametric test equipment, circuit simulation testbench data needs to be formatted to the appropriate test description file for the parametric tester. Close coordination with the foundry's support team is required.

- **Array repair**—The SoC methodology team faces an implementation decision related to the process technology options available for fuse programmability and memory array repair. The team needs to collaborate with the foundry to assess the SoC array requirements, anticipated defect densities, and yield impact with and without repair. The available redundant row/column resources incorporated by the array IP provider, combined with the fuse programmability features at production test, are also integral to the cost trade-off decision.

From an EDA perspective, the tools used to generate and grade the production array test patterns must also assist with the diagnosability of an array test fail and the definition of a subsequent fuse programming sequence. Upon detection of the set of failing array patterns, the diagnosis of the failing location(s) determines whether the array is reparable and, if so, which combination of row and/or column replacements is required. The SoC production test sequence needs to be optimized to efficiently retest the array after the corresponding repair fuses are programmed.

It should be noted that the DFT methodology should ideally include the capability to independently read the programmable fuse values.

It is also feasible to implement *soft fuses*, in which case a specific value is scanned into configuration registers from an external source during the reset sequence; this approach avoids the additional cost of the fabrication module for fuses. However, this is a volatile method compared to the non-volatile data in a fuse register. Increasingly, high-end SoC designs are also including specific identification and internal configuration information on-die, using additional fuse locations. As a fuse block may be needed for other (non-volatile) requirements, the additional external overhead associated with soft fuses becomes less attractive.

The fabrication of fuses involves unique physical layout design rules to separate the fuses from other circuitry for the application of the fuse programming sequence and to accommodate displacement of the fuse material after programming.

The majority of fuse applications can be met during wafer-level production testing, where the automated test equipment supports the programming step. The most prevalent technology in use today utilizes fuses that can be blown through a sequence of electrical pulses (i.e., *e-fuses*). Prior to the development of e-fuse process technology, laser-blown fuses were available. However, they required a separate set of equipment and, therefore, additional test overhead costs compared to an inline e-fuse programming sequence. There is also an option to include an additional e-fuse programming step during subsequent package testing. The SoC architecture and methodology teams need to review the requirements and costs for package-level programmability with the OSAT provider and assess whether to make this functionality accessible at the package pins.

### 2.7.2 Voltage and Temperature Test Options

Finally, the SoC design and methodology teams need to evaluate the requirements for production testing at different applied voltage and temperature conditions. Wafer-level test equipment may provide background temperature control at the wafer chuck, representing the temperature of the thermal interface material (TIM) between final die and package. The pattern activity factor during testing is typically much lower than functional operation; to achieve a transistor temperature representative of an operating extreme, the selected chuck temperature may be outside the normal environmental range. Test pattern measurements that are extremely sensitive to temperature conditions may need additional time to allow for adjustments to the chuck temperature if multiple test passes will be used in support of static leakage screens, path delay tests, analog parametric measures, and so on.

### 2.7.3 Summary

Establishing the DFT methodology for a complex SoC ultimately comes down to cost. To achieve a target product quality level, manufacturing defect screening methods can be applied at wafer and package levels, with different overhead cost adders. The efficient application of test pattern stimulus and response observability involves critical up-front planning of the SoC clock and register architecture, built-in test controllers, fuse repair programmability, and special test signal multiplexing to access unique IP (e.g., analog IP blocks, wrap test logic around embedded cores). These methods have area and performance impacts that result in additional costs.

EDA tools utilize various fault models and coverage metrics to aid in pattern generation, prioritization, and failure diagnosis. Production tester resources are a significant percentage of the total part cost; these tools are critical to optimizing the test time per die/package.

In the infancy of the VLSI industry, a DFT architecture may not have been included. A set of functional simulation testbench patterns were simply selected (and a reset sequence prepended). Die area and circuit performance overhead for production testing was minimized. As the complexity of VLSI designs grew and evolved into modern SoCs, the need for a comprehensive DFT methodology became paramount. Successful projects incorporate the DFT trade-offs as an integral part of initial project planning, implement the selected DFT architecture(s), and develop fault coverage measures to ensure appropriate product quality levels.

## 2.8 Design-for-Manufacturability (DFM) and Design-for-Yield (DFY) Requirements

Section 2.2 introduces DFM and DFY design implications—specifically how the chip physical implementation and physical design rule verification flows are developed using foundry PDK data. The main comments to add in this section relate to where and when flows that address these requirements through physical layout updates are exercised and where and when the updates are subsequently verified.

The methodology team needs to address where in the overall physical model hierarchy that DFM physical data (e.g., fill shapes) are added to selected layout cell(s). The scope of the foundry's uniformity design rules provides some physical guidelines (i.e., the stepping window and the acceptable percentage range of mask layer utilization). The addition of a DFM layout cell instance to an existing physical design needs to ensure a high success rate in meeting these rules. Selecting an area that is too small or too large in the physical design hierarchy detracts from the likelihood of passing these DRC rules. The electrical impact of DFM data on the physical circuit implementation is a factor, as well, although the size of individual fill shapes and the spacing to design data is selected to minimize the capacitive influence. Nevertheless, DFM data needs to be present for detailed parasitic extraction and thus is part of the electrical analysis.

The stability of the SoC physical hierarchy is a consideration as well, and it helps to define when DFM data should be included. If the physical implementation of a design block is very fluid, adding DFM data (and investigating DRC errors) would be an unnecessary distraction.

For mask layers subject to multipatterning litho/etch steps, the DFM layout pattern can hopefully be developed so as to have minimal impact on mask data decomposition. If the layout incorporates decomposition coloring designations to achieve specific electrical behavior, subsequent addition of DFM shapes will hopefully not invalidate the existing color assignment.

Similar considerations apply to DFY flows (i.e., where in the physical layout hierarchy these optimizations are applied, when the flows should be exercised during the physical implementation phase). In the case of DFY updates, however, there is commonly a more significant electrical influence on circuit performance; for example, redundant vias between interconnect route segments reduce the series resistance in the net. Further, DFY optimizations also have a more direct impact on available routing resources. As a result, the selected DFY methodology may apply these flows as an integral part of physical implementation, as opposed to a less frequent, more opportunistic post-routing approach.

## 2.9 Design Optimization

The SoC methodology includes numerous opportunities to inject optimization algorithms into the overall functional and physical design flows. Generally, flows include "switches" to enable/disable these algorithms in various EDA tools. The CAD team establishes which algorithms are on/off by default, based on prior results on representative design examples, specifically from the tool evaluation and trailblazing activity described in Section 2.4. For various tool optimization features, the evaluation results include assessment of compute resources and runtime versus the design quality of results (QoR). Algorithms that offer minimal QoR benefit yet add to the required IT resources may be off by default in the released flows. If these optimizations incur additional license cost to the base EDA software product, the number of add-on license features is likely limited—and less than the base product license quantity. License availability often dictates which features are off by default and which are selectively enabled for specific blocks and/or in specific project design phases.

Examples of functional optimizations include:

- RTL-based power estimation analysis, providing recommendations for RTL clock gating and micro-architectural updates
- RTL *linting*, which is semantic analysis of the RTL to identify and recommend corrections to problematic statement constructs and hierarchical instances, such as:
  - Incomplete case statement conditions
  - Incomplete if-then statements with no else condition
  - Sourceless/sinkless nets
  - Uninitialized signal values and undesired "U-value propagation"

  RTL model behavior typically does not include inferred sequential state through the concurrent expression logic semantics. The machine state is represented by capturing logic expression values in clocked register signals. Linting is an effective method, separate from functional validation and synthesis, for finding anomalous RTL code.

- RTL logic optimizations incorporated into netlist synthesis algorithms, including:
  - Redundant and "dead" logic identification and removal
  - Parallel logic insertion (*cloning*) for performance and apportioning fan-out loading
  - Insertion of DFT architecture logic

A key methodology flow to exercise after these model updates is logical equivalency checking (LEC) between the pre- and post-optimization models (see Chapter 6, "Characteristics of Formal Equivalency Verification").

Physical optimizations occur frequently throughout implementation flows (e.g., cell/IP placement, interconnect routing, clock buffer selection, global power and clock distribution tuning, DFM/DFY shape modifications). For physical design, the corresponding methodology flow to functional equivalency between two logic models is layout-versus-schematic (LVS) checking. Although one model reference is commonly denoted as the schematic, typically the comparison between the physical layout connectivity and the netlist is directed to stop at various instances in the netlist hierarchy (e.g., library cells, hard IP blocks).

Regardless of whether physical updates are introduced by flow algorithms or through manual ECO edits, the checkin of a new model revision to the design data management system needs to include exercising LEC (functional) and/or LVS (physical) flows as a prerequisite.

## 2.10 Methodology Checks

The flow that verifies that design blocks have adhered to the overall methodology is an integral part of SoC project management. Methodology checking comprises both flow results analysis and reporting (commonly denoted as *scoreboarding*). Flow reports are parsed for pass/fail status, plus any key metrics that indicate model quality. These results are then posted on a project scoreboard for the blocks in the SoC model hierarchy that are being tracked. The scoreboard covers functional validation, testability, physical verification, and electrical analysis flows.

### 2.10.1 Functional Validation Status

Functional validation progress is typically tracked in terms of the number of:

- Failing simulation testbench cases ("directed" tests and pseudo-random test sequences)
- Pseudo-random instruction testcase length (and coverage of various instruction interactions)
- RTL design model coverage measures (e.g., line, statement, condition, net toggle coverage)
- Open bugs and bug severity (failing testcases where triage has identified the root cause and assigned a priority to fix)

A feature of the methodology manager is an *issues tracking* application, more commonly known as a *bug tracker*. Failing testcases result in a new record in the bug database. Design and validation engineers add information to the issue database entry until the bug is closed. If project-wide validation progress is being impeded, the issue is a high priority. If the testcase conditions are more isolated, a lower priority may be assigned. The bug may certainly be associated with the testcase itself, rather than with the functional model, if it is determined that the testcase does not correctly reflect the SoC specification.

As the root cause is being analyzed, an owner is assigned in the issues database. The owner is responsible for providing a target resolution date when a model revision will be released. The project management team can then assess how to best proceed until a release is available (e.g., determine what testbench changes are appropriate and/or what temporary functional model workarounds may be needed).

### 2.10.2 Testability Status

Although the production test pattern set is not completed until after the final tapeout netlist is available, it is beneficial to exercise testability analysis on individual blocks during the design phase. For block-level netlist testability analysis, the flow setup is typically easier, as the clock definitions and serial scan-shift chains are more straightforward. Conversely, the full SoC incorporates global logic to generate clocks, to multiplex scan and BIST functionality, and to define array repair sequences, requiring more complex and detailed setup for production test pattern set definition.

Test pattern generation on a block-level netlist provides fault coverage percentage measures for each fault type. As the block inputs and outputs would be regarded as directly controllable and observable, the coverage percentages are maximum compared to the embedded block instance in the SoC model. Thus, networks with poor fault coverage at block level do not subsequently improve and warrant early attention to untestable faults to improve test visibility. The project scoreboard highlights any blocks whose testability percentages are below limits; the block-level criteria are more stringent than the overall SoC target.

### 2.10.3 Physical Verification Status

Physical verification flows provide a multitude of potential results for each block in the model hierarchy to include in the project status scoreboard. The methodology and SoC project management teams need to determine what flow data is most important to pull from output log files and report in the project scoreboard.

Physical layout verification flows such as DRC and LVS offer several options for results data. The total number of layout errors can be misleading, as an error in a highly replicated layout cell inflates the count and does not accurately reflect the time required to fix the design; it is best to reduce the error count to unique cases within the block. LVS errors may also require specific interpretation (e.g., net shorts to power/ground or open/incomplete nets). The presence of power/ground shorts may generate many individual errors that may be addressed by quickly correcting the distribution grid definition (and blockage map) used by signal routing algorithms. On the other hand, the presence of a large number of incomplete signal nets may be a more critical indication that there may be insufficient routing resources available in the allocated block area in the floorplan.

### 2.10.4 Electrical Analysis Flow Status

Electrical analysis flows provide a wealth of data, and assessment is needed to determine what information is most informative to include in the project status scoreboard. For example, the block-level static timing analysis flow would offer the following results:

- Total negative timing slack (TNS) summed across all timing path test points
- Worst negative timing slack (WNS) for any single timing path

- Slack histogram data for all paths
- A breakdown of the timing path data results into the following categories: primary input-to-internal test point, internal launch-to-primary output, and internal timing paths
- Internal network nodes with negative timing slack that are part of multiple failing timing paths

The primary input arrival time and primary output expected time constraints at the IP block level require coordination between design teams to allocate timing budgets and thus may need specific designation in the project scoreboard.

Similar considerations on the amount of analytic data to gather for project reporting apply to the other electrical analysis flows. In addition, there are correlations between analysis fails from different flows that offer insights into the highest-priority errors to address; for example, a weak local power grid may result in both high electromigration reliability risk and I*R voltage drop margin fails.

The goal of the scoreboard (and bug database) is not to reward or penalize a block for its current status but rather to provide the project management team with insights into where team resources may need to be rebalanced and/or where block power budgets and performance/area targets need to be reviewed and potentially revised.

### 2.10.5 Tapeout Audit

All the methodology manager features are exercised as the project enters the final tapeout phase with a frozen design database. The scoreboard is now populated with the results of full-chip flows. Further, attention is given to verifying the specific versions of foundry PDK techfiles and flows to which the design data have been subjected. Specific flow settings are also captured from post-analysis reports and checked to ensure consistency with project-wide constraints.

A final audit of the full-chip scoreboard and bug database, including version consistency, is conducted as part of the tapeout release review. The audit team may be faced with several key trade-offs:

- **Functional issues may still be open in the bug database.** To avoid schedule impacts, some functional bugs may be acceptable—for example, those with low priority or with software workarounds—especially for a first-pass tapeout. The intent would be to address these issues in a subsequent tapeout (perhaps with spare gates).

- **The full production test pattern set may not yet be fully assembled.** Often, the final production test set is prepared in the interval between tapeout release for fabrication and silicon wafers available. The tapeout audit reviews the test scoreboard status to confirm (with high confidence) that the full-chip pattern set can achieve fault coverage targets.
- **Analysis flow errors may be waived.** Any remaining analysis flow issues are reviewed and evaluated to determine whether waivers should be applied in the scoreboard. Timing data-to-clock setup failures suggest that silicon bring-up could proceed with a reduced clock frequency. Power dissipation estimates above chip targets may not adversely impact bring-up and could be waived. Electromigration and device aging circuit margin issues pertain to production volume lifetime failure predictions and could potentially be deferred to be addressed after first-pass tapeout.

The results of the tapeout audit review are recorded by the methodology manager. Any DRC waivers are included in the tapeout report, along with the acceptance sign-off from the foundry customer support team.

## References

**[1]** Ren, H., et al., "Intuitive ECO Synthesis for High Performance Circuits," Design, Automation & Test in Europe Conference (DATE), 2013, pp. 1002–1007.

**[2]** Chang, N., et al., "Efficient Multi-domain ESD Analysis and Verification for Large SoC Designs," 33rd Proceedings of the Electrical Overstress/Electrostatic Discharge (EOS/ESD) Symposium, 2011, pp. 300–306.

## Further Research

### Layout-Dependent Effects

Describe some of the layout-dependent effects from neighboring structures that impact device characteristics.

Describe the considerations for a layout context cell to surround a design cell prior to executing the parasitic extraction flow for worst-case and best-case cell performance.

### Antenna Rules

Describe the fabrication steps that result in accumulated electric charge on metal wires in a net. Describe how the antenna rule check is formulated, in terms of metal layers, metal area, and connected device gate area.

Describe how and where diode layout cells could be inserted and connected to the metal antenna and the potential impact on route congestion.

### Pattern Matching

Efficient methods to describe a complex layout pattern, and subsequently to check for the presence of that pattern in a layout view, are required for DRC, DFM, and DFY applications. An industry standard for pattern definition is emerging. Describe the Open Pattern Analysis for Layout (OPAL) modeling language.

### Inverse Lithography Technology (ILT)

For several process generations, optical proximity corrections have been applied to drawn mask data to improve the exposure and post-etch fidelity to the drawn design. More recently, these corrections have been displaced by a more exhaustive ILT analysis.

Describe the OPC design modifications originally developed and the scope of their impact on the final patterned shapes. For example, review the history of serifs on mask shape corners, (alternating and attenuating) phase shift masks, and sub-resolution assist features added to the mask data.

Describe the origin of "forbidden pitches" for the width plus space dimensions on design data.

Describe the ILT algorithms currently used to derive mask data and the exposure illumination pattern.

### EDA Tool Evaluation and Trailblazing

Describe the (general) criteria that are pertinent to the evaluation of a new EDA tool, both in terms of new features added for productivity or technology requirements and for competitive assessment against an existing tool used in a flow. Describe the quality of results (QoR) criteria that would be commonly defined for simulation, synthesis, automated placement and routing, and test pattern generation tools.

Describe the potential resource costs and schedule impact of migrating from an existing EDA tool.

Describe the characteristics of an optimal trailblazing block design used to evaluate new EDA tool features and/or qualify new CAD flow releases.

### Configuration Management

Describe how the features of a configuration management application developed for (large) software development projects are applicable to SoC design.

### Bug Tracking Tool

Describe how the features of a bug tracking tool used for software development are applicable to SoC design. Describe the typical SoC functional validation bug evolution, from identification through the various stages of triage to final closure. Describe how the bug status and information recorded in the tracking tool would be linked to the configuration management of the SoC data.

CHAPTER 3

# Hierarchical Design Decomposition

## 3.1 Logical-to-Physical Correspondence

As illustrated in Figure 3.1, two hierarchical models are maintained for an SoC design: a logical hierarchy and a physical hierarchy. The functional validation environment is developed around the logical hierarchy, with references to model registers and array storage that reflect the instance path concatenation. In addition to the top node of each model representing the full-chip SoC, there are correspondence points throughout the two hierarchies. Below the correspondence nodes in each model, the hierarchical decomposition diverges.

A pair of hierarchical correspondence nodes share the same pin definitions. The correspondence nodes represent the same functionality; logic equivalency flows are exercised between the RTL and the netlist/schematic of the correspondence pair. The initial step in equivalency checking is to read the two models and establish the common primary inputs, outputs, and internal state points. The default matching algorithm looks for identical pin names, although designers can provide a configuration file with logical-to-physical

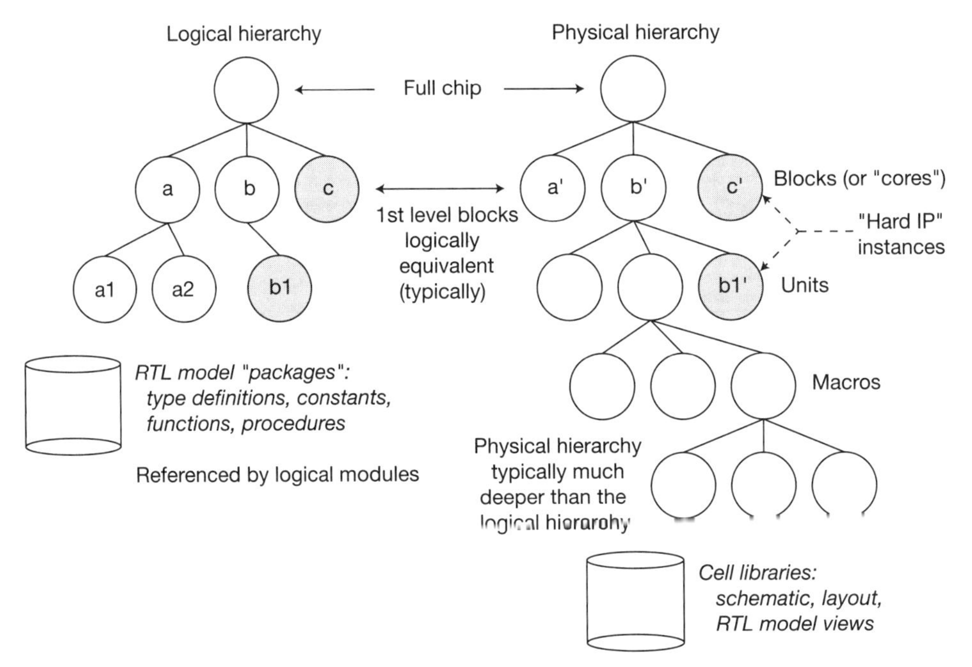


**Figure 3.1** Illustration of logical and physical hierarchies for an SoC model (similar to Figure I.3). The leaf nodes of the physical hierarchy are hard IP blocks and IP library cells. Hard IP blocks in the physical hierarchy are also equivalency points in the logical hierarchy, using the HDL model from the IP provider.

pin-pair overrides. Internal state point match candidates are selected using multiple approaches:

- Common RTL signal and netlist node names
- Common input signals to the *logic cone* terminating at the state points
- A user-provided configuration file with state pair recommendations

The logic equivalency checking (LEC) flow is exercised for correspondence nodes in the hierarchy. When confirming equivalency at the full-chip SoC level, the correspondence nodes need not be reevaluated and can be regarded as black boxes in the SoC configuration file.

It is common, but not necessary, for the correspondence points to be the second-level nodes in the physical hierarchy. The SoC physical floorplan is typically defined by cores/blocks at a single level of physical implementation below the full chip, with matching logical models. Global logic and global circuits (e.g., repeaters, buffers) can be consolidated into a pair of correspondence nodes in the model hierarchies or simply flattened into the SoC top node. An issue with consolidating these functions into a physical hierarchy node is that the implementation will be present across many disjointed areas across the die; floorplanning tools usually associate a physical node with a single (rectilinear) area boundary definition.

Hard IP blocks are naturally correspondence nodes in the hierarchies, as each one represents a fixed logical and physical set of models. Soft IP designs are candidates for correspondence. However, it is also common to integrate multiple soft IP RTL models into a larger physical block for implementation to reduce the floorplan complexity and leverage logic synthesis optimizations across flattened model interfaces. Indeed, EDA vendors of synthesis tools are constantly working on extending the model capacity of their algorithms while maintaining suitable runtime and IT resources.

Hierarchical model correspondence is being emphasized due to the potential impact on the project schedule if frequent changes are made to either of the hierarchies during the SoC design. As highlighted earlier, the logical hierarchy is an integral part of the functional validation environment. Testbenches reference internal logic signals by their hierarchical path. Some flow(s) exercise simulation testcases on the physical netlist, as well. For example, power dissipation analysis needs to apply representative testcases on the detailed physical model to accurately calculate net switching activity and the resulting capacitive charge/discharge energy. The output of the logic equivalency flow provides the state point and pin matches between logical and physical models. For simulation testcases it is (relatively) easy to adapt the logical model references to the equivalent physical net when initializing or querying a model value. If the correspondence node pair definitions are in flux, maintaining the validation environment across both hierarchies becomes more difficult. At the start of a new SoC project, the architecture and project management teams review the design specification and establish the first few levels of model hierarchy to enable functional validation and physical implementation planning to proceed. As the design evolves, the lower hierarchical levels of logical and physical decomposition diverge; however, the correspondence nodes should ideally remain stable.

### 3.1.1 Physical Implementation Details and the Logical Model

The key issue for the methodology team relative to model hierarchy and correspondence pertains to how the test and physical implementations could influence the logical model. If test logic and serial scan connectivity are inserted during logic synthesis of a soft IP block, the additional pins and functionality are not present in the original RTL model. Equivalency checking is directed to disregard the inserted logic. Validation testcases are needed to confirm the test sequence operations on the netlist model, not the RTL. The team must assess the merits compared to the difficulties of attempting to represent test function in the logical model. (Representation of the serial scan chain from the physical implementation back to a different logic hierarchy is especially complex.)

In addition to test functionality, a similar trade-off between RTL functionality, physical implementation, and validation exists for SoC reset sequencing. Validation testbenches can readily "force" an initial logical model state prior to simulation, avoiding the overhead of simulating a reset preamble. The physical reset implementation could take several different forms and could potentially take many clock cycles. The methodology team must evaluate the overhead in representing reset behavior in the RTL logic model compared to pursuing separate reset validation efforts on a more detailed netlist and should confirm that the physical reset matches the forced initial RTL simulation state.

Finally, a similar trade-off is present when considering where to represent the power management and power state sequencing functionality. In this case, however, the EDA industry has defined a separate power format file standard. The intent is that a detailed description separate from either the logical or physical model captures the full power management functionality, from power state transitions to the inactive behavior of a gated block. EDA tools have been specifically developed to synthesize and validate the implementation against the power intent description. As a result, there is little direct value in representing this behavior explicitly in the RTL model. (Note that the maintenance of this power format file is also simplified if the logical-to-physical hierarchy node correspondence is stable.)

The methodology team needs to evaluate the IT resources needed for validation of functionality that is present only in the physical netlist, not reflected back in the logical model. The netlist detail implies that software simulation executes more slowly and that debugging is more difficult. If simulation acceleration technology is available, the netlist model validation workload is likely to be directed to that platform.

## 3.2 Division of SRAM Array Versus Non-Array Functionality

When developing the hierarchical decomposition of the SoC logical and physical design, a key consideration is the definition of on-chip SRAM memory arrays. The hierarchical node for a memory array is commonly a correspondence point between the logical and physical models. The methodology flows for arrays are unique from the flows for digital logic networks; the success of the array flows is strongly dependent on the exclusion of unrelated logic from the array definition.

### 3.2.1 Array Compilers

Array IP may be produced by a software *array compiler* provided by an IP vendor. Using an array configuration description, the compiler generates a full suite of logical and physical models. In this case, the logical and physical hierarchical nodes are defined by the generated array ports in the RTL model and the physical layout cell output from the compiler.

### 3.2.2 Array Methodology Flows

If an array is being designed without benefit of a compiler, the array functionality is presented to specific test, validation, and electrical analysis flows. These unique flows include the following:

- **Logical-versus-physical equivalency**—The array physical implementation starts with a schematic netlist. EDA vendors have adapted equivalency algorithms to work with the unique nature of array schematics (e.g., evaluating row and column decoders, interpreting the dotted bit cells along bit lines, resolving the differential sense read amplifiers and bit line write drivers, evaluating the match circuitry of a cache address tag).
- **Timing model generation**—The characterization of array timing involves a topological analysis of the internal storage cells, read sensing circuitry, and bit cell write behavior; detailed circuit simulations; recording of specific simulation measures that define the array operations; and generation of the clock/address/data timing model and constraints. This flow is significantly more complex than timing characterization for logic cells.
- **Test pattern generation**—For large arrays, EDA vendors offer products to generate MBIST logic to integrate around the array. For smaller array macros, EDA vendors provide tools to generate test patterns for the array in isolation and to confirm the SoC test mode conditions to apply these patterns to the embedded array from controllable points in the design.

The EDA tools used in the array flows have inherent limitations on the complexity of supported array topologies, especially logical-versus-physical equivalency (using symbolic simulation, discussed in Section 6.7). The methodology team should review the EDA tool features and confirm whether support is provided for advanced custom array implementations, including:

- Pre-charged bit lines
- (Self-timed) signal pulse generation within the array schematic
- Multi-stage address pre-decode
- Latched versus unlatched output data
- Hierarchical memory arrays (with an array block selector and global bus)
- Multi-port bit cells (multiple, mutually exclusive word lines for a write operation; potentially, multiple active word lines for a read operation)

As mentioned previously, the array logical and physical models should exclude logic unrelated to data read/write operations to better align with the array flows.

### 3.2.3 Register Files

The definition of array functionality also applies to register files. A register file has a similar architecture to a memory array but is usually implemented with a bit storage cell that more closely resembles a flip-flop, without the need for differential sensing. The flows for register files may be unique, as with SRAM arrays. Alternatively, it may be feasible to implement and model the register file as composed of conventional logic cells, without the schematic topologies listed earlier that require special analysis algorithms. If conventional flows apply for register file arrays, the constraints for division of array versus non-array logic functionality will be relaxed.

## 3.3 Division of Dataflow and Control Flow Functionality

The previous section highlights requirements for the definition of array functionality in the model hierarchies to ensure that the unique flows for array models and analysis are successful. A similar methodology consideration may also apply to the hierarchical decomposition of logic networks: distinguishing control flow from dataflow. Generally, *dataflow* refers to the (wide, multi-bit) logic and registers associated with operations on information resident in data cache and main memory. *Control flow* encompasses the fetch, decode, and sequencing of signals associated with instruction execution. Often, a synchronous finite state machine (FSM) is the modeling approach used to describe a set of control flow signal sequences. The outputs of FSMs are inputs to the dataflow networks, serving as multiplexer selects to steer data through the correct logic operators. These selects are typically high fan-out signals, spanning the data width of the overall architecture. As a result, the circuit performance of these control signals is also typically in critical timing paths. The physical implementation of the dataflow network—and especially the control signal interfaces—is crucial to achieving clock cycle time performance targets. The SoC methodology may offer specific physical design algorithms for dataflow implementation, which may necessitate attention to functional decomposition in the logical and physical model hierarchies.

### 3.3.1 Dataflow Library Cell Design

One implementation consideration is whether a unique cell library specifically for dataflow networks is warranted. As illustrated in Figure 3.2, the optimal logical fan-in of dataflow functions may be higher than other, more random logic. The number of data steering select signals may suggest that larger AND*OR (NAND*NAND) cells are available in the circuit library.

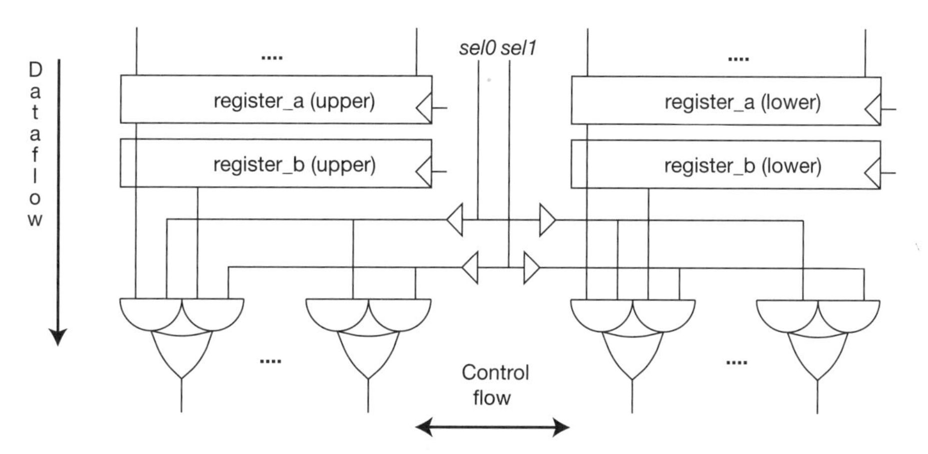


**Figure 3.2** A separate dataflow-centric cell library may be offered by the IP vendor, with wider fan-in logic functions. A modified cell template for dataflow library cell placement may be optimal.

Also, the pin locations on the cells should facilitate an optimal signal route across the dataflow width (e.g., AND*OR select input pins should be on different routing tracks to enable direct routes and minimize congestion). A set of dataflow cells with a larger track height may be beneficial. The tradeoff would be the overhead to develop and support a modified cell placement image and local power/ground distribution grid for dataflow networks within a block throughout the physical composition flows.

Another dataflow library design consideration would be to try to *pitch match* the logic operations to the cell width of the register bits. As depicted in Figure 3.2, the overall dataflow execution runs orthogonally to the control flow routing. The optimum dataflow physical implementation should avoid widely varying widths of operand slices. Figure 3.2 also illustrates the buffering of control select signals for the high fan-out of the data width. Unique buffer cells that have pin locations and drive strengths for the dataflow fan-out may be preferable. For a larger architectural data width, the preferred dataflow physical implementation may be to split the design into smaller subwidths, with the control select signals physically interspersed throughout the dataflow operators. The resulting decision from an assessment of these physical tradeoffs is that the circuit library team may indeed be tasked with development of specific cells for dataflow design.

### 3.3.2 Dataflow Cell Synthesis, Netlist Placement, and Routing

The example in Figure 3.2 suggests that there may be specific methodology flows for dataflow networks. Logic synthesis flows could be directed to a unique dataflow-specific cell library. During synthesis, RTL expressions should be interpreted to maintain common logic functions across the data operand width in the output netlist. During cell placement, several physical constraints should be observed:

- Relative placement of the cells in the data operand slice should be aligned (by bit position).
- Control signal repowering buffers also require specific cell placements.
- (Control) signal routes should be as direct as possible, facilitated by the cell placement.

EDA vendors have added features in their design tools to support dataflow implementations. For example, during logic synthesis cell mapping optimization, interconnect load estimates added to the fan-out gate loading would be reduced to reflect the directed routes. In preparation for placement, a set of netlist cells could be identified as a *relative placement group* in an input configuration file to the flow; these cells effectively become a single (1 × N) placement object. High fan-out control select signals would be assigned increased priority during routing. The separation of dataflow functionality in the physical hierarchy enables these flow features.

### 3.3.3 Merged Control Flow and Dataflow

Unlike the array functionality distinction discussed in Section 3.2, where unique circuit topologies required very different analysis approaches, the dataflow design described in this section is based on library cells. Electrical analysis methods are the same for dataflow and control flow networks using (pre-characterized) cells in the netlist; only the synthesis and physical design algorithms would be unique.

Further, it is often both difficult and somewhat impractical to try to separate control flow from dataflow in the RTL and physical hierarchies. Small FSMs may be integral to the functionality of a block and are tedious to maintain as separate nodes in the physical hierarchy. The physical separation of interdependent control logic may result in timing paths that are difficult to close. Ideally, the SoC methodology would support a general combination of control (and glue) logic with dataflow expressions and registers, in the same corresponding logical and physical hierarchical model.

Unique methodologies have been developed for a general block model comprising both dataflow and control logic descriptions to target specific IP cell library elements, with a small impact on RTL coding style.[1] As an example, a subset of an RTL model could be surrounded with (non-functional) pragma directives in comment statements, which would be interpreted as directing synthesis and physical design algorithms to the unique implementation techniques described above. This technique is similar to the pragma statements in Figure 2.1, where the goal is to bypass a set of HDL statements for synthesis altogether. In this case, the intent is to direct synthesis to a specific mapping algorithm for the set of statements. The RTL code snippet in Figure 3.3 illustrates the concept.

```
sel0_vec(31 downto 0) <=  (others => sel0);
sel1_vec(31 downto 0) <=  (others => sel1);

-- synthesis  map_as_written

xyz:  a(31 downto 0) <=  (rega(31 downto 0) and sel0_vec)
   .                       or  (regb(31 downto 0) and sel1_vec) ;  // placement group "xyz", 32 2-2 AO cells
   .
   .


-- synthesis map_off
```

VHDL coding style example with a "map_as_written" synthesis directive

**Figure 3.3** Example of a pragma added to an RTL model for a "direct map" synthesis algorithm.

The remaining RTL functionality in the model would be synthesized and optimized together with the direct map network(s) and would be part of the complete block netlist used during physical design and electrical analysis. If the methodology can support dataflow-specific synthesis and physical design optimizations within a larger block, the design constraints on separating control flow are significantly reduced. If a set of dataflow library cells uses a unique template, physical design flows need to support multiple images within a block, with placement restrictions for the dataflow cells in the netlist to specific rows.

Any discussion of identifying dataflow and control flow in an SoC design for optimum (library cell based) implementation needs to include mention of complex functionality that may not fit this approach and that perhaps requires special flow features. An example that presents unique difficulties would be a data multiplier operation. The natural block shape for a multiplier would be a skewed orientation rather than the orthogonal data/control signal routing depicted in Figure 3.2. For these special functions, additional methodology support may be needed, extending dataflow implementation support.

## 3.4 Design Block Size for Logic Synthesis and Physical Design

When establishing the hierarchical decomposition of an SoC design, there are trade-offs related to the sizes of the blocks that are presented to various methodology flows. Specifically, there are two key considerations:

- The blocks in the logical hierarchy using logic synthesis
- The blocks in the physical hierarchy that will be maintained in the physical floorplan

### 3.4.1 Logic Synthesis and Block Size

The evolution of fabrication process scaling has enabled tremendous increases in the effective logic gate count available for SoC integration. To keep the requisite engineering design resources in check, there is a corresponding need to increase the block sizes presented to synthesis flows. EDA synthesis tool vendors have responded by enhancing their products to support larger designs, including the following:

- Efficient data structures for representation of unmapped logic operations
- Multi-threaded algorithms for logic reduction, logic factoring, and library cell mapping with timing/power optimizations
- Efficient in-memory models for cell electrical characterization data to support initial cell *mapping* and post-mapping optimization *cell swaps*
- Physical guidance during cell selection, using block floorplan information to improve wiring-based calculations of estimated interconnect loading and routability

For the last feature listed here, incorporating physical data during synthesis ideally results in fewer post-routing electrical issues and/or fewer floorplan/synthesis/physical design iterations. Note that this feature assumes a logical node-to-physical node hierarchy correspondence for the synthesized block and thus requires the synthesis flow to support large models. If multiple synthesized netlists are subsequently merged into a single floorplan block for physical design, the value of physical guidance in synthesis is lost.

The features listed above have enabled block designs to realize netlists with multiple millions of logic gates. The support for larger synthesis block sizes has several advantages:

- Fewer block input/output pin constraints to maintain (as per "Rent's rule"[2])
- The potential for improved logic reduction and factoring, with fewer partial logic paths to/from block pins and timing test points
- The potential for improved local DFT logic insertion, such as adding more comprehensive test compression logic for the block (see Section 19.4)
- The potential for improved high fan-out signal (clock) repowering tree distribution in the larger block if a greater number of sinks are present

There are disadvantages to adopting larger synthesis blocks, as well:

- Synthesis flows (typically) support a single power domain and power state management strategy. The coding of a power intent description is based on the logical model hierarchy; a single power gating implementation is usually assumed during synthesis. A larger design block may use multiple architectures for power management, which may be ill suited for synthesis.
- Managing pin constraints is easier for a single (root) clock domain. Similarly, it is more straightforward to manage clock domain crossings (and the insertion of synchronizing cells) at block interfaces. A larger design block may incorporate multiple (architectural, unrelated) clock domains, which may be ill suited for synthesis.
- Larger blocks have more functional revisions, necessitating more synthesis iterations. This is the greatest consequence of using larger logic hierarchy blocks for synthesis. As functional bugs are uncovered during validation, RTL models are updated and new revisions released. A "minor" RTL change can have a major impact on the synthesized netlist, invalidating existing physical design results from a previous logical model. If a larger design block incorporates a sub-block that is still fluid in functional definition, the confidence level in the overall floorplan, netlist routability, and electrical analysis closure may be lower. The SoC project management team might decide to identify smaller, more stable design blocks for synthesis (and, potentially, physical floorplan correspondence) that can be pursued independently, with fewer expected iterations.

EDA synthesis tool vendors have added features to identify the logic cones that differ between RTL revisions and excise those cones, re-synthesize, and insert a reduced cell netlist. Ideally, this reduces the resources required to iterate on the synthesis flow for a larger block. However, the relative success of such features is difficult to predict due to the extensive logic and cell mapping transformations that are applied to an RTL model.

### 3.4.2 Low-Effort Synthesis

A methodology approach that has proven to be extremely useful during the initial project design phases, which are focused on RTL definition and validation, is to execute a *low-effort synthesis* (*LES*) flow (see Section 7.11). A reduced set of logic synthesis algorithms is applied, essentially stopping right after initial cell mapping (perhaps to a generic Boolean library, without detailed electrical models). The LES output netlist can provide useful analytic information such as the following:

- Estimated logic gate count (lower bound as no timing-driven repowering is exercised)
- Identification of very high fan-out nets (which may subsequently need high fan-out repowering algorithms)
- Logic gate path length distribution

If the target clock cycle time for the block is "N technology library gates," an LES netlist with path lengths much greater than N indicates subsequent timing closure issues; required micro-architectural changes can be identified early. Similarly, a high number of LES short paths indicates additional delay cells (and resources) that are subsequently needed to address hold time issues at timing path endpoints. The LES netlist is also useful to the SoC test engineering team, for analyzing the logic stuck-at fault coverage measures for the block.

### 3.4.3 Physical Design and Block Size

Many of the same considerations for selecting the logic hierarchy block size for synthesis also apply to the physical design and analysis flows. Larger blocks facilitate simpler floorplan and pin management and provide more opportunities for cell placement and routing algorithms to balance area utilization and interconnect resources. Larger physical design blocks also provide more flexibility for the location of embedded hard IP macros within the blocks.

The disadvantages of larger physical blocks are also similar to those in logic synthesis. Physical implementation is simpler if the design block represents a single clock and power domain. Generating multiple clock grids and power grids (potentially with power gating) in a single physical block is certainly more complicated.

As with synthesis, the stability of the design is the most significant factor in reducing the number of resource-intensive iterations. For a larger physical block, the biggest impact pertains to the stability of the floorplan pins. The definition of the pin location, pin width, pin metal layer (with multi-pattern coloring), and global signal route planning to the pin are all detailed tasks. If the physical hierarchy is being revised frequently, the updates required to existing pin definitions and global route plans are extremely resource intensive. Whereas synthesis block size selection is impacted by RTL stability, physical design blocks are more impacted by pin partitioning.

As mentioned earlier, it is feasible to merge multiple synthesized netlists into a single physical block. To support synthesis iterations in this case, EDA tool vendors have added features to allow edits to a netlist sub-block within an existing physical implementation (e.g., delete all the sub-block netlist cells, place a sub-block netlist revision, and rip up/reroute existing net segments to update the physical design).

The SoC logical and physical hierarchical decomposition involves crucial decisions on block size to present to synthesis and physical design flows. Current EDA tools enable models with multiple millions of logic gates to be efficiently managed. Thus, the determining factors in developing the hierarchical description are primarily driven by the SoC architecture and the expected stability of the functional model.

## 3.5 Power and Clock Domain Considerations

As mentioned briefly in the previous section, the hierarchical decomposition of the logical and physical models for an SoC design is influenced by the power domain and clock domain architecture and implementation. Methodology flows that insert (and verify) power gating functionality typically assume that a single domain is present:

- A single set of VDD, VSS, and gated power distribution grids are used in the cell library image.
- A single set of "deep sleep" gating transistor connections are added to the image between the VDD and VDD_gated (or VSS and VSS_gated) rails.
- Power domain level-shifter cells have two specific power rail requirements and are typically present only at floorplan block interfaces, not placed throughout the block.

In addition, development of the physical hierarchy needs to be cognizant of the estimated power dissipation of the design block. A high level of switching activity within the block leads to a local *thermal hot spot*, potentially resulting in failing circuit characteristics and/or accelerating reliability degradation. A thermal sensor inserted within the block may be needed to throttle the activity (e.g., suppress clock cycles) with the corresponding impact on performance. An alternative approach would be to review the logic micro-architecture to evaluate whether a different physical hierarchy could better distribute the dissipated power.

Similarly, the logical and physical hierarchy decomposition needs to consider the various SoC clock domains. Physical implementation flows typically assume a single clock distribution network—whether using a grid, balanced repowering tree, or grid/tree hybrid. For logical-to-physical correspondence nodes, typically using synthesis for netlist generation, a single fundamental clock domain is typically assumed. A consideration for logical decomposition is the functional validation strategy that will be used for the SoC. Various techniques available for accelerating simulation throughput partition the functional model, using the existing logical hierarchy as a starting point. Both parallel software simulation on multi-core compute servers and SoC model prototyping on field-programmable gate array-based systems can benefit from RTL logical decomposition with single clock domain endpoints.

## 3.6 Opportunities for Reuse of Hierarchical Units

A major factor in the development of the logical and physical hierarchies is the potential for reuse of existing models from a previous design. Project resources for functional design, functional validation, and physical implementation are greatly reduced when existing, proven blocks can be reused. As defined in Section 1.1, a logical block candidate for reuse is denoted as "soft IP." An existing RTL model, potentially licensed from an external provider, is reused, and the corresponding node in the SoC logic hierarchy is created.

Although the RTL source may be carried forward, the actual functionality may certainly differ between designs. Configuration registers within the block could be initialized with different values, resulting in significantly different behavior. The RTL model could include a rich set of features, only a subset of which are required for a specific SoC design. Unused logic present in the RTL model could be tied to inactive logic levels (and optimized during synthesis). A more elaborate RTL coding style may utilize parameterizable *generate* statements, where input values defined in the parent model are passed to the soft IP model instance. The elaboration of the model applies these generate values to set the specific bit width of dataflow expressions and/or to define a specific number of unit instances within the hierarchy of the soft IP. The planning of the logical hierarchy should leverage available soft IP for reuse.

The physical hierarchy is also influenced by the availability of existing physical IP for reuse. As introduced in Section 1.1, "hard IP" is by definition a physical reuse implementation candidate, with a corresponding logic model.

In general, it is much more difficult to reuse a physical block from another design than reusing a (parameterized) RTL logical model. The new SoC design usage may have different PPA targets than were applied for an existing physical block. Floorplan aspect ratios allocated for the block may differ between various applications. And, most commonly, the new SoC design may simply be using a newer fabrication process technology, with more aggressive physical design rules; an older physical layout may be non-optimal or simply invalid in the newer technology.

"Firm IP" corresponds to an existing cell netlist for reuse. This mitigates the floorplan aspect ratio concerns and (perhaps) simplifies the transition to a newer process technology. A detailed physical implementation flow with new PPA targets is needed.

Nevertheless, if the opportunity for (hard IP) physical block reuse is available, that should be a major consideration as a resource and cost savings when planning the SoC physical hierarchy.

## 3.7 Automated Test Pattern Generation (ATPG) Limitations

Sections 3.1 and 3.2 mention some of the DFT architecture constraints when developing the logical and physical hierarchies, such as differentiation of memory and non-memory functionality for the generation of embedded MBIST test logic. This brief section expands on that discussion, focusing on optimizing production testing throughput.

The SoC project management and product engineering teams develop cost models for volume production. A key contribution to the final product cost is the (wafer and package) tester time, along with the resources associated with any early life *infant failure* stress testing/screening, commonly denoted as *burn-in*.

The tester time is dependent on several factors that affect the model hierarchy decomposition (and the overall global design planning):

- **Number and length of serial-shift scan test chains**
- **Scan-shift cycle time**
- **Wrap test features incorporated with an IP block**—A wrap test chain integrated with an IP block is similar to a serial scan-shift DFT architecture but may have used unique clocking and test mode select sequences from the rest of the SoC to apply and capture pattern data. In other words, it is likely not a candidate for integrating with other chains and may require a specific test duration.
- **MBIST logic insertion**—The addition of on-chip pattern generation and response signature compression logic for arrays in the logical hierarchy significantly reduce tester time, at the cost of additional die area (and EDA tool/flow licensing).
- **I/O boundary scan**—The SoC pad-level I/O circuits also include their own unique serial scan-shift architecture, known as *boundary scan* (or the industry-standard term, *IEEE JTAG standard 1149.1*[3]). The boundary scan chain test sequence is similar to that of wrap test in terms of setup and clocking. Testing logic paths between internal blocks and SoC I/Os requires coordination between internal scan-shift and boundary scan clocking and pattern generation. In addition, parametric electrical tests need to be applied at the I/Os, which also contributes to overall tester time.

For the traditional scan-shift approach, a scan chain may be a concatenation of chains from multiple hierarchy nodes. Or, for large blocks with a high register count, multiple chains may be present within a single hierarchy node. The DFT architecture includes a tester sequence to transition between serial test pattern stimulus shift-in, functional clocking, and response pattern shift-out. In this sequence, the pattern application time is dominated by the overall scan-in and scan-out intervals. The number and length of chains is a critical design hierarchy optimization consideration; ideally, the length of individual chains should be balanced. The physical implementation of the scan_enable select signal or the scan-shift clock distribution, as well as the interconnect loading on the scan_out to scan_in for each bit in the chain, defines the achievable pattern shift cycle time, which is a critical factor in the calculation of the overall tester time. (Section 19.4 presents a novel *test compression* scheme, where the DFT architecture uses many short scan chains to test a logic network, whose chain outputs are connected to a *signature register* to compact the overall response. This architecture improves the tester time considerably over the traditional scan chain design through all flops and registers.)

In general, the chip I/Os do not have a strong influence on the SoC hierarchy; they are often present in the physical model as flat instances in the top model or perhaps grouped by their side/orientation on the die.

Testing of analog IP is also a key contribution to tester time and cost.

A test engineer is challenged with determining the optimal ordering of the various test pattern methods. Typically, a simple shift of 1s and 0s to validate scan chain and clocking integrity should precede any logic test pattern application. A specific challenge for a test engineer is to determine which pattern stimulus/capture/observe sequences can be executed concurrently at production test.

Burn-in stress testing presents additional logical and physical hierarchy constraints. The goal of burn-in testing is to accelerate early fail mechanisms and exclude infant fail packaged parts from product volume shipment. One approach is to apply an elevated supply voltage with static SoC inputs at an elevated burn-in chamber temperature; defect mechanisms sensitive to electric fields and/or temperature are magnified. A more advanced method is to apply representative pattern sequences at reduced clock rates, with the elevated supply voltage and temperature, resulting in widespread switching activity within each packaged part. The goal of dynamic burn-in testing is typically not to verify functionality at these environmental conditions, as circuit behavior at these extremes is not necessarily guaranteed. The chamber electronic controls may be able to measure each part's pattern response on a limited number of pins; nevertheless, the typical flow is to apply stimulus patterns to multiple socketed parts on the burn-in board in parallel without measures. After stress, the parts are retested at normal conditions.

The key issue with the design of the burn-in board relates to the total current draw and power dissipation of the parts under test. For throughput, a high number of parts screened in parallel is desirable; however, each part is at high power dissipation due to the elevated supply and temperature. The dynamic burn-in pattern rates are relatively low, with low active power; the leakage power is substantially larger. A trade-off exists between the number of parts and their power dissipation to remain within the limits of the burn-in equipment. To maximize the number of parts in the chamber, this trade-off is often biased toward reducing the power dissipation of each part. It may be preferable to develop a dynamic burn-in strategy consisting of a series of tests, where only a subset of the SoC blocks is powered for each test. Burn-in patterns would be developed to exercise a subset of SoC blocks, striving for high toggle coverage for nets within the blocks. A series of burn-in tests would ultimately span the full SoC physical implementation. In this case, the SoC model hierarchy may play a role. The physical hierarchy may need an implementation with separable burn-in power domains, up to and including (groups of) package power pins that connect only to each burn-in block power subset. The logical hierarchy is used by functional validation and test engineers to identify and excise patterns that can be applied to SoC inputs, with switching activity coverage measured for each burn-in subset.

## 3.8 Intangibles

When defining the SoC hierarchy, in addition to the technical considerations mentioned in this chapter, the project management team may also have to deal with non-technical issues.

The size and expertise of individual design groups has a major impact on the scope of logic architecture, (block and full chip) functional validation, and physical design tasks assigned to the group. A logical-to-physical correspondence node assigned to the group should be a fit to their skills and resources, to maintain a project schedule timeline comparable to the timelines of other groups.

Similarly, the engineering management organization to which the development groups report may influence how the block and full-chip responsibilities are assigned. This is especially prevalent when the engineering team is dispersed geographically. It would be uncommon to split a hierarchical node for development across multiple geographic locations if it is feasible to co-locate the team. Although the bug tracking database serves as the principal means of communicating project issues, the co-location of engineers working on a hierarchical block across functional definition, validation, and physical implementation tasks inevitably accelerates design closure.

Note that there are many approaches to the organization of the SoC engineering team. The preceding comments assume that an engineering manager is responsible for both the functional and physical design of an SoC hierarchy node, with engineering resources that include a wide breadth of expertise. Alternatively, an engineering organization may consolidate similar skills together (e.g., all functional designers in the same management reporting chain, separate from the physical design reporting chain, merging at a corporate executive level). An SoC hierarchy node would then be developed by engineers assigned from each organization; their collective progress would be tracked in a "matrix management" style.

Both organizational approaches offer advantages and disadvantages. It is perhaps easier to re-allocate resources between SoC nodes (or between SoC projects) from an organizational pool of engineers sharing common expertise. Conversely, this matrix approach requires more project management coordination and cross-organizational communication to track each SoC node to the overall project schedule.

It is more common to maintain only the functional validation and test engineering resources in a consolidated pool. The project schedule is typically defined by the duration of the full-chip SoC functional validation phase. Resources from the pool can be pulled (temporarily) for development of node-level simulation testbenches, while the balance of the validation team is reviewing the SoC specification and preparing the full-chip validation strategy. Once the RTL model for each node is integrated, the validation team focuses on full-chip functional simulation, with related formal property definition and model prover activities.

Regardless of the organizational structure, the allocation of resources to sub-chip development tasks needs to balance the scope and complexity of each hierarchical node to the expertise of the specific engineering teams.

## 3.9 The Impact of Changes to the SoC Model Hierarchy During Design

The goal of this chapter is to highlight the significance of SoC logical and physical hierarchies on the project planning and execution phases. The SoC methodology influences the decomposition of each model and the correspondence points between the two hierarchies. This section attempts to further that goal by briefly describing the impacts when the hierarchical model changes during the detailed design phase of the project schedule.

Functional simulation testbenches, assertions, and invariant properties are developed by the validation team to exercise against the logical model. These statements refer to signals, registers, and memory storage within the model. Each of these references consists of a net name prefixed by the concatenation of hierarchical instance identifiers. If the functionality associated with validation testpoints moves within the logical hierarchy, these testbench references are invalidated. Any resources devoted to maintaining the validation infrastructure are not really advancing the project; ideally, these hierarchical changes can be minimized.

Physical hierarchy changes relatively deep in the hierarchical tree typically have little impact on project schedules. The physical design and layout engineers may identify implementation opportunities to reuse physical layout components (e.g., sets of devices and local wiring, interconnect routes between circuits). By introducing new physical macrocells in the hierarchy, the bottom-up layout assembly productivity can be improved. Conversely, changes to the top-level physical hierarchy definition may have a major impact on the overall SoC schedule and PPA objectives due to global bus plan changes, floorplan aspect ratio and pin changes, and updates to power management states and controls. The floorplan is developed to align with the pin definitions of each (top-level) node; changes to these pins affect the validity of the floorplan. In addition, global bus planning occurs concurrently with the block-level floorplans. Hierarchical changes to bus sources and/or fan-outs have a pervasive impact to the floorplan and global track planning. The ripple effect of global route plan allocation changes is likely to affect the clock and power grid distribution as well.

The power domain hierarchy is the most far-reaching; changes to the definition of the power domains, the methods for temperature sensing, and/or the power state transition control signals impact not only the floorplan and global power grid but the package design as well.

In all these cases, embedding top-level physical hierarchy updates requires significant project design and logical-to-physical correspondence resources and may adversely affect assumptions on die area and cost.

There is an additional consideration when making changes to the top-level hierarchical model—specifically, the impact on the functional and physical pin properties that are input to SoC methodology flows. Timing analysis utilizes input pin arrival times and slews, with output pin required times and loads, for pin-to-timing path endpoint checking. Multiple flows use relationship properties among input pins, such as mutual exclusivity (e.g., one-hot, zero-hot, pairwise inversion). Frequent hierarchical pin definition changes require considerable design resources to analyze and maintain valid functional and electrical property information.

During the initial chip design definition phases, some degree of hierarchical decomposition changes is inevitable. The SoC methodology team can assist with minimizing the resources required to maintain validation testbench and physical floorplan data by providing model revision *difference checking* utilities to succinctly inform designers of critical model hierarchy changes in successive version releases.

## 3.10 Generating Hierarchical Electrical Abstracts Versus Top-Level Flat Analysis

The hierarchical design decomposition of an SoC enables block-level definition, functional validation, physical design, and electrical analysis to progress using smaller individual engineering teams. As detailed physical block models evolve and are available to release to full-chip integration, there is a methodology question to address: Which models should be used for full-chip electrical analysis flows?

As VLSI process scaling has progressed and the number of available interconnect layers has increased, the corresponding electrical model data volume (with layout extracted parasitics) has increased tremendously. At the same time, EDA vendors are addressing this issue by adapting their algorithms to execute in a multi-threaded manner. The methodology team needs to assess which analysis flows could (or should) utilize a flat full-chip model and which require the generation of a suitable electrical abstract for individual blocks to reduce the full-chip dataset size input to the flow. As an example, consider the I*R voltage drop analysis flow to measure the local supply and ground voltages at individual cells and ensure that they are within the assumed circuit characterization margins. The flow utilizes (saturated device) current measures during an output switching transient for each cell from its characterization file (see Section 14.1). The methodology team needs to address whether this flow could be successful using a flat full-chip model or whether a flow to generate an I*R current abstract for the block is required. The I*R flow needs extracted parasitic data for the power/ground grids, a much smaller dataset size than for signal interconnect parasitics; as a result, the flat extracted model for power and ground nets is more manageable. (Note that the model for internal power and ground grids usually consists of only RC segments; a more detailed RLC extracted model is required for the power and ground-to-I/O circuits, for a simultaneous switching output [SSO] transient noise analysis flow.) The methodology team also needs to evaluate whether the accuracy of an abstracted model is sufficient for full-chip evaluation. In the case of I*R analysis, a continuous global grid at top metal levels may be present over multiple blocks. As a result, there is no specific demarcation of the current distribution to a specific block, as the switching activity in neighboring blocks on the same grid will have an impact on the I*R drop results.

Whereas I*R analysis may be a candidate for flat full-chip models, other electrical analysis flows simply require too much data to consider exercising on a flat model; in these cases, generating and promoting a block-level abstract is a requirement. For example, static path timing analysis requires detailed extracted parasitics for signals to merge with the cell library delay models. Block-level timing closure uses timing constraint assumptions for pin arrival times, input slews, and output loads. One approach to full-chip timing analysis would be to treat each block as a black box, maintaining the timing constraint values used during block timing closure. A more accurate approach would be to treat each block as a black box but derive updated timing constraints, pin loading, and output drive strength measures from the detailed block-level timing analysis results; in this case, the abstract is denoted as an *extracted timing model* (*ETM*). The most accurate block abstract for the full-chip timing methodology would excise detailed cell and interconnect networks from within the block that provide full pin-to-flop electrical models; this abstract is known as the *interface-level model* (*ILM*). The examples in Figure 3.4 illustrate the ETM and ILM approaches to block-level timing abstract generation. The EDA static timing tool vendor typically provides block abstract generation features for either modeling method.

For the ILM abstract method, the resulting full-chip timing model is presented with detailed networks to/from each timing test point for delay calculation and propagation, forgoing the pin property assumptions of the black-box approach. The more detailed abstracts result in a larger, yet hopefully manageable, full-chip dataset.

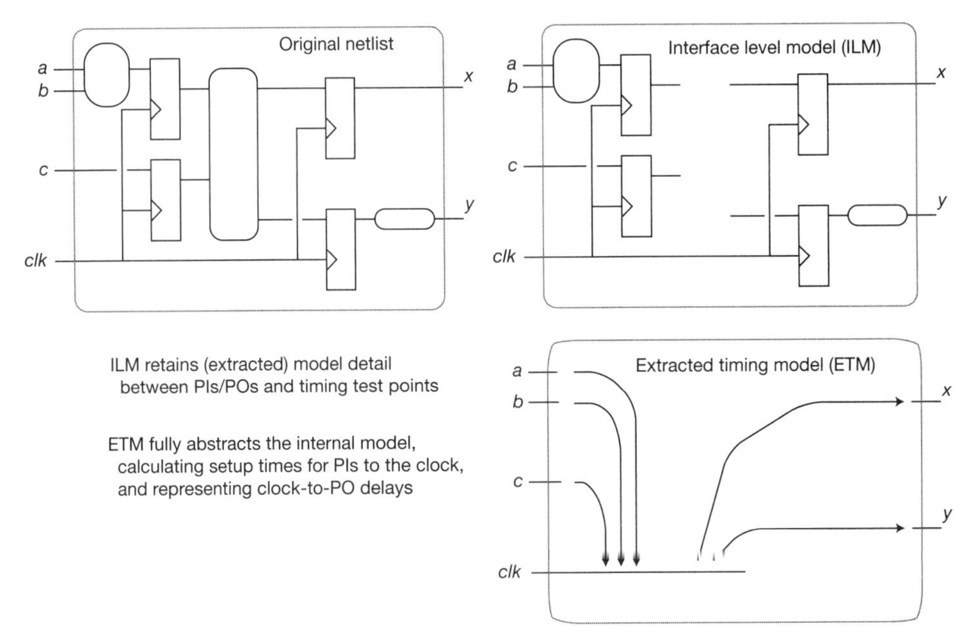


**Figure 3.4** Block-level timing abstraction methods. An extracted timing model (ETM) and interface-level model (ILM) abstract representation for a block are depicted.[4]

The methodology team needs to ensure the optimal utilization of flat or block-abstracted electrical models for each full-chip analysis flow. In each case, a trade-off decision is required on the compute resources, the flow runtime, and the abstract model accuracy.

## 3.11 Methodologies for Top-Level Logical and Physical Hierarchies

Section 1.1 defines the term *glue logic* to describe the SoC functionality not included with block-level IP. This could encompass control flow for miscellaneous chip states, logic for reset initialization, logic defining various clocking and test modes, power state management, and so on. A common glue logic function is a flop used as a repeater for global interconnect pipelining. In addition to the repeaters, there are signal buffering cells to meet required inter-block timing and signal slew constraints if an additional pipelining clock cycle is not required.

The methodology team has several options for representing this logic and signal buffering in the logical and physical hierarchies. One alternative is to locate the glue logic at the top level of the logic hierarchy, not assigned to any specific block. The physical floorplan would include channels between blocks for placement of these cells. The floorplan designer would be responsible for providing a suitable power and ground distribution to these cells. (Fortunately, glue logic and signal buffers are typically not power gated, simplifying the power and ground distribution to these cells.) The floorplan must also distribute clock(s) to the pipelining repeaters and any finite state machines. The corresponding top level of the physical hierarchy would include the floorplan block abstracts, the cells in the channels, the top power, ground, and clock grids, and global routes.

Another alternative is to physically place (some or all) the glue logic within the area allocated to floorplan blocks; as illustrated in Figure I.5, the optimum path for a pipeline repeater may be within a block. This approach requires several methodology features:

- The global floorplan needs to be able to disrupt a block design by inserting (banks of) repeaters.
- The local clock distribution within the block needs to accommodate a separate clock to the repeaters (which likely are not on the same clock domain as the block).
- Similarly, the power distribution to the global cells within the block may be on a distinct domain; the local power and ground grids also need to accommodate unique global grids.
- Logical-to-physical hierarchy verification needs to isolate the cells that do not belong to the block.
- Signal routes to the embedded physical repeater cells need to be part of the extracted model for the block to include their coupling capacitance for block signal timing and noise analysis.

The complexity of embedding top-level glue logic within physical block-level floorplan areas is certainly higher than a methodology utilizing physical channels between blocks. Yet the criticality of addressing the performance impacts of additional repeater pipelining cycles may necessitate pursuing the additional flow features required.

The preceding discussion on placing global physical cells within a block floorplan area mentions the significance of the coupling capacitance parasitics for routes to these cells on local signal electrical analysis. In general, the block-level parasitic extraction methodology must include the effects of all top-level global routes over the block. The methodology team needs to define a physical top-level overlay cell to merge with the block for extraction. The overlay may initially be a (conservative) estimate of routes occupying tracks over the block. After the initial SoC integration phase, subsequent iterations of the block-level extraction flow may use more precise physical *cutout overlay* cells that include power/ground (non-switching) grid wires and occupied signal route tracks, as depicted in Figure 3.5.

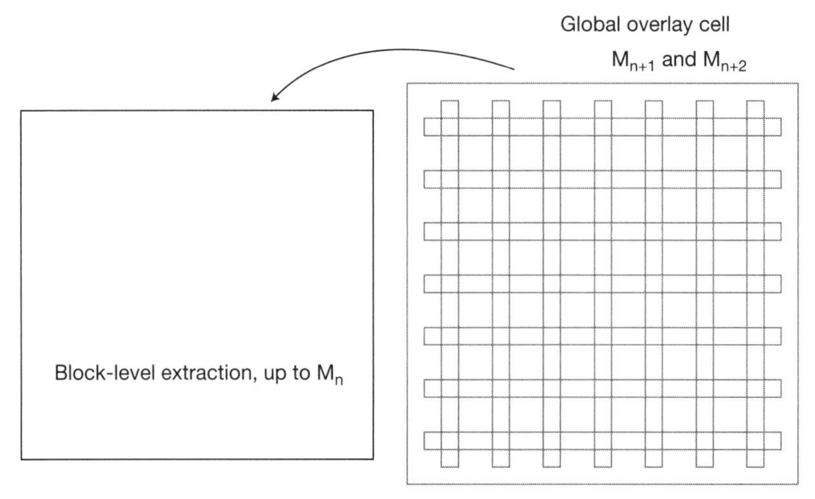


Block-level parasitic capacitance extraction includes a global overlay cell (initially a conservative design, perhaps re-extracted with the actual "cutout" overlay).

**Figure 3.5** Block-level signal extraction involves addition of a global overlay cell, which may evolve from an initial (conservative) design to a “cutout” cell from the global physical integration.

## 3.12 Summary

This chapter highlights some of the key considerations for establishing the logical and physical hierarchical decomposition of an SoC design project. The correspondence points in these two hierarchies represent models where several block-level flows are exercised. The stability of the pin definitions at the correspondence points has a strong impact on the design resources to maintain the additional project data used by the block-level flows. The additional glue logic residing at the top level of the logic hierarchy may be reflected in the physical floorplan in different ways (e.g., located in channels, embedded within blocks).

The specific methodology approach pursued may require additional features of the floorplanning and block-level analysis flows to implement and model this logic.

## References

[1] Damiano, R., et al., "Method for Mapping in Logic Synthesis by Logic Classification," U.S. Patent number 5,537,330.

[2] For a review of the original work by E.F. Rent in 1971 and an updated interpretation applied to more recent electronic component design, see Lanzerotti, M., et al., "Microminiature Packaging and Integrated Circuitry: The Work of E.F. Rent, with an Application to On-chip Interconnection Requirements," *IBM Journal of Research and Development*, Volume 49, Issue 4.5, July 2005, pp. 777–803.

[3] IEEE Std 1149.1: "IEEE Standard for Test Access Port and Boundary-Scan Architecture,"https://standards.ieee.org/findstds/standard/1149.1-2013.html.

[4] Daga, A., et al., "Automated Timing Model Generation," *Proceedings of the 39th Design Automation Conference*, 2002, pp. 146–151.

## Further Research

### Rent's Rule

Describe the relationship between logic gate partition size and pin count, denoted as Rent's rule.

Although Rent's rule was originally developed in the 1970s, describe how this relationship can be extended to block partitioning and floorplanning for a current SoC design targeting an advanced process node.

TOPIC II

# Modeling

"All models are wrong, but some are useful."

—*George Box*

The design methodology for a complex SoC is based fundamentally on a diverse set of models used to represent the logic functionality and the electrical response of the fabricated silicon. These models are provided to EDA software tools that simulate and analyze the design behavior. The functional models originate from the SoC design team and the IP library vendor. The circuit electrical models are derived from the basic device and interconnect models provided by the foundry in the PDK release. The electrical models are commonly abstracted to the level of the library cell as the output of the library *characterization* flow. This part of the text describes key features of the models used in the SoC methodology.

CHAPTER 4

# Cell and IP Modeling

## 4.1 Functional Modeling for Cells and IP

Functional models are written in a specific hardware description language (HDL), for which EDA vendors have developed corresponding simulation tools. The semantics of an HDL model consist of a set of *concurrent sequential processes* (CSPs). In short, a set of model processes (or procedures) are compiled. Each process has an input sensitivity list of signals; a transition on a signal in that list results in the execution of the statements in the process. Evaluation of the statements in the process/procedure proceeds sequentially to completion (or until a *wait statement* clause is encountered). All model processes are pending concurrently, and once active, they may execute in parallel (i.e., all starting at the same reference point in simulation time).

As described in the following sections, several levels of modeling abstraction are used to represent the logic functionality of an IP block and individual library cells using an HDL.

### 4.1.1 Behavioral Modeling

A behavioral model represents the (complex) functionality of a large IP block using programming language semantics typically associated with sequential statement execution. Procedures, processes, and functions are written using a combination of logical and arithmetic operators and statements that execute sequentially, following the CSP paradigm. The HDL sequential statements available for modeling include the control flow typically available with conventional programming languages, including if-then-else, (conditional and fixed iteration) loops, and case statements. As mentioned previously, the process is invoked when a change in value is assigned by the simulator to a signal in the input sensitivity list of the routine, as illustrated in Figure 4.1.

Examples of the "Sensitivity List" in HDL Semantics (Verilog and VHDL)

```
always @(posedge clk)         p1: process ( rising_edge(clk) )
  begin                         begin
  ....                            ....
  end                           end process p1 ;

always @(select)              p2: process ( select )
  begin                         begin
  ....                            ....
  end                           end process p2;

always @*                     procedure p3( a: in Boolean ; b: in bit ;  c: out std_logic )  is
  begin                         begin
  ....                            ....
  end                           end p3 ;
```

A sensitivity list is used in several statement types — e.g., process, procedure, assertion.

**Figure 4.1** An HDL model consists of a set of concurrent, sequential processes. An example of a process and its sensitivity list are illustrated. Both Verilog and VHDL examples are depicted.

Although similar statement types are used, several aspects of the behavioral HDL model distinguish it from a conventional programming language, as discussed in the following sections.

### *Representation of Simulation Time*

The statements used in behavioral HDL models reflect the times at which the results of statement evaluation are reflected by the simulator. As illustrated in Figure 4.2, a statement may have an additional clause that provides an explicit delay, which the simulator posts to a future *event queue*.

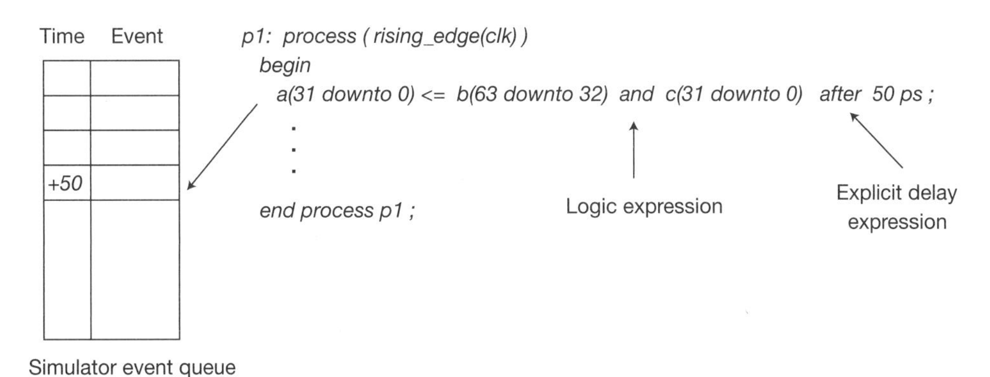


**Figure 4.2** Illustration of an HDL statement with an explicit time delay until the assignment is to be applied to the simulation model. A pending update is posted on the event queue.

The HDL language semantics also include an *immediate assignment* statement. The result of this statement execution is an immediate assignment to the left-hand side variable, without advancing simulation time. Subsequent statements in a sequential procedure use the updated values, as illustrated in Figure 4.3.

The representation of simulation time also includes the concept of an infinitesimal *delta delay*. An assignment of a new signal value from executing a deferred statement without a delay clause is not reflected immediately. Rather, the deferred assignment value is placed on the simulation event queue, to be evaluated at a delta time in the future, after all sequential statements in all active processes are complete. The explicit simulation time is not advanced until the delta event queue is empty and the next pending event explicitly advances the time base (using an "after n psec" clause). Figure 4.4 illustrates the evaluation flow.

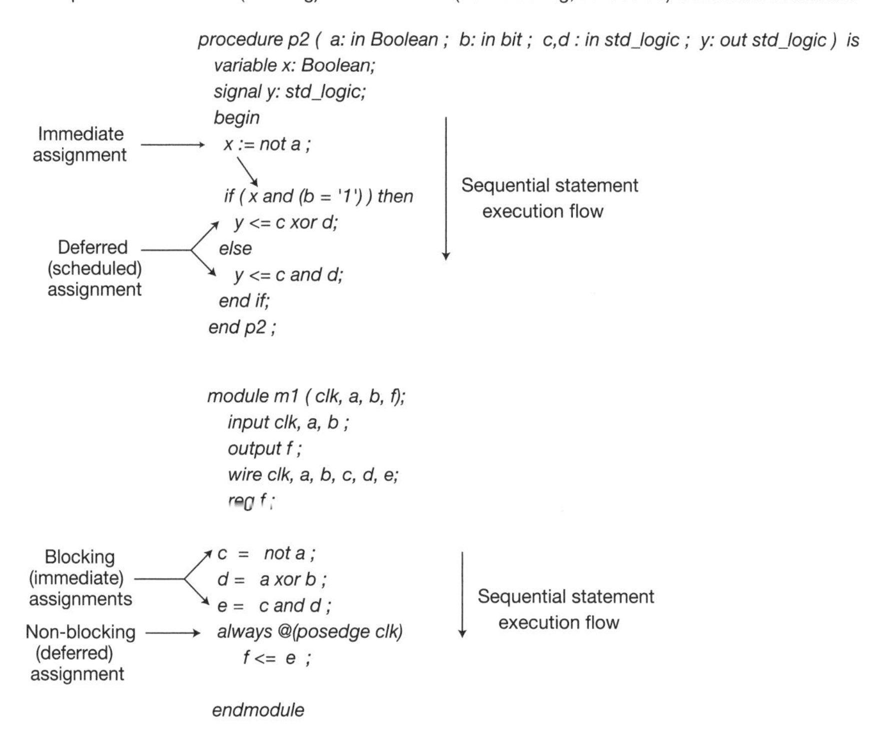


**Figure 4.3** An immediate assignment statement executed in a CSP updates the left-hand variable value directly after evaluation of the right-hand side expression rather than posting to the event queue.

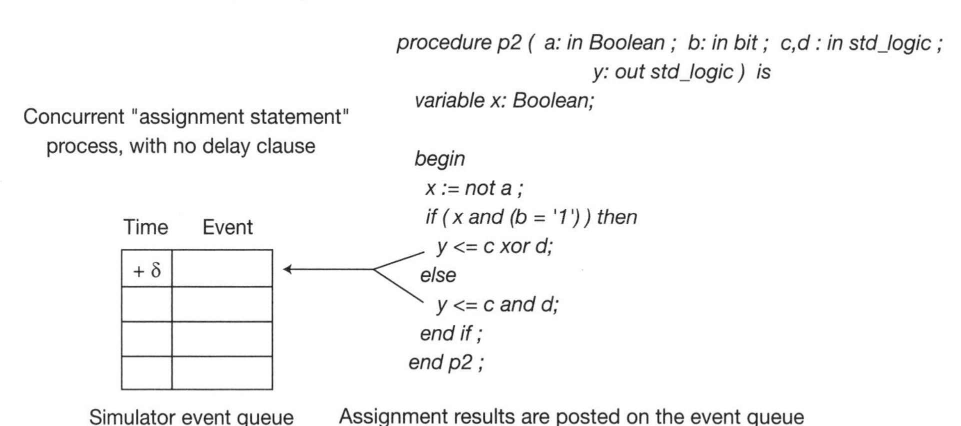


**Figure 4.4** A non-immediate assignment without an explicit delay clause is posted to the event queue in an infinitesimal "delta delay" after the current simulation time. After all active processes complete, the event queue is evaluated, and new processes (and delta time assignments) are active. Simulation time does not explicitly advance until there are no pending events on the delta time queue.

### *Variables and Signals (VHDL), Wires, and Regs (Verilog)*

The declaration of HDL model identifiers differentiates between the left-hand side targets of immediate (also known as *blocking*) and deferred (*non-blocking*) assignments. An immediate assignment is made to identifiers declared as variables (VHDL) or wires (Verilog). A deferred assignment is made to signals (VHDL) or regs (Verilog).

### *Resolution Function*

In the specific case in which multiple assignments to the same signal from different statements are pulled from the event queue at the same (delta) time, a *resolution function* is invoked by the simulator to calculate the value to be assigned to the signal. The specific resolution function is associated with the type declaration of the signal (discussed shortly).

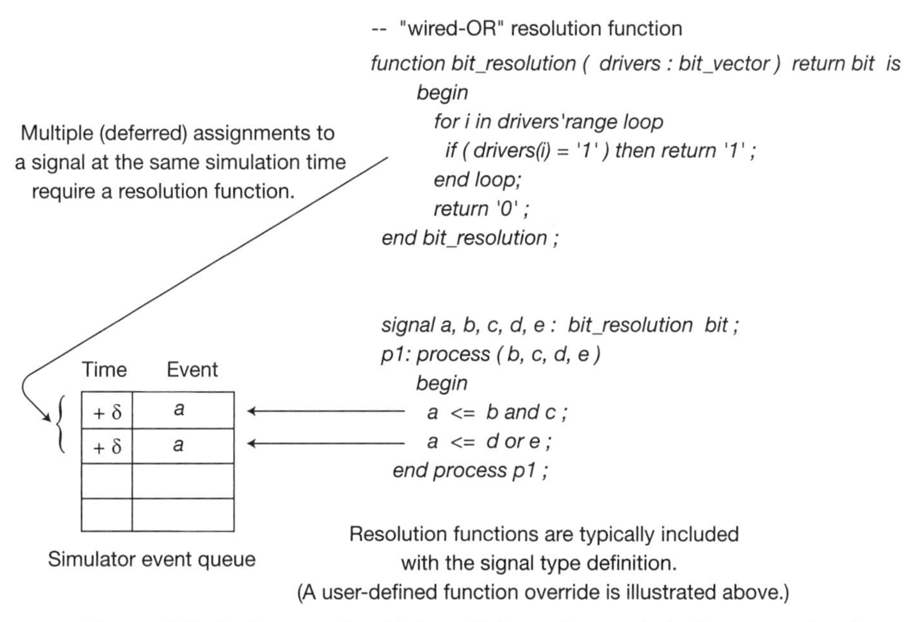


**Figure 4.5** In the case in which multiple assignments to the same signal are present on the event queue at the same time (originating from one or more processes), the simulator evaluates the resolution function corresponding to the signal type to determine the assigned value. A "wired-OR" resolution function description in VHDL is shown.

### Data Structures

A conventional programming language typically provides data structures that are based on collections of variables, such as vectors and multidimensional arrays, records, and linked lists with pointers (where variable values are references as address pointers to other variables). The HDL design model is typically presented to a behavioral synthesis flow; data structures that are "non-synthesizable" are not typically included in the model. Note that HDL testbench models compiled with the design model for functional simulation validation are typically able to utilize the breadth of HDL statements, including non-synthesizable constructs. A hardware simulation accelerator used in the validation methodology may restrict the testbench modeling style if all or part of the testbench stimulus and monitoring features are to be incorporated on the accelerator.

### Types

Like programming languages, HDL behavioral models associate type declarations with each variable and signal. The compilation and elaboration of a simulation model check the validity of the right-hand expression as an appropriate result to assign to the type of the left-hand statement identifier. Typical HDL language types for logic signals define the enumeration of discrete values that are valid assignments for Boolean operators—for example, '1', '0', 'U' (uninitialized), 'Z' (high-impedance), 'D' (don't-care), and 'X' (undefined, typically due to an error condition detected within the model). Other levels of modeling abstraction may utilize additional type values, as discussed shortly.

At the start of simulation time t = 0_minus, the type's default (or *initialization*) value is assigned by the simulator to signals and propagated throughout the model to reach an appropriate initial state. Subsequently, the simulator advances to t = 0, where the testcase stimulus is applied. An uninitialized signal present in the simulation model well after t = 0 identifies logic paths that may not have adequately propagated the desired initial (or reset) condition. To expedite simulation runtime, a specific model state can also be provided to the simulator to avoid having each testcase exercise an initialization preamble.

The type definition also includes the resolution function to be evaluated when multiple assignments to a signal are active at the same simulation time.

Each type definition can typically be expanded to allow vectors and arrays (*vectors of vectors*) to be easily defined. The indices to these structures are typically integers, and the definition would support ascending or descending index ranges.

A behavioral model commonly utilizes additional types beyond Boolean logic definitions. For improved simulation execution throughput and coding productivity, the behavioral model is likely to also include integer, floating point, and enumerated types, the latter being especially advantageous for describing machine state. The integer and floating point types have corresponding arithmetic and relational operators in HDL semantics.

For the synthesis of the behavioral model to a detailed hardware description, the integer type declaration for a signal needs to be bounded when the identifier is declared. The valid bounds ultimately define the number of bits required to represent the identifier in the synthesized implementation. These bounds are also used in the simulator's real-time range checking. The IEEE-754 standard provides the definition for 32-bit and 64-bit floating point value representation (and arithmetic evaluation) for behavioral synthesis.

### *Inferred State*

The sequential statement execution within an HDL process may result in a signal that is not necessarily assigned in each sensitized execution of the process. The example in Figure 4.6 depicts an if-then-else conditional statement in which the set of assigned signals differs in the potential statement execution paths. If a signal present on the left-hand side of an assignment statement is not explicitly updated during process execution, the current value of the signal is retained in simulation.

Example of "Inferred" State in a CSP Model

```
p1: process ( a, b, c, d )
 begin
   ....
   if (( a = '1') or (b = '0')) then
 // no assignment to signal f
1     e <= not c ;
   else
 // no assignment to signal e
2     f <= not d ;
   end if ;
   ....
 end process p1 ;
```

Distinct statement execution paths through the CSP do not share the same set of assigned signals, implying a retained value between successive process evaluations.

EDA tools will build a statement traversal graph and signal assignment list for each CSP to check for inferred state.

**Figure 4.6** A signal that is not assigned during the execution of a CSP implies that the signal value is retained (i.e., an inferred state).

As a result, when synthesizing the behavioral model to a logic network, the value of the signal needs to be retained between evaluations of the process: An *inferred state* is implied. The resulting synthesized netlist includes a register, clocked by any of the input sensitivity list of the process, retaining its current value when the synthesized logic bypasses a new assignment to the signal.

Note that a loop construct in the behavioral HDL code denotes an explicit state machine.

### *Modules/Entities and Configurations*

The HDL behavioral model also differs from a programming language with the additional support for a logical hierarchy. The model functionality is allocated to hierarchical elements denoted as *modules* (Verilog) or *entities* (VHDL), as illustrated in Figure 4.7.

```
entity c1 is
  port ( a : in std_logic ;
         b : in std_logic ;
         c : in std_logic_vector(31 downto 0) ;
         d : out std_logic ) ;
  end entity c1 ;

  architecture arch of c1 is
    begin
      -- CSP statements
     ....
    end arch ;

entity parent is
  ....
 end entity parent ;

architecture arch2 of parent is
  signal d, e, f, x, y, z :  std_logic ;
  signal  i, j:  std_logic_vector(31 downto 0);
   ....
-- component declaration for c1 included
  component c1 ...  ;
begin
 -- port connection by position
inst1:  c1 port map (d, e, i, f);
-- example of explicit port connection
inst2:  c1 port map(a => x, b => y, c => j, d => z ) ;
  ....
end architecture arch2 ;
```

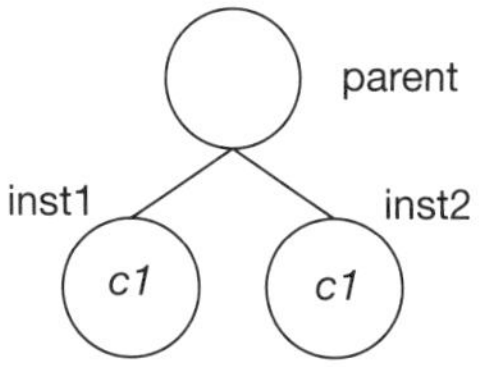

```
module c1 (a, b, c, d );
  input a, b, c[31:0] ;
  output d ;

// CSP statements
   ....
 endmodule ;

module parent
 ...
wire d, e, f ;
wire x, y, z ;
wire i[31:0], j[31:0] ;
 ...
// port connection by position
inst1  c1( d, e, i, f );
// explicit port connection
inst2  c1( .a(x), .b(y), .c(j), .d(z));
 ...
endmodule ;
```

parent
inst1 c1
inst2 c1

**Figure 4.7** A hierarchical model is constructed using the definition of an entity (VHDL) or a module (Verilog) and then adding a component instance in the parent model.

Using VHDL semantics as an example, the connectivity to an instance in the logical hierarchy is through the ports of the entity. Signals defined in the parent connect to the ports of the child entity instances within the body of the parent model. During model compilation and elaboration, the signal-to-port connectivity is verified to ensure consistency of type and range declarations.

To provide increased flexibility in modeling styles, compilation may utilize a *configuration specification* (not to be confused with the DDM configspec described in Section 2.5, although they are similar in concept). This specification identifies a specific *body* that is to be associated with the entity/module definition when compiling the model. In this manner, different body code can easily be inserted, maintaining the same (invariant) connectivity throughout the logical hierarchy. For example, an RTL- or netlist-level model could replace a behavioral model at a subsequent project phase of functional validation by modifying the configuration specification. This model build approach simplifies the task of aligning the simulator used for each validation phase with the optimal model coding style.

### *Scope*

The CSP execution paradigm includes the definition of the scope of an identifier, similar to a general programming language. A declaration of a local identifier implies that the value is not visible outside the process; it cannot be referenced by other (concurrent) processes. This enables independent HDL model development between blocks, as well as the integration of external IP models, without concern for identifier collisions. The EDA vendor simulator tool is likely to offer features that bypass the scoping rules of the HDL standard; for example, internal values in a process are visible to query during interactive simulation, *breakpoints* could be set to pause execution upon an internal value condition, and so on. In these cases, the simulator uses the hierarchically qualified instance prefix to access an internal identifier.

The behavioral abstraction typically simulates efficiently, with model compilation leveraging many of the optimizations developed for software programming languages. This style is commonly used by hard IP providers. Complex memory array and processor core IP are more readily modeled using the full HDL language semantics. Behavioral modeling is emerging as an attractive abstraction level for design IP, as well, especially for signal processing applications that perform arithmetic manipulations on streaming data packets. The numeric data types, arithmetic operators, and loop statements enable concise model coding and, thus, reduce the likelihood of design bugs. EDA vendors have also contributed to the growing adoption of behavioral modeling, with support for an increasing set of high-level HDL semantics in their behavioral synthesis tools.

### 4.1.2 RTL Modeling

The most prevalent HDL modeling style in SoC design uses an abstraction denoted as register-transfer level (RTL). All storage elements are explicitly coded; thus, all registers are readily identifiable in the HDL source code, as illustrated in Figure 4.8.

```
module example ( clk, reset, d, q );
  input clk, reset. d;
  output reg q ;

always @(posedge clk or posedge reset) begin
  if (reset)
   begin
     q <= 1'b0 ;
   end  else
    begin
     q <= d ;
    end
endmodule
```

```
entity example is
  port ( clk, reset, d : in std_logic ; q : out std_logic ) ;
end entity example ;

architecture arch of example is
begin
  p1: process ( rising_edge(clk) or rising_edge(reset) )
   if (reset = '1') then
     q <= '1' ; else
     q <= d ;
   end if;
  end process p1 ;
end architecture arch ;
```

**Figure 4.8** The register-transfer level (RTL) coding style incorporates explicit register assignment process statements, where the sensitivity list consists of a clock signal.

An RTL model of a design block includes a smaller set of HDL statement semantics than a behavioral model. No inferred state through the execution of conditional statements in a CSP is allowed (i.e., all branches through the clauses in if-then-else and case statements require assignments to a complete and consistent set of signals). A common coding style is to express combinational logic as stand-alone CSP statements (see Figure 4.9); the signals on the right-hand side become the sensitivity list for statement execution.

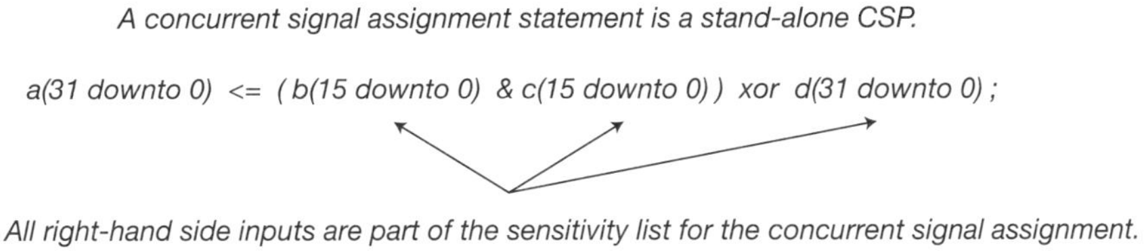

*A concurrent signal assignment statement is a stand-alone CSP.*

*a(31 downto 0) <= ( b(15 downto 0) & c(15 downto 0) ) xor d(31 downto 0) ;*

*All right-hand side inputs are part of the sensitivity list for the concurrent signal assignment.*

**Figure 4.9** A concurrent, deferred assignment statement is equivalent to a CSP. The right-hand-side expression inputs are the elements of the sensitivity list. VHDL semantics are shown in the figure. Verilog uses a different method, with combinational logic represented by immediate assignments and coded outside the always statement.

The RTL coding style typically uses a limited set of signal types. The EDA industry has established de facto standards for type declarations, supported signal values, and operator evaluation for these types.

Simulation of an RTL model proceeds by calculating the combinational logic values to "transfer" to register signals each successive clock cycle. The RTL statements rarely include detailed delay information associated with any expression; the simulation timebase is advanced by the periodic clock signal input stimulus from the testcase. The RTL style enables simulation optimization during compilation. A *cycle simulation* tool assumes that only register signal values need to be recorded each cycle, in an output *trace file* for subsequent debug. Combinational signal values are not recorded (under the assumption that their values can be recalculated during debug from register values). The compilation of the cycle simulation model *levelizes* the RTL assignment statements between registers—no combinational loops are allowed—and optimizes the resulting compiled code. (Note that the cycle-optimized simulation execution requires special consideration for design blocks on different clock domains to achieve fastest simulator performance.)

Logic synthesis support for RTL model abstraction has been an established toolset from EDA vendors for decades. The synthesis flow identifies the registers in the design, constructs the combinational networks, and exercises optimization algorithms prior to technology mapping (e.g., constant value propagation, redundant logic removal, common sub-expression factoring). And, as mentioned previously, the RTL synthesis flow initially confirms that combinational network loops and statement clauses with inferred state are not present in the model.

As with behavioral coding, RTL models include a structural hierarchy of modules/entities. An SoC typically includes a mix of behavioral models for hard IP instances and RTL for design blocks.

Power domain information for the SoC design is typically not represented in either the behavioral or RTL model descriptions. The specific implementation of domain power gating (and internal state retention) for a sleep state operating mode is captured in the separate power format description rather than the RTL model. During RTL synthesis, consistency checks are evaluated to ensure that the power format description aligns with the RTL model.

### 4.1.3 Netlist

The most detailed HDL model representation is a netlist, consisting solely of structural instances of elemental cells from a library. No assignment statements are present. The netlist model could be generated by the logic synthesis flow or translated from a graphical schematic consisting of library cells, as depicted in Figure 4.10. The detailed text data in a netlist are rarely keyed in directly.

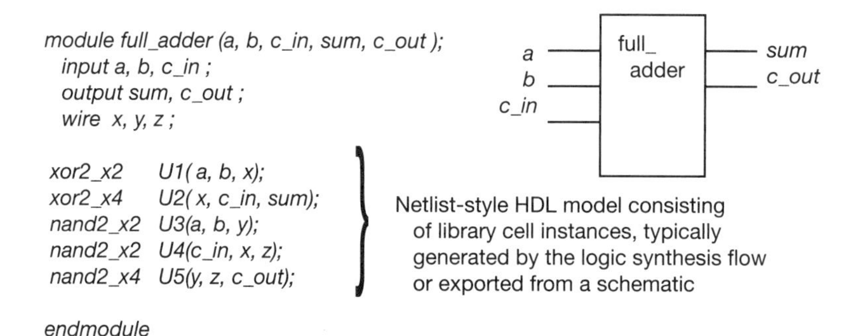


**Figure 4.10** A netlist-style description is generated by the logic synthesis flow consisting of library cell instances. A netlist could also be exported from a schematic entry tool that uses graphic symbols for library cells. (A custom macro circuit-level netlist would be exported from device-level schematics.)

The use of schematics to represent a netlist may reflect the desire to build a block (or sub-block) from the bottom up, where the critical performance of a digital (or mixed-signal) function necessitates technology-specific cell selection and early timing simulation rather than using logic synthesis driven by timing constraints. Due to the rather poor simulation performance of a detailed netlist model, an RTL model for the schematic function is typically developed as well. The schematic-based netlist is subsequently presented to the logic equivalency checking (LEC) flow against the RTL model.

Note that the technology-specific netlist contains cell instances that do not add to the logic functionality of the model (e.g., parallel/serial repowering trees for high-fan-out signals). The netlist also includes the cells inserted to match the power format specifications.

### 4.1.4 Additional Functional Model Considerations

Regardless of the language description style used for each node of the logic model hierarchy, there are additional considerations for successful simulation:

- **Initialization**—An HDL signal type has a default initial value. In addition, HDL statement semantics allow an overriding initial value to be coded in the model, as illustrated in Figure 4.11.

```
signal a:  std_logic_vector(31 downto 0) := X"CAFEFEED" ;
```

Initial value with declaration, overriding default for signal type

**Figure 4.11** Initialization values in the HDL model override the default init value for the signal type.

  To simplify coding and improve testbench efficiency, EDA simulators support the use of an external initialization file to establish the time t = 0_minus values on (a subset of) signals in the design. Note that these values are logically propagated at t = 0 in delta time. The simulation methodology flow needs to support use of an external init file, the capture of init state from a separate testcase that performs model reset(s), and a check to confirm consistency of the init file with the reset testcase results.

- **Test model views**—The netlist model is used for test pattern generation. For logic primitive cell instances, the HDL representation of the cell is used directly by the EDA test pattern generation and fault simulation tools. However, for more complex cells, the functional description could be expanded to better reflect the circuit-level faults that contribute to the test pattern coverage measures. Figure 4.12 shows an example of a netlist instance in which the cell model could include two HDL *views*: one for simulation/equivalency flows and one for test analysis. The methodology team works with the cell library modeling and test teams to define how these alternative views are presented to the model compilation step of different flows.

```
// functional model
module ao22 (a, b, c, d, z );
  input a, b, c, d ;
  output z  ;

 z <= (a and b) or (c and d) ;

endmodule

// test model
module ao22 (a, b, c, d, z );
  input a, b, c, d ;
  output z  ;
  wire x, y ;

 nand2_x2   U1(a, b, x);
 nand2_x2   U2(c, d, y);
 nand2_x4   U3(x, y, z);

endmodule
```

```
entity  c1  is
  port  (  a, b, c, d : in std_logic ;
              z : out std_logic ) ;
  end entity c1 ;

  architecture func of c1 is
    begin
      z <= (a and b) or (c and d) ;
    end architecture func ;

  architecture test of c1 is
     signal x, y  :  std_logic ;
   -- component declarations included
      ....
   begin

     U1:  nand2_x2 port map (a, b, x);
     U2:  nand2_x2 port map (c, d, y);
     U3:  nand2_x4 port map (x, y z);

  end architecture test ;
```

Multiple model views of cells are developed with varying detail:
- optimized for functional simulation platform throughput, and
- with circuit-level detail for fault modeling in test pattern generation.

**Figure 4.12** A cell (or IP block) could utilize multiple netlist views for different flows. A simulation view and a test view are illustrated.

The netlist also includes the full structural detail of the DFT architecture on the SoC (e.g., serial scan chain connectivity, test clocking).

- **X- and don't-care modeling**—The behavioral and RTL models are likely to utilize don't-care designations when writing vector-based comparisons in conditional expressions, as illustrated in Figure 4.13. The use of the don't-care designation makes the model coding easier and more readable; simulation compilation and logic synthesis disregard these bit comparisons.

```
module case_stmt ( x, y );
  output  y ;
  input [3:0] x ;

  wire z ;

 always @(x) begin
   case ( x )
   4'b0100   : z = 1'b1 ;
   4'b1???   : z = 1'b1 ;
   4'b00??   : z = 1'b1 ;
   default   : z = 1'b0 ;
 ....
 endmodule
```

```
architecture func of c1 is
  constant compare : std_logic_vector(31 downto 0) := X"FEED--FE" ;
  begin
   p1: process ( rising_edge(clk) )

   if ( b(31 downto 0) = compare ) then
     ...
   else
     ...
   end if ;
  end process p1;
 end architecture func ;
```

Don't-care values are used in comparison expressions to enhance model readability — these are optimized out during model compilation for simulation and synthesis.

**Figure 4.13** Don't-care values are commonly used in HDL comparison expressions to improve code readability.

An additional model value designation—commonly, an 'X'—is used to indicate that an erroneous condition has occurred, as part of a behavioral, RTL, or netlist model. For example, if an attempt is made to concurrently write to the same address from multiple write ports on a register file, an invalid value should be recorded (see Figure 4.14).

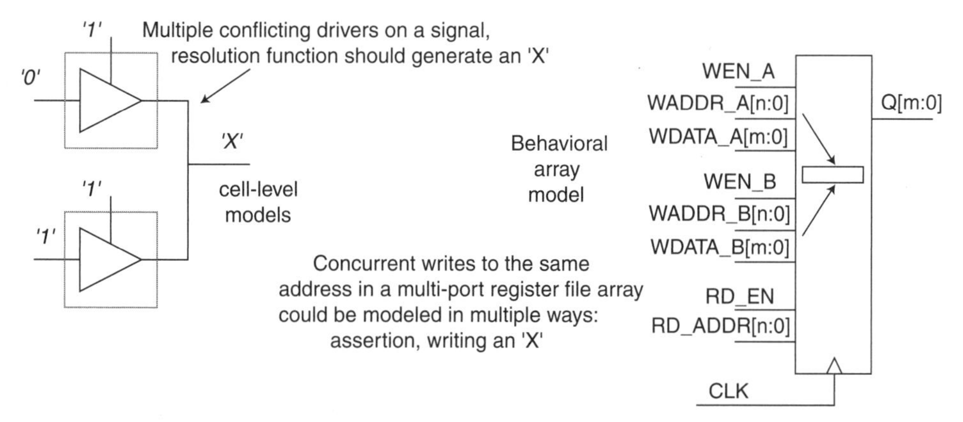


**Figure 4.14** An unknown or undefined 'X' value may be assigned to a signal to represent an anomalous condition during simulation.

The propagation of an 'X' value on a signal in simulation usually expands throughout the network and can readily be detected. However, the evaluation of HDL statements with an input error value may provide unexpected results, as the behavioral or RTL model may not be coded to respond to unexpected conditions, as shown in Figure 4.15.

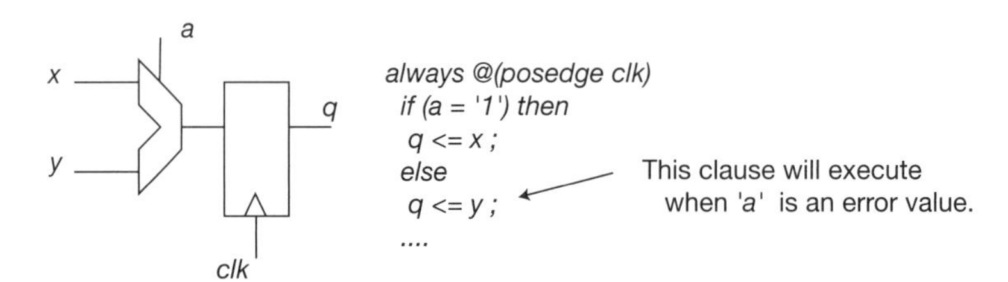


**Figure 4.15** HDL models may not expect an 'X' value to be present at statement inputs and thus might not propagate the (internal) error condition. A methodology review of 'X' generation and propagation coding styles is needed to ensure consistency throughout the model.

If the 'X' signal value is to be regarded as an unknown value, rather than as indicating an error condition, a different simulation algorithm may be invoked. To enable testcase evaluation to continue, EDA simulation tools have implemented algorithms to set the 'X' signal value to both '1' and '0', evaluate the model twice, and merge the results. The methodology and functional validation teams review the modeling conditions that generate an 'X' value and determine what simulation tools settings are appropriate.[1,2]

Rather than rely on X signal value propagation, a more precise approach to handling model error conditions during simulation would be to add functional *assertion statements* to the model. In general, an assertion statement defines an invariant "true" expression and can be regarded as another concurrent process during simulation execution. (Sequential assertions embedded within a CSP are also available and would be evaluated only during sequential statement execution in the process.) Figure 4.16 shows an example of an assertion statement that includes an additional *severity* parameter.

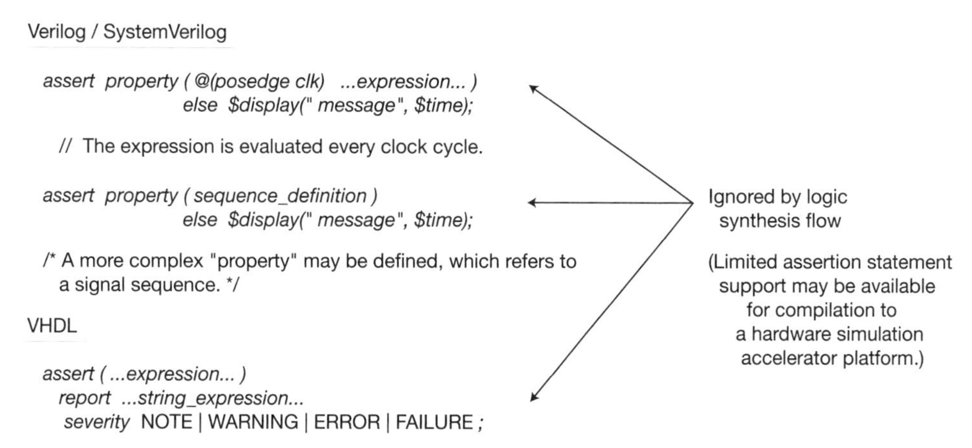


**Figure 4.16** An example of an assertion statement to continuously monitor the simulation model for a specific condition. A severity clause is included, and it can be used to control how simulation proceeds if the assertion condition becomes false.

The EDA simulation tool includes settings that define what execution steps to take when as assertion statement of a specific severity is *fired* (i.e., when the "always true" assertion evaluates to false). An example of this simulation setting would be "stop_on_ERROR, continue_on_WARNING." Rather than rely on the evaluation and propagation of 'X' signal values during the testcase, assertion-based validation enables more precise detection of anomalous simulation behavior. Both the simulation control setting and assertion output messages enable improved debugging. Further, EDA vendors now provide tools to analyze the model functionality to *formally prove* whether an assertion is always valid, without depending on functional simulation; if the assertion could be invalidated, the tool provides simulation counter-examples that fire the expression.

An assertion is a non-functional statement in an RTL or behavioral model. Assertions are skipped by logic synthesis tools when generating netlists. There is one exception: The synthesis of a model for a hardware simulation accelerator may be able to compile simple assertions into equivalent accelerator primitives with an output error signal, thus including the intent of the assertion in the accelerated simulation execution.

## 4.2 Physical Models for Library Cells

The physical model for each cell is based on the abstract view. The abstract defines the cell area and pin locations. Typically, the cell abstract includes power/ground pin definitions, as well, for coverage by the power and ground grids in the block physical design once the cells are placed. (It is uncommon for cells to include grid segments within the cell layout.) In addition to the abstract view, additional physical cell properties are required for subsequent flows:

- **Equivalent pin groups**—Routing algorithms include features to deviate from the as-provided netlist, swapping signal connections among the input pins in an equivalent group to alleviate congestion.
- **Cell orientation options**—Current fabrication processes require that all devices adhere to a single (vertical) orientation. As a result, the valid cell placement orientations are limited to mirror and flip operations. In addition, I/O cells have a strict distinction between left/right and top/bottom chip edge legal placements. The introduction of multipatterning decomposition among metal interconnect layers implies new placement orientation restrictions, such that the cell pin is consistent with the corresponding interconnect wiring track color, as illustrated in Figure 4.17.

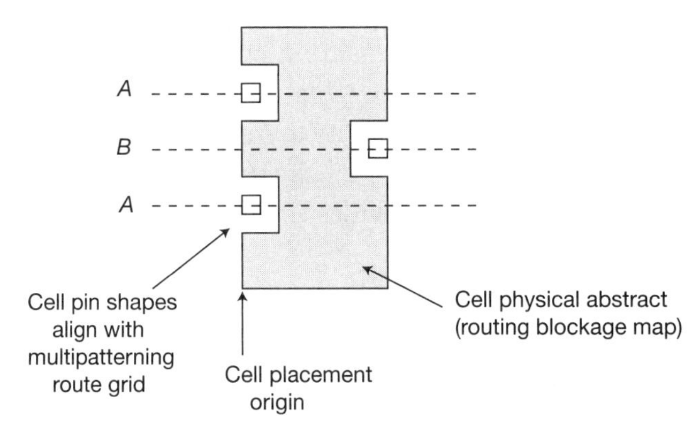


**Figure 4.17** The physical locations of cell pins align with the multipatterning track color assignment for advanced process nodes.

## 4.3 Library Cell Models for Analysis Flows

### 4.3.1 Cell Models for Synthesis and Testability Analysis

The netlist model of an IP block on an SoC includes instances of library cells. The functional HDL model for each library cell may provide different views specifically developed for different flows. The library release includes information with each cell HDL model that enables the logic synthesis flow to select the correct functionality during technology mapping of the post-optimization RTL model. Specifically, the models for sequential cells (e.g., flops, latches) require a specific representation to be selected by synthesis, as these cells may have multiple clock/data ports, various (synchronous or asynchronous) set/reset input conditions, etc.

The testability fault model for the cell needs to be consistent with both the EDA tool capabilities and the prevalent manufacturing defect mechanism information from the foundry. Figure 4.12 provides an example in which the HDL model and Boolean gate-level model for a complex cell necessitate release of different views. Traditionally, test tools have utilized *stuck-at* pin faults on the gate-level cell model (i.e., test patterns are derived to generate and propagate values to demonstrate gate pins were not s-a-1 or s-a-0). This approach sufficed to cover the primary manufacturing defects, commonly localized to transistor operation. Note that CMOS logic circuits use complementary nFET and pFET transistors for each logic gate pin; the pin stuck-at fault model proved to be sufficient for early CMOS process technologies, merging the defect mechanisms of the two device types. More recently, resistive defects in the fabrication of contacts, vias, and local metal interconnects have led to the use of cell test models that may include greater detail. Test pattern generation tools may also accept a table of user-defined test pattern sequences for the cell that augment pin faults to exercise specific connections in the cell physical layout.

### 4.3.2 Cell Delay Models for Static Timing Analysis

Initially, cell delay models consisted of tables corresponding to input pin-to-output pin *arcs*. Each delay value in the table for each arc was a function of input pin slew and output pin (effective) capacitance load. Static timing tools would interpolate between these table entries to calculate the arc delay for each instance and flag cases where the network parasitics required extrapolation beyond the min/max slew and min/max load index values. Characterization of library cells would fill in these tables with measures from many circuit simulations, incorporating extracted circuit layout parasitics in the simulation model for each cell. Tables were provided for rising and falling input transitions for each pin. (Arc delay measurements are typically made between the 50% signal crossings of input pin and output pin waveforms.) The methodology team would trade off the number of table entries and the slew/load index range with the corresponding characterization effort for the library. Sequential cells required additional simulations for setup and hold time tables. Output pin waveform slew tables were also generated by the characterization flow, over the same input slew and output load index ranges as used to calculate the cell delay arc. The output pin slew for each cell delay arc was then used as part of the interconnect delay calculation, from the cell output to fan-out pins in static timing analysis.

The requirement for additional cell delay accuracy has recently led to the adoption of timing model formats that represent much greater detail and cell complexity. Pin-to-pin delay arcs may be conditional upon the logical values at other input pins, known as a state-dependent timing arc. The output signal slew table information proved to be inadequate for both interconnect delay calculation and the noise analysis flow. Current cell models replace each output slew table entry with a detailed (voltage or current) waveform representation. Interconnect delay calculation uses this waveform to determine the signal arrival and slew measures at fan-out pins.

Cell timing models are provided by the characterization flow for specific process tolerances, the supply and ground voltages (at the cell), and the (local) temperature values. The process tolerances span variations in fundamental device parameters, as well as a range of local interconnect dimensions after fabrication. The cell models are represented by a number of distinct parasitic extraction settings, each producing a unique netlist of resistance and capacitance elements between devices. Each extraction setting is selected to produce a netlist that biases the element values, such as max_R, max_C_total, max_RC, and max_C_coupling. The contribution of coupling capacitance is further complicated by the assignment of adjacent wires to multipatterning colors, with additional coupling variation due to the mask overlay tolerances.

The fabrication process parameters needed for characterization circuit simulations at each PVT corner require collaboration with the foundry engineering team. To maximize production yield, a set of worst-case (WC) process parameters is normally used for timing setup tests, while a best-case (BC) process would be used for hold tests. However, the definition of WC circuit simulation transistor and interconnect process values for circuit delay is not straightforward; setting all parameters to their statistical extremes from the foundry manufacturing measurement data requires establishing very conservative design performance targets. Rather, a set of WC characterization parameters is selected to represent n-sigma performance (e.g., 3-sigma performance from nominal). A set of representative circuit simulations is run, using sampled values from device and interconnect parameter statistical distributions. The resulting distribution of simulation measurements is analyzed to set an n-sigma process parameter sample that can be used across the library characterization flow.

The selection of the n-sigma parameters for library characterization assumes that the sampled delay distribution is Gaussian. The statistical delay for a representative library cell may have a different distribution shape, especially at low VDD supply values. As a result, the n-sigma performance target may need additional statistical analysis when selecting the characterization parameters.

Several simplifying assumptions are made when providing the n-sigma WC/BC cell delay model, for each operating voltage and temperature condition. The characterization values are used for all instances of all cells in the SoC netlist; the actual devices have some fabrication variation across each SoC die. A static timing analysis flow may introduce a unique margin for logic path and clock path cell delays to reflect the *on-chip variation* (*OCV*) when performing WC/BC setup and hold timing endpoint tests. (Section 11.4 discusses the use of *derating factors* in delay calculations.)

The use of n-sigma cell delay corner models is generally accepted for logic path timing analysis but may not be applicable to other IP blocks. For example, an array may incorporate such a large number of devices that a single n-sigma characterization approach cannot adequately represent the statistical variation within the array. A single device outside the WC process parameter set may result in an *array weak bit* that would be a significant yield detractor. (A single outlier device in a logic cell essentially results in a WC/BC delay anomaly that is only one constituent of a total path; its impact is less than an outlier present in an array, where no path averaging applies.) The array characterization must therefore be done to a high-sigma statistical process parameter set. The brute-force approach would be to run a large number of sampled Monte Carlo simulations on the array netlist, which would be computationally expensive. EDA vendors have developed unique "importance sampling" algorithms to reduce the number of simulations required to confirm high-sigma array behavior with high confidence levels.

Application-specific markets may also require high-sigma logic cell characterization, where the performance path averaging is not regarded as an acceptable trade-off, and there must be a high statistical confidence level in the calculated timing path delays.

Note that the cell delay model characterization reflects a single input transient, from which the pin-to-pin arc delay is measured; all other input values are assumed to be static. As illustrated in Figure 4.18, a multiple-input switching (MIS) event affects the actual circuit response.

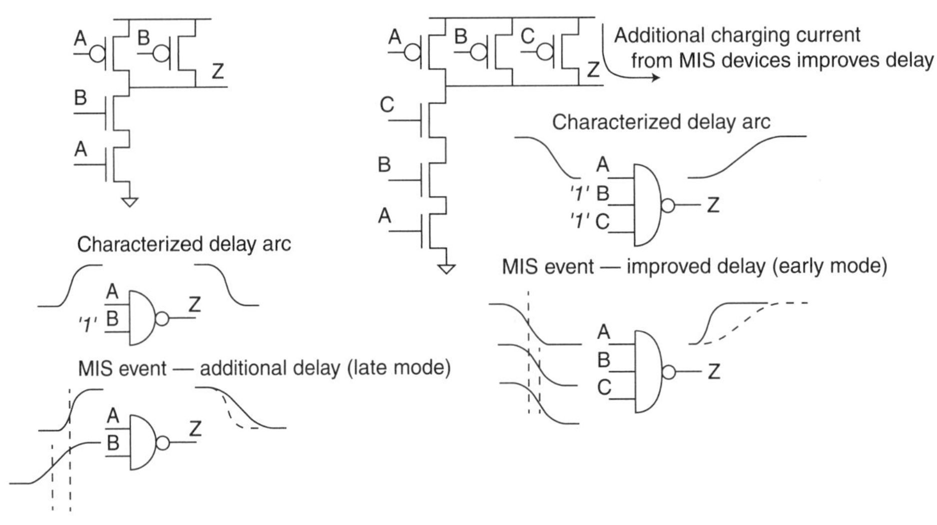


**Figure 4.18** Example of a multiple-input switching (MIS) event at the cell inputs. Cell delay arc characterization currently assumes static values on other input pins. Both late-mode and early-mode arrival time propagation due to the multiple-input switching arrivals are depicted.

The static timing analysis (STA) algorithm adds path delays to calculate an arrival time and slew at each input pin. In Figure 4.18, for late-mode timing analysis, the arrival at input pin B occurs before the arrival at input pin A; although B is still transitioning when A arrives, the STA algorithm uses the A-to-Z delay arc to propagate the Z-falling arrival time, characterized with B at a static value. Thus, there is a degree of optimism in using the characterization delay to calculate the output arrival.[3,4] Conversely, for early mode timing analysis, multiple input transitions arriving shortly after the initial arrival could accelerate the delay transition; the arc characterization delay measured with static side inputs would be optimistic for hold time. The methodology team needs to review what MIS-based cell delay calculation derating features are provided by the EDA tool and to what extent those features should be applied in the static timing analysis flow to "margin" path delay calculations.

### 4.3.3 Power Analysis Characterization

Cell model characterization data is required for calculation of SoC power dissipation. There are three contributors to power dissipation associated with an individual cell (see Figure 4.19):

- A switching transient on the output capacitive load
- The cross-over current during a switching event, as the complementary devices in a CMOS circuit transition from active to off
- Static (sub-threshold) leakage current

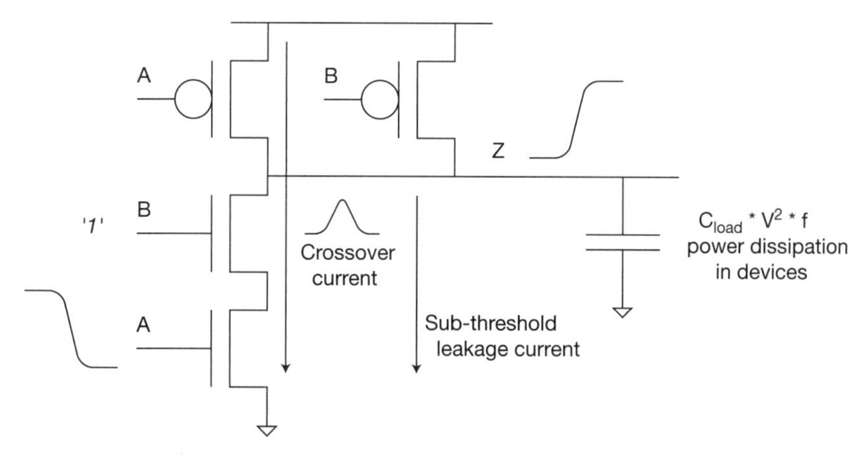


**Figure 4.19** Contributions to cell power dissipation: subthreshold leakage current, cross-over current during a switching transition, and dissipation due to charging/discharging the capacitive load through the pullup and pulldown networks, equal to $C_{load}$*(V**2)*f, where f is the activity factor.

The power dissipation as a result of charging and discharging the output capacitive load is equal to:

$$P = (C_{load}) * (VDD**2) * f) \quad \text{(Eqn. 4.1)}$$

where $f$ is the frequency of the output signal transitions—the signal *activity factor*, in simulation terminology. The dissipated energy due to the cross-over current is a strong function of the input pin slew and can be measured during characterization. The sub-threshold leakage current is also easily determined, although specific characterization simulations are required. (For simulation efficiency, sub-threshold device current calculations are normally disabled during delay characterization.)

In addition to the data for cell power dissipation, the cell power model includes similar information used for the I*R voltage drop analysis flow for the power and ground distribution grids. A simplifying assumption is typically made for I*R analysis modeling. During a switching event, the saturated device current value is used for the duration of the transient, injected into the grid at the cell location, as depicted in Figure 4.20.

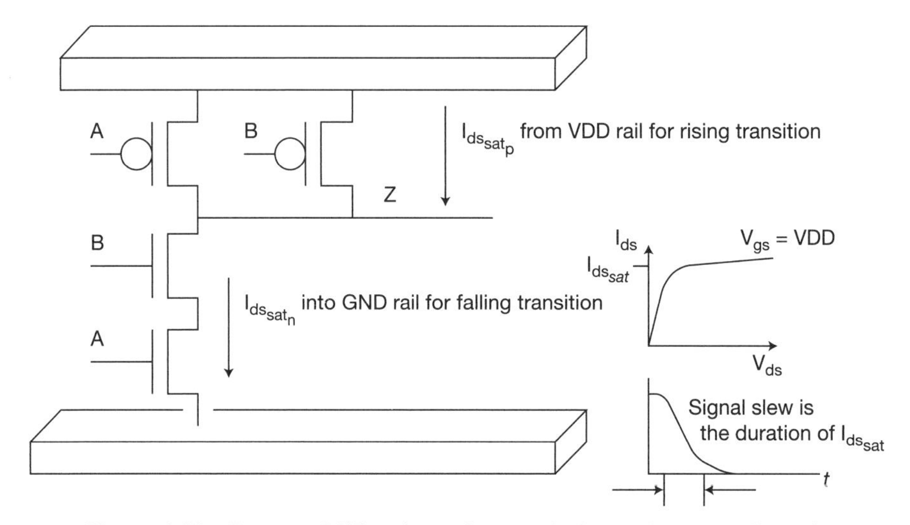


**Figure 4.20** Power rail I*R voltage drop analysis requires modeling of the cell as a current source connected to the power rail for a switching transient.

The cross-over current in the other supply rail is typically neglected for the switching event. The cross-over current is indeed included in the cell power dissipation calculation, but its magnitude and duration would not contribute significantly to the dynamic I*R voltage drop in the other rail.

### 4.3.4 Cell Input Pin Noise Sensitivity

The noise analysis flow ensures that a capacitive-coupled transient from an *aggressor* does not result in an erroneous response on the circuitry associated with a *victim* signal and its fan-out cells, as depicted in Figure 4.21. The figure illustrates a "low-up" transient, representing the victim signal original value and the direction of the aggressor capacitive coupling. The driving cell will ultimately restore the signal to the correct logic value. However, the transient injected on the net will propagate at the fan-out cells, potentially resulting in an invalid network state.

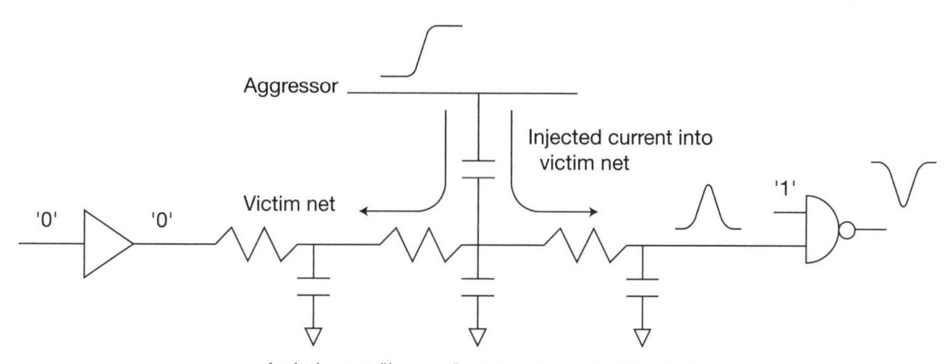


**Figure 4.21** Model for a noise coupling event, from aggressors to a victim net.

The noise characterization of the library cell input pin determines the output transient due to an input noise pulse. The noise analysis flow determines how the coupled energy from the aggressor is dissipated through the victim net RC network and the pulse arriving at each cell input pin on the net. This input pulse at each victim net fan-out pin results in a perturbation at the output pin of the fan-out cell, to be further propagated by the noise analysis flow.

The characterization of the cell for input pin noise response requires specific features:

- The interconnect extraction corner for characterization should maximize C_coupling.
- The other (static) cell input values during characterization should be selected to maximize the gain of the input pin devices for maximum signal swing.
- A set of magnitude/duration noise responses is required for each input pin (e.g., high_down, low_up). For circuit reliability, the characterization methodology should also establish high_up and low_down magnitude and duration data.

Figure 4.22 illustrates a simple input pin magnitude and duration curve from characterization (for a specific cell output load).

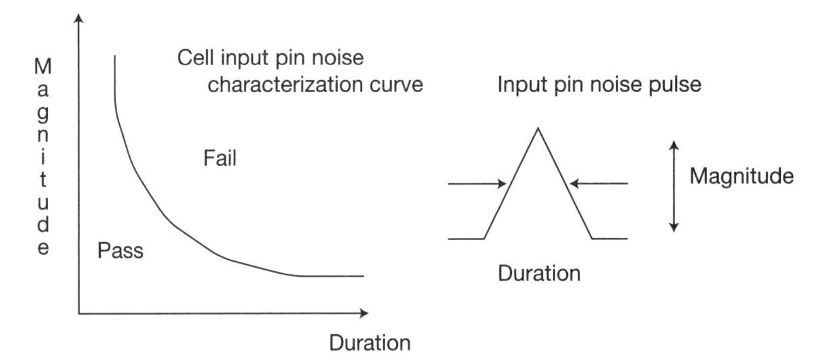


**Figure 4.22** Cell noise characterization data. A simple pass/fail noise rejection curve is depicted. Characterization simulations sweep the input pin magnitude and duration for a range of output capacitive loads. The pass/fail delineation is based on an interpretation of the cell output response during characterization. At advanced process nodes, the actual noise pulse at the cell output pin is recorded for each simulation for the noise analysis flow to propagate this pulse through the network.

At advanced process nodes, the relative contribution of the coupling capacitance to the total interconnect capacitance on a net has increased. To maintain a suitable sheet resistivity with the scaling of line width, the aspect ratio of metal wires has been increasing in successive process nodes. As a result, the adjacent wire coupling capacitance is a greater percentage of the total wire capacitance. The graph in Figure 4.22 depicts a simple pass/fail curve from characterization for an input pin noise pulse. With this method, the noise analysis algorithm evaluates the arriving input pin pulse from a noise coupling event on a net and makes a direct pass/fail interpretation against the characterization noise curve. As discussed in Chapter 12, "Noise Analysis," this restrictive pass/fail method is no longer suitable at advanced nodes, with the increased contribution of noise coupling from aggressors. Instead, cell input pin characterization exercises a simulation sweep of the input transient pulse magnitude and duration over a range of output loads and records the noise pulse measured at the cell output. The noise analysis flow now calculates the arriving noise transient at the input pin and subsequently propagates the characterization noise pulse from the cell output pin through the network. Additional aggressor events add to the noise pulses in the network simulation. The noise analysis flow determines the magnitude, duration, and arrival time of a noise pulse at a flop input to determine the risk of a network state upset error.

### 4.3.5 Modeling for Clock Buffers and Sequential Cells

A specific subset of the library cells is used in the distribution of clock signals. These cells consist of a limited number of logic functions, such as, inverters, buffers, and clock gating functions (refer to Figure 1.36). Characterization of these cells is slightly different than for the remainder of the logic library. Ideally, the rising and falling clock-to-output pin arc delays (RDLY, FDLY) at each corner should be equal, over the range of characterization loads and slews, to minimize clock phase jitter. The characterization range for clock-related cells may be more limited than for general cells; each drive strength option for these cells is typically designed for a rather precise load.

The noise characterization of these clock cells also involves more stringent magnitude/duration input pin limits, as the allowable output pin transient pulse is extremely limited. The fan-out set of these cells are other clock distribution circuits and sequential cell clock pin inputs, associated with very high-gain devices.

The characterization of sequential cells includes additional modeling requirements:

- **Setup and hold constraints apply at each corner.** The measurement of setup and hold requires a number of circuit simulations, sweeping the arrival of the data input transition relative to the clock arrival, over the allowable slew range of each of these two input pins. The characterization flow team needs to collaborate with the library engineering team to determine what circuit simulation node measurements are appropriate to establish robust operation and what (maximum) deviations in those nodes are allowable when selecting setup/hold constraints (see Figure 4.23).

The most direct setup and hold measurement criteria relate to the allowed increase in the clock-to-q delay. The clock-to-q arc delay is measured when the data input is stable. During setup and hold sweep simulations, as the data transitions near the clock, the clock-to-q arc delay increases. The maximum percentage increase is commonly used to define the setup and hold timing constraints for each of the input pin slew and output load data points.

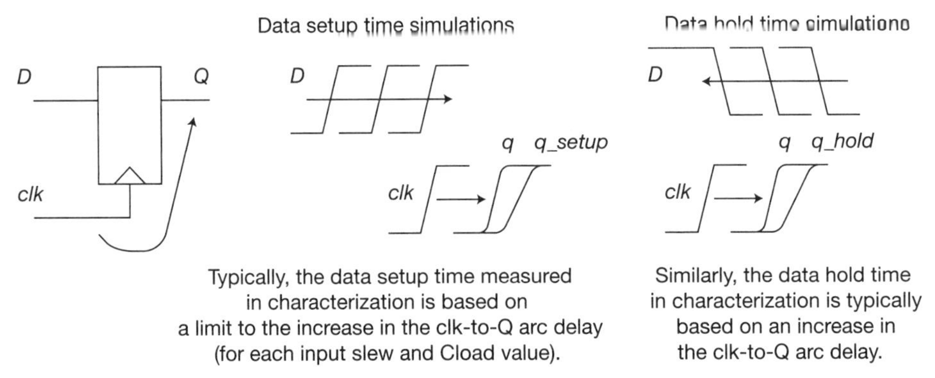


**Figure 4.23** Sequential circuit setup and hold characterization measurements in circuit simulation.

- **Power characterization may require additional data, associated with input events that do not result in an output transition.** For (single-stage) logic cells, if the output pin does not transition, there is little active power dissipation. Power calculations using SoC functional validation testcases to derive the activity factor on signals include only the contribution of switching logic gates. However, there may be internal power dissipation in a sequential cell when the output is unchanged. For example, consider the internal clock inverter power dissipation for the master/slave topology in Figure 4.24 when the data input is unchanged in successive cycles.

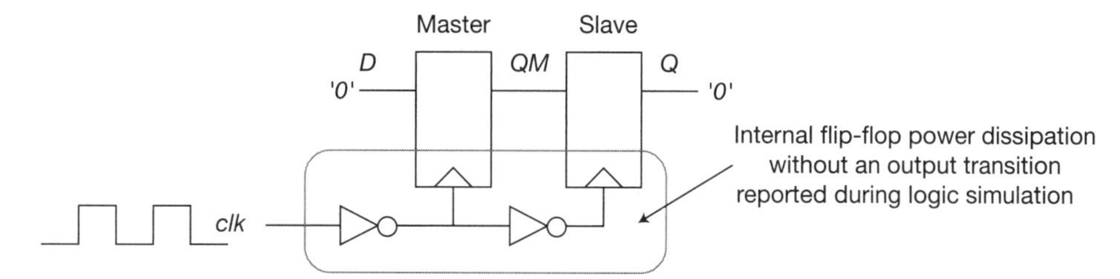


**Figure 4.24** Internal cell power dissipation without a change in output value and, thus, no contribution to the signal-switching activity calculation.

The characterization of sequential cells and the power calculation flow require additional features to include the contribution from events that are not directly associated with functional validation testcase signal activity.

- **Noise characterization requires awareness of clock transitions.** For combinational logic cells, noise characterization on input pins assumes static values on other inputs. The noise sensitivity of the data input pin of a sequential cell is maximized when the clock input pin is transitioning. The noise characterization for the data input should be thus simulated while the clock is in transition, using internal node measures for flop stability similar to those used for setup/hold simulations.

A special set of sequential cells is added to the circuit library for asynchronous interfaces. The characterization of these cells includes additional internal node measures during the data versus clock sweep circuit simulations. The calculation of metastability for these cells requires time constant measures for internal circuit nodes.[5,6]

Latches are not commonly present in an IP cell library. A design using alternating clock phase-based logic with latches (with potential path slack *time borrowing* in successive phases) requires special static timing analysis algorithms. If latches are present, cell characterization requires additional features—not only for setup/hold tests at path endpoints but also for the data-to-output timing arc for "clock transparent" mode value propagation.

## 4.4 Design for End-of-Life (EOL) Circuit Parameter Drift

The discussion on cell IP characterization in the previous section focuses on the circuit simulation testcases and measures used to provide models for electrical analysis flows. These circuit simulations utilize device and interconnect parameters from the foundry, based on process qualification wafer measurement data (with statistical distributions). Some device mechanisms during SoC operation, such as the following, lead to *parameter drift*:

- **Device $V_t$ shifts due to "negative" bias temperature instability (NBTI for pFETs) and "positive" bias temperature instability (PBTI for nFETs)**—During device operation where a high electric field is present between gate and conducting channel, a small population of free carriers enters the gate dielectric, filling available *trap states*, which results in an effective change in the device threshold voltage. For pFET devices, the direction of the electric field is from channel to gate, with free holes injected into the gate oxide; this is denoted as *negative bias temperature instability (NBTI)*. For nFET devices, the direction of the electric field is from gate to channel, with free electrons injected into the oxide. As a result, for both pFET and nFET devices, the $|V_t|$ increases over the operating lifetime. This mechanism is partially reversible during device operation, with the opposite gate-to-channel electric field direction. Also, the $|V_t|$ parameter shift ultimately saturates.
- **Hot carrier effect**—During device operation in saturated mode, where a high lateral electric field is present between the drain and the conducting channel, a flux of energetic carriers is also injected (locally) into the gate oxide near the drain node. The principal device model parameter impact is a reduction in the effective channel carrier mobility.

The magnitude of these parameter changes is provided by the foundry reliability engineering team. These effects are very much dependent upon local device temperature. Note that these mechanisms are especially prevalent in circuits with static DC bias currents, such as those used in mixed-signal IP.

The model characterization flows do not reflect these parameter drift mechanisms. Instead, special circuit simulations are separately pursued, typically using (high-sigma-sensitive) SRAM arrays and mixed-signal IP. The circuit sensitivities to these parameter drift changes are analyzed to assess the performance impacts over the SoC lifetime. To help with this assessment, SoC methodology flows provide additional data:

- **Switching activity data from functional validation tests and signal slews from static timing analysis**—The duration and frequency of device operation in the high electric field modes is used to estimate the rates at which parameter drift and recovery occur.

- ***Thermal map* generated to estimate the local device junction temperature**—The BTI and hot carrier effect parameter drift activation rates are highly dependent on the device temperature. The WC/BC cell characterization simulations use a junction temperature extreme. Product lifetime calculations commonly adopt an approach in which (spatial and temporal) temperature estimates are used, rather than an operating temperature extreme for the full lifetime. A temperature map of the SoC integrates the power dissipation flow calculations with the thermal resistance model of the die and package environment. The evolution of FinFET and silicon-on-insulator processes has introduced more complex local *self-heating* thermal resistance paths.

## 4.5 Summary

This chapter provides a brief introduction to cell and IP functional modeling, as well as the circuit characterization methods used to generate the model data required for SoC analysis flows. The EDA industry has established a detailed (and evolving) library cell modeling format that encompasses all this information.[7] This standard has enabled EDA vendors to release library characterization products to IP developers, whose output models can then be accepted by tools from any EDA vendor.

## References

**[1]** Greene, B., "Catching X-Propagation Related Issues at RTL," Tech Design Forum, February 26, 2014, http://www.techdesignforums.com/practice/technique/catch-x-propagation-issues-rtl/.

**[2]** Baddam, K., and Sukhija, P., "Challenges of VHDL X-Propagation Simulations," Design and Verification Conference (DVCON) Europe, 2015. (Presentation slides available at https://dvcon-europe.org/sites/dvcon-europe.org/files/archive/2015/proceedings/DVCon_Europe_2015_TA5_2_Presentation.pdf; full paper available at https://dvcon-europe.org/sites/dvcon-europe.org/files/archive/2015/proceedings/DVCon_Europe_2015_TA5_2_Paper.pdf.)

[3] Lutkemeyer, C., "A Practical Model to Reduce Margin Pessimism for Multi-Input Switching in Static Timing Analysis of Digital CMOS Circuits," TAU Workshop, 2015. (Presentation slides available at http://www.tauworkshop.com/2015/slides/Lutkemeyer_TAU15_PPT.pdf.)

[4] Kahng, A., "New Game, New Goal Posts: A Recent History of Timing Closure," 52nd Design Automation Conference, 2015, http://ieeexplore.ieee.org/document/7167187/.

[5] Veendrick, H.J.M., "The Behavior of Flip-Flops Used as Synchronizers and Prediction of Their Failure Rate," *IEEE Journal of Solid-State Circuits*, Volume 15, Issue 4, April 1980, pp. 169–176.

[6] Horstmann, J.U., et al., "Metastability Behavior of CMOS ASIC Flip-Flops in Theory and Test," *IEEE Journal of Solid-State Circuits*, Volume 24, Issue 2, February 1989, pp. 146–157.

[7] Liberty modeling format; open source licensing is available from Synopsys: https://www.synopsys.com/community/interoperability-programs/tap-in.html

[8] Sun, S., et al., "Fast Statistical Analysis of Rare Circuit Failure Events via Scaled-Sigma Sampling for High-Dimensional Variation Space," *IEEE Transactions on Computer-Aided Design of Integrated Circuits and Systems*, Volume 34, Issue 7, July 2015, pp. 1096–1109.

[9] McConaghy, T., et al., *Variation-Aware Design of Custom Integrated Circuits: A Hands-on Field Guide*, Springer-Verlag, 2013.

## Further Research

### Behavioral Modeling

An emerging behavioral modeling language for hardware simulation is SystemC. Describe how SystemC differs from both C-language and HDL semantics.

Describe the unique features required for synthesis of a SystemC model with a logic IP cell library.

### Parameter Sampling for n-Sigma Characterization

The statistical distribution of a (dependent) circuit measurement due to fabrication process variations is required for characterization. The traditional method is to pursue random sampling (also known as *Monte Carlo* sampling) from the independent input parameter distributions and re-simulate a sufficient number of the same testcase stimuli with these different sample sets to plot a distribution of the circuit measurement. Of greatest interest are the n-sigma extremes of the measured value. More efficient methods are being pursued to reduce the number of simulations to estimate the high-sigma "tails" of the measured distribution (with high confidence).

Describe how *importance sampling* is applied to circuit simulations from independent parameter distributions.

Other sampling methods have also been proposed, including "scaled-sigma" sampling[8], worst-case distance, and high-sigma Monte Carlo[9]. Describe how these methods differ from importance sampling.

TOPIC III

# Design Validation

The evolution of VLSI process nodes enables system architects to integrate an increasing number of functional blocks into a single SoC chip design. The complexity and diversity of the IP blocks is growing, as well. There are a greater number of concurrent operations in flight at any time, resulting in blocks contending for shared memory resources and bus interface bandwidth. To address product markets requiring low power dissipation, the SoC design must support a growing number of clock and power domains with multiple power modes and intricate power state transitions. As a result, the validation of the SoC functional model is a demanding task.

The resources applied to preparing the functional validation plan, composing and executing testcases on the model, measuring testcase coverage, and debugging testcase failures is the largest contingent of the SoC development expense. It is crucial that the productivity and efficiency of the validation team is optimized. This part of the text describes the functional validation methodology steps used throughout the SoC design schedule. In addition, the flow steps to prove functional equivalence between two models are discussed.

There are ongoing industry-wide initiatives to provide standards for testcase structure and to release the corresponding software libraries into the public domain. The primary goals are to reduce the time required to develop the testcase suite and to enable EDA tool vendors to optimize simulation model compilation and execution. The detailed nature of these object-oriented programming class-based testcase generation software functions is beyond the scope of this text; the reader is encouraged to review the references and future research recommendations at the end of Chapter 5.

CHAPTER 5

# Characteristics of Functional Validation

## 5.1 Software Simulation

The evaluation of hardware description language (HDL) model functionality is based on the paradigm of *event-driven* simulation. As mentioned in Section 4.1, an HDL model consists of a set of concurrent sequential processes. The evaluation of a process is triggered by an event in the sensitivity list for the process, typically a transition on a signal in the list. The result of the process evaluation is the assignment of a new value to variables and signals; the new signal values are scheduled in future time on the event queue maintained by the HDL simulator. When all active processes complete evaluation at the current time, the top of the event queue is queried, and time is advanced corresponding to the next set of pending transitions. The event simulation time may advance in irregular intervals, from an infinitesimal delta time to an explicit delay included with an assignment statement.

The functional model stimulus is provided by a *testcase*, as described in the next section. The collection of testcases used to validate the block or SoC functionality is denoted as the *testbench*. Typically, the testbench includes assertion and monitor functions that are incorporated as part of each testcase.

The event simulation throughput is a key metric used by the SoC project management and methodology teams to estimate the EDA simulator license quantity and compute resources required to complete execution of the testbench. The throughput is varied, as it depends on the signal activity written to the event queue; average estimates are commonly used for resource estimation.

Another factor that strongly influences simulator throughput is the degree to which event activity is logged to assist with post-simulation debugging; writing a trace file of signal event/time values (and other HDL-based messages from monitors and assertions) slows the evaluation speed. Often, testcases are developed to periodically probe the HDL functional model for expected values without requesting a trace file. If the testcase fails the self-checks, simulation flow scripts resubmit the testcase with (a subset of) signals selected for tracing.

Event-driven simulation includes the unique requirement to manage multiple potential future assignments to the same signal. Depending on the signal properties, the simulator may need to cancel a future event from the queue if a new assignment overrides a pending future event, as illustrated in Figure 5.1.

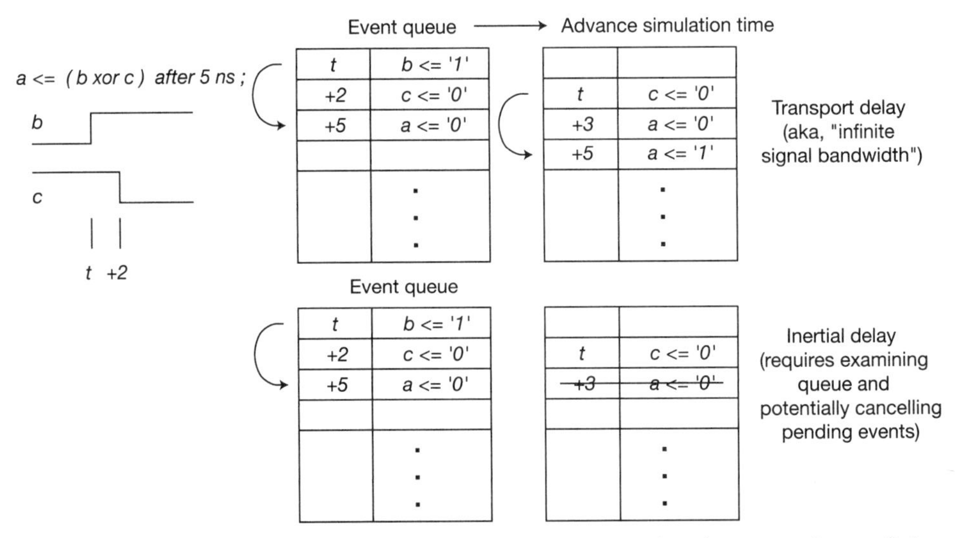


**Figure 5.1** Inertial and transport signal properties related to managing multiple transitions on the event queue.

A signal with the transport delay property would retain all pending events on the queue (i.e., an "infinite bandwidth" response). A signal with the inertial delay property could result in cancellation of pending events to prevent short-duration pulses from propagating.

EDA vendor HDL simulator tools include several features to assist with model and testcase debugging, as described in the following sections.

### 5.1.1 Interactive Mode

The bulk of the testbench suite is (repeatedly) run in batch mode on the various mainline and checked-out active versions of the functional model. Failing testcases may be resubmitted with additional tracing and message output, or they may be evaluated using a model compiled for interactive execution. Interactive simulation utilizes additional user input provided at the simulator shell command line to control the testcase. The user can specify simulation time execution increments or enter a new logical signal expression to serve as an interactive "breakpoint." The user can (temporarily) override a current signal value by entering a "force" command (followed by a "release" command) to determine how the model responds to a value that supersedes an HDL assignment statement.

### 5.1.2 Waveform Display

EDA vendors provide graphical display tools to show signal transition/time trace information in the familiar waveform format. This interface is used both with a post-simulation trace file and during interactive mode execution. Users can select which signals to display, move back and forth in time, and query for a specific signal value (or breakpoint expression). The adoption of more complex data types in HDL behavioral and RTL models has introduced some challenges to EDA vendors. The traditional waveform display of binary '1' and '0' levels needs to succinctly reflect other bit values ('X', 'Z', 'U'), as well as values from enumerated, integer, and floating point type ranges. Note that all signal identifiers used in the waveform display are qualified by the logical hierarchy instance names; an additional debug tool aids the user link from a signal waveform identifier back to the corresponding HDL source.

### 5.1.3 HDL Source/Configuration Cross-Reference

The waveform trace graphical display is useful for debugging, but a functional validation engineer needs to correlate simulation results to the original HDL and testbench environment source code. Advanced EDA vendor debugging tools compile the HDL model into a database that represents the configuration hierarchy and signal assignment dependencies. Validation engineers leverage this database in conjunction with the waveform display:

- Selecting a signal in the source adds the waveform to the display window.
- A database *dependency traceback* on a waveform displays the set of signals that are present in assignment expressions to the waveform signal.

The HDL source-to-waveform display cross-reference interface manages multiple instances of the same model IP when users traverse the signal database.

As mentioned in Section 4.1, the RTL model abstraction typically does not include explicit event delay values. Register states are updated when the CSP is sensitized by a transition on the corresponding clock input signal. Although HDL semantics are inherently based on event simulation, the compilation of the model for simulation may include cycle-based optimizations. The resulting simulation executable executes more quickly if intermediate combinational signals need not be explicitly retained. The resulting trace file is reduced to the register values each cycle. Advanced debugging tools that link HDL source and trace results also include the capability to dynamically recalculate an intermediate combinational signal value, using the register values at the inputs to the signal's logic cone.

## 5.2 Testbench Stimulus Development

As SoC complexity has increased with each process node, so has the complexity of the testcases in the testbench suite. Indeed, detailed methodology references have been published, along with a recommended set of object-oriented programming class libraries for developing *transaction-based* stimulus/response testcases that connect to (internal and external) SoC bus interfaces.[1,2] These approaches are beyond the scope of this text, but the following sections provide a simple introduction to the types of testcases in the testbench suite that utilize these methodologies.

### 5.2.1 Directed Testcases

The simplest testcase uses a *directed set* of stimulus vectors in a sequence over simulation time, applied to the inputs of the model (also known as the *device under test* [DUT]). The directed testcase includes the expected response on the DUT outputs, either monitored each cycle or monitored only at the end of the vector set. Directed tests often reflect the requirement for compatibility to prior implementations of an IP core in the SoC micro-architecture and are commonly derived using existing functional traces from a previous design. The directed testbench suite is intended to provide a degree of coverage over the instruction set and/or the command set of each IP block. This suite is typically used as a prerequisite for checking of any HDL model revisions for promotion to the version management mainline. At a minimum, the quality of a model revision must pass the (subset of) directed testcases selected as a *release gate*.

### 5.2.2 (Constrained) Random Testcase Generation

Although the directed testcase suite may provide basic coverage, it does not thoroughly exercise potential interactions between successive instructions and their register/memory references. In addition, an existing set of (embedded or external) program code examples do not anticipate potential future compiled code dependencies. To provide more exhaustive coverage, a random testcase generator is often developed for the SoC micro-architecture.[3,4] The generator code utilizes a separate "golden model" of the architecture to calculate the expected simulation responses. The generator includes numerous settings to constrain, or "bias," the stimulus for each testcase. For example, vectors could be selected to exercise specific architectural features, especially those that require unique handling (e.g., cache and branch prediction misses, queue "full," interrupts, "lock/unlock" references in shared memory transaction processing).

The methodology team is faced with the dilemma of estimating the EDA simulator licenses and compute resources to apply to random testcases from a generator, which are essentially unbounded in number and duration. Often, the workload allocated to random simulation testcases consumes the available license/server capacity; the goal is to maximize the license and IT resource utilization at all times. Random testcases continue to be submitted to the compute server job dispatcher at a low priority. When a user submits a directed simulation job request, it is placed on the job queue at higher priority (even potentially suspending existing random simulation jobs).

A critical element of random testcase execution is to develop a coverage goal to evaluate when the functional model is of a sufficient quality for tapeout (although simulation jobs will no doubt continue after tapeout, until silicon debug commences). The functional validation team reviews the bias settings and instruction/address/data interdependencies with the SoC architecture team to define a tapeout-quality coverage metric.

For execution performance, random testcases do not request a signal trace file. A testcase that fails the predicted response from the golden model is logged, along with the bias settings and the random seed; it is subsequently resubmitted with tracing, using the same generator parameters.

As mentioned earlier in this section, successful execution of the directed testcase suite is a common quality prerequisite for the release of a new model version. Attempting random testcases too early is likely to result in a significant percentage of failures. As a result, random testcases are not usually part of a release gate—at least not initially. Directed tests are more thoroughly documented and, generally, easier to debug. Examining a large set of failing random tests on early model versions could be problematic.

Note that the random testcase generator requires additional validation resource to prepare and maintain the golden model response calculation. For a new SoC architecture, the code developed for the golden model from the specification also requires debugging. The golden model is typically coded in a behavioral abstraction for efficiency. To eliminate the need to develop a separate golden model, the SoC industry is actively researching techniques to capture the salient micro-architectural features in an *executable specification*.[5]

### 5.2.3 Assertion-Based Validation (ABV)

As described in Section 4.1, both behavioral and RTL abstraction models are enhanced with the addition of non-functional (concurrent or sequential) assertion statements. The assertion expression defines an invariant property of the model. If this expression evaluates to "false" during simulation, the output message is written, and any directive clause of the assertion is active. In addition, EDA vendor tools are available to attempt to formally prove assertion expressions, avoiding the simulation execution overhead.

The value of model assertions is extremely high. Model debug is improved, as the assertion pinpoints the failing condition without the need to examine waveform values.[6,7] However, the SoC validation team needs to address ABV trade-offs. The validation engineering team is typically not sufficiently familiar with the design to add detailed assertion expressions. This team's expertise is focused on external stimulus/response interaction with the DUT as a black box. The design engineer is therefore often asked to add assertions while coding the functional RTL model; however, completing the functionality is the highest priority, and the attention to defining assertions is secondary. Another complicating factor in pursuing assertion-based validation is the complexity of the signal relationships that would be valuable to track. Designers work within the scope of an HDL model within the overall logical hierarchy; an assertion expression requiring signals outside this scope cannot be readily coded and compiled. (Separate "property-specification" languages have been developed, offering richer semantics and scope than HDL assertion expressions. EDA vendors have developed tools to input the HDL model and the property specifications to submit to a formal prover algorithm.[8]) Despite these limitations, ABV offers significant opportunities to accelerate simulation debugging and reduce testbench volume through separate formal analysis.

### 5.2.4 Reset Validation, Uninitialized Signal Propagation, and Power Supply Sequence Validation

For efficiency, the testbench "slams" an initial state into the functional model for each testcase at time "t = 0_minus," from which the (clock) stimulus commences. A separate class of validation tests is developed to reflect the functional model reset sequence. Model reset could take several forms and include several unique features:

- "Soft" reset, from a valid existing state to a new reset condition
- "Hard" reset, where all signals start with unknown/uninitialized values (such as would be representative of initial power-on)
- Additional built-in self-test operations after reset (e.g., to ensure proper operation of on-die memory arrays before the conclusion of the reset interval)
- Exercising (embedded) software initialization routines to put the SoC into a specific reset state

The engineering team responsible for reset validation is faced with the issue that the duration of the testcases is very long, especially with self-test and software initialization sequence evaluation. The validation and methodology teams may find that EDA software simulation tools do not provide sufficient throughput; an investment in hardware simulation acceleration resources may be appropriate.

The reset validation group also needs to close the loop with the rest of the functional validation team and ensure consistency between the initial "slam" model state used by functional testcases and the reset state determined from separate simulations.

### 5.2.5 Power Supply Sequence Validation

A small yet extremely significant aspect of reset validation relates to an electrical characteristic of the SoC. Prior to the application of (hard reset) clock and input signal stimulus, the SoC may have additional requirements for the sequencing of power supplies (typically from voltage regulator models [VRMs]). A "power good" signal from a preceding supply is required before the next supply ramp occurs. The SoC engineering teams responsible for reset validation and the global circuit implementation need to coordinate how to confirm power sequencing.

### 5.2.6 Power State Sequence Validation

The SoC power format file description captures the requisite information on how IP cores may transition to/from various power management states. The validation team needs to develop testcases based on this description. A complication arises because the functional HDL model for the core is likely to include minimal detail about the electrical isolation features; that is, the use of sleep FETs and state retention flops are represented in the core's power format file but not the RTL description. As a result, distinct power format analysis tools need to be used to evaluate the electrical interface between IP cores in different power management states, and the functional validation team must confirm the power state sequencer design and the clock cycle duration between state transitions.

### 5.2.7 Bad Machine Error Injection Testcases

The functional validation team collaborates with the SoC architects and methodology team to assess the testcase strategy for injecting errors into the simulation model and the error recovery response that is expected in the RTL.

EDA simulation tools typically provide the capability for testcases to "force" and "release" overriding values onto selected signals, which are values that differ from the logic model calculations. In this manner, an error could be (statically or intermittently) applied during a running simulation. The testcase needs to comprehend the SoC error recovery features to monitor the expected SoC response. Some errors are relatively straightforward to inject and observe (e.g., a memory bit error for an array read access where the architecture provides for a parity checking error interrupt). Other potential injected errors must be more intricate to create and confirm the correct SoC behavior (e.g., a non-responsive bus transaction sequence triggering a retry request). Although there are several methodology approaches to testbench coverage measurements for the functional model, to date, the sufficiency of error injection testcases has relied more on engineering judgment.

### 5.2.8 Visibility into Hard IP Models

An IP vendor providing a hard core for SoC integration includes a functional HDL model as part of the design enablement kit. The vendor wants to protect the proprietary details of the IP implementation and is likely to provide a binary model for linking into the compiled simulation executable. However, the validation team needs access to specific signals in the model in addition to the input and output ports (i.e., for initialization/reset, signal value assignment and error injection, and internal monitoring). To address the conflicting requirements of proprietary IP information protection and (limited) testbench visibility, the EDA industry has developed a standard methodology for the IP vendor to identify specific internal signals/arrays that are accessible to testbench developers.[9] Figure 5.2 illustrates a *model manager* software layer between the simulation application and the compiled IP model. A set of functions defined for this interface controls the value and timebase communication between the simulator and model. In addition to the input/output ports, additional internal signals (or "viewports") are registered by the IP model provider for testcase visibility.

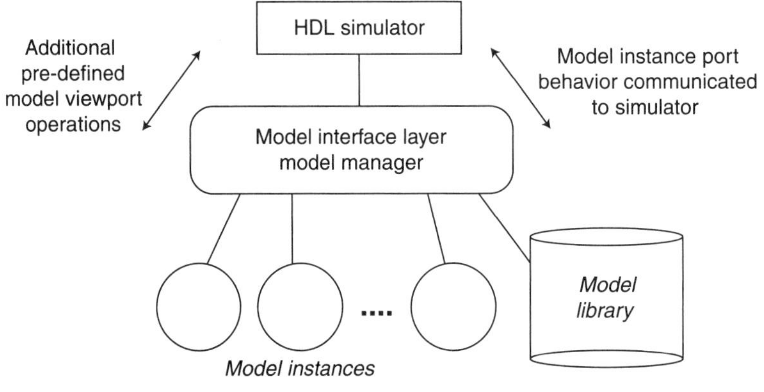


**Figure 5.8** A model manager software layer in the EDA simulation tool allows integration of proprietary models. The model provider is thus able to protect the intellectual property. The model includes "viewports" to define internal signal visibility during execution of the testcase.

### 5.2.9 Summary of Testbench Development Approaches

This section reviews the varied approaches to testbench development. The SoC project management and methodology teams need to plan how to allocate validation resources to developing the complete testbench (and individual testcases) using a combination of these techniques. Subsequent sections of this chapter discuss the EDA tools available for functional validation, their throughput/cost/scope trade-offs, and the appropriate subset of the testbench suite to submit to these platforms.

## 5.3 Hardware-Accelerated Simulation: Emulation and Prototyping

With the increasing size and complexity of SoCs in current process nodes, EDA software simulation tools (using a concurrent sequential process and event-driven paradigm) are unable to provide sufficient throughput. The duration and signal activity of full-chip testcases often necessitate accelerating evaluation using a hardware system specifically designed for HDL simulation. There are two general classes of hardware acceleration products from EDA vendors: emulation and prototyping.

### 5.3.1 Emulation

An emulation system comprises several identical application-specific processor chips. The fundamental architecture of these processors utilizes a wide microcode instruction (see Figure 5.3). Each instruction to be executed contains fields representing the following:[10,11]

- The addresses of the logic gate input signal values
- The address of a logic gate input value not resident in the local data memory of the processor but from the crossbar switch provided for inter-processor signal communication
- The address of the logic truth table for the logic gate
- The address to store the logic gate output value in the data memory
- An interrupt mask to trigger a change in normal execution flow, based on the calculated output signal value

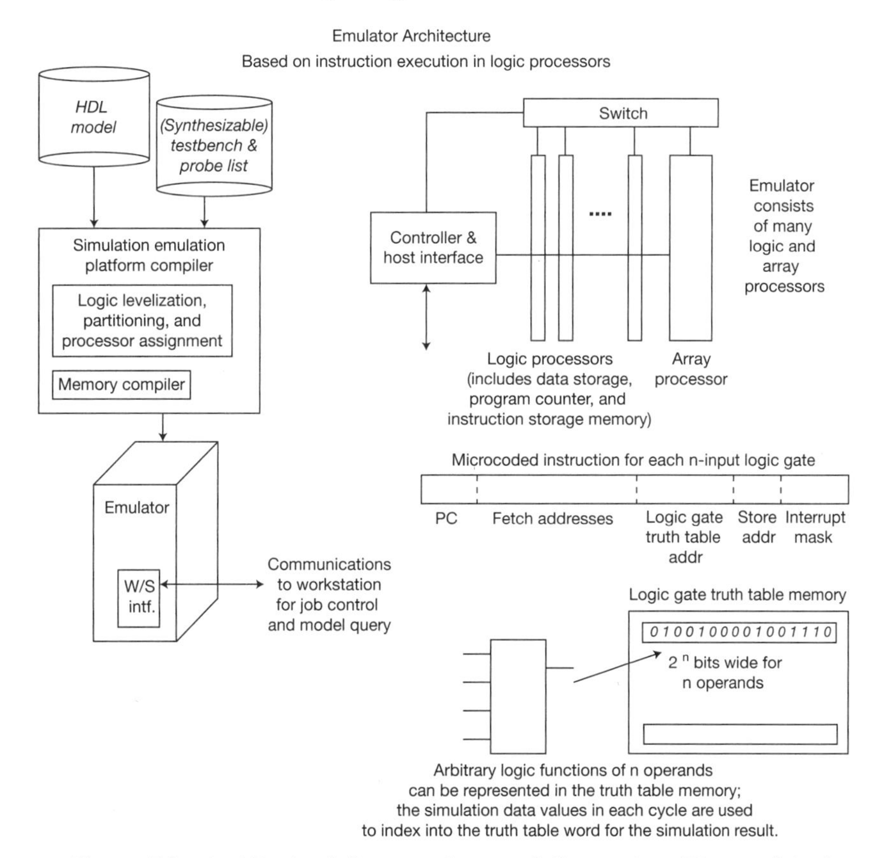


**Figure 5.3** Architectural diagram of an emulation system. The emulator is composed of a large number of application-specific processors that execute a wide microcode instruction to evaluate each "logic gate." The HDL model compiler synthesizes the gates, generates the microcode, and manages the assignment of the gates to a specific processor address space. A crossbar switch provides for inter-processor communication.

The logic simulation model is represented as "one microcoded instruction per gate." An RTL model is compiled into gate-equivalent functions, in a manner similar to logic synthesis. The functional execution unit in each processor evaluates a sequence of microcode instructions to calculate the result of the logic model each successive clock cycle. Compiled logic gates between registers are levelized, as illustrated in Figure 5.4. The order of microcode instruction execution follows the level rank order.

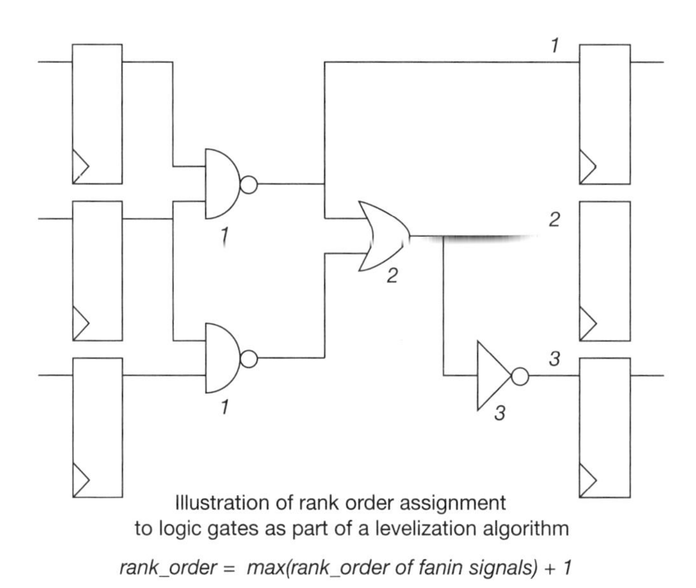


**Figure 5.4** Levelization of logic gates between registers defines the emulator microcode instruction execution order.

The processors run in lockstep, each executing its specific sequence of microcode instructions for the logic assigned to the processor during model compilation. If the processor execution time is N psec for the complete operation—that is, microcode instruction fetch, operand access, function table lookup, and operand store—the RTL model executes at (N * rank_order_length) psec per RTL clock cycle. Note that every "gate" is evaluated on every RTL clock cycle, regardless of the event activity. All operand values are read from the (multiport) data memory for every instruction. Nevertheless, the efficiency of emulation execution on this specialized processor architecture is much greater than when using a general-purpose microprocessor to execute an event-driven simulation algorithm (with the additional event queue management overhead).

The emulation architecture includes memory for each gate's functional lookup table. The number of operands in the microcode instruction (e.g., a four-input function) determines the (maximum) number of different tables required and the size of each table. Model compilation generates these arbitrary logic function tables; they are arbitrary in the sense that there is no fixed Boolean target library. The compiled "gates" could provide any permutation of '1' and '0' output values for the (2**(number of operands)) input values. Compilation includes assignment of each logic network to a specific processor, cognizant of the location of each signal's data memory, whether resident locally or accessed through the crossbar switch connected to the data memory of a different processor.

The emulator may use multiple bits in the data memory and functional table lookup to represent each signal. Two bits would enable each signal to reflect a four-valued simulation (i.e., '0', '1', 'X', and 'Z'). The interrupt mask might be used to halt simulation upon a signal transitioning to the 'X' value, for example. The emulator includes additional memory storage (and microcode instructions for memory access) to represent arrays that are part of the functional model. Compilation and testbench development are constrained to map memory addresses in the model to the available memory footprint of the emulation system.

The emulator may include additional features to assist with validation debug:

- General visibility to register values, which can be cached and written out each RTL model clock cycle, after completing the rank-order sequence of microcode instructions
- Recording of switching activity of signals each RTL cycle for use with SoC power estimation

The testcases executed on the emulator are typically based on lengthy (embedded or external) software code examples, for which average and instantaneous power estimates are desirable.

- Multiple clock domains on the SoC are managed during compilation and microcode instruction execution control.
- Multiple models and/or multiple testcases could be running concurrently on separate partitions of the emulator if the overall emulation logic gate capacity exceeds the RTL model size.

- If the testbench transactions and monitors are judiciously coded, compilation may be able to integrate a significant percentage of each testcase as part of the emulation microcode.

For maximum efficiency, the requisite signal communication between the emulator and host system should be minimized. Ideally, testcases running at this level of verification can be developed to be self-contained and self-checking.

Note that model compilation for an emulator is likely to require some modifications to the RTL and testbench code originally developed for event-driven simulation. Memory address space may need to be tuned for emulation, as mentioned earlier. Power gating functionality needs to be "stubbed out" from compilation, and so do any mixed-signal models (e.g., in I/O communication protocol modules). Emulation systems also typically include a feature to inject errors into the model from a host system, forcing and releasing values in the data memory.

The SoC methodology and validation teams need to assess what value emulation capabilities will offer to the project and at what cost. Emulation systems are expensive, as measured in "cents per effective gate capacity" as well as additional system simulation memory. The overhead of RTL and testbench tuning, model compilation, and host-to-emulation startup initialization implies that emulation is applied primarily during the more stable full-system validation phase. To avoid the cost of purchasing an emulation system for the (relatively short) duration of full system validation, it may be possible to lease the resources of a remote system, hosted by an EDA tools cloud provider.

### 5.3.2 Prototyping

In addition to emulation, the other class of hardware-accelerated simulation products is a prototyping system. Whereas an emulator uses a number of custom processing units, a prototyping system utilizes field-programmable gate array (FPGA) modules. Commercial FPGAs are widely used in electronic products for low-volume offerings and/or field-upgradable functionality. The logic gate capacity of a commercial FPGA is below that of an SoC design (at the same process node); thus, the implementation of a hardware prototype of an SoC or a system model typically necessitates using multiple FPGAs. Prototyping systems are available in a wide range of capacities and physical footprints, from a plug-in card with a small number of FPGAs inserted in an existing computer system slot (e.g., PCIe) to full stand-alone server rack hardware with many FPGA parts.

The SoC model compilation flow for FPGA-based prototyping is similar to that of a commercial-based FPGA design, with the added requirement that the synthesized netlist be partitioned across multiple FPGAs. This partitioning needs to satisfy the physical constraints of each part in terms of (routable) gate count, partition I/O signal pin count, and clocking domains. Register files and cache arrays in the SoC model can potentially be problematic in terms of mapping to available FPGA memory resources. After partitioning, the individual FPGA netlists are placed, routed, and timed, using commercial FPGA implementation tools.

The prototyping throughput is thus based on the timing optimizations and timing analysis results from the (slowest) FPGA physical implementation. In addition, some cycle time margin is allocated to the global clock distribution from the central clock source to the multiple parts/boards in the prototyping system.

The physical nature of the FPGA-based prototype enables the system to readily include a *hardware attach interface*. If a subset of the overall target product design is already available in hardware and can be down-clocked to interface with the prototype model at its clock frequency, a very efficient hardware prototype implementation for system validation can be realized. Note that the role of the traditional testbench is diminished when pursuing a prototyping solution; ideally, I/O traffic to/from the SoC model can be physically provided by the hardware attach card(s) rather than through a much slower interface to a host controller. In lieu of hardware interface attach, an alternative approach would be to compile testbench transactors into FPGA gates, much as with the goal of compiling testbench features into emulation system processor microcode.

Compared to emulation, prototyping offers the potential for much faster throughput once the model is successfully partitioned and the physical implementation is complete. Execution of boot firmware and operating system–class testcases is feasible with reasonable turnaround time. The key trade-off is that visibility into the model and signal trace capture for debugging are more limited for prototyping systems. Commercial FPGAs offer some features for internal observability, albeit from a predefined signal set. The ability to debug an error is thus hampered, requiring iterative signal set definitions, model recompilation (hopefully to a limited subset of FPGA partitions), and testcase re-execution to capture additional signal trace information. In addition, EDA vendors are pursuing "custom" FPGA-based prototyping systems, replacing commercial FPGA parts with a unique programmable logic design that includes greater internal signal observability.

As with emulation technology, the SoC methodology and validation teams need to assess the value of prototyping for system validation, applied near the end of the project schedule. Prototyping products are also effectively priced at cents per gate. The additional complexity of model partitioning and physical implementation constraints for FPGA parts, testbench preparation, and testcase debugging needs to be weighed against the achievable prototyping throughput, which enables much longer software testcase execution. Indeed, the value of prototyping acceleration may also accrue from providing software/firmware programmers with access to the prototype model for their code development and testing once a stable system model is available.

### 5.3.3 Applications of Emulation and Prototyping

An application in which either class of hardware simulation acceleration platform is extremely beneficial is system reset validation. The duration of a system reset sequence may be problematic for software simulation throughput, especially given the high percentage of signal event activity (which is immaterial to acceleration systems).

Both types of platforms are extremely useful during the post-silicon debug phase, as well. Functional errors uncovered in product testing of silicon prototypes from the foundry are re-created on the accelerated system model, using the product software trace as a testcase. (A separate analysis is pursued in parallel to determine how the error escaped pre-tapeout validation coverage.) Given the need to minimize the debug cycle in preparation for a subsequent SoC tapeout, and the likely complexity of the silicon bring-up testcase, utilization of an accelerated platform is extremely valuable.

## 5.4 Behavioral Co-simulation

At the opposite end of the validation spectrum from hardware acceleration systems, the SoC validation team may envision an environment with a mixture of RTL modeling and high-level testbench code. Section 5.1 highlights the fact that current HDL semantics include behavioral-level functional modeling constructs; the application of logic synthesis in the physical implementation flow may confine the extent to which these statements are used in the SoC model. However, the testbench environment around the model needs to utilize a productive coding style. A number of different semantics may also be supported by the EDA vendor simulation tool, outside the scope of the HDL language reference definition (e.g., C, C++, and SystemC, which is a unique language definition derived from C, with specific hardware simulation features).

The compilation of an executable from the SoC model and testbench enables a variety of high-level, behavioral functions to be integrated into the simulation. A specific validation application is the co-simulation of the SoC, the testbench, and a (C-language-based) SoC performance model. Preliminary system performance estimates may utilize a separate, lightweight model of system throughput, with parameterized functions for the system cycles required for resource access and task execution. System architecture optimization utilizes the performance model with appropriate program code traces to make trade-offs in instruction pipeline definition, cache array size and associativity, out-of-order instruction prediction success (and mispredict penalty), and so on. A subset of the detailed SoC validation testbench subsequently needs to co-simulate with the performance model to confirm the accuracy of the performance model and its system throughput estimates. The methodology, validation, and performance modeling teams need to evaluate how to best build this environment for the EDA software simulator and plan the scope and project timeline for this validation task.

## 5.5 Switch-Level and Symbolic Simulation

This section describes a different type of validation than the event-driven semantics used with an HDL simulation tool. In addition to the task of exercising the HDL (or netlist) model against a suite of testcases, it is also necessary to verify that the device-level schematic for each library cell and hard IP block is functionally equivalent to the corresponding HDL model released in the library. Figure 5.5 depicts the verification task; note that this is really a test of the equivalency of two separate models rather than a simulation test. Nevertheless, the proof of "full and complete" equivalency may indeed involve exercising unique testcases against the models.

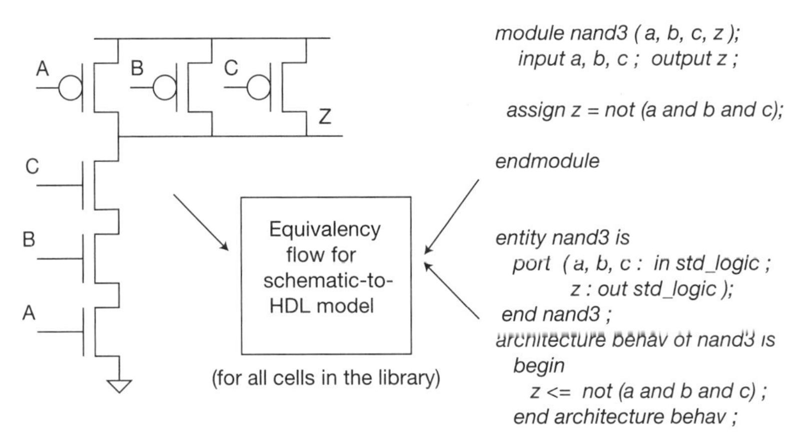


**Figure 5.5** Flow to demonstrate equivalence of a library cell schematic and its released HDL model.

### 5.5.1 Switch-Level Simulation

One approach to verifying that the device-level schematic for each library cell and hard IP block is functionally equivalent to the corresponding HDL model released in the library would be to develop HDL-like models for an individual nFET and pFET device and to exhaustively apply input patterns to both the schematic-based netlist and the corresponding functional HDL description. Figure 5.6 illustrates this *switch-level* simulation methodology.

This verification approach to equivalency is productive only for simple cells with few inputs. Note in the figure that the number of suggested patterns is n**4, where *n* is the number of cell inputs, and the testcase includes '0', '1', 'X', and 'Z' values. A smaller number of vectors is certainly feasible, with knowledge of the cell's function; for example, for a NOR2 gate, vectors ('1', 'X') and ('X', '1') cover several other vectors. Also, note that the 'U' (uninitialized) value is not considered to be required. This approach actually would only apply to combinational logic cells and schematics. Any cells with one or more storage bits would require a more sophisticated generator of test vectors.

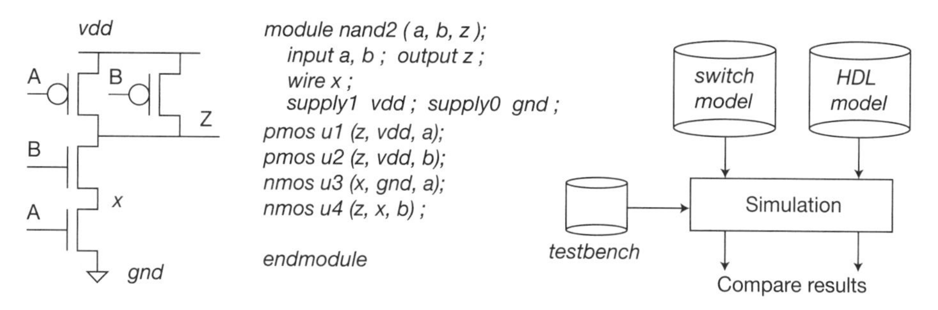


*nFET device model output truth-table*
*nmos ( out, in, gate );*

| gate \ in | 0 | 1 | x | z |
|---|---|---|---|---|
| 0 | z | z | z | z |
| 1 | 0 | 1 | x | x |
| x | L | H | x | x |
| z | L | H | x | x |

z = high-impedance
x = unknown ('0' or '1')
L = '0' or 'z'
H = '1' or 'z'

**Figure 5.6** Switch-level simulation of an nFET and pFET device netlist. Device behavior is reduced to binary "on" and "off" conductivity, based on the logical input. For circuits with feedback, the switch-level simulation algorithm incorporates a resolution function comparing weak conductivity and strong conductivity.

Note that a switch-level model could also utilize 'weak' and 'strong' properties on signal output values, as derived from device widths, for use with sequential circuits when overwriting the stored value; the sophistication required of the simulation testcases increases significantly for this type of simulation.

### 5.5.2 Symbolic Simulation

To address the limitations of switch-level simulation for schematic-to-HDL model equivalency verification—especially for memory array IP designs—EDA vendors have implemented a specific algorithmic approach. The general field of model equivalency verification through *symbolic simulation* incorporates two key algorithms. First, detailed (switch-level) models are abstracted to equivalent Boolean functions. Then, rather than applying testcase vectors with patterns of 1s and 0s (and, potentially, other bit values), the reduced model is simulated using input symbols, which are propagated functionally. Figure 5.7 depicts a greatly simplified example of the symbolic simulation concept.

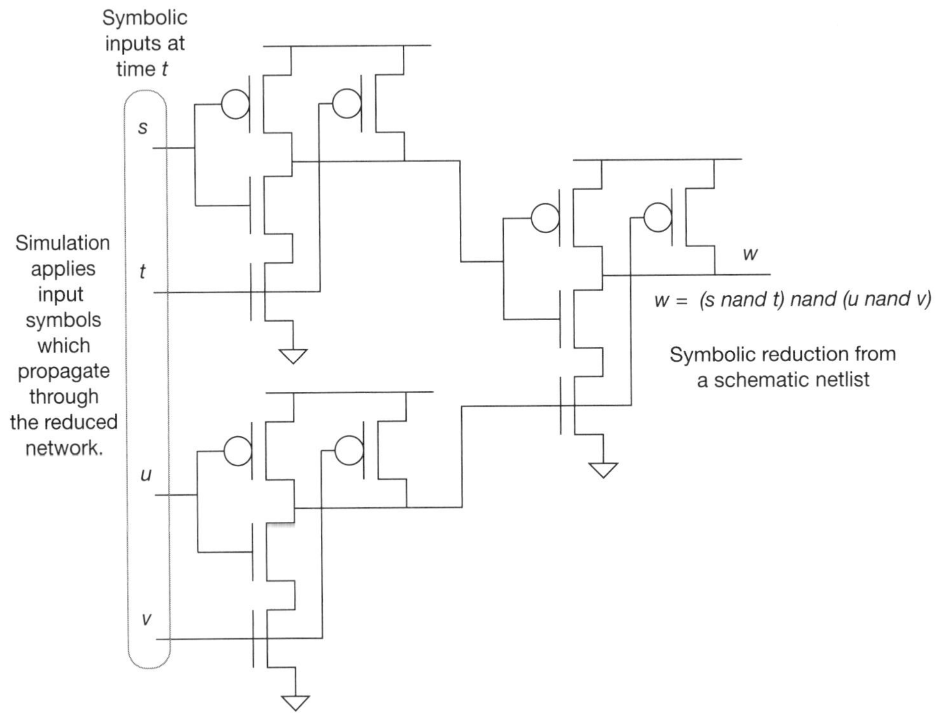


**Figure 5.7** Illustration of symbolic simulation. Symbols replace the traditional representation of signal values. Symbols (and operators) are propagated during simulation.

The first step in preparation for symbolic simulation for equivalency is schematic reduction. The schematic netlist is decomposed into subsets consisting of device *channel-connected regions* (*CCRs*), as illustrated in Figure 5.8. The specific nature of static combinational CMOS logic results in complementary (series/parallel) topologies of nFET and pFET devices, which leads to relatively straightforward deduction of the Boolean function. Also, note that a reduced schematic network could reflect multi-bit values represented by a single symbol.

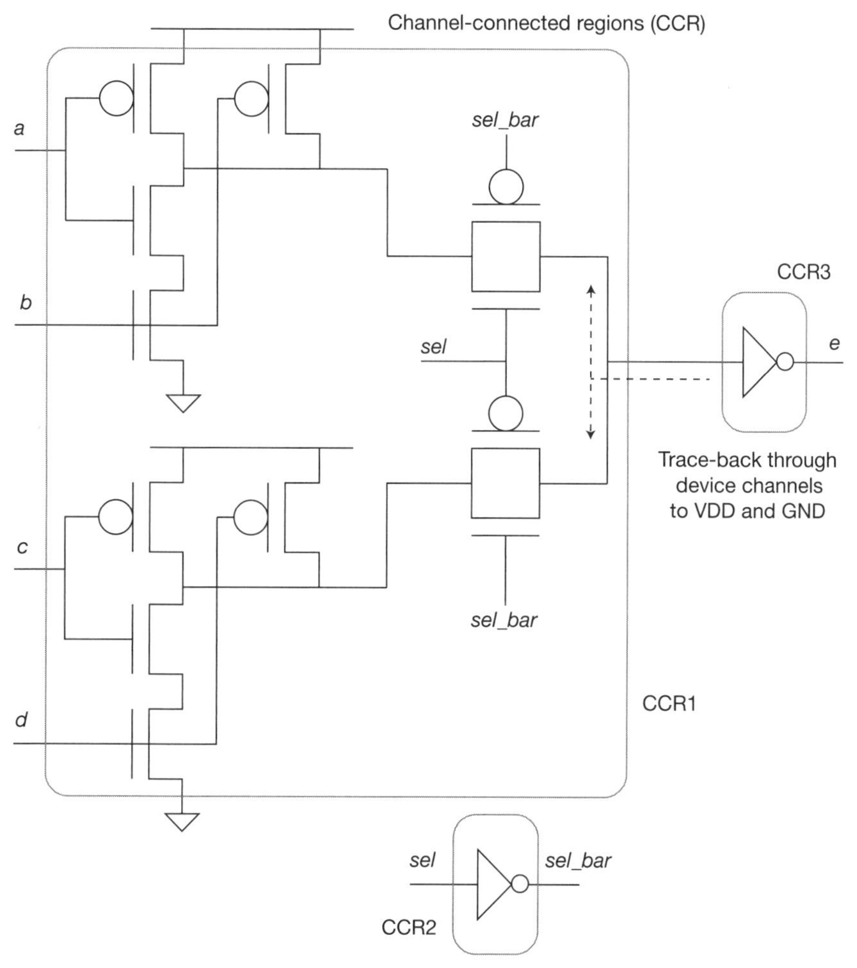


**Figure 5.8** Reduction of the device netlist model evaluates the connectivity of circuit nodes through channel-connected regions (CCRs).

Any non-complementary topologies, such as clocked flop circuits (with feedback) require more sophisticated reduction algorithms that include the temporal behavior. Specific EDA tool features are required to represent flip-flop circuit identification and reduction. The reduction of memory array storage bit cells, which are dotted connections to array bit lines and sense amplifiers and thus represent very large CCRs, presents a unique challenge. Special reduction techniques, typically using circuit topology pattern matching algorithms, are used for these complex CCRs.

A memory array circuit design commonly includes several internal clocked elements that are active over successive cycles. The symbolic functional representation for signals includes the temporal behavior over the testcase duration. As the size of the propagated symbolic function grows with each clock cycle, accumulating symbol values from prior and current network states, there is a realistic limit on the number of cycles for any symbolic simulation testcase. For the specific application of memory equivalence verification, the number of cycles associated with read and write operations is fortunately small. (Note that there is a unique challenge of schematic-to-RTL verification for test scan chains, such as for registered storage in a memory IP design outside the bit cell array that is connected in a scan chain.)

Since the memory array may also include internally generated clocks to provide specific pulse timing to sensing circuits, the functional reduction and symbolic simulation approach may include a "unit delay" timebase for propagation, within a larger clock period associated with the symbolic testcase input.

In summary, the specific task of verifying the equivalency of memory schematics to the corresponding HDL model commonly uses symbolic simulation testcases. The setup to run these simulations involves identification of:

- Pattern-matching schematic topologies
- Input pin groups for address, data, R/W enables, and clocks
- Unit-delay propagation requirements
- Device sizing guidelines for weak drive strengths that would be overwritten by other dotted drivers

The memory array IP verification team needs to develop the symbolic testcase suite; given the use of symbolic representations, functional coverage of the memory address and data space is provided by a vastly reduced number of tests. Specific attention needs to be given to the number of clock cycles in each testcase to avoid generating an unmanageable size for the symbolic propagation result.

If an SoC design project is anticipating developing memory array blocks, the methodology and equivalency verification teams need to incorporate symbolic simulation tools into the overall project flow. As the memory RTL description is likely to precede the availability of the schematic netlist to support functional SoC validation, the timeline for array verification with symbolic simulation is deferred until schematics are available. The project duration allocation for symbolic verification is under pressure to be completed as quickly as possible. Although preliminary array physical layout work may have been undertaken to provide block area estimates, final layout cannot commence until the RTL-to-schematic equivalency is proven.

If the SoC design integrates existing memory array IP, the methodology team should review the results of the array equivalency flow used by the IP provider. A schematic-to-RTL equivalency review should also be done for the cell library models from an IP provider to evaluate the symbolic or switch-level simulation methods used.

## 5.6 Simulation Throughput and Resource Planning

The SoC project management and methodology teams review the validation strategies at each level of the overall design hierarchy (i.e., block/core, SoC, and system levels). Associated with each validation phase is an estimate of the number and duration of testcases in the corresponding testcase suite when submitted to a software simulator. The project management team will have incorporated a target turn-around time for *regression* of the testcase suite (or a subset thereof) on a proposed model revision. For example, a bug fix at the block level may be targeted to "complete *n* directed tests in *x* hours." Or, a full SoC model revision may be expected to complete a set of release-gating testcases overnight so that the design and validation teams can focus on addressing any failing testcases the next day. If the regression suite passes, the new revision is released, and the SoC team can focus on developing and executing additional testcases to enhance model coverage.

The key metric for the methodology team is the throughput of the software simulator on the various model sizes, commonly denoted as "testcase clock cycles executed per second of compute server time." For an event-driven software simulation algorithm, this metric is strongly dependent on several factors:

- Internal model switching activity level during each testcase
- Characteristics of the testcase stimulus transactors and monitors
- Degree of simulation output messaging and signal tracing enabled during each testcase (or any other external file I/O)

For planning purposes, an average number of testcase cycles per server second is typically used. The number of EDA software licenses and the number of compute servers required for regression testing can thus be calculated, and the corresponding costs can be provided to the SoC project team.

The testcase cycles per server second measure is thus a common benchmark comparison used to evaluate an HDL software simulator from different EDA vendors. Model compilation time, HDL/C/C++/SystemC language semantics support, and post-simulation debug features are important, to be sure, but the key metric that most directly equates to the project cost for functional validation is the simulator throughput. EDA tool vendors have implemented multi-threaded HDL software simulation algorithms to leverage the multi-core CPU architecture prevalent in most compute servers. The throughput benchmark is therefore also dependent on the potential for model execution to be directed to a multi-core resource. Several different approaches have been taken toward multi-core implementation of event-driven simulation, from partitioning different tasks (e.g., simulation assertion evaluation, coverage statement data collection) to user-assisted model partitioning. The goal is to ensure that the event queue operations are as efficient as possible.

EDA simulator benchmark results may be quite varied. For model regression testing over a target duration, if the multi-threaded speedup with additional cores is sub-linear, it may be more worthwhile to simply run all regression testcases in single-thread mode. The methodology team needs to evaluate the trade-offs in testcase suite throughput and license/server cost between single-threaded and multi-threaded execution. The licensing of multi-threaded applications by EDA vendors uses a variety of different approaches, ranging from the simple "one base license per thread" to more complex combinations, such as "one master license plus *n* slave licenses." The license purchase negotiation with the EDA vendor should include an assessment of the extent to which multi-threaded execution will be used, especially given the unique cost structure between a master license and a slave license.

Section 5.2 describes an emphasis on assertion-based validation, in which a portion of the total validation coverage workload can be presented to formal (static analysis) algorithms in lieu of software simulation. The pursuit of ABV reduces the software simulation workload; that savings is weighed against the resources associated with incorporating the associated HDL coding style and the costs of the formal model prover EDA tool licenses. Section 5.2 also describes a validation strategy that utilizes a pseudo-random testcase pattern generator (PRPG) to provide stimulus to the model. The real-time generator would be biased in different simulation jobs to focus on specific SoC micro-architectural features. Unlike directed testcases, PRPG-based testing does not necessarily include a fixed endpoint. SoC validation resource planning requires specific attention to PRPG testing in terms of:

- What degree (and duration) of biased PRPG tests are included in a release gate regression test (at the IP core and the SoC level)
- How many PRPG tests will be running continuously, consuming EDA tool licenses and compute servers
- What compute server job dispatch policies are provided, to enable a PRPG job to be halted/checkpointed and the tool license released, so that a higher-priority (validation debug) job can be initiated

The last item in this list applies more generally to any high-priority job submitted to the compute server farm. It is common to keep the compute servers as busy as possible, executing validation jobs—especially PRPG-based jobs—assuming that sufficient simulation licenses are available. This is true even during the project validation phase in which hardware simulation acceleration resources may be used. When a critical job in another discipline is submitted, job dispatch policies are needed to assess what server(s) should be allocated to the high-priority job and what steps are appropriate for suspending/terminating existing jobs of lower priority (commonly the PRPG simulations).

The compute server policies may even designate different classes of servers to different job types. The computer memory images required for the compiled simulation model executable are typically much smaller than the model size associated with electrical analysis or physical layout design rule checking tasks. As a result, the required "server memory per core" is less for functional validation. Significant cost savings may be realized by populating the compute server farm with a "validation class" of servers with less memory. A smaller number of big-memory servers would be associated with other job types. Although the core utilization rate for the large memory footprint servers may be more sporadic if validation jobs are excluded, their availability for critical tasks from other SoC engineering teams may be improved. Compute server farm resource planning and management is directly related to the cost and schedule constraints addressed by the SoC project management.

The validation resource planning also needs to ensure that appropriate EDA tool licenses and compute servers are allocated to the debug task.

Ideally, functional error debugging is enabled by post-simulation investigation of messages, assertions, and simulation trace values. (A failing testcase without tracing enabled for highest execution throughput is typically resubmitted with selective signal tracing.) EDA simulation tool vendors commonly provide additional products to view and search waveform trace files and to cross-reference trace signals to the corresponding HDL model source statements. The number of these stand-alone waveform viewer application licenses anticipated to be used is part of the overall SoC project cost. In addition to post-simulation trace analysis, validation debug also involves running testcases in an interactive mode. A validation engineer needs to resubmit the testcase from a command line interface. The simulator executes commands that specify the time increment to advance or, alternatively, to continue until a conditional signal expression is satisfied. When the simulator is paused, signal values and the event queue can be examined. Extensive signal tracing is commonly enabled, as well, to explore waveform data generated during each simulation time increment. To assist with debugging, the simulator may include commands for the engineer to override calculated signal values with force/release settings to explore the testcase simulation results with minor "quick fix" modifications to the model.

The validation resource planning should include a number of simulation tool and waveform viewing licenses for interactive debugging use. A specific set of compute servers may also be designated for interactive debugging jobs, as the workload characteristics are different than for a testcase regression; that is, interactive jobs are not compute-intensive and may require larger memory footprints for real-time trace analysis.

## 5.7 Validation of Production Test Patterns

The production SoC test pattern set may comprise a combination of automatically generated vectors and manually defined sequences. The SoC test model is based on a detailed netlist of logic cell and IP models. Although the confidence level in the accuracy of the automatically generated stimulus and expected response pattern values is high, it is nevertheless mandatory to separately validate the test pattern set against the (tapeout release-level) functional model.

Specifically, the test vector set includes a preamble and a postamble sequence that uses the scan connectivity between flops to serially shift in the pattern stimulus and, after suitable functional clocking, serially shift out response values. The scan chains are commonly very fluid during the design implementation phase; during physical design, the placement of flop cells is updated frequently. The goal of scan chain physical design is to both reduce the wire loading and congestion for the scan_out to next scan_in pin connections and to balance the number of flops in individual scan chains (for the most efficient shift time during production test). Validation of the full test pattern set against the netlist model is required. The SoC project management and methodology teams need to assess whether this validation milestone is a prerequisite to tapeout or whether it could be completed during prototype fabrication (with the pattern set updated as required).

Two characteristics of this validation task require methodology focus:

- The netlist model is used, rather than the RTL model.
- The number of scan and functional clock cycles in the test vector set is large.

The number of simulation cycles for embedded array IP—whether exercised by an on-chip built-in self-test or through IP macro isolation and external pattern application—is large. The throughput of an event-driven software simulator is slowed by both the netlist detail and the relatively high activity level, especially during scan shifting. The test pattern set may need to be divided into smaller, isolatable testcases submitted concurrently to complete functional validation. Alternatively, a hardware acceleration platform may be a more suitable resource for this validation task, especially if this is indeed regarded as a tapeout prerequisite. Ideally, a validation testbench for acceleration would instance the SoC netlist and mimic the behavior of the production tester, as well, from loading the tester's pattern buffer to application of the test vectors and clocking sequences.

Note that the network switching during scan is immaterial to the validity of the test pattern. Nevertheless, it is still important to measure and analyze this activity, as it directly relates to the power supply source current required from the tester.

### 5.7.1 RTL Modeling of Scan

A key methodology decision is whether scan chain connectivity is to be reflected in the RTL model of the SoC. By definition, an RTL model explicitly defines all sequential states, with signal names declared for all registers. (If a high-level hardware description model is used for the SoC, where behavioral synthesis identifies and maps inferred sequential state, scan connectivity could not be represented.) Depending on the DFT architecture adopted by the SoC, it may be feasible to annotate an RTL model with the scan logic and chain connectivity, using a structural approach to register definition (see Figure 5.9).

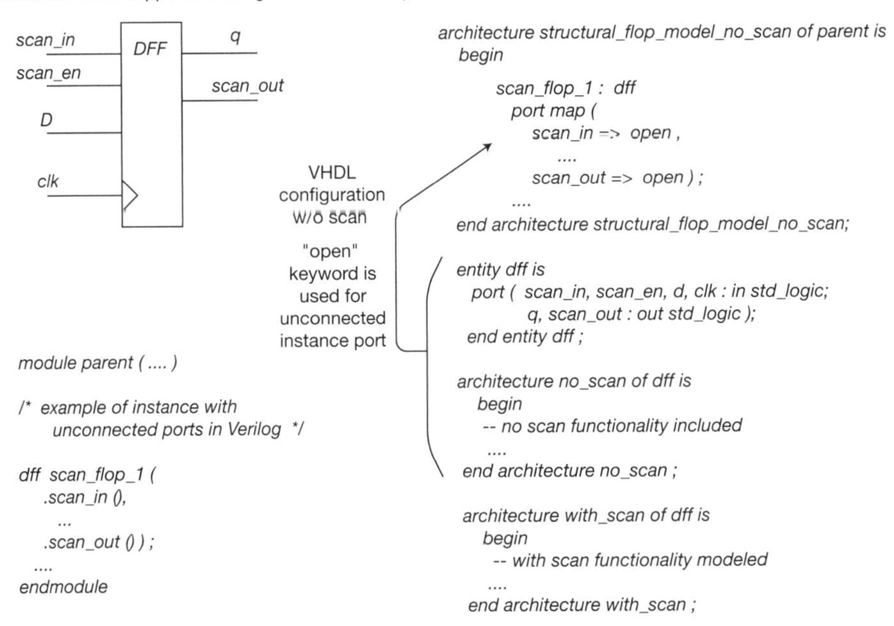


**Figure 5.9** Illustration of a "structural" RTL coding style for registers that could support subsequent annotation of a scan chain.

The motivation to include scanning in the RTL model is two-fold: (1) Scan validation of test pattern data executes more quickly and more significantly than when using the tapeout netlist, and (2) some designs use scan chains as part of the SoC functional operation. For example, an initial functional state can be established by scanning in reset values. Figure 5.10 illustrates using logical inverters inserted in the scan chain to initialize machine state to the reset pattern of 1s and 0s with a static scan chain input.

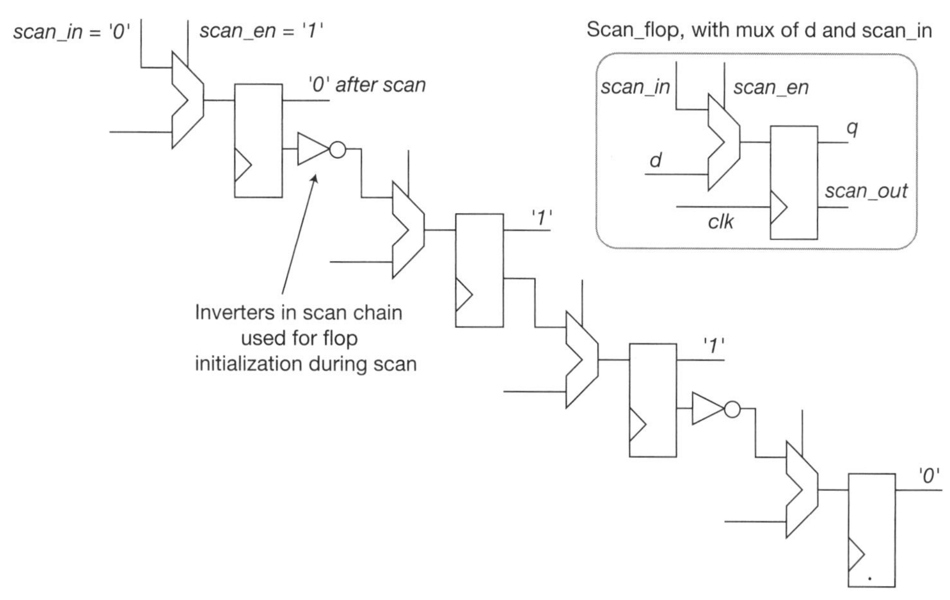


**Figure 5.10** Illustration of inverters in the scan_out to scan_in connectivity, used for functional reset.

There are significant difficulties associated with RTL scan annotation, however. SoC designs incorporate multiple scan chains, whose final implementation is driven by physical design constraints (and routing optimizations). As a result, scan chains will likely not reflect the logical RTL hierarchy. Whereas Figure 5.9 depicts a single set of (unconnected) RTL module ports to annotate with a single scan chain, the actual implementation will be much more complex. Rather than attempt to represent scan in the RTL, the netlist-based validation testbench is typically used instead.

SoC test mode operation requires unique validation approaches for production parts, as well. Reliability screening using accelerated temperature often uses test pattern subsets to exercise internal logic, known as "dynamic burn-in." Scan-shift pattern sequences are typically very efficient, capable of achieving extensive signal switching activity with a simple interfacc to the (parallel) SoCs loaded on a burn-in board. When preparing dynamic burn-in patterns, the validation and product engineering teams need to measure the (instantaneous) switching load and corresponding power delivery requirement to ensure that the burn-in chamber supply current limits are observed.

## 5.8 Event Trace Logging

The testbench developed by the validation team needs to include additional constructs to quickly pinpoint errors detected during simulation:

- Assertions (directly incorporated into the HDL model)
- Bus interface monitors
- Results comparison to predicted values:
  - Either pre-calculated or from a parallel testbench high-level model running concurrently
  - Either comparing individual transaction-based results during the test-case or using compaction for a final result at the end of the testcase

Despite efforts to develop a self-checking testbench with descriptive assertion and monitor messaging, debugging of failing testcases often requires examination of detailed signal values throughout the model over many clock cycles. Event-driven simulation tools incorporate an input list of signal values that are associated with a subsequent output trace file, written during testcase execution. Upon detection of a failure in a self-checking testcase (e.g., a fatal model assertion raised during simulation) the testcase is commonly resubmitted with selected signal tracing for stand-alone debugging.

Note that EDA simulation tools may have different methods for accepting the signal trace list and/or generating detailed signal values. Model compilation may invoke optimization algorithms to reduce the resulting executable size and improve throughput (e.g., reduction of combinational signals in an RTL model to a register-based cycle time period representation). A detailed debug signal trace input list may necessitate recompilation. Alternatively, a post-simulation debug platform may be able to utilize register-only trace values and calculate combinational signal results from the RTL description "on the fly" for each cycle.

The EDA industry has adopted two de facto standards for the output trace file format—one optimized for viewing/searching events by time increments, and one optimized for viewing/searching by signal values. These file formats enable multiple EDA vendors to offer waveform display tools with various differentiating features:

- Waveform scrolling by time or by signal value conditions
- Cross-referencing signal waveforms to RTL and testbench source code

- Efficient signal waveform representation of bus values (and complex signal types)
- Specific source-code debugging features for a SystemC- or C/C++-based testbench (e.g., representation of advanced data structures, presentation of class libraries and function inheritance relationships)
- Concurrent display of digital and analog signal values for mixed-signal simulations (typically with cell-based delay model data annotated to a gate-level netlist)

As testcase debugging time is in the critical schedule path for completing the various block/core and SoC validation phases, the productivity features of the trace file debugging toolset have a high value to the overall project.

The SoC methodology team needs to collaborate with the validation and IT management teams to estimate the resources to allocate to the debug effort, including:

- The number of waveform debugging licenses (e.g., one debug session per validation engineer)
- The compute servers where simulation testcases with tracing will be submitted (e.g., potentially requiring large local storage allocated to trace file output)
- Servers for interactive debugging job sessions, supporting a light computational workload and high memory per user

The progress of debugging efforts is recorded and monitored in the project's bug tracking database.

The goal of debugging is efficiency (i.e., a prompt bug closure rate). Resimulation of a failing testcase with tracing enabled would ideally record only the necessary signals and duration to enable the functional failure to be diagnosed, minimizing the simulation runtime impact of tracing, and optimizing the trace file size for exploratory searches. However, as described in the next two sections, in some applications, full signal trace analysis is required.

## 5.9 Model Coverage Analysis

The most common (and, perhaps, most difficult) questions posed by the SoC project management team to the validation team are: "How much simulation is enough?" and "How do you know when you are done?" Although the testbench is likely to include a set of testcases to ensure that a new SoC design maintains compatibility with existing application software, this set of tests would likely not sufficiently exercise new features in the design. SoC design teams and EDA vendors have attempted to address these questions to measure the *model coverage* of the testbench suite.

The assessment of model coverage by a testbench is the cumulative measurement from individual testcases. A single test is written to focus on validation of a specific product feature or operating condition and is thus by itself not very comprehensive. A coverage measurement database application accumulates results from individual tests into an overall summary, highlighting areas where additional tests may be warranted to fill coverage gaps.

From a project schedule planning perspective, coverage measurement is not an initial focus of the validation team. Early RTL and testbench environment release versions are likely to have a high bug rate, and debugging turnaround is the highest priority. As the design stabilizes, and the bug identification rate starts to fall, the validation team adjusts resources to address model coverage and the need to compose additional tests.

The SoC project management team may choose to measure coverage and apply target metrics at block/core levels of the SoC model hierarchy as a gate to release and promotion to the full-chip validation environment. However, the most stringent coverage targets are applied during full-chip validation and, ultimately, will serve as tapeout readiness review criteria.

Model coverage data can be measured and recorded during simulation of the testcase. HDL language semantics have incorporated non-functional statements to enable designers to explicitly describe coverage conditions of interest (e.g., the *covergroup* construct in the SystemVerilog language, consisting of *coverpoints* to record the occurrences of explicitly defined expression values of interest). In addition, EDA simulation tools often include model compilation options that add specific measurements to the compiled executable and output a coverage report after completing the testcase. A related EDA vendor utility merges the individual testcase reports into the coverage database.

Alternatively, coverage measurements can be made after simulation by analyzing the signal activity in a trace file. The advantage of post-simulation analysis is that the validation team can develop unique coverage algorithms when querying the trace file. Specific conditions of interest across the SoC design can be investigated, beyond the measures included in the simulation executable. The disadvantages of post-simulation coverage analysis are the simulation trace runtime overhead and the search/query complexity of working with very large, full SoC model trace dump files.

### 5.9.1 Types of RTL Model Statement Coverage

HDL coverage statements included by the designer or validation engineer pinpoint specific conditions of interest within the design. In addition, there is a set of commonly applied algorithms for coverage measurement that are independent of the SoC functionality and that examine only RTL statement execution. The general acceptance of these algorithms enables the EDA simulation tool vendor to incorporate them into model compilation.

#### *Signal Toggle Activity*

The most direct measure is a count of the number of binary ('0' to '1' and '1' to '0') transitions for each signal in the model during execution of the testcase. The merit of this approach is its simplicity: If a signal never toggles in the entire testbench suite, the surrounding logic network has not been exercised. Conversely, despite a signal having a high toggle count, there is no guarantee that the surrounding logic network propagates this signal value to an observable register, which would provide higher confidence that the signal functionality is correct; in this case, more sophisticated coverage algorithms are needed.

#### *Expression and Conditional Coverage*

RTL logical assignment statements typically consist of multiple compound expressions. An expression coverage measure would analyze the logical operations in the statement and count the number of signal output transitions associated with individual expressions, as illustrated in Figure 5.11.

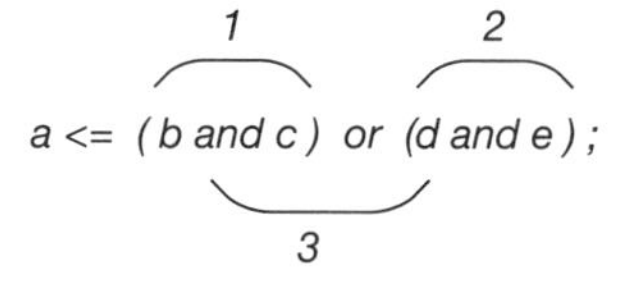


**Figure 5.11** Example of expression coverage measurements for an RTL assignment statement. The assignment statement is analyzed semantically to identify each sub-expression and the necessary expression values so that each sub-expression influences the assigned value.

This algorithm improves on simple signal toggle coverage, measuring how individual logical operators uniquely contribute to the assigned simulation value. Similarly, conditional coverage examines RTL statements with if-then-else semantics, measuring whether all branches have been exercised, as depicted in Figure 5.12.

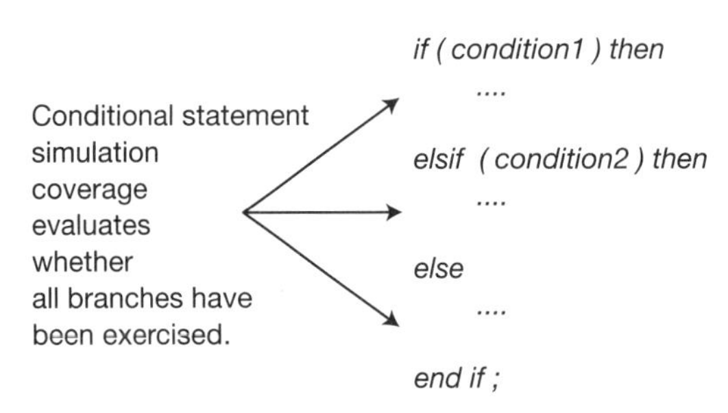


**Figure 5.12** Example of conditional statement branch coverage measurements.

For more complex conditional statements (e.g., a case statement) coverage measures would be recorded for all values of the case expression.

#### *Memory Array Coverage*

Various types of arrays are integrated on SoC designs (e.g., [multi-ported] register files, caches, stacks/queues, general randomly accessible data storage). The validation team needs to review the scope of coverage appropriate for each type. For stacks, the testbench suite needs to exercise stack full write ("overflow") and stack empty read ("underflow") condition handling. Cache read miss and cache write-allocate/write-through operations need to be exercised (specifically looking for read-after-write handling issues). Ideally, the testbench suite should span all storage addresses and exercise all read/write register file ports for coverage of address decode logic; however, the practicality of array address coverage is likely to lead the validation team to adopt a reduced approach. (The production test vectors for the array extensively cover address and data signals to detect fabrication defects and circuit sensitivities; the functional validation coverage criteria are less stringent.)

#### *Path Coverage*

An extension to the expression and conditional statement coverage algorithms applies the semantics of the HDL concurrent sequential process to develop more complex coverage criteria. For each CSP, a control-flow graph (CFG) is constructed to enumerate all the possible execution paths that could be evaluated sequentially when the process is active. (With an RTL model coding style, note that for every potential path in the CFG, the same signals are assigned values; otherwise, a signal that is missing from a path implies inferred state.) A coverage measurement target might select a representative subset of the potentially many possible start-to-end paths in the CFG; for example, coverage could be deemed satisfactory if each CFG testcase exercises an edge in the graph not present in other tests and, cumulatively, all edges are covered. The selection of CFG paths (and, for that matter, all the coverage criteria selected) is a heuristic approach to try to identify likely sources of design errors to guide ongoing testcase development.

The nature of the CSP model paradigm is that a very large number of processes may be active concurrently; as a result, it is difficult to extend path coverage to represent the interaction among processes.

Note that a simple, single concurrent signal assignment statement is itself a CSP. In this case, the expression coverage measure is effectively the same as the path coverage.

#### *Bad Machine Path Coverage*

A fundamental aspect of SoC validation is to inject (permanent or intermittent) errors into the design model and testcase stimulus to confirm SoC behavior. Much as the choice of CFG paths for coverage involves a heuristic approach, the scope of "bad machine" error injection criteria requires judicious selection of appropriate CFG paths.

As with array coverage, bad machine path coverage for functional SoC validation differs from production test pattern set development. Manufacturing faults are sensitized, propagated, and observed during application of test patterns specifically developed for defect detection, whereas bad machine paths and coverage are selected to verify functional behavior and error recovery features during field operation.

#### *Finite State Machine (FSM) Coverage*

The complexity of the control flow logic in an SoC design is great, and validation of the potential sequences of control state is difficult. The most sophisticated coverage measurement algorithms analyze RTL finite state machine coding styles, build the state machine graph, and subsequently record the state transitions exercised during a simulation test. The identification of the state machine values and input variables controlling the state transition arcs could be guided by designer input or, potentially, automatically extracted from the RTL description by an FSM coverage tool. (The tool may also be able to readily identify issues with the state transition graph without simulation—for example, unknown initialization state or unreachable or unrecoverable states.)

Complex control flow logic is often implemented using multiple concurrent interacting finite state machines. A more sophisticated algorithm for deducing coverage targets is thus required, and it involves collecting coupled pairs of transitions between pairs of state machines.

#### *Analog Mixed-Signal Coverage*

An SoC validation model that includes the interfaces between analog and digital IP requires special consideration for coverage measurements at the interface pins. For digital-to-analog input interface conversion, the full range of analog response behavior needs to be evaluated during mixed-signal circuit simulation. However, the field of analog circuit validation coverage measurement techniques remains an active area of research.

### 5.9.2 Strengths and Weaknesses of Coverage Measurement

The costs of EDA tool licenses for measuring and accumulating validation coverage data for SoC testbenches is considerable, as are the project resources applied to examining the coverage database and deciding what additional testcase development is warranted. The value of this investment has been proven empirically to find design errors that have escaped other criteria. Manual design reviews of the completed RTL code among members of the architecture and validation teams are resource intensive, require deep knowledge of the SoC micro-architecture, and are quite prone to missing functional defects. Biased random testcase generators require extensive simulation to exercise interactions and interdependencies between functional blocks, and they require considerable insight to ensure that the breadth of bias settings is sufficient. The active bug detection rate is often used as a qualitative measure of functional validation, but it can be extremely misleading: A bug identification rate of zero is definitely not a sufficient measure of design quality. The coverage measurement techniques described previously have proven to uncover RTL design defects that these other methods have missed.

However, coverage methods do have deficiencies. As mentioned in the description of the simpler measures, the data are based on local excitation of a logical expression or condition; the observability of the propagated active condition is not measured. The path coverage measure attempts to improve the observability, but realistically, the measurement targets are heuristically defined. The evaluation of FSM transition coverage reports may require assessment of the difficulty in writing additional directed testcases to exercise hard-to-reach state transitions; in this case, a manual RTL design review and/or random testcase generator setting may be a viable alternative. In addition, the overhead of simulation-based coverage execution or post-simulation trace file analysis impacts the cost and throughput of completing each validation phase of the SoC project schedule.

Nevertheless, the overall cost of functional bug discovery and design fixes increases throughout the SoC project, and it jumps tremendously if a bug is uncovered in prototype silicon. Each coverage measurement described in this section has unique merits, especially at different levels of design maturity. As a result, validation teams tend to apply all these coverage techniques at some point to accelerate bug finding. Still, the question posed at the beginning of this section—"How much simulation is enough [prior to tapeout]?"—remains elusive to answer, either definitively or empirically.

## 5.10 Switching Activity Factor Estimates for Power Dissipation Analysis

A critical collaboration between the SoC physical implementation and validation teams is to calculate the (average and peak) power dissipation for the blocks in the SoC physical floorplan. The physical design team seeks to confirm that the local power grid distribution in each block and the connections to the global grid are sufficient to satisfy the active switching current. In addition, the (local) power dissipation data are used to calculate the thermal map for the SoC die, as required for the calculation of (maximum) device temperature and, subsequently, for the electrical analysis algorithms that are strongly temperature dependent.

The switching activity of netlist-level logic signals in the validation testbench is a (linear) factor in the calculation of the device power dissipation of the driving logic gate, as illustrated in Figure 5.13.

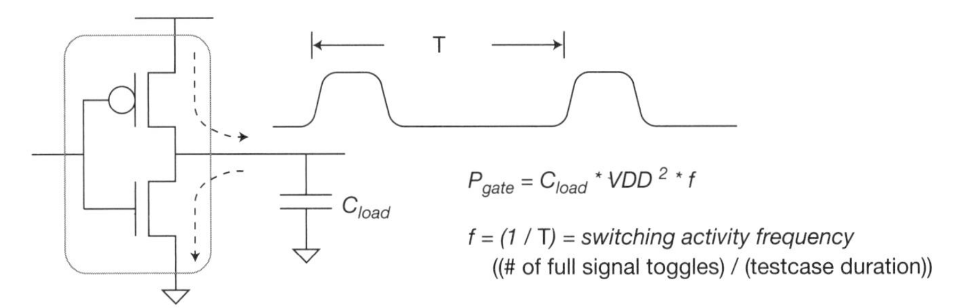


**Figure 5.13** Switching activity factor used in power dissipation calculation.

As mentioned in the previous section, a common coverage measurement technology is the signal toggle count; this count divided by the duration of the validation testcase equates to the switching frequency used in the figure. Technically, *activity factor* is a more accurate term than *frequency*, as the time duration between signal transitions in any representative power dissipation measurement testcase varies widely; the activity factor simply averages the toggle count over the testcase duration of interest.

The SoC validation team needs to develop a strategy for selecting testcases (and, likely, specific intervals within each testcase) in which signal toggle data recording is enabled. The toggle count for power calculation involves a full signal set of the model. The physical implementation provides the extracted total capacitance on each signal in the netlist for the power dissipation calculation.

Detailed power calculation involves rerunning selected validation testcases on the gate-level netlist model. However, the initial physical floorplan development also requires power dissipation estimates for each block/core when planning the power and ground grids. An emerging area for EDA tool development is to provide an early power estimation methodology, based on the RTL model and RTL validation testcases. EDA vendors of hardware simulation acceleration platforms are also incorporating switching activity RTL power estimation features, enabling validation teams to leverage much longer and more representative testcases.

## 5.11 Summary

This chapter briefly reviews the various methodology approaches to validation of the SoC functional model. The resources applied to validation are typically much larger than any other aspect of the SoC design project. Project management and methodology teams seek to find the optimum investment in software simulation and hardware-accelerated throughput, testbench development, and debug strategies while striving for thorough model evaluation coverage. Another critical facet of SoC validation not discussed in this chapter is the prototype silicon bring-up activity (see Chapter 21, "Post-Silicon Debug and Characterization ("Bring-up") and Product Qualification"). Successful SoC projects are characterized by predictable schedules and accurate resource forecasts, which rely on a thorough functional validation methodology and execution plan.

## References


[1] The Verification Methodology Manual (VMM), https://www.vmmcentral.org/grg.html.

[2] The Universal Verification Methodology (UVM), maintained by the Accellera industry consortium, http://www.accellera.org/downloads/standards/uvm.

[3] Aharon, A., et al., "Test Program Generation for Functional Verification of PowerPC Processors in IBM," *Proceedings of the 32nd Design Automation Conference (DAC)*, 1995, pp. 279–285.

[4] Wagner, I., et al., "StressTest: An Automatic Approach to Test Generation Via Activity Monitors," *Proceedings of the 42nd Design Automation Conference (DAC)*, 2005, pp. 783–788.

[5] Gajski, D.D., et al., "A System-Design Methodology: Executable-Specification Refinement," European Conference on Design Automation (EDAC), 1994.

[6] Litterick, M., "Assertion-Based Verification Using SystemVerilog," SystemVerilog Users Group (SVUG), 2007. (The presentation slides are available for download at http://www.verilab.com/files/svug_2007_abv_litterick.pdf.)

[7] Jose J., and Basheer, S., "A Comparison of Assertion Based Formal Verification with Coverage Driven Constrained Random Simulation, Experience on a Legacy IP," Design-Reuse.com, https://www.design-reuse.com/articles/18353/assertion-based-formal-verification.html.

[8] IEEE 1850-2010: "IEEE Standard for Property Specification Language (PSL)," https://standards.ieee.org/findstds/standard/1850-2010.html.

[9] IEEE 1499-1998: "IEEE Standard Interface for Hardware Description Models of Electronic Components," https://standards.ieee.org/findstds/standard/1499-1998.html.

[10] Pfister, G.F., "The IBM Yorktown Simulation Engine," *Proceedings of the IEEE*, Volume 74, Issue 6, June 1986, pp. 850–860.

[11] Beece, D.K., et al., "The IBM Engineering Verification Engine," *Proceedings of the 25th Design Automation Conference (DAC)*, 1988, pp. 218–224.

## Further Research

### Trace File Standards

Describe the characteristics of the standard Value Change Dump (VCD) trace file format for post-simulation waveform display and search (part of the Verilog HDL standard, IEEE 1364).

Describe the characteristics of the Switching Activity Interchange Format (SAIF).

### Portable Test and Stimulus Standard (PSS)

The emerging Portable Test and Stimulus Standard is intended to allow validation engineers to specify testcase stimulus and coverage characteristics for SoC blocks and interfaces at an abstract level. Subsequently, an EDA tool would interpret the PSS descriptions to generate testbenches, applicable across a range of simulation platforms (see https://accellera.org/downloads/standards/portable-stimulus).

Describe the key features of the PSS semantics, how PSS descriptions are applied across the SoC hierarchy, and how (pseudo-random) testcase data are generated. Describe how coverage measurements are defined in a PSS model.

Describe how a PSS description could be applied for reuse of (configurable) IP across multiple SoC designs.

### Property Specification Language (PSL)

Describe the key features of the IEEE PSL for advanced temporal descriptions of model assertions (IEEE 1850-2005, available at http://ieeexplore.ieee.org/document/4408637/).

Describe the features of model prover software tools that evaluate the PSL property specifications.

### Prototyping

Prototyping platforms for accelerated system-level validation offer unique capabilities to attach existing hardware as part of the system model.

Describe how hardware modules are connected to a prototype platform. Describe the specific steps required to "bridge" the frequency differences between the platform and the interface to the hardware module.

CHAPTER 6

# Characteristics of Formal Equivalency Verification

## 6.1 RTL Combinational Model Equivalency

A critical SoC project checkpoint at multiple phases of the development schedule is to prove functional equivalency between the RTL and gate netlist implementation model throughout the SoC model hierarchy. An RTL logic synthesis flow is expected to remain equivalent through the various stages of logic reduction, factoring, repowering, technology library mapping, and clock/test insertion algorithms. Yet for gate netlists that are manually captured or, especially, for ECOs applied directly to the netlist, model equivalency between RTL and netlist must be proven. Functional validation testbenches applied to both RTL and netlist is a possible approach to evaluating equivalency, although such an approach is not exhaustive. EDA vendors offer a toolset to formally prove (combinational) logical equivalency between two models: RTL-to-RTL, RTL-to-gate, or gate-to-gate netlists).

The equivalency methodology utilizes the concept of a *logic cone*, as illustrated in Figure 6.1.

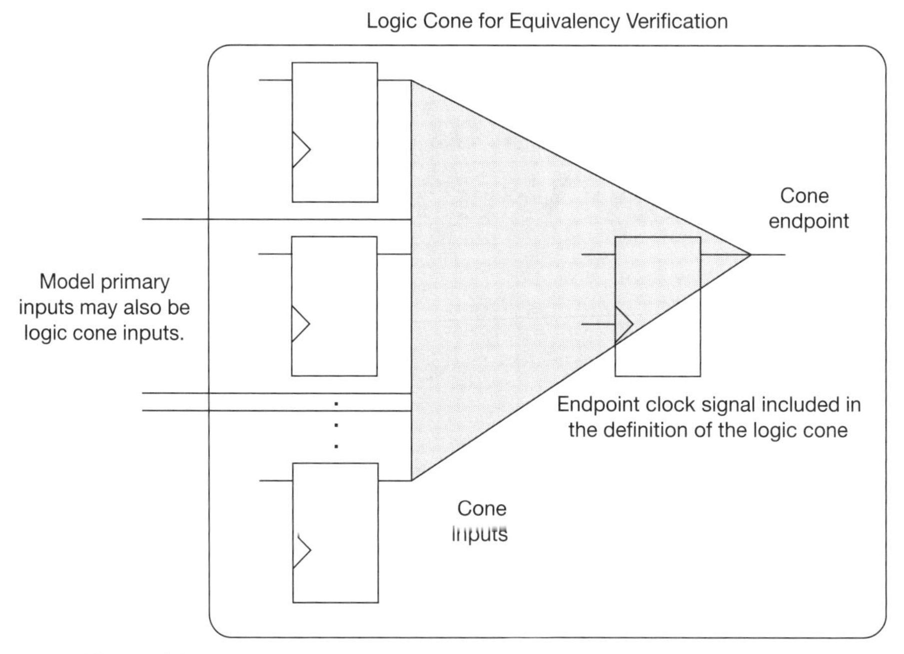


**Figure 6.1** An endpoint signal for a logic equivalency test (or a path timing test) is driven by a set of inputs that define a logic cone.

The endpoint of a logic cone, or equivalency *testpoint*, is a state element output or primary output of the model. The logic cone inputs are the comprehensive set of fan-in signals to the testpoint, traversed back through combinational logic to state element outputs and model primary inputs. The equivalency flow begins by constructing the logic cones in each of the two models through identifying the testpoints in the model and tracing backward to construct the logic network, stopping at the cone inputs. Once a correspondence is established between all test points in the two models (as described in the next section), algorithms are applied to compare the logic cone functionality.

Note that combinational logic equivalency is based on the register-transfer level (RTL) paradigm. State elements are explicitly defined in the HDL model description (and no combinational feedback loops are present). This enables the direct identification of testpoints and a logic cone traversal algorithm that results in a directed acyclic graph (DAG) with precise levelization.

## 6.2 State Mapping for Equivalency

As mentioned in the previous section, once the sets of logic cones are constructed for the two models, the next step in the equivalency flow is assigning correspondence between each testpoint in one model to a testpoint in the other. This *state mapping* step is heuristic in nature, employing a number of algorithms to investigate the two models (e.g., common testpoint signal names, cones that share the same input signal list). Note that state mapping is iterative: Confirming that two cones have the same inputs for testpoint correspondence requires prior correspondence of the fan-in signal list. To expedite state mapping, it is recommended to maintain RTL-to-gate testpoint signal name correlation wherever possible and use common primary input/output port names between the two models. EDA tools also accept a user input file, in which designers can provide recommendations for state correspondence to guide the mapping algorithms.

The user input file to the equivalency flow also identifies instances in the RTL and netlist hierarchies that are black boxes, not part of the logic cones. As illustrated in Figure 6.2, the pins of a black box introduce additional cones for analysis. This method would be applied to exclude arrays and any mixed-signal IP. It would also be used in support of a hierarchical equivalency methodology, in which child nodes in the two parent models have already been proven equivalent.

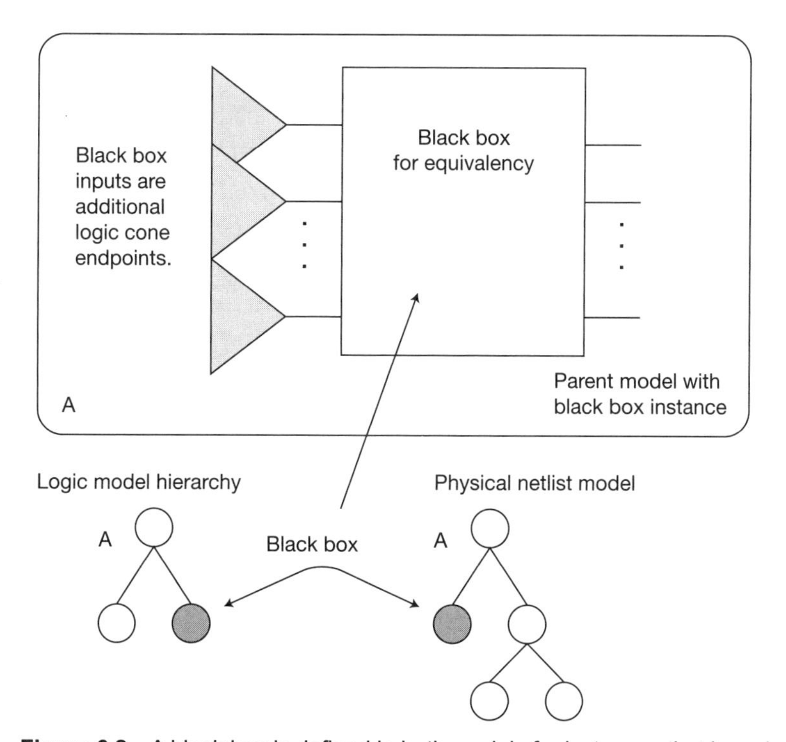


**Figure 6.2** A black box is defined in both models for instances that have been previously verified for equivalency. The input pins of the black box are added to the set of endpoint tests. The output pins of the black box serve as additional logic cone inputs.

The normal situation is that a 1:1 state mapping correspondence is established between the two models. A special case arises if the implementation model has made a parallel copy of the testpoint, as depicted in Figure 6.3. The EDA tool would need to support a 1:n state mapping and equivalency proof.

Equivalency Flow Support for 1:n State Point Mapping between Logical and Physical Models

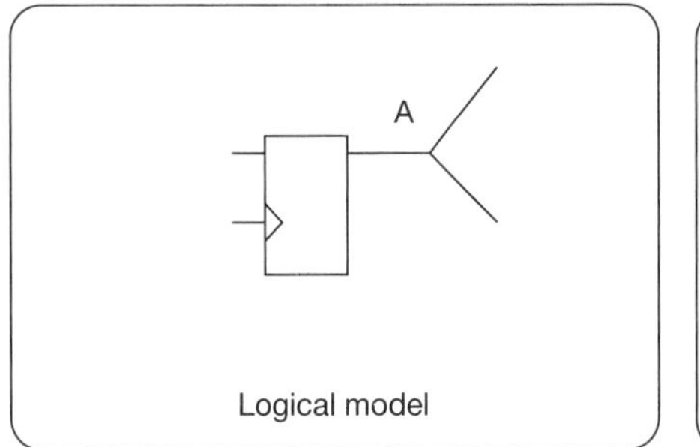


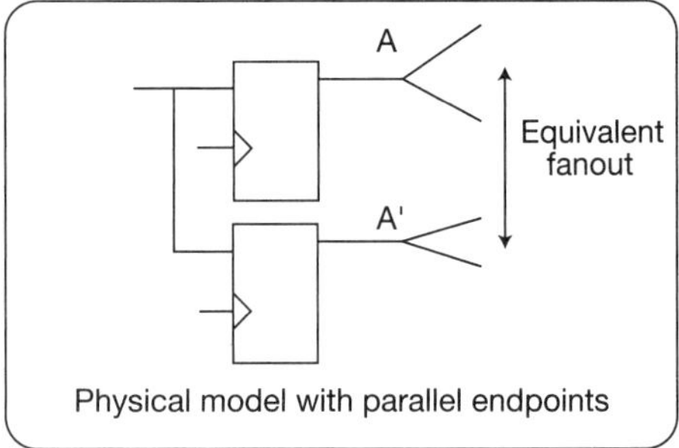


**Figure 6.3** Parallel copies of an endpoint (in the physical model) require a 1:n mapping description to be provided to the equivalency flow.

Note also that there will be connections in the gate netlist that are not present in the RTL model; for example, the result of scan test logic insertion will alter the logic cone inputs with signals such as "scan_enable". In addition to a user input description with state mapping guidance to the equivalency flow, a set of logic signals and their Boolean values to apply during equivalency will be needed to disable logic not pertinent to the equivalency comparison. (Section 5.7 discusses the focus on scan validation at the netlist level and the difficulties in representing the physical scan implementation back to the RTL model.)

One additional difference that may arise during state mapping is a logic inversion between the RTL and gate models, as illustrated in Figure 6.4.

The EDA tool may support recognition of the two cones in Figure 6.4 as *inverse equivalent* and propagate the logic inversion property for the cones where the testpoint is in the fan-in list.

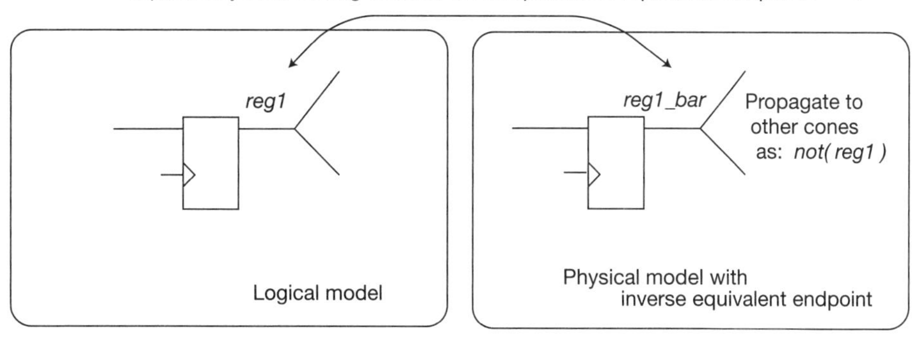


**Figure 6.4** An inverse equivalent endpoint in the two models could be accepted by the EDA tool, which will propagate the inversion property as the input to other logic cones.

## 6.3 Combinational Logic Cone Analysis

In general, the equivalence-proving algorithms make no prior assumptions on any similarities between internal logic cone signals in the two models. Also, the execution runtime increases with the input size of the cone. As a result, the formal proof may be subject to a "time limit exceeded" for a very large cone, a setting provided to the EDA tool by the equivalency flow. The EDA tool may also include a number of heuristic algorithms to look for potential additional (combinational) correspondence points within the two cones to reduce the effective size. Equivalence checking of different versions of RTL-to-RTL or gate-to-gate netlist models leverages this internal correspondence and is thus relatively straightforward. RTL-to-gate equivalency is more difficult to prove for large cones, although the adoption of RTL logic signal names within the gate netlist when feasible is certainly beneficial.

### 6.3.1 Equivalency Debug

The results of the equivalency flow fall into several categories:

- Incomplete logic cone endpoint mapping (If the initial testpoint correspondence step in the flow is not completed, additional user input to provide mapping recommendations will be required.)
- Cone too large, timeout during comparison (Again, additional input guidance to suggest [internal] cone signal correspondence between the two models will be needed.)

- Cones equivalent
- Cones provably different

If there is a provable difference between the cones of the two models, the EDA tool provides additional information to assist with further investigation. At a minimum, one or more vectors applied at the inputs to the cones are generated, highlighting the different calculated testpoint values. Additional tool debug features may present visual representations of the logic in the two cones, with the evaluated results from an applied vector annotated to signals within each cone. A key diagnostic feature would be to identify logic present in the "overlapping cones" of failing testpoints, as depicted in Figure 6.5.

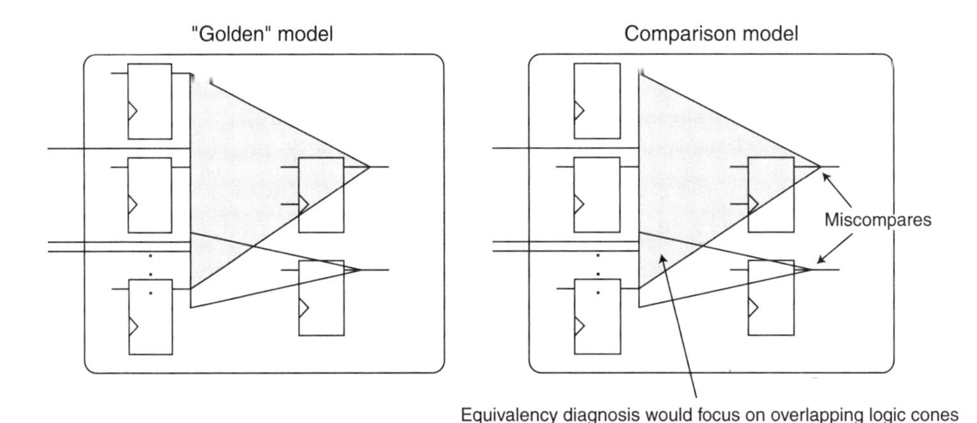


**Figure 6.5** A debug aid for an equivalency error is to identify overlapping logic in failing cones, reducing the network size within which to search for a logical difference.

Assuming that one model is "golden," the overlapping logic in failing cones in the other model is a likely source of the logic discrepancy. In general, the diagnosis of the root cause of RTL-to-gate equivalency mismatches is difficult, especially for large cones.

## 6.4 Use of Model Input Assertions for Equivalency

The previous sections in this chapter highlight some of the user input information that can help accelerate the equivalency flow. Another special case relates to the additional use of logic assertions. Figure 6.6 illustrates a simple example in which an assertion between netlist input signals is required to establish correspondence and evaluate equivalence.

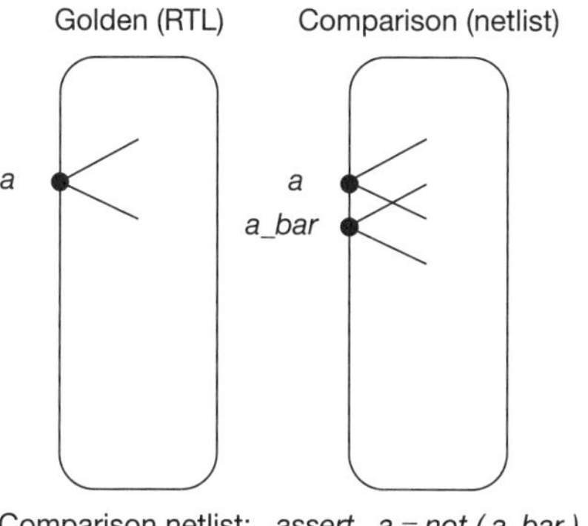


**Figure 6.6** To support equivalency verification of models with different pinouts, assertions may be provided to the flow. The figure illustrates that the netlist model for a block uses both polarities of a primary input port. These assertions need to be independently proven.

For completeness, any model assertions used in the equivalency flow need to be (formally) proven separately, such as with an EDA property prover tool.

## 6.5 Sequential Model Equivalency

Section 4.1 described the use of behavioral logic synthesis when the SoC model incorporates sequential process statements with inferred signal states. A key factor in the adoption of sequential synthesis is the corresponding availability of sequential logic equivalency checking tools from EDA vendors. These tools utilize methods that are related to model prover algorithms, as they must represent the behavior of a model over some number of clock cycles.

The general problem of sequential equivalence is illustrated in Figure 6.7.

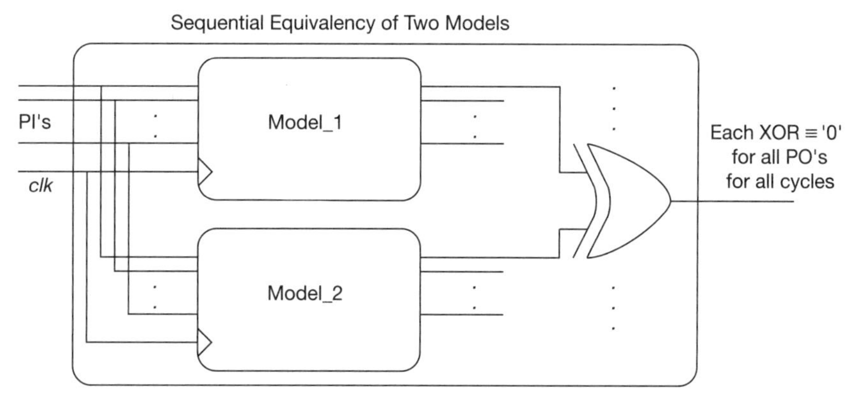


**Figure 6.7** General methodology for demonstrating sequential equivalence for two general logic networks. (Note that a common initial state between the two models is assumed.)

Some facets to sequential logic equivalence are significantly more complex than combinational equivalence. Figure 6.8 illustrates how logic is moved across state testpoints as part of sequential path timing optimization.

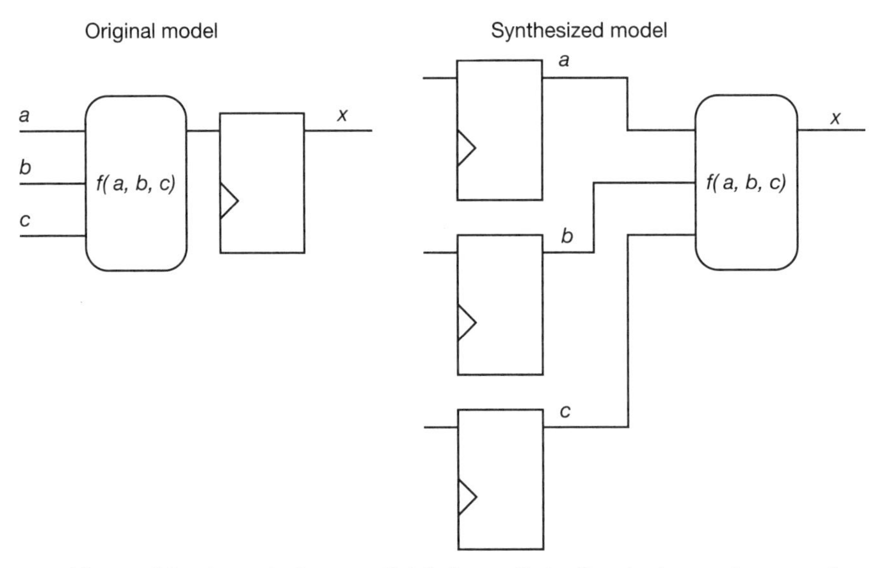


**Figure 6.8** As part of sequential timing optimization, logic may be moved across state points. Sequential equivalency verification is significantly more complex than a combinational logic cone proof.

Another example of a complex sequential optimization addresses power reduction using sequential clock gating, as illustrated in Figure 6.9.

The finite state machine implementations in the behavioral HDL and sequential synthesis output models may differ significantly, as with the sequential power gating topology in Figure 6.9. An algorithmic approach to sequential equivalency would begin in an initial state for the combined network of the two models, as depicted in Figure 6.7. Analysis would proceed by computing the set of *reachable* states over multiple clock cycles. The set of reachable states grows in cycle n+1, based on the current state in cycle n plus a specific vector of inputs. The sequential equivalence algorithm ends when the reachable state set is unchanged in successive cycles.[1,2,3]If the calculated model_1 and model_2 outputs do not differ throughout the reachability analysis, sequential equivalency is demonstrated.

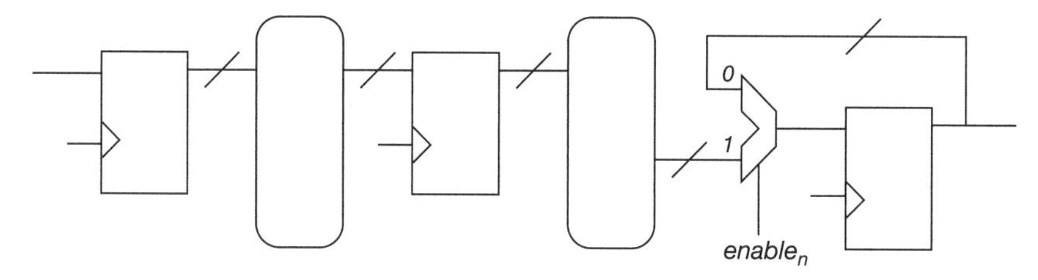


Pipeline Design with "Free-running" Clock (multiplexing data feedback)

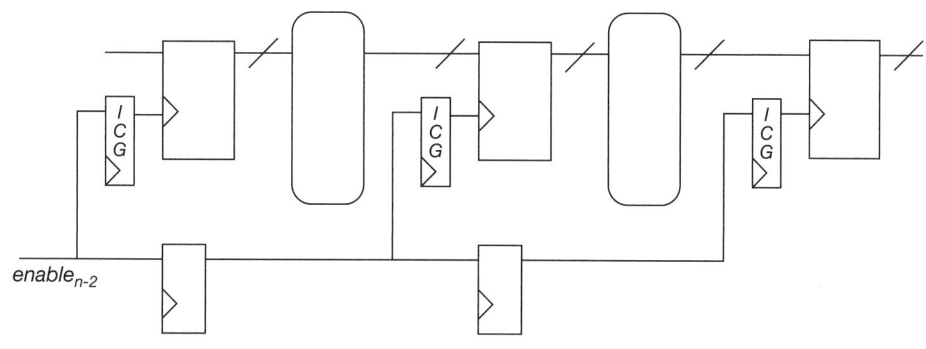


Pipeline Design with Multi-stage Sequential Clock Gating (power optimization)
ICG: Integrated Clock Gate

**Figure 6.9** Example of sequential clock gating.

The general complexity of sequential equivalence is reduced when the application is to verify a behavioral model to the corresponding RTL model generated by behavioral synthesis. The task of transitively checking the equivalence of a gate netlist logic ECO to updated RTL (using combinational equivalence) and ultimately back to an updated behavioral model is more involved.

## 6.6 Functional and Test-Mode Equivalence Verification

Section 6.4 briefly discusses the user inputs to the functional equivalence flow, including Boolean assignments to exclude netlist logic and signal connectivity that is not part of the HDL model. The typical application of these input assertions is to exclude test logic, which would commonly be inserted during the logic synthesis flow. Another area to exclude would be any netlist-level functionality associated with power gating control that is inserted during synthesis using the power format input file description. A unique area where significant logic is likely to be added to the netlist that may not be present in the HDL is the on-chip e-fuse array controller. (The fuses are programmed during production test to insert redundant SRAM resources for manufacturing defects detected on the die, after executing memory BIST pattern sequences.) In all cases of input assertions applied to modify the netlist (or HDL) model for equivalence, the methodology team needs to identify the validation strategy to evaluate the functionality of the excluded logic. Assuming that the constraints are primarily associated with the netlist model, a suite of gate-level simulation testcases is required.

The platform for simulating the netlist model, combined with the expected testcase workload, requires estimating the EDA tool licenses and compute resources to be allocated. Fortunately, the frequency of running gate-level simulation tests is typically low. The expectation is that test and power gating logic insertion algorithms will be error free, and MBIST repair-focused testcases need only be run infrequently. For any simulations used to augment functional equivalency verification, special attention needs to be given to the coverage criteria. As the netlist model is used, the EDA tools that analyze RTL-based line/expression/path coverage would not apply; manually developed coverage analysis is needed.

## 6.7 Array Equivalence Verification

Section 5.5 introduces symbolic simulation technology, which is typically used in conjunction with reduction of a transistor-level netlist to a logic model, and the reduction of (bus) signals to symbol representations. Symbols are propagated through the model during testcase execution. The complexity of simulation results increases with each successive clock cycle, which limits the general applicability of this technology. One area to which symbolic simulation is well-suited is the evaluation of memory array behavior in both an RTL model and a reduced transistor netlist.

A special class of equivalence verification compares the behavior of an RTL IP model to the device netlist used for physical implementation. For simple IP cells, such as the elements of a standard cell library, demonstrating model equivalence is relatively straightforward, using one of the following methods:

- Exhaustive testcase simulation of the RTL logic and transistor switch simulation models
- Reduction of channel-connected regions (CCRs) to an equivalent logic model for input to equivalency verification
- (Simple) symbolic simulation testcases

Figure 6.10 illustrates these equivalency options for a logic cell.

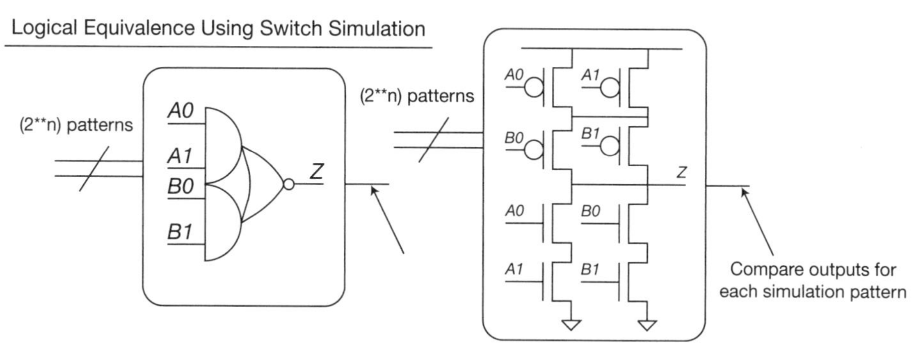


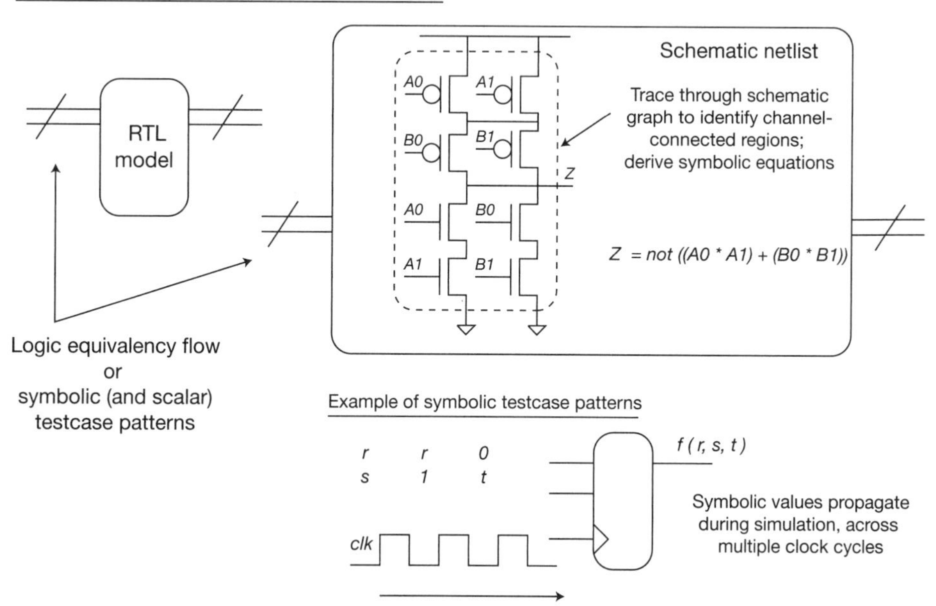


**Figure 6.10** Illustration of switch-level simulation, channel-connected region (CCR) identification and model generation, and symbolic simulation for a logic cell.

The demonstration of the equivalency of the RTL and netlist models for an SRAM array IP offering strictly uses symbolic simulation. A set of testcases is written for each array operation and applied to the RTL and device netlist models. As the number of clock cycles required to complete an array operation is typically few, the complexity of the symbolic simulation results (and the model output comparisons) is manageable.

EDA vendors offer specific products for array equivalence verification, embedding a symbolic simulator into a framework for testcase development, simulation results visualization, and discrepancy debugging (see Figure 6.11).

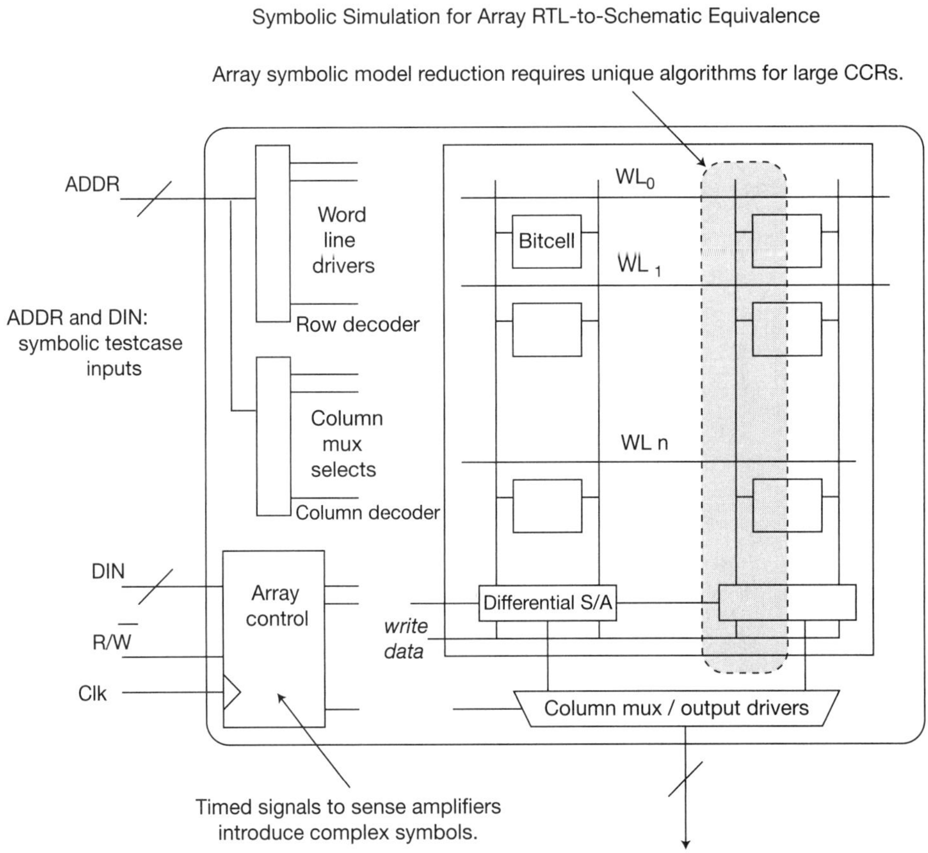

**Figure 6.11** Illustration of symbolic simulation requirements for an SRAM array. Unique topology identification algorithms are required to manage the (large) CCR topologies for dotted bitcells. Timed signals generated internally to the array result in complex simulation symbols.

To help ensure the success of this approach, the EDA vendor makes recommendations on the functionality to include/exclude from the array. The hierarchical boundary for array equivalence should exclude random logic, if at all possible, which would quickly add to the internal simulation symbol complexity. The array internal model may include registered input/output data values; although the clock cycle depth in testcases will be longer in this case, the complexity of the simulated results does not increase substantially, as the symbolic data value is not altered. Array circuits often include transistor size ratios that enable "strong" active devices to overwrite "weak" devices (e.g., weak bit line pullup devices versus the active devices driving bit lines in the array bitcell). The EDA vendor also enables the designer to define how device size ratios should be modeled as part of the initial transistor array netlist reduction.

## 6.8 Summary

The SoC methodology team must present a comprehensive strategy for demonstrating the equivalence of the (tapeout) netlist to the HDL model used for functional validation. Manually designed IP requires equivalency to be demonstrated between the implementation schematic and the corresponding HDL model released as part of the IP library. By definition, the output of synthesis flows should be logically equivalent to the HDL. A critical collaboration between the equivalency and functional validation teams is required to examine additional test and power gating logic inserted during synthesis that is not present in the HDL source. Most importantly, the insertion of late (functional or non-functional) ECOs directly to the netlist to expedite implementation, enabling minimal disruption to the existing physical design, must be reflected in the HDL model and quickly verified for equivalence.

The SoC project management team needs to decide when equivalency is required to be demonstrated during the validation and physical design schedule. It is a prerequisite for hard IP promotion to higher levels of physical integration, as well as a tapeout checklist gating item. Additional equivalency milestones may be appropriate, especially during the physical ECO implementation phase. Section 3.1 describes the development of the logical and physical model hierarchies. Correspondence points between these hierarchical models are valid candidates to submit to the equivalency flow. Full-chip equivalence can be demonstrated by equivalence of hierarchy correspondence points, which can then be black boxes when analyzing equivalence at parent nodes.

Finally, the SoC team may also seek to develop specific expertise for debugging equivalency discrepancies. Finding the root cause of logic cone differences between netlist and RTL is not easy. The skills required include detailed knowledge of RTL coding semantics, IP libraries, and synthesis flow algorithms. (Debugging symbolic simulation discrepancies between two models is an additional skill.) The documentation policies used by the SoC project to describe changes in successive revision releases of RTL model versions and (especially) the application of ECOs certainly help pinpoint where differences could have arisen. Nevertheless, the SoC design team may want to develop expertise in the equivalency flow and allocate design engineering resource to assist the validation team during the phases of the project schedule focused on model equivalence verification.

## References

[1] Kuehlmann, A., and van Eijk, C., "Combinational and Sequential Equivalence Checking," Chapter 13 of *Logic Synthesis and Verification*, S. Hassoun and T. Sasao, Editors, Kluwer Academic, 2001.

[2] Ravi., K. and Somenzi, F., "High-Density Reachability Analysis," *Proceedings of the IEEE International Conference on Computer-Aided Design (ICCAD)*, 1995, p. 154–158.

[3] van Eijk, C., "Sequential Equivalence Checking Based on Structural Similarities, "*IEEE Transactions on Computer Aided Design of Integrated Circuits and Systems*, Volume 19, Issue 7, July 2000, pp. 814–819.

## Further Research

### Array Equivalence

Hard IP memory arrays may include internal data/address registers. Describe techniques for verifying the equivalence of the scan chains through these registers between the HDL and schematic models for the array.

### Equivalence Verification with Array Repair

Section 19.3 describes how redundant rows and/or columns incorporated into a memory array may be multiplexed into the array to replace defective bitcells after die production testing. Commonly, on-die fuses are (permanently) programmed after testing to alter the array configuration. This chapter briefly describes (good) array HDL-to-schematic equivalence verification using symbolic simulation. However, methodologies for verifying the full model of redundant array topologies with fuse array programmability are currently rather ad hoc.

Research and propose methods for improving the verification robustness of the full memory array and fuse array model (with shifting of fuse values to array multiplexers).

### Data Structures for Equivalence Verification

EDA research has pursued techniques for efficient representation of combinational logic networks for equivalence verification. Many EDA tools use data structures based on the concept of a binary decision diagram (BDD).

Describe the BDD data structure and the advantages/disadvantages of this logic representation for equivalence verification.

TOPIC IV

# Design Implementation

This section of the text describes the methodology flow steps associated with the translation of the functional model of a VLSI design block to a completed physical implementation. Chapter 7 focuses on the logic synthesis of the functional description to a target technology cell library. Logic optimization algorithms are applied to the functional model to optimize the resulting cell netlist. Additional optimization algorithms are employed that evaluate the cell-level model against performance, power, and clocking constraints provided as inputs to the synthesis flow. Logic synthesis will iterate on the cell selection and signal fan-out repowering topologies to address these constraints.

Chapter 8 discusses the placement of the netlist cells on the floorplan image allocated for the design block.

Chapter 9 focuses on the completion of the netlist interconnect routes between cell pins. Throughout these design implementation flows, the estimation of interconnect electrical parasitics is a crucial aspect of the optimization algorithms, and is discussed in detail.

CHAPTER 7

# Logic Synthesis

## 7.1 Level of Hardware Description Language Modeling

This section briefly reviews the topics introduced in Section 4.1. There are three principal semantic styles for HDL coding: behavioral, register-transfer level (RTL), and structural cell-level netlist.

Behavioral-level models utilize the concurrent sequential process (CSP) paradigm for time-based model simulation. Behavioral models allow a sequential statement flow within an active process evaluation that assigns a new signal value on a subset of the possible execution paths; as a result, there is an inferred signal state that retains its value when not explicitly assigned. The RTL coding style explicitly defines all data registers and finite state machine elements, and the CSP execution flow includes no inferred state. A cell-level netlist HDL model is rarely entered manually but rather is typically generated as an output option from a schematic capture EDA tool, using the schematic symbol for each cell released as part of the IP library data.

All three coding styles can be submitted to a set of algorithms incorporated into a logic synthesis flow. For the behavioral model, the (intermediate) output is an RTL model, with additional state registers and FSM state encoding added. For the RTL model, the synthesis flow output is a cell-level model, after a set of library mapping and power/performance tuning algorithms have been applied. Logic optimizations during synthesis include Boolean constant propagation, logic redundancy removal, and common sub-expression logic factoring. For an initial cell-level netlist input model, the synthesis flow output is another cell-level netlist, with instances of the same (or possibly retargeted) technology cell library to which similar power/performance optimization algorithms have been applied, as during RTL synthesis.

It is certainly feasible to include a cell-level model as an instance into an RTL model hierarchy to override the synthesis mapping algorithm with a specific, manually optimized implementation. The synthesis flow includes the capability to exclude this netlist from further optimizations (e.g., a "don't touch" directive) or present the cells and mapped RTL together to subsequent algorithms.

Chapter 6 discusses the importance of demonstrating the functional equivalency of the original model and the output of the logic synthesis flow, especially as additional test and power-gating logic are to be inserted into the final output netlist.

## 7.2 Generation and Verification of Timing Constraints

After technology cell mapping from an optimized logic model, the logic synthesis flow attempts to meet performance targets through a set of algorithms that select drive strengths of a logic cell in the library and/or perform local logic network modifications (see Section 7.4). Interconnect loading estimates are incorporated into the netlist model to try to reflect logic stage delays that approximate what will be realized during detailed physical design.

The SoC project input to the synthesis flow describes a set of *timing constraints* specific to the block that represent the timing optimization targets. Figure 7.1 illustrates some of the information contained in the constraints file input:

- Primary input pin expected arrival time, estimated input capacitance, and target (max) input signal slew
- Primary output pin required arrival time and estimated output pin loading
- Internal clock arrival skew budget (physical design target)
- Any timing modes
- Any timing "don't cares"

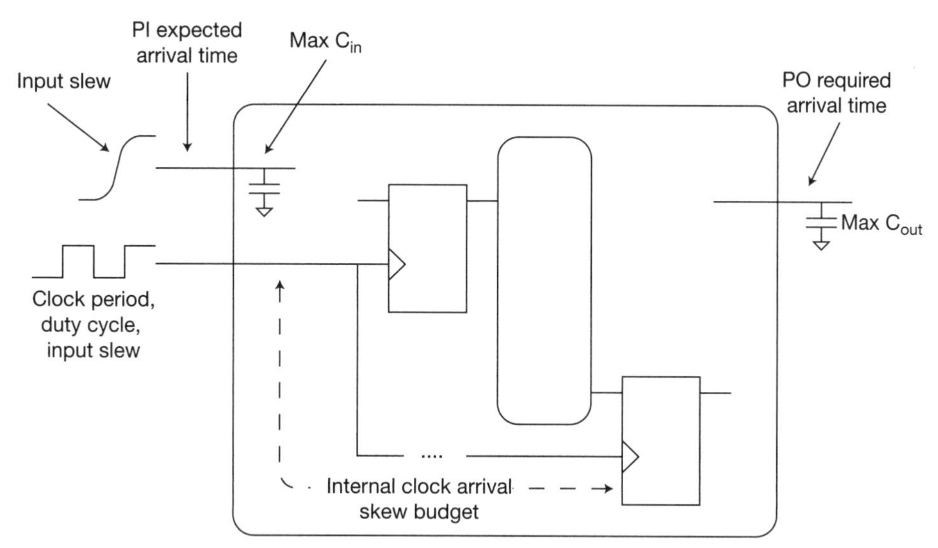


**Figure 7.1** Example of timing constraints provided as input to the logic synthesis flow.

Note that these constraint values are required for each timing corner for which the synthesis flow will be directed to calculate cell and cumulative path delays. Timing corners could reflect different voltage and temperature operating environment conditions, as well as a set of fabrication process variation parameters, commonly known as the *PVT corner*. In addition, constraints may also identify different functional operating modes, for which the synthesis algorithm should evaluate the technology performance. For example, a synthesized block may have distinct timing constraints associated with active and power-saving modes. Also, after test logic and test signal connectivity are inserted, performance targets are associated with the scan shift operations in a test mode.

Note that the terms *early mode* and *late mode* path delay calculations during synthesis are also often used. Technically, these are not functional operating modes. Rather, these terms represent the synthesis algorithms used to check for hold time and setup time calculations against the values in the cell library model for a clocked element, to be evaluated at each PVT corner for each mode. Note also that the synthesis algorithms need to incorporate suitable interconnect loading estimates for each PVT corner. Perhaps the most intricate optimization algorithm applied across the corners and modes relates to the construction of a suitable clock repowering tree network; this high fan-out distribution strives to satisfy stringent arrival latency and skew targets across a range of conditions.

With the advent of wider application markets for SoC designs, the number of environments and operating mode conditions has increased. EDA synthesis tool vendors have focused on concurrent multi-corner, multi-mode (MCMM) execution, with appropriate internal data structures and timing optimization algorithms to keep the compute resources and runtime in check.

Another possible synthesis constraint is to identify acceptable *multi-cycle paths*. Rather than attempt to realize a logic path in one period of the clock, a constraint may direct the timing optimization algorithms to apply setup timing tests at the path endpoint using a multiple of the clock cycle time. One potential implementation of this feature would be to associate a *delay adjust* constraint at a specific signal in the network, as illustrated in Figure 7.2.

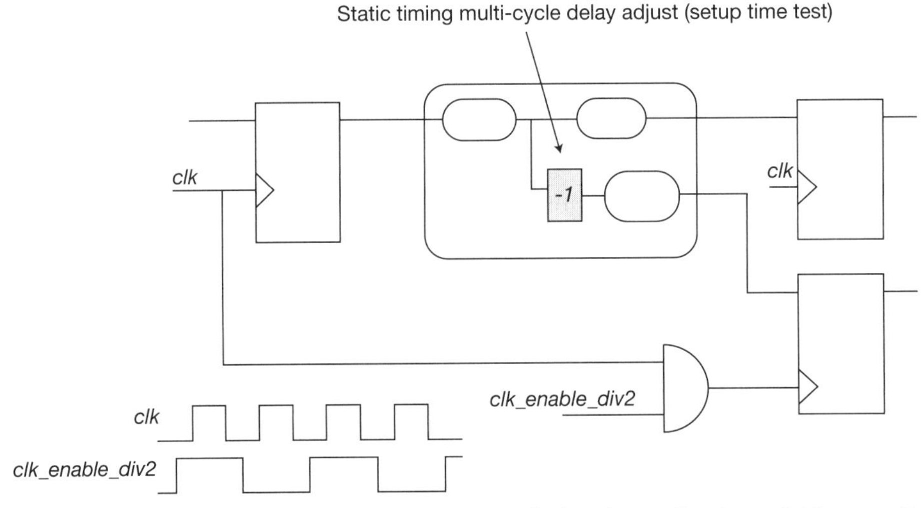


**Figure 7.2** A delay adjust directive is applied to the synthesis model for a multi-cycle timing path. The use of a delay adjust assigned to a signal allows the timing analysis tool to apply a "single cycle" algorithm throughout the network.

Rather than apply a delay adjust to a signal in the network, a more elaborate multi-cycle constraint implementation utilizes an enumerated "from-to" path description. Note that the hold check test at the multi-cycle path endpoint differs from the setup test; to ensure data integrity, the hold check is based on the common cycle clock arrival skew.

An extremely powerful synthesis flow constraint is to identify timing optimization "don't cares" (also commonly known as *false paths*) within the design block. This constraint removes paths from timing tests and thus should be used very judiciously (and reviewed thoroughly). These constraints could arise due to multiple signals being connected by a path in the network, but a transition on one cannot logically result in a transition on the other. (Recall that a cell's input-to-output pin arc and interconnect delay calculations used in path delay measures are independent of specific logic values.) Alternatively, a constraint could identify a signal that defines an operating mode, whose timing transition is not critical to the other path delays. Figure 7.3 illustrates simple examples of the need for a false path constraint.

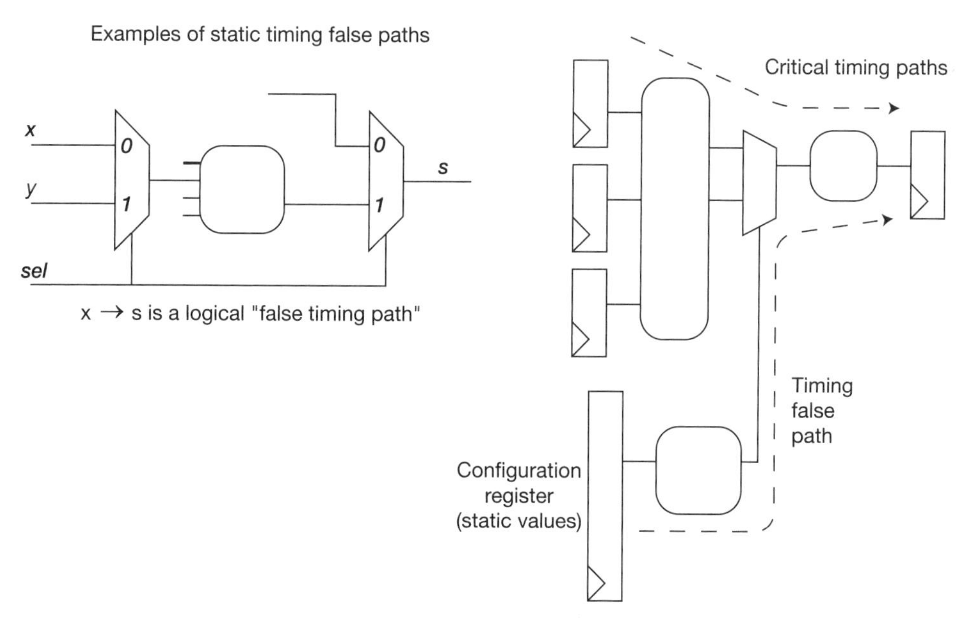


**Figure 7.3** Illustration of logic networks where a false path timing constraint directive to the logic synthesis flow would be applied.

Note that the data interface between asynchronous clock domains is also effectively a timing false path (and requires the addition of a set of synchronizing flops).

Speaking of clock domains, each of the figures used to this point to illustrate timing constraints for synthesis depict paths with a clocked element as a timing reference point:

- Primary input pin-to-register (input-to-capture path endpoint)
- Register-to-register (launch-to-capture path endpoint)
- Register-to-primary output pin (launch-to-output)

It is feasible, but more intricate, to include combinational paths between primary input and primary output pins in the synthesized block, without a clocked element in the path. As the SoC design evolves, each block's timing constraints are adjusted. A failing register-to-PO timing test in one synthesized block needs to allocate a greater percentage of the cycle time, requiring blocks in the fan-out of the PO to relinquish some of their PI-to-register timing margin. The SoC project management team maintains a timing report scoreboard that tracks the current timing results and highlights where timing constraint values need updates. If a block contains purely combinational paths, the initial timing constraint budgets and subsequent re-allocation decisions are considerably more difficult.

There may also be synthesis flow constraints that set relative priorities between area and performance when exercising optimization algorithms.

As mentioned earlier in this section, the accuracy of synthesis timing optimizations can be improved by incorporating physical design information into the interconnect loading assumptions. Floorplan and metal routing layer characteristics allocated for the synthesized block are provided to guide *physically aware logic synthesis*.[1] Physical synthesis starts with the initial mapping of the (minimized) RTL logic to a cell library—either a general Boolean library (with area estimates) or cells from the target technology library with default threshold voltage and drive strength. The resulting intermediate cell netlist from this initial mapping is placed in the floorplan area, typically using simple, fast cell affinity criteria. Interconnect routes are estimated for the placed cells to provide a representative capacitive loading and RC delay. Technology library cell selection now proceeds (iteratively), using timing optimization algorithms. The resulting output of a physically aware synthesis flow is the cell netlist, timing reports, (potentially) power estimation reports, and a starting cell placement definition. The placement is likely not legal, as there may be cell overlaps and abutment violations. A detailed placement legalization step is the first algorithm to be applied in the subsequent placement flow.

Physical synthesis enables greater accuracy in cell delay estimates due to the availability of representative routes for each net. (Prior to the availability of physical synthesis tools from EDA vendors, more rudimentary "wireload per fan-out" estimates were used for delay calculation.) Physical synthesis also offers additional insights prior to running detailed physical implementation flows. The set of cell pins in a net defines a physical *bounding box*. The route segments within this box used for loading estimates also provide insights into potential local route congestion. The physical synthesis results may necessitate a review of the floorplan area and/or the development of additional placement optimizations for more efficient routing track utilization (e.g., *clustering* of individual cells into multi-bit library offerings or provision of user-defined *relative cell placement* directives). Similarly, the physical synthesis data may highlight block floorplan primary input/output pin placements that are significant contributors to path delay or route congestion; feedback to the global integration team from block physical synthesis on pin assignments is critically important. Andrew Kahng's "Physical Synthesis 2.0" provides experimental data on the impact of clock pin location on the resulting clock distribution performance after physical synthesis[2].

There are physical synthesis limitations, as well. The routing estimates are improved over basic wireload per fan-out assumptions, but the capacitive loading calculation for these estimates does not include the detailed coupling capacitance associated with adjacent routes. Also, for block floorplans that utilize several metal routing layers, the RC delay characteristics improve up the metal stack, where thicker/wider layers are used. The routing estimates in physical synthesis are likely to use a characteristic R*C per length metric that is an average value over the available metal layers and is subject to change during detailed (timing-driven) route optimization. To try to absorb these differences between physical synthesis and physical implementation, SoC designers may provide synthesis timing constraints that are more aggressive than the actual performance specification (e.g., provide a clock period constraint that is shorter than the actual design target). At first glance, it would appear that this margining approach would reduce the difficulty in closing timing during physical design. However, it might result in a final synthesized netlist that has incorporated significant signal repowering to meet the more aggressive cycle time, at the expense of area and power. In short, to be applied effectively, constraint margining during synthesis requires engineering judgment.

Finally, there is a risk that the clock tree distribution in the output netlist from physical synthesis may not subsequently satisfy the target clock latency and clock skew constraints provided. EDA vendors incorporate numerous features for the optimization of clock tree buffers and (balanced) fan-out routes during physical design to mitigate this risk.

### 7.2.1 $C_{total}$ and $C_{eff}$

There is an additional, extremely significant, characteristic of capacitive constraints provided as part of the synthesis flow input. Figure 7.4 illustrates the concept of an *effective capacitance load,* $C_{eff}$, which differs from the *total capacitance load,* $C_{total}$. The figure illustrates a net with several fan-out segments on resistive metal layers.

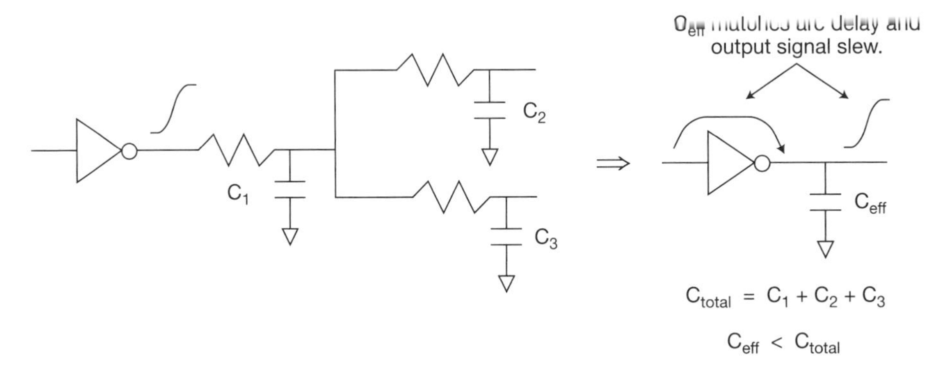


**Figure 7.4** Illustration of the effective capacitance load ($C_{eff}$) on a cell output pin compared to the total capacitance, $C_{eff} < C_{total}$.

Due to *resistive shielding*, the effective load on the driving cell pin is less than $C_{total}$. The cell input-to-output arc delay and signal slew rate at the output pin are in response to $C_{eff}$, not $C_{total}$. From the output pin driving waveform, the individual R*C interconnect delay adders and waveforms to the input pins of fan-out cells can be derived to determine the arrival time and slew rate. O'Brien and Savarino wrote a landmark technical paper that defines effective capacitance and the division of total output delays into cell and interconnect contributions[3]. The reduction of complex fan-out topologies to a single $C_{eff}$ is fundamental to the cell delay characterization methodology for library release.

From the perspective of synthesis constraints, the required arrival time and target loading at block primary outputs should be based on a $C_{eff}$ calculation for the global net. The SoC integration team is responsible for allocating global R*C delays between blocks, as depicted in Figure 7.5.

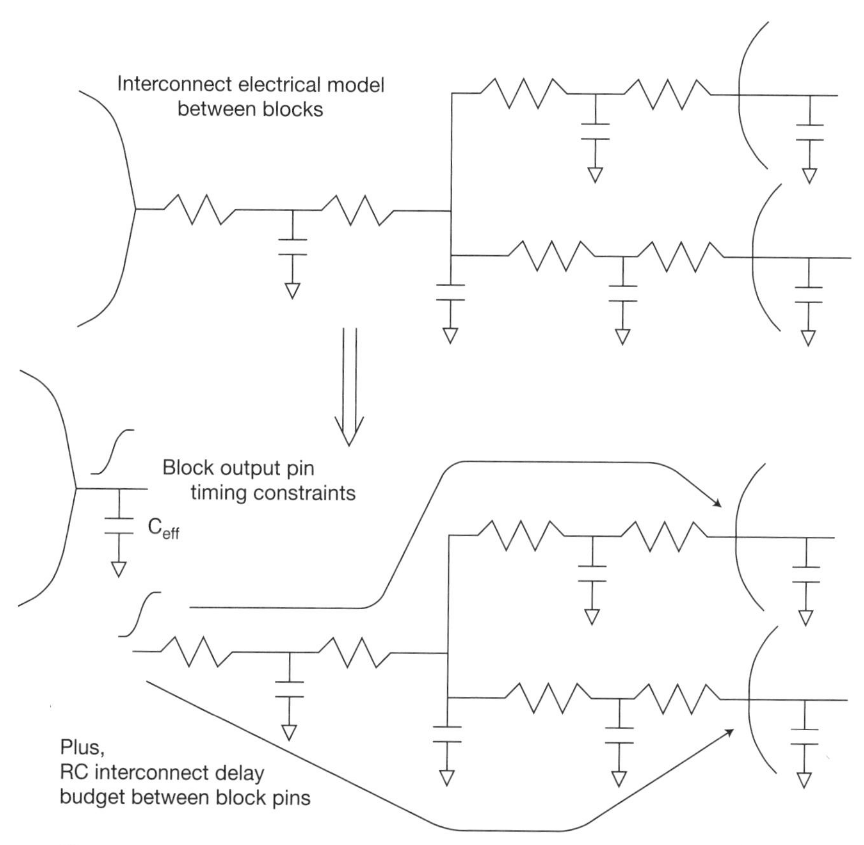


**Figure 7.5** Synthesis timing constraints require budgeting for global interconnect delays between blocks.

Note that there is an implicit assumption for synthesis constraints that the R*C interconnect delay from a primary input pin to the fan-out pins within the block is small. The global timing model in Figure 7.5 relies on a fixed capacitive block input pin load, based on the synthesized cell input pin capacitances. A significant R*C interconnect from the primary input pin to an internal cell would invalidate the timing constraint model to some extent (see Figure 7.6).

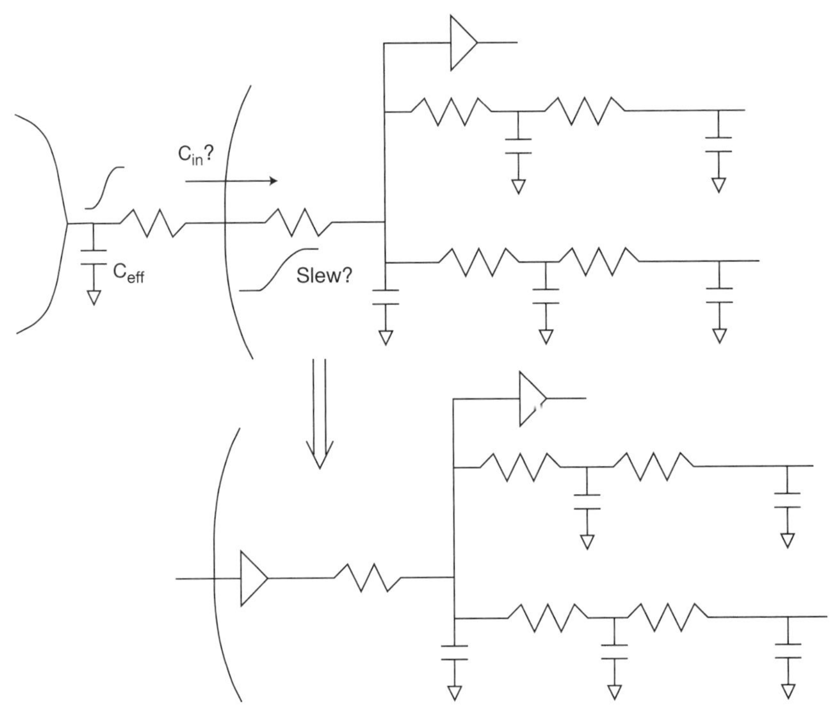


**Figure 7.6** A distributed R*C interconnect tree from a block primary input pin is inconsistent with the timing constraints used in the synthesis flow.

To maintain the accuracy of the timing and capacitive constraint assumptions, cells connected to PI/PO pins are strongly weighted in the placement flow to be assigned a location close to the pins. Buffering circuits are commonly inserted in synthesis to reduce the total R*C interconnect network between a pin and internal cells.

## 7.3 Technology Mapping to the Cell Library

After logic minimization of the RTL model in the synthesis flow, the resulting logic expressions are mapped to the target technology library (often with an intermediate Boolean library mapping as the initial structural netlist model). As an aside, there are several strategies for RTL coding that direct synthesis to a specific topology during mapping (e.g., embedded functions and logic assignment statements with "map as written" directives).

The complexity of the library varies from technology to technology. Early bipolar and CMOS circuit families had a minimal number of logic cells (e.g., a few dozen), whereas current CMOS libraries are much richer. One of the key technology library characteristics is the maximum fan-in count for a series stack of transistors, as illustrated in Figure 7.7.

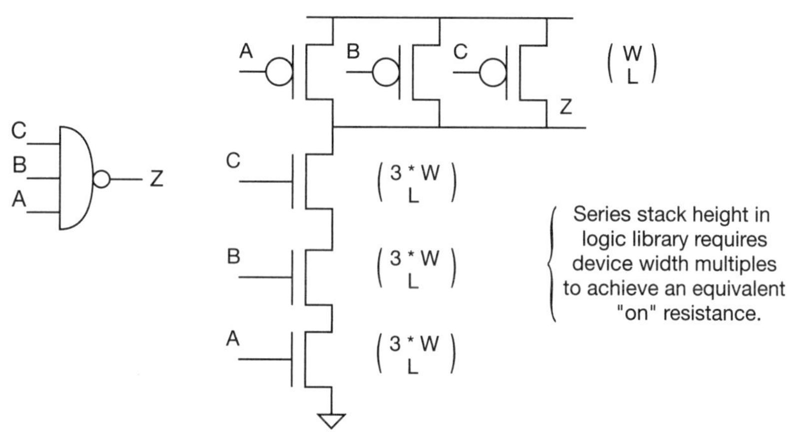


**Figure 7.7** Illustration of the maximum device stack height (and logic gate fan-in) for the cell library.

Each logic cell is typically designed to have a pin-to-pin arc delay comparable to a simple inverter. Indeed, a technology library performance metric that is often used is the stage delay of an inverter driving a fan-out of four identical inverters, or the *FO4 delay*. (The FO4 measure is derived without an R*C interconnect delay adder; it is primarily a reference to the device performance, not the full device plus metallization stack process capabilities.) To achieve comparable performance to an inverter, higher fan-in cells require commensurately wider series devices, with a significant impact on cell area. Also, the series device connection to the cell output suffers from the device *body effect*, which adversely affects performance. As described in Section 1.2, current FinFET and FD-SOI process options have a reduced body effect (and, for FinFETs, higher transistor drive current per unit area), suggesting that higher series fan-in cells may be attractive, especially if the final synthesized netlist results in fewer logic stages in critical timing paths. As a result, current cell libraries also include a rich set of AND*OR*INVERT logic primitives, which are relatively efficient to implement in a CMOS technology. Multiplexing logic (MUX) cells are typically a large subset of the library; a *transfer gate* is a unique CMOS circuit that can result in efficient select and data steering logic. Full adders, half adders, and exclusive OR/NOR (parity) functions are also commonly available.

Cell libraries commonly do not include dynamic logic (e.g., precharge/evaluate clocked "domino" logic) or high-impedance (tri-state) functions. These are difficult to map from an RTL definition, and there are extensive methodology flow requirements to ensure that the corresponding circuits are electrically robust in a general physical block design.

There are multi-stage circuits in the target cell library. Primitive CMOS logic circuits (e.g., NAND, NOR, AOI) are by nature inverting functions. The addition of an output inverter to the cell provides an opportunity to map directly to the (more common) non-inverting operators in logic expressions and case statements in the RTL model. For high fan-out signals, the addition of an inverter in the multi-stage cell offers an efficient means of minimizing the interconnect loading between stages and providing multiple drive strength variants (see Figure 7.8).

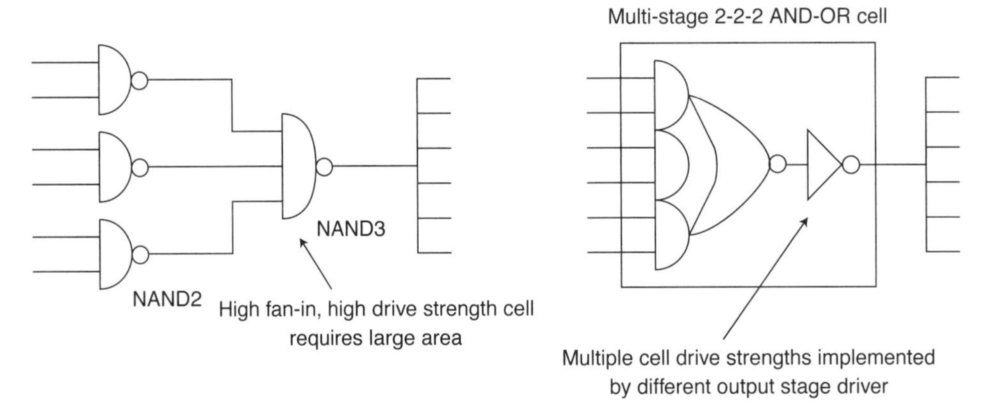


**Figure 7.8** Illustration of multi-stage logic cells in the library to efficiently provide multiple drive strength options.

Some cell libraries may include unique logic definitions, where the inputs are not swappable (symmetric), as illustrated in Figure 7.9.

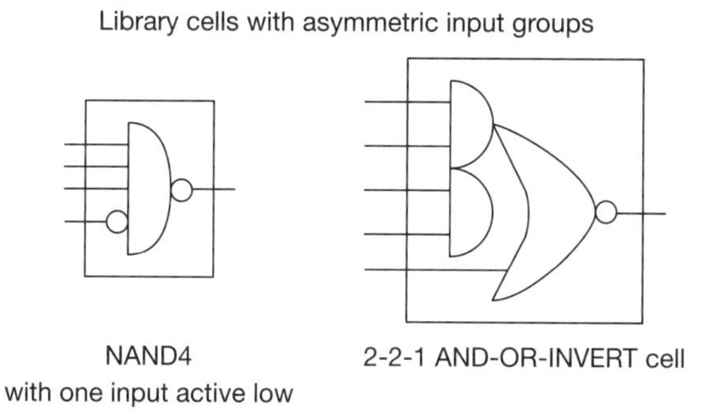


**Figure 7.9** Illustration of logic functionality embedded in logic cell inputs.

These functions are more difficult to map directly in the initial stages of the synthesis flow but may subsequently be selected by a logic compression algorithm.

Buffering cells (inverting and non-inverting) of a wide range of drive strengths are available during subsequent timing optimizations but typically are not used as part of the initial mapping. The exception is the insertion of clock buffers—and especially clock gating cells—to the mapped model as part of the clock tree synthesis step.

Logic network mapping and clock tree synthesis are closely interdependent, as depicted in Figure 7.10. For logic conditions where a flop retains its value at the next cycle, the mapping algorithm needs to decide whether a free-running or gated clock topology is warranted.

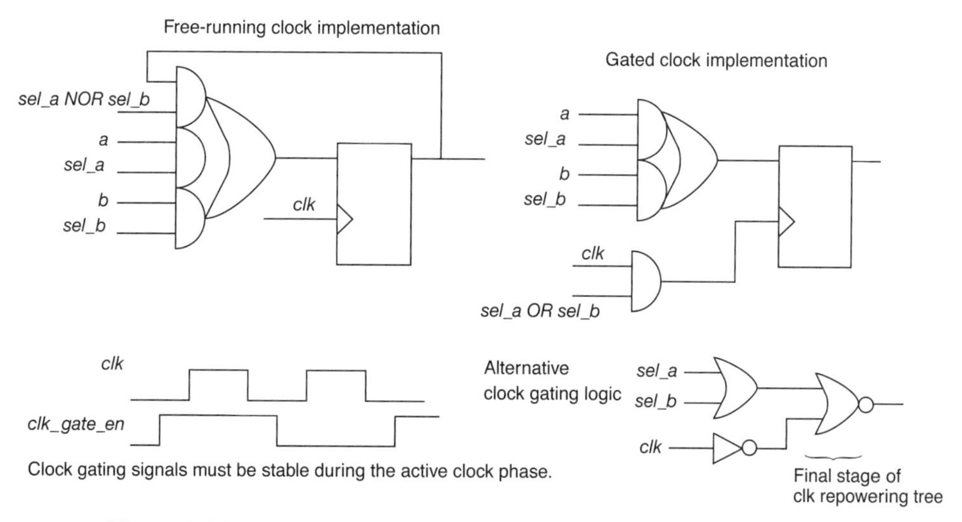


**Figure 7.10** Illustration of free-running versus gated clock logic inputs to a flip-flop.

In addition to containing clock buffers, the library also contains clock gate cells in a wide range of drive strengths. The decision to map to a free-running or gated clock topology is an intricate one. If the number of flops with a common clock gate signal is large, there is an area savings to implementing a gated clock topology. If the percentage of cycles during which the clock is disabled is large, the gated clock topology results in significant power reduction, which is a critical synthesis optimization objective for SoC designs targeting battery-powered applications. In the simple clock gate design in Figure 7.10, the correct functionality requires stringent arrival time requirements of the enable condition; the enable must transition only during the inactive clock phase, and it must satisfy a setup timing test to the active clock transition. An *integrated clock gate* (ICG) cell would be added to the library, and a latch element is included to ensure that the proper enable timing conditions are met (see Figure 7.11).

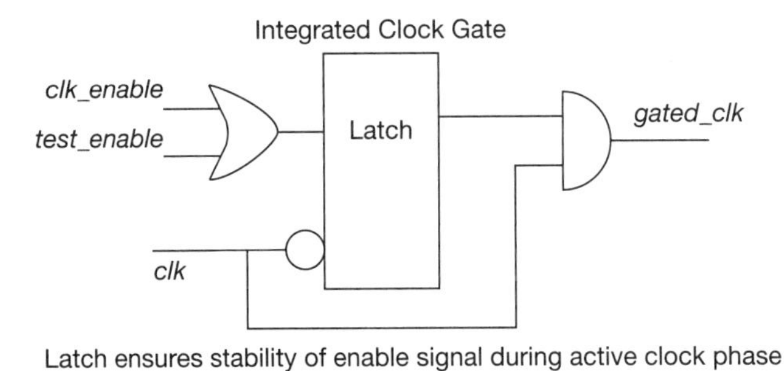


**Figure 7.11** Integrated clock gate (ICG) cell with a latch added to simplify the clk_enable timing constraints.

The ICG cell in the figure includes a test_enable input, as well, to allow direct control of the functional clocking during testing.

The mapping of RTL state signals to flops is in some sense straightforward because the RTL model has explicit definitions of all internal states. However, the (final) mapping is indeed more intricate, due to the following:

- Translation of initial values to (asynchronous or synchronous) flop set/reset logic
- Active high and active low clock assignments in the RTL, translated to the available library flops

- Insertion of test logic and scan chain connections
- Flop cells that offer both Q and $Q_{bar}$ outputs
- Flop cells that integrate a logic stage into the data input
- Multi-bit flops

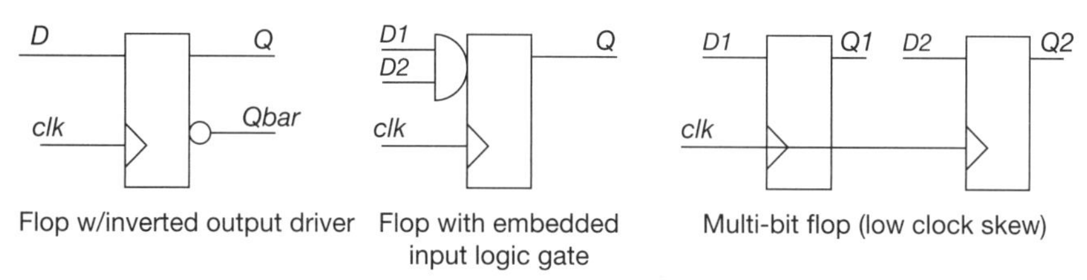


**Figure 7.12** Various flip-flop types commonly provided in the logic cell library.

Figure 7.12 illustrates some of these complex flop types. The last three flop cell types in the list above offer unique options. Subsequent to the initial technology mapping, a logic compression algorithm may seek to subsume a mapped cell into another complex cell that integrates that function. In some CMOS flop circuits, a logic function can be efficiently added to the data input. (Cell characterization now references setup and hold timing tests to the cell input pins with the logic gate embedded in the circuit simulation model.) It may also be circuit efficient to provide both polarities of the registered data as circuit outputs. Providing both Q and Qbar flop outputs needs to be pursued cautiously. For some circuit types, the output load on the clock-to-Q delay arc also strongly affects the clock-to-Qbar arc; the delay calculation tools in synthesis and timing analysis flows may not support this interdependence. The cell library may offer multi-bit flops, providing an overall reduction in clock tree loading, at the risk of potentially higher local route congestion by consolidating multiple flop input and output pin connections into a single cell.

If a power-gating implementation is inserted in the synthesized block, a request for internal state retention in the power format file results in mapping to the flop cell with integrated retention storage, as depicted in Figure 7.13.

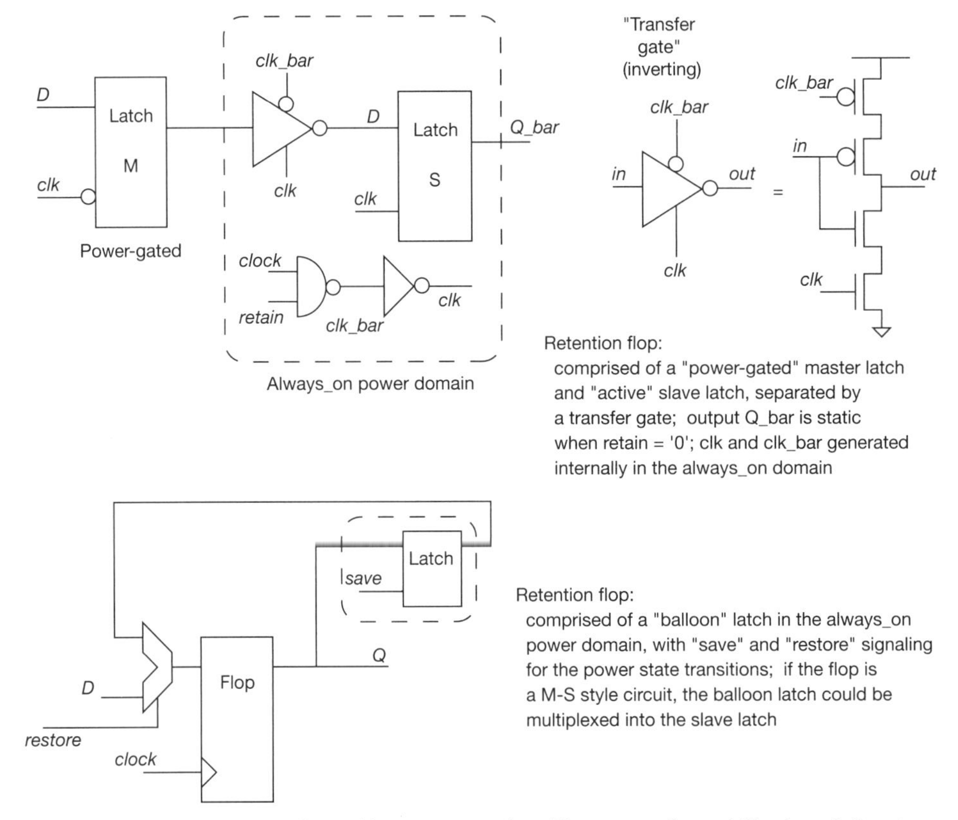


**Figure 7.13** Flip-flop with state retention. The power format file description for the block indicates which internal states require retention when the block is power gated.

The flop value is saved during the power-gating transition, and it is restored upon the transition to the active state. The retention flop requires an internal cell placement image with access to both the power-gated and always_on domain within the block.

The initial technology mapping of the RTL logic model is typically directed by a user input to a default $V_t$ threshold voltage variant of the logic function. Although the standard $V_t$ (SVT) cells would be the common choice, performance-critical block designs may default to low $V_t$ (LVT) cells, and power-constrained designs might default to high $V_t$ (HVT) offerings. As the goal of clock tree synthesis is to provide low-skew arrival latency across the block rather than performance, the clock buffering and gating cells used are commonly the SVT cell variants. These clock cells also have improved statistical variation over other cell $V_t$ options, because the SVT device is the "natural" device threshold (i.e., without additional channel impurity implants). The initial cell drive strength assigned during mapping is selected based on the physical synthesis route loading estimates and signal fan-out.

After technology mapping, the timing constraint inputs to the flow are evaluated against the physical synthesis model, and a series of timing and power optimizations are applied. Cell selections are changed to modify the drive strength and/or $V_t$ properties. Path endpoints that easily meet their timing goals introduce opportunities for power reduction for cells in the path; they can swap to lower drive strength and higher $V_t$ variants. Conversely, paths not meeting their timing constraints require higher-performance cells or a network modification using a repowering strategy (discussed in the next section). The notion of the *timing slack* on a signal denotes its criticality for all paths through that signal (see Section 11.4). The slack directs the optimization algorithms where to focus to improve timing closure.

## 7.4 Signal Repowering and "High-Fan-out" Net Synthesis (HFNS)

Two general classes of algorithms are used in logic synthesis for managing performance requirements for very high fan-out signals: those that involve construction of balanced repowering trees to achieve a low-skew arrival at the fan-out pins and those that involve construction of (unbalanced) buffering networks, prioritizing the fan-outs that are part of path delays that require timing optimization.

### 7.4.1 Construction of Balanced Repowering Trees

An algorithm that involves construction of balanced repowering trees to achieve a low-skew arrival at the fan-out pins proceeds as follows:

1. Selecting buffer cells
2. Analyzing the physical synthesis cell placement to group the signal fan-out into subsets ("clusters")

3. Placing a buffer at an optimal location within each cluster and updating the netlist for each inserted buffer
4. Recursively constructing the repowering tree using the inserted buffers back to the signal source pin

Figure 7.14 illustrates the result of High Fan-out Net Synthesis (HFNS).

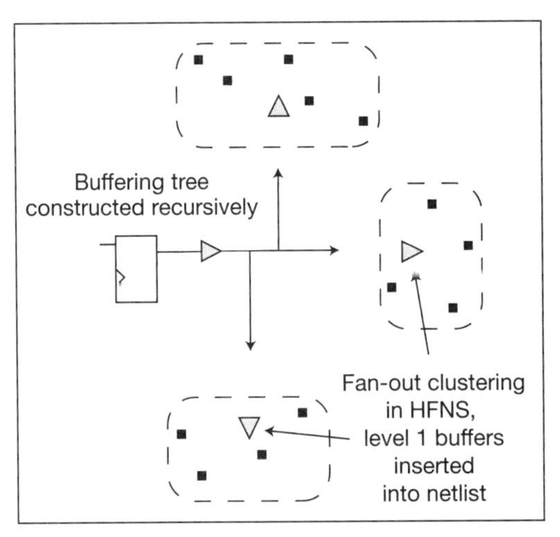


**Figure 7.14** Illustration of the balanced repowering buffer tree topology for high-fan-out net synthesis.

The key step is the selection of the individual fan-out clusters to provide a topology that can meet arrival skew targets after final placement, routing, and buffer size tuning. (Section 7.9 describes an extension of the balanced tree HFNS insertion method specifically for clock signals, denoted as clock tree synthesis [CTS].)

### 7.4.2 Construction of Buffering Networks for Timing Optimization

A variety of series/parallel repowering algorithms are used in an attempt to improve critical path timing. Figure 7.15 depicts the most basic series and parallel repowering approaches; many more combinations are often used (in conjunction with $V_t$ and drive strength selection for the inserted cells). These repowering algorithms incorporate performance, area, route congestion, and power dissipation trade-offs when evaluating a repowering netlist update to the mapped model.

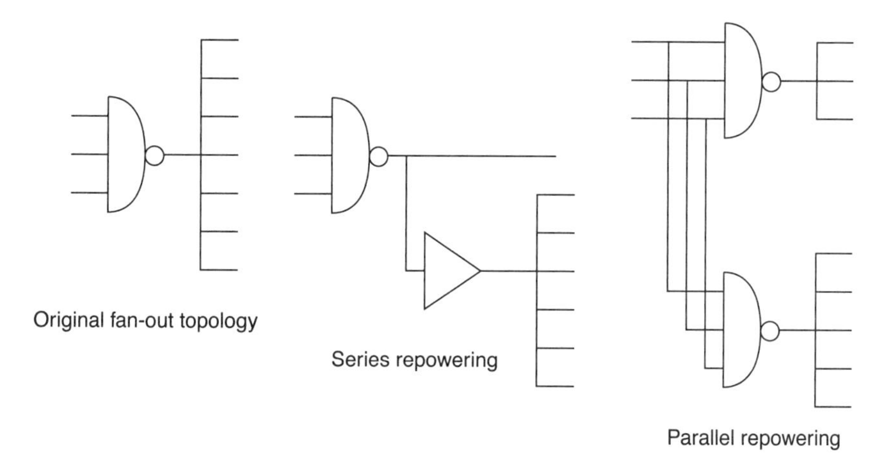


**Figure 7.15** Illustration of series and parallel repowering topologies for signal buffering.

Note that these series/parallel repowering model transformations are local in nature. Cells that are part of failing timing paths are prioritized and selected to evaluate different repowering implementations. A proposed network update impacts the path delay over a limited number of logic stages. To keep the runtime in check for a potentially very large number of repowering evaluations throughout a model, the physical synthesis tool must support *incremental* recalculation of cell delays and path timing.

A key methodology decision pertains to the allowed repowering strategies for flop cell outputs with high fan-out—specifically, whether parallel repowering of flops will be enabled in the RTL synthesis flow, as illustrated in Figure 7.16.

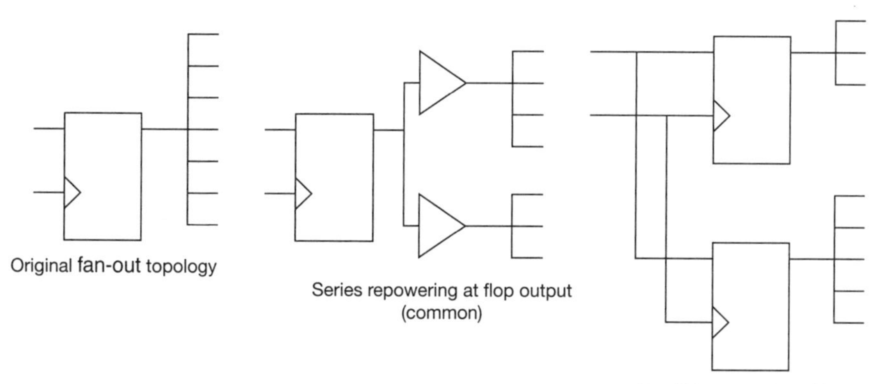


**Figure 7.16** Illustration of high fan-out flop cell repowering strategies. Parallel flop repowering is also known as "flop cloning."

To allow flop cloning in synthesis, a number of methodology flows need to be addressed to maintain the functional state correspondence between the RTL, test, validation, and physical models between iterations of the block synthesis flow. Commonly, the repowering of high fan-out flop signals uses buffers at the flop output rather than cloning of the (large area) flop cells and increased clock power.

Note that some EDA tools can analyze the RTL model and make design recommendations for power optimization that incorporate a flop cloning approach, somewhat related to HFNS of the flop signal. Consider an example where a flop output is referenced in assignment statement expressions in multiple clauses of an RTL case statement. As only one clause of the case logic would typically be active, the high fan-out of the flop to all clauses results in wasted power dissipation. Assume that the logic conditions for each clause are available in the previous clock cycle. It may be more power efficient to make copies of the flop and allocate the fan-out of each to individual case statement clauses and apply clock gating to each flop using the clause condition value as the clock enable. The EDA tool provides RTL design recommendations for power optimization, using flop cloning and clock gating logic restructuring.

Signal repowering for path timing optimization is undoubtedly the most intricate set of algorithms used in synthesis flows due to the wide variability in potential repowering topologies and cell selection options. Each alternative needs to be evaluated against area, routing, and power constraints. In addition, it is quite easy for an SoC architect to be unaware of high "logical fan-out" signals in the RTL model and the difficulty that may ensue due to close timing. A common example is illustrated in Figure 7.17, where a timing-critical signal is present in a very large number of logic expressions. When the model is mapped, this signal is present in the initial logic levels of the network graph. As the signal is timing critical with high logical fan-out, an ill-structured repowering tree and significant path timing fails are likely to result. Recall that repowering transforms are rather local in nature; a critical signal may need to be re-architected in the RTL logic model to promote its usage to deeper levels in its timing paths.

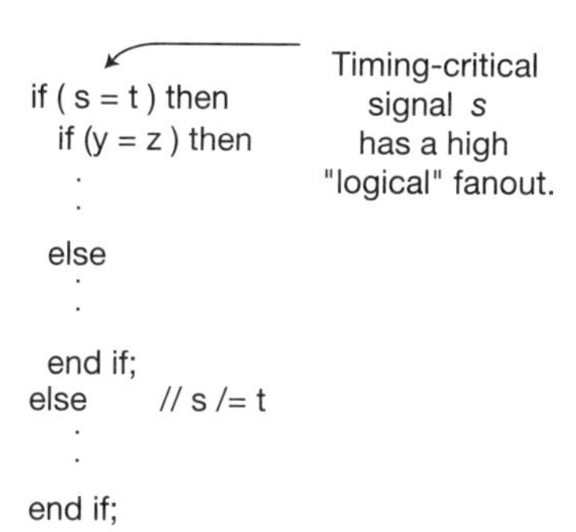


**Figure 7.17** Illustration of a high "logical fan-out" RTL signal. If the signal is timing critical, timing optimization algorithms may have difficulty generating the major logic network restructuring required.

Section 7.11 describes a low-effort synthesis flow that may be a useful addition to the SoC methodology to provide a quick check of potential RTL modeling issues prior to detailed physical synthesis.

## 7.5 Post-Synthesis Netlist Characteristics

Several identifiers in the netlist output from logic synthesis can provide useful information for subsequent analysis and validation flows. The following sections examine them in turn.

### 7.5.1 Cell Instance Name

The name assigned to a cell instance can provide insight into the specific synthesis algorithm invoked. For example, designers may wish to query the output netlist to see what specific repowering strategy was used for timing critical signals. Cells whose $V_t$ type was changed to save (leakage) power while still meeting path timing also have informative instance names in the netlist. Identification of cells added during clock tree synthesis and test insertion also provides useful data to help quantify the area overhead of these features.

### 7.5.2 Netlist Signal Name Correspondence to the RTL

The initial synthesis network model, prior to logic reduction and factoring optimizations, reflects signal names originating from the RTL. As logic and technology mapping transformations are applied, these signal names may be removed, subsumed into other functions, cloned during parallel repowering, and so on. It is often beneficial to know how synthesis algorithms apply changes to model signal names during these transformations—specifically, whether a functional correspondence exists between a common RTL and netlist signal name present in both models. This RTL-to-netlist correspondence is beneficial to the equivalency flow as it simplifies generating the state mapping input file. It is also helpful to the netlist-level simulation flows when exercising existing RTL testcases that monitor specific signals. Ideally, the RTL-to-netlist signal correlation should be well-defined for model PI/PO ports and flop outputs.

Note that there are some subtle characteristics associated with signal name correlation between RTL and netlist models:

- **RTL (1D and 2D) vectored signals**—The translation of vector indices from the RTL model to the netlist may introduce different nomenclature to the netlist signal.
- **Elaboration of the RTL model to collapse hierarchy for synthesis**—The RTL model may have a deep hierarchy, with multiple instances of the same entity/module and/or duplicate signal name identifiers in the scope of separate entities. RTL model elaboration for a single synthesis block flattens this hierarchy. The resulting synthesis model prior to optimization needs to systematically derive a signal name that reflects the hierarchy prior to elaboration.

The SoC methodology team needs to be aware of the RTL-to-netlist model translations used during synthesis and provide the tool utilities to aid with the derivation of correspondence data for other flows.

## 7.6 Synthesis with a Power Format File

The functional HDL representation does not reflect the specific power domains within the SoC design, nor the various operating conditions that these domains may reach as they sequence through different power states. To enable logic synthesis to implement the power state functionality, a separate power format file is provided as input to the synthesis flow. (This file is also often described as a *power intent* file.) Two separate formats evolved from EDA vendors; the IEEE has since taken on the standardization activity for the Unified Power Format (IEEE1801).

There are three major SoC design features related to power management:

- Multiple supply voltages distributed on-chip, which are different between blocks
- Changes to the supply voltage reflecting different operating modes (e.g., normal, "boost") using dynamic-voltage frequency-scaling (DVFS) techniques
- Block-level power gating, which involves disconnecting the domain supply (or ground) from logic circuits with the addition of power switches (also known as *sleep FETs*)

The power format file correspondingly describes these design characteristics:

- Definition of the power supply and power gating domains
- (Legal) power states for each domain and the (valid and invalid) sequences between states
- Internal registered signal state retention when in a sleep power state
- Output signal behavior during a sleep state (i.e., the *isolation* value)
- Input signal interface requirements during an active state (i.e., the need for *level shifter* cells between domains operating at different supply voltages)

These power format semantics span the entire SoC hierarchy. Each statement that declares a power domain and related power state has an associated scope within the model hierarchy. The power format file for a synthesized block references its own scope, to be subsequently included in the SoC description. The block power format file includes commands that define both the *configuration* and *implementation* of power management features, such as the following:

- create_power_domain < domain_name >
- create_supply_port < supply/ground_port_name > -direction in
- add_power_state -state < state_name > { logic_expression } (As the RTL model submitted to the synthesis flow does not include the definitions of power domains, the power format file must create new power ports for each domain. The supply voltage value for the domain may vary in different power states, as indicated in the power format file.)

- create_power_switch (This command includes the supply rail names for the input and gated output of the power switch and the corresponding power state names for the active and sleep operating modes.)
- create_logic_port, create_logic_signal
- describe_state_transition

As with the addition of power ports to the RTL input model to the synthesis flow, the power format file defines any new block signals/ports that will be used in the logic expression for each power state. These additional block ports and signals will be incorporated in the synthesis output model.

Power-aware simulation tools from EDA vendors have been enhanced to interpret the power format file commands when simulating the RTL model. These tools incorporate the new ports and signals defined in the power format file and evaluate power state transitions. Any invalid transitions would result in corrupted signal values. State retention/restore and sleep state isolation are also simulated, using the expression clause in the following power format file command examples:

- set_retention -domain < domain_name > -retention_power < port_name > -retention_ground < port_name >
- set_retention_control -domain < domain_name > -save < signal_name > -restore < signal_name > (Figure 7.13 illustrates a library flop cell with an additional internal state retention circuit.)
- set_isolation -domain < domain_name > -clamp_value < 1, 0 >
- set_isolation_control { expression } (Figure 1.34 illustrates the "clamp" of all block outputs to a non-floating (logic '0') value when the internal block logic is in a sleep state.)
- set_level_shifter -threshold < voltage_difference > -applies_to < inputs, outputs, both > (Figure 1.33 illustrates how a level shifter cell would be added at block inputs, when the sourcing logic '1' voltage would be less than the block's supply voltage by a value greater than the specified threshold.)

There are additional power format implementation-style statements to direct logic synthesis to map to specific retention flops, isolation cells, level shifter cells, and power switch cells. The insertion of power switch cells during synthesis requires attention to controlling the magnitude of the current transient when switching from the sleep state to the active state, as illustrated in Figure 7.18.

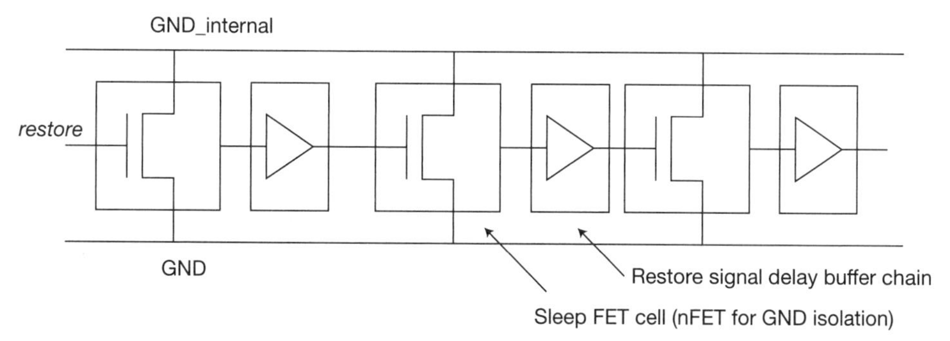


**Figure 7.18** Multiple power switch cells are inserted for power gating, with a distributed delay for the power gate enable signal to control the current transient.

The most complex aspect of power format synthesis is the addition of the finite state machine logic that reflects the power state transitions. A validation testbench would need to be exercised on the synthesized netlist to confirm correct behavior. Note that such simulation involves compiling a netlist with full power/ground connectivity and with cell functional models generating and propagating 'X' values when disconnected from power in a sleep state.

## 7.7 Post-Technology Mapping Optimizations for Timing and Power

After the initial technology mapping and cell placement of the physical synthesis flow, the algorithmic focus shifts to optimization of path delays and power dissipation.

### 7.7.1 Performance Optimization

The constraints description submitted with the RTL model to the synthesis flow includes the performance targets for the network, including the following:

- Input clock period and jitter tolerance
- Input pin capacitance and signal slew limits
- Output pin capacitance load estimate
- Block physical implementation clock skew target

- Timing modes (for multiple operating configurations)
- Don't_care (non-timed) paths
- Multi-cycle paths

As described in Section 7.4, there are multiple synthesis algorithms available to improve path timing, starting with the selection of the drive strength and threshold voltage variant of each mapped technology cell to the selection of signal fan-out repowering topologies. The conventional approach is to initially map all functions to a single $V_t$ selection—typically the standard SVT cell—and select the drive strength based on the estimated output loading. Delay calculation is performed, followed by static path timing analysis. Failing paths are assigned negative slack values to individual cell pins (see Section 11.4). Optimization algorithms focus on the signals with the most critical slack first, where cell delay improvements can have the greatest impact on the network.

A substitution of a low-$V_t$ (LVT) or ultra-low-$V_t$ (ULVT) cell variant accelerates the circuit turn-on time and improves the output drive strength, reducing the cell's input pin-to-output pin arc delay. The input pin loading capacitance on the sourcing cell is minimally impacted. The disadvantage of this optimization is the significant increase in sub-threshold leakage current when the cell is quiescent. Figure 1.3 illustrates a (nominal) $I_{on}$-versus-$I_{off}$ log-linear curve for a single device, with the device thresholds highlighted. Due to the significant increase in leakage power, the synthesis optimization algorithms typically limit the assignment of LVT/ULVT library cell variants to a low netlist percentage (e.g., <5% of the netlist cells). With the drive strength selected to match the estimated loading and with the (limited) $V_t$ cell swaps applied, most of the timing optimization effort in logic synthesis flows is focused on signal repowering strategies.

The post-mapping timing optimizations in physical synthesis are relatively constrained in scope. The netlist restructuring modifications impact only a few gate levels around a critical path logic network node. After these optimizations, the synthesis flow timing calculations for the output netlist may identify remaining negative slack paths. The timing reports include information such as *worst negative slack* (*WNS*)—that is, the network node with the most critical late arrival timing versus the required time—and the *total negative slack* (TNS) summed for all network nodes. These data offer insights into paths where RTL model restructuring may be warranted (e.g., moving logic across registers, rewriting logic expressions to promote late arrival signals). An assessment is made by the SoC methodology and project management teams of the TNS and WNS data reports after synthesis timing optimization. The outcome of this review determines whether RTL updates are needed or whether subsequent optimizations within the physical design flows will likely be sufficient. Physical synthesis uses interconnect loading estimates; the addition of routing directives, such as the preferred use of upper metal layers for critical nets, may allow the physical design flows to close timing without requiring an RTL model revision.

To offer RTL designers quick feedback on the logic path depth in the model, Section 7.11 describes a low-effort synthesis (LES) flow. This flow provides a path length distribution graph after a direct (quick) mapping to a generic logic library. The SoC architects typically associate a technology-specific path length with the target clock period. For example, a high-performance processor may be characterized as a "14-stage path length," using a representative FO4 technology-specific stage delay. A less aggressive architecture might use a clock period with FO4 path lengths in the mid to upper 20s. If the technology target path length is "n stages," path lengths much greater than this target in the LES data are indicative of significant difficulty in achieving timing goals for the subsequent physical synthesis flow. Initial logic reduction and refactoring, as well as subsequent post-mapping timing optimization algorithms, are not likely to be able to compress the RTL input model to ~n technology-specific stages.

### 7.7.2 Power Optimization

The optimization of performance during physical synthesis is typically the highest priority; subsequent to path timing netlist updates, post-mapping algorithms focused on power optimization are invoked. The simplest approach is to evaluate path slack data after timing optimization, looking for candidate cells where a lower power cell swap could be made to the netlist model. As mentioned earlier in this section, the initial technology mapping typically uses a default cell $V_t$ selection. Cells that remain at the default with significant positive slack after timing optimization are candidates to swap for a variant with a higher $V_t$ to reduce leakage power (e.g., an SVT-to-HVT or SVT-to-UHVT cell swap). (Note that there is an implicit assumption that cell $V_t$ variants can be freely swapped, with little subsequent impact on the physical design. In advanced process nodes, the lithography design rules associated with threshold voltage implants and multipatterning mask decomposition may introduce cell placement adjacency restrictions; as a result, cell variant swaps may have an area utilization impact.)

In addition to reduction of static leakage power, there are opportunities to optimize active power dissipation. Active power is a function of net switching activity and capacitive loading. Estimates of activity factors for signals in the synthesized network originate from the RTL model and may be applied through the logic optimization, logic factoring, and technology mapping steps. The RTL signal activity data can be calculated from simulation testbenches (*vectored*) or estimated from a stand-alone analysis of the RTL model without simulation (*vectorless*). For nets with high estimated activity, synthesis optimizations may restructure the local mapped network model to try to reduce the switching probability or loading; an example is illustrated in Figure 7.19.

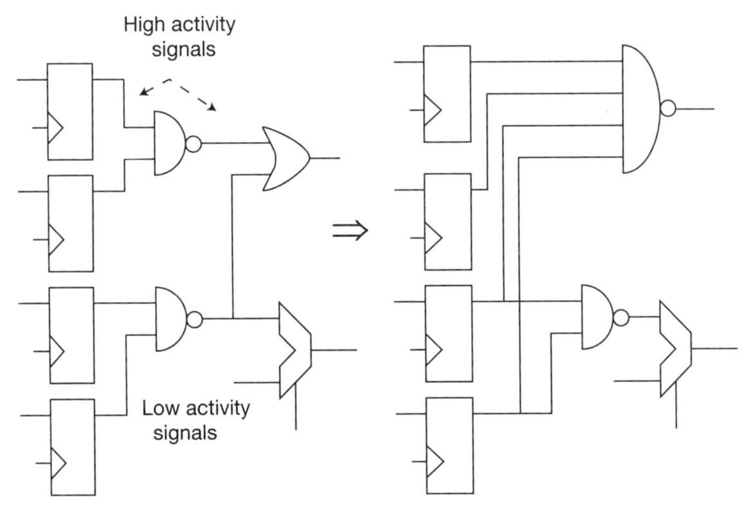


**Figure 7.19** Power optimizations in the synthesis flow may locally restructure the logic network to reduce switching activity or loading.

Note that the breadth of logic functions provided in the library has a strong impact on the active power dissipation. A higher fan-in set of functions may offer reduced switching probability. The SoC methodology team needs to establish the relative priorities between active power, total cell area, and path performance during power-focused optimizations in the synthesis flow.

An additional static leakage power optimization is potentially available; it is usually deferred until later in the overall physical design flow rather than being implemented during logic synthesis (see Chapter 20, "Preparation for Tapeout"). The sub-threshold leakage current of a device is dependent on the gate length dimension as part of both the (W/L) device sizing ratio, and the $V_t(L)$ dependency (also known as the *short channel* $V_t$ *effect*). Figure 7.20 illustrates an algorithm to selectively add to the gate length within a cell layout. The layout design is developed in collaboration with the foundry to allow reduced gate-to-gate and gate-to-contact spacing and non-uniform gate lithography in support of an extended gate length.

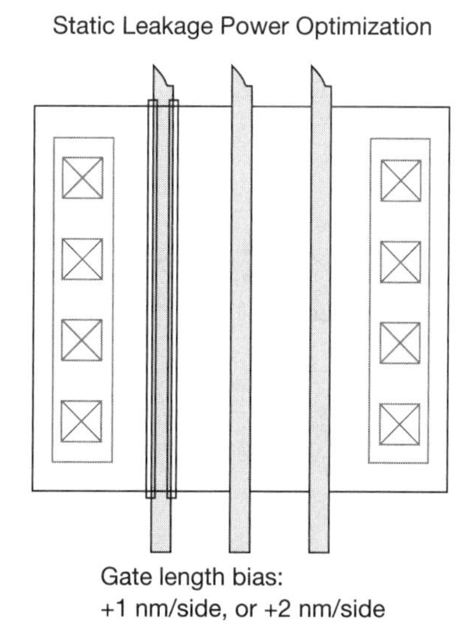


**Figure 7.20** A small gate length bias could be added (in a physical overlay cell) to reduce leakage power dissipation.

Note that the gate length bias approach is a very selective optimization:

- The cell is not recharacterized with the longer gate length (of a few nm). The assumption is that the impact on path performance is slight.
- The base library cell layout is not modified; rather, a physical cell overlay with the gate length bias shapes is added, typically late in the design schedule. The existing device $V_t$ used in the cell layout is maintained.

This ad hoc power optimization is becoming less applicable at advanced process nodes, as the gate length lithography options offered by the foundry are more limited.

## 7.8 Hold Timing Optimization

The discussion of timing optimization has focused on improving setup (late) timing checks. A methodology decision is needed to assess the extent to which the physical synthesis flow should address the risk of hold (early) timing checks, as depicted in Figure 7.21.

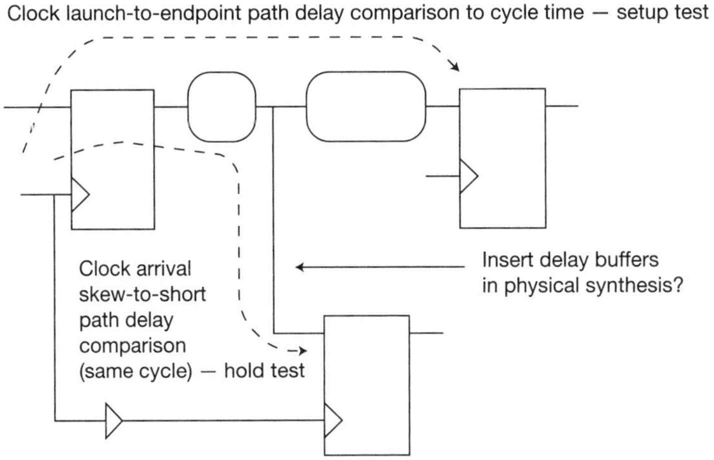

**Figure 7.21** Synthesis flow optimization of hold time check fails, with cell insertion to extend path delays.

Any path length between registers of very few gates needs to be evaluated against the (anticipated) maximum arrival skew at the register clock inputs. If the skew could potentially exceed the path delay, an error has occurred. The endpoint register will have captured source data for clock cycle 'n+1' at clock cycle 'n'. A delay padding algorithm is required to extend the short logic path to exceed the clock arrival skew, which is evaluated at all PVT environment corners. The methodology team needs to evaluate whether to invoke (delay) cell insertion algorithms during the timing optimization phase of physical synthesis or defer the hold timing correction steps until physical implementation. The advantage of deferring is that skews can be better defined after the detailed clock repowering network is implemented. If the actual skew is less than the physical synthesis estimates, unnecessary padding can be avoided. Also, deferring hold timing fixes until physical design also allows the use of *scenic route* segments to address hold timing errors. An increased R*C interconnect delay may be selected rather than inserting delay cells into the netlist. Alternatively, if a significant number of delay cells are anticipated to be required (i.e., if the low-effort synthesis path length distribution shows a high number of short paths), it may be preferable to ensure that these cells are added during synthesis so that the subsequent physical design flows are presented with a more representative cell area and routing utilization for the block floorplan. The insertion of delay padding cells during synthesis should ideally extend short paths without adversely impacting late paths converging at the same endpoint, as depicted in Figure 7.22.

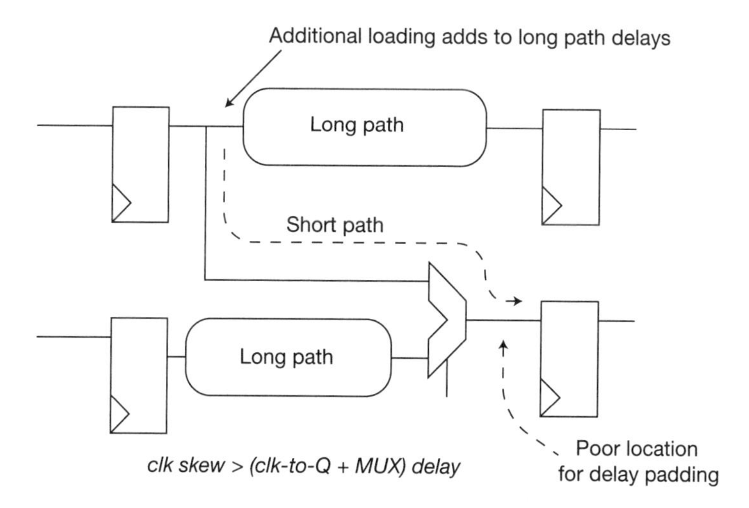


**Figure 7.22** Hold time delay padding needs to minimize the impact to critical setup timing paths; selecting the optimum network location to insert delay cells involves a complex algorithm.

The hold time padding algorithm in physical synthesis requires sophisticated decision-making features to select the appropriate delay cell(s), the optimum network insertion connection, and the preferred cell location.

## 7.9 Clock Tree Synthesis (CTS)

As described in Section 7.4., the synthesis flow invokes a specific signal repowering algorithm on high fan-out nets within the design block, where the objective is to provide a low-skew arrival time at the endpoints of the repowering tree. Although similar in nature to HFNS, clock signal repowering algorithms are invoked by specific synthesis flow CTS commands and involve both logic insertion and physical implementation. The goal of this subflow is to build a symmetric repowering tree with balanced loading at each branch for each level of the tree. The clock "tree" may use a variety of physical topologies, such as an *H-tree* or a *fishbone* (see Figure 7.23).

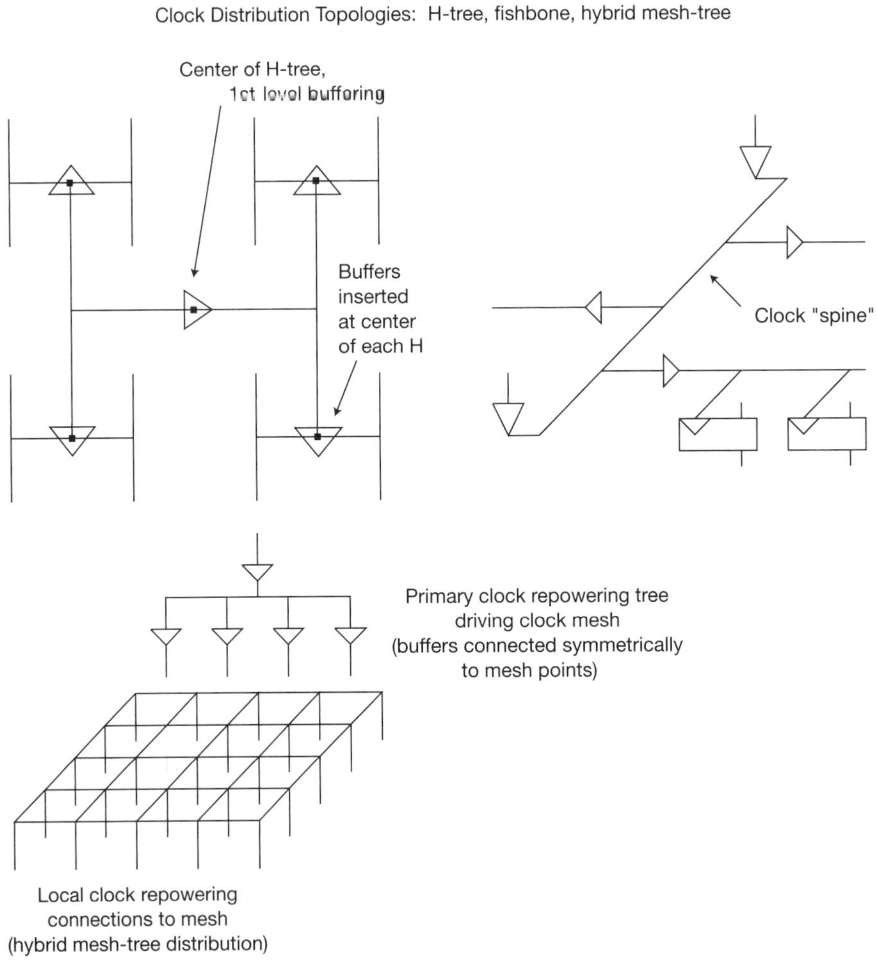

Clock Distribution Topologies: H-tree, fishbone, hybrid mesh-tree

Center of H-tree,
1st level buffering

Buffers
inserted
at center
of each H

Clock "spine"

Primary clock repowering tree
driving clock mesh
(buffers connected symmetrically
to mesh points)

Local clock repowering
connections to mesh
(hybrid mesh-tree distribution)

**Figure 7.23** The clock tree synthesis algorithm is directed toward a specific physical implementation. Examples of an "H-tree" and "fishbone" topology are illustrated.

The clock H-tree (recursively) selects a buffer size and location for inserting a clock driver cell, for a balanced load at each branch of the tree at the same level. The fishbone utilizes a predefined clock spine, from which balanced repowering trees (of reduced depth) are constructed.

In addition to the CTS commands to insert clock drivers and construct the repowering network, synthesis flow constraints on the maximum allowable clock signal slew rates control power dissipation. The clock arrival skew within the block is only part of the CTS timing equation; there are also constraints on the latency from the clock primary input connection(s) to any endpoint in the block so that paths that cross blocks sharing the same clock domain have a common timing reference.

The CTS steps use specific library buffer cells that are designed for balanced rising and falling pin-to-pin delay arcs (RDLY, FDLY) to maintain a 50% pulse duty cycle. Section 7.3 describes how the CTS buffer tree is modified to provide gated-clock topologies for power reduction.

An additional physical synthesis input constraint for CTS commands relates to the choice of a hybrid clock grid-tree topology with an upper metal clock grid overlay throughout the block. The resulting grid provides a low arrival skew at frequent access points within the block. The local clock tree for the hybrid grid-tree design style is very few buffer stages in depth.

Rather than optimize the synthesis of clock trees to a low skew arrival difference at flop input pins, an alternative approach, commonly described as *useful skew*, may be implemented. The goal of a useful skew synthesis algorithm is to build asymmetric repowering trees, with endpoints arriving at staggered times within the overall period. Logic path endpoints with negative slack are assigned to later-arriving clocks, effectively extending the period in an attempt to satisfy path timing. Conversely, paths originating from an extended skew clock would be subject to a reduced period to fan-out endpoints. The success of assigning useful clock skews during synthesis depends on the ability to find variable logic path lengths in the network model and the confidence in physical synthesis interconnect estimates. An additional, secondary result of applying useful skew algorithms is a reduction in the simultaneous switching current that originates with the clock transition: The power and ground grid distribution has less dynamic resistive losses and inductive noise.

The CTS discussion in this section has assumed the use of clock edge-triggered (flop) timing endpoints. If the design utilizes transparent latches for half-cycle paths, the CTS specification and setup/hold timing tests are significantly different, as depicted in Figure 7.24.

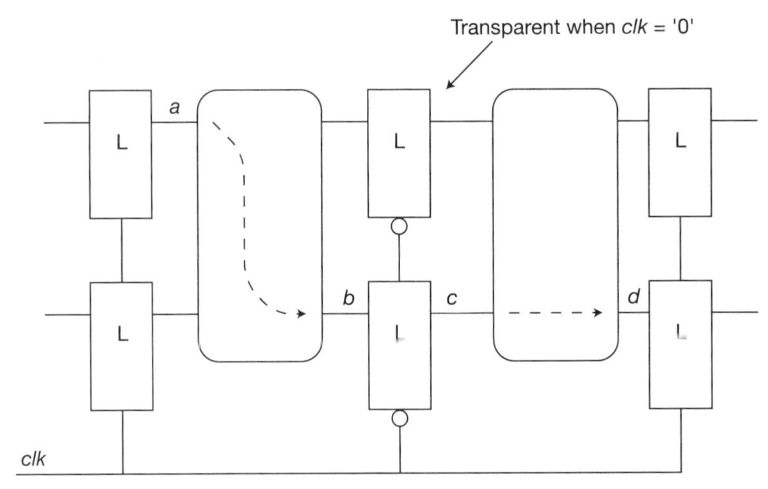


Paths can be > 1/2 cycle — "time-borrowing" from the next phase.
Path delay arcs at latch inputs may be either *clk*-to-Q or D-to-Q.

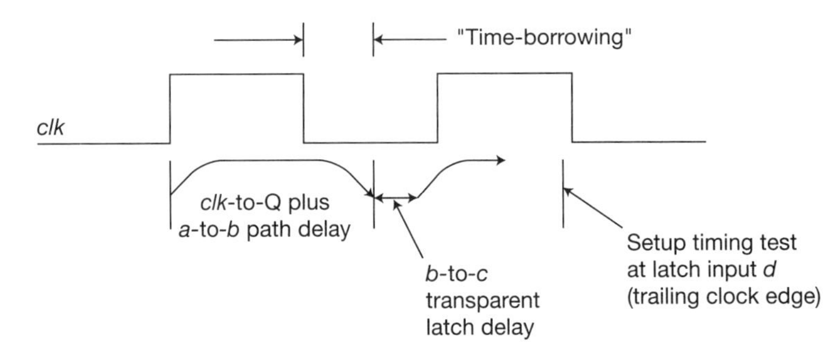


**Figure 7.24** Synthesis for latch timing uses a different set of setup and hold timing tests.

The design of clock repowering trees is a key facet of the overall synthesis flow. The subsequent physical design flows must treat the placement and interconnect routing of these netlist cells with special algorithms to address skew, slew, latency, and signal noise constraints. Non-default metal routing wire widths, unique wire spacing (and adjacency to power and ground grid rails), and optimum metal layers for clock tree route segments are all part of the physical implementation. The synthesized netlist includes specific instance labels on the cells inserted by CTS to guide these physical design algorithms.

## 7.10 Integration of Hard IP Macros in Synthesis

The synthesized block model may include instances of hard IP macros, such as a register file array or microcontroller core. The physical synthesis optimization algorithms incorporate the timing model for the IP into the path timing checks. The SoC methodology team needs to assess whether the IP macro should be preplaced within the block floorplan or may be free to be assigned during physical synthesis. The macro placement has an impact on several characteristics of the synthesis model:

- **Routability**—A macro with a high local pin density may require sparse cell population in the local neighborhood to provide sufficient route access to the macro.
- **Critical timing paths to/from/through the macro**—Automated placement algorithms utilize several metrics to evaluate the quality of the overall solution. A common, highly weighted measure is the total (estimated) wire length, which is used to ensure subsequent routability for the block. A hard IP macro with low pin density has a small contribution to this metric compared to the large number of logic cells in the synthesis model. As a result, placement algorithms tend to assign hard IP macros toward block edges, leaving more internal area for cells. However, the timing paths associated with the logic around the macro are likely critical; an assigned placement for the macro internal to the block may improve synthesis timing closure, as illustrated in Figure 7.25.
- **Clock tree construction**—The hard IP macro integrates an internal clock distribution. The CTS algorithm utilizes the clock specification for the IP to build a (potentially truncated) tree to the macro clock pin. The macro placement has a significant impact on how the tree is constructed and on the success in optimum placement of clock buffer cells to the remaining logic endpoints. A hard IP placement internal to the block may increase the difficulty in placing and routing a balanced clock tree.

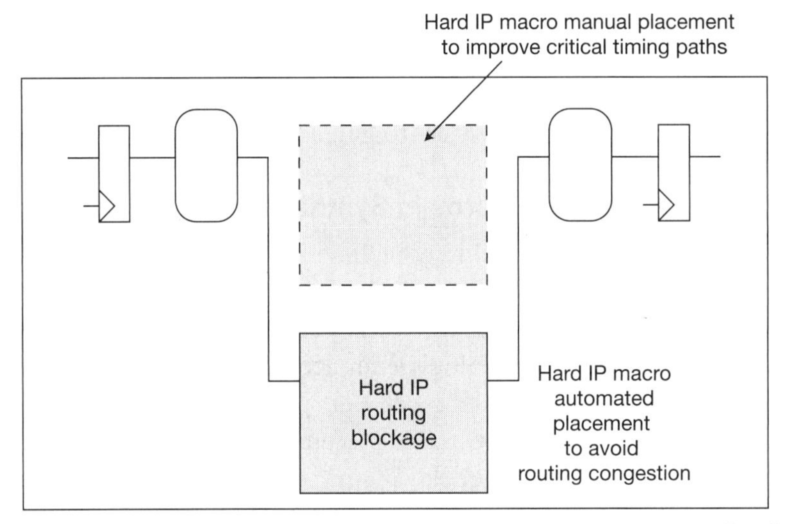


**Figure 7.25** Placement options for an embedded hard IP macro within the synthesis block. Manual placement may be required to optimize timing paths through the macro.

## 7.11 Low-Effort Synthesis (LES) Methodology

The previous sections of this chapter review the steps associated with the synthesis of a hardware description language (HDL) model to a netlist composed of logic cells from a specific technology target library. The following chapters discuss the remaining steps in the physical implementation of the SoC—namely, the placement of the netlist cells and the design of the routing interconnections between cells.

The project flow for closing on the physical implementation is likely to require multiple iterations through logic synthesis before proceeding with detailed physical design, for several reasons:

- Experimentation with the synthesized block's physical area and aspect ratio and corresponding routability and timing results
- Logic updates to the HDL model to fix bugs found in functional validation
- Refinement of the block primary input/output timing and loading constraints, as more implementation detail is available for other blocks with connectivity to the synthesized design

The appeal of logic synthesis is that the iteration time is relatively short. However, synthesis does not necessarily reduce the iterative nature of physical design closure. The disadvantage of synthesis iterations versus manual-refinement engineering changes is the tendency for the new synthesized netlist version to differ substantially in cell count, logic net names, and so on.

As mentioned in Chapter 2, "VLSI Design Methodology," the SoC project schedule requires support for concurrent engineering tasks. The SoC methodology needs to provide various engineering teams with access to a netlist-based model to proceed with their own analyses during the block physical synthesis and physical design phases; this netlist must be functionally equivalent but need not necessarily reflect cell updates for timing and noise closure.

A methodology option for a flow to execute in parallel with physical synthesis is to provide a low-effort synthesis (LES) netlist model. The HDL design is input to synthesis, as usual. The conventional synthesis flow is interrupted after the logic minimization, test insertion, and target library mapping steps, and a model netlist is exported. The target library cell models are modified to remove any fan-out limits, as no repowering algorithms are invoked. This abbreviated synthesis flow is depicted in Figure 7.26.

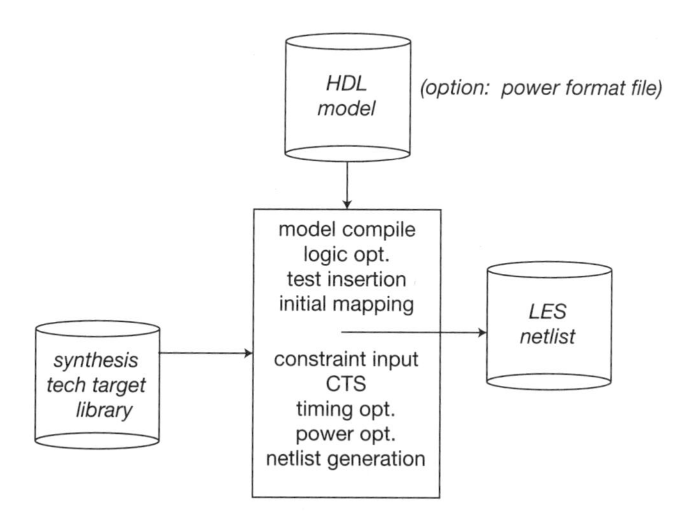


**Figure 7.26** Low-effort synthesis flow.

The characteristics of this netlist are as follows:

- It is functionally equivalent to the original HDL model.
- Ideally, netlist signal names are very close to those in the original HDL model for ease of correlation.
- Test connectivity is present (e.g., scan-chains, built-in self-test functionality).

Because the LES flow does not include logic factoring or repowering algorithms, the RTL signal names are typically carried forward to the output netlist. The scan chain order in the LES netlist is not likely to match the final physical implementation, which may include scan reordering algorithms; for the design flow steps that utilize the LES netlist, this is typically not an issue.

The target synthesis library for the LES mapping step could be the same as the physical implementation, or it could be a "generic" Boolean cell library (of rich logic content and high fan-in). As mentioned previously, the fan-out limits in the LES library are removed. The specific choice of the target library depends on the other flows that use this netlist.

### 7.11.1 Design for Testability Analysis (DFT Design Rule Checks)

The test engineering team requires an "early" netlist that reflects the testability features of the SoC design. Prior to exercising automated test pattern generation (ATPG), the SoC model is first subjected to a set of design checks to confirm that the structure of the design fits the underlying assumptions of the ATPG algorithms. For example, functional and test-clocking modes need to be explicitly defined and externally controllable. The connectivity and any logic functionality in logic scan chains need to be analyzed for consistency with the ATPG tool. The test pattern application sequence for scan input shifting, functional data capture, and scan output shifting needs to be validated. The operation of built-in self-test functionality also needs to be confirmed.

The LES netlist is ideally suited to enable testability analysis on the design and provision of early feedback to the SoC design team on DFT design rule checking (DRC) results. The fact that the LES netlist contains a "dummy" scan chain order, which would not reflect any reordering optimizations during physical implementation, does not substantially alter the DFT DRC analysis of test model control and pattern application.

The SoC test engineering team could use either a generic Boolean or technology cell library for the LES netlist target for DFT DRC analysis, as long as the behavior of the mapped (scannable) sequential elements is the same. If the DFT DRC results are clean, the test engineering team may want to continue with ATPG pattern generation on the LES netlist. If the fault coverage results were poor in some local logic network of the LES model, it is likely that the coverage would be equally suspect in the final physical implementation netlist, as well. Although the ATPG patterns generated would not be directly applicable to the final design, the percentage fault coverage data on a similar netlist would provide early feedback to the SoC design team.

### 7.11.2 Path Length–Based Timing Model Analysis

In the early planning stages of an SoC design project, performance goals are established by the architecture and performance modeling teams. These goals are then translated into a target cycle time for clock domains by the SoC logic design, circuit analysis, and global clocking/integration teams. After accounting for clock skew and jitter margins, the cycle time is translated into a representative logic path length, using a typical cell delay measure. The de facto standard for representing a cell library's typical delay is denoted as the inverter FO4 metric, as introduced in Section 7.3. This measure enables different cell libraries to be readily compared. More importantly, it offers logic and circuit designers insight into the logic path length and micro-architectural pipeline definitions required to hit the target cycle time.

An issue that is generally raised with HDL-based modeling and logic synthesis is that the HDL design can easily become removed from the constraints of the physical implementation. The LES netlist, when mapped to the technology target library (without fan-out repowering) could be used to quickly provide path length results. If there are paths in the LES netlist that far exceed the number of stages corresponding to the cycle time–based goals, it is very unlikely that the post-timing netlist optimization will resolve the issue. As highlighted in Section 7.4, timing constraint-driven optimizations typically result in additional serial and/or parallel signal repowering networks on high fan-out and timing-critical nets, which adds to the path length. If the LES path length results already far exceed the SoC target, an architectural change in the HDL model is warranted.

One caveat is that many micro-architectural critical paths involve array accesses (e.g., read, write, read/modify/write, associative cache lookup) whose delays are not well represented with a simple path length counting algorithm applied to an LES model. An integer instance-based path length measure from the LES netlist is useful for highlighting potential issues in achieving the target cycle time, but it by no means covers the space of potential micro-architectural timing problems.

The distribution of short path lengths in the LES netlist model is indicative of the magnitude of the hold time delay padding that is subsequently required. If a significant percentage of paths are likely to result in hold time errors, the SoC design implementation team should invoke the delay padding algorithms in physical synthesis rather than attempting to resolve hold time issues in the subsequent physical design flows.

### 7.11.3 Functional Simulation Using the LES Model

The LES netlist has applicability for functional validation. As briefly mentioned in Section 5.3, the validation team may have access to hardware simulation acceleration resources. The actual model to be loaded on the hardware may require a gate-level equivalent representation of the HDL design, executing the testcase stimulus on programmable gate-level logic modules. Rather than using the output model from the "HDL compiler" provided by the accelerator's EDA vendor, the LES netlist model may be preferable for simulation. The potential advantage to using the LES model for simulation acceleration depends in large part on the testcases to be used in the validation methodology. One key consideration is whether any of the testcases require the addition of test functionality. For example, if the accelerator is to be used to verify (typically very lengthy) scan and BIST sequencing testcases, the LES model offers suitable connectivity.

Another consideration is the validation strategy for any power-format functionality inserted during synthesis. Section 7.6 discusses the power format file used to represent the power state sequencing in the HDL model. Although the methodology for using power format data is likely to focus on using formal verification techniques to confirm the added power management functions, there may be simulation testcases added to the validation suite as well. The LES netlist may serve as the model for power sequence simulation. (The power state sequencing involves transitions to/from high-impedance 'Z' and unknown

'X' signal values; the LES simulation platform may not be able to support these additional signal values, necessitating translation of the cell model outputs to a reduced Boolean value set.)

Another, somewhat related, potential validation application of the LES netlist model would be to derive representative switching activity data. HDL simulation traces provide signal activity information that is used for power estimation. However, the potential advantage of the LES model is its closer correspondence to the final physical implementation, having been subjected to logic minimization, plus power management and test logic insertion prior to mapping. Although the LES model is not suitable for detailed power calculation, switching activity "hot spots" identified from validation testcases executed on the LES model are similar to the results from the final physical design netlist. These results provide early insight into potential changes required in the physical design (e.g., the addition of more local decoupling capacitance, more local power grid strapping, and the addition of a thermal sensor circuit for local temperature measurement).

This section described a methodology option to generate a low-effort synthesis netlist model from the HDL design and provide that model as an initial cell-level representation to other flows. The availability of the LES model enables concurrent analysis by multiple teams while the detailed physical implementation is progressing. The results of this analysis provide valuable early feedback on the suitability of the HDL design for test, timing, and netlist-level functional validation requirements.

## 7.12 Summary

This chapter described some of the algorithms used to synthesize the HDL model description of a block into a netlist of cells from a target technology library, with optimizations applied to address path timing constraints and to reduce estimated power dissipation. A critical facet to logic synthesis is the incorporation of a (preliminary) physical model of the cell netlist, to provide an estimate of the interconnect route parasitics prior to detailed physical implementation. The synthesis flow incorporates steps to insert the design-for-testability architecture into the output netlist. A power format file is also provided to the flow, to define the functionality to be inserted for the power domain switching states in the block. The synthesis of the clock tree network utilizes specific algorithms for the selection and insertion of clock repowering buffer cells, to achieve clock latency and skew targets. A low-effort synthesis flow option provides an output netlist functionally equivalent to the HDL model, with netlist characteristics suitable for testability analysis and validation of the inserted logic.

## References

[1] Alpert, C.J., et al., "Techniques for Fast Physical Synthesis," *Proceedings of the IEEE*, Volume 95, Issue 3, March 2007, pp. 573–599.

[2] Kahng, A., "Physical Synthesis 2.0," Keynote presentation at 24th International Workshop on Logic & Synthesis (IWLS), 2015. (Slides available for download at http://www.iwls.org/iwls2015/physical-synthesis-2.0.pdf.)

[3] O'Brien, P.R., and Savarino T.L., "Modeling the Driving-Point Characteristic of Resistive Interconnect for Accurate Delay Estimation," *Proceedings of the IEEE International Conference on Computer-Aided Design (ICCAD)*, 1989, pp. 512–515.

## Further Research

### Hardware Description Language Modeling Using a Truth Table

A potential HDL model coding style replaces a set of assignment statements with a truth table (sum-of-products) format. For example, the Verilog HDL standard includes the option of coding a "user-defined primitive" (UDP), which allows a truth table for a single output signal to be instanced in a model. A more general truth table could also be supported: An HDL preprocessor could convert an embedded truth table into the equivalent (large) case statement.

For a combinational network with m inputs and n outputs, a truth table would consist of (m + n) columns. The number of rows in the truth table would be equal to the number of product terms in the logic network. All input column entries in the table would be '1', '0', or 'don't_care'. The entries for the output columns would be '1' or '0', depending on whether the product term is used or absent in the equivalent sum-of-products output equation. The HDL preprocessor would add a row with the "default" (or "when others") clause to the generated case statement to cover all other product terms, with all outputs set to '0' (although an 'X' simulation output value for the default case clause may also be a consideration).

Describe the advantages and disadvantages of using a truth table (large case statement) format for logic synthesis.

Describe the advantages and disadvantages of using a truth table/case statement format for other RTL-based flows (e.g., logic simulation, assertion-based verification, simulation model coverage, simulation debug, power estimation).

### HDL Preprocessor Opportunities

In addition to translating a combinational truth table to a large case statement, describe other opportunities for an HDL preprocessor that may offer productivity benefits to the micro-architecture designers (e.g., conversion of state transition diagrams to equivalent HDL statements implementing the finite state machine, graphical entry of the top decomposition level of the SoC logical hierarchy).

With the use of a preprocessor, the original (graphical) representation is of a different file format than the rest of the HDL and testbench models, and it is often drawn using unique EDA tools. Thus, a methodology trade-off is introduced. These new file formats and tools could be incorporated as part of the project design data management and configuration management system. Alternatively, the methodology could disregard these external files and incorporate the HDL source output from the preprocessor into the project data management scheme, consistent with the remainder of the SoC and testbench models. Describe the advantages and disadvantages of these two methodology approaches.

### Timing Constraints

A methodology flow that provides useful feedback to block designers (and the SoC project managers) evaluates block-level logic synthesis timing constraints for consistency.

Describe block PO-to-block PI global path timing constraint checks that would be informative (assuming that the blocks share a common clock domain).

### Hold Time Padding

The synthesis algorithm for determining optimal network model locations to insert hold time correcting cells without adversely impacting setup time is extremely complex. The insertion of delay cells is exacerbated by the requirement to insert delays at a "fast corner" to fix hold time issues; the corresponding cell delay arcs are much greater at the "slow corner" used for setup tests.

A unique situation arises for global hold time padding. Block IP designs often incorporate registered inputs/outputs, resulting in short logic gate path lengths globally. In addition, the clock arrival skew estimate for paths between blocks may be higher due to the disjointed CTS trees in separate blocks. (A global clock grid spanning the blocks will reduce the arrival skew.) Describe a methodology flow for (early) identification of global paths requiring hold time analysis and how the results of this early analysis impact floorplanning, global/local clock skew budgeting, and block timing constraint generation.

CHAPTER 8

# Placement

## 8.1 Global Floorplanning of Hierarchical Units

Prior to detailed placement of cells in block netlists (and the global cells at the top of the SoC hierarchy), a physical floorplan of the chip design is required. As briefly described in Section 3.1, the floorplan typically represents the first level of the SoC model hierarchy; it is uncommon to further develop a "floorplan within a floorplan" for the physical design of subsequent levels of the SoC hierarchy. The glue logic functionality at the top hierarchical level is commonly allocated to *channels* between block floorplan boundaries. An alternative methodology would be to define abutting block floorplan regions and insert global glue logic within various blocks. The advantage of the reduced channel area is offset by the additional dependency of global cell and route data on block-level physical verification and electrical analysis.

The physical floorplan data include the global power and ground grids and global clock distribution, typically originating from a PLL hard IP macro that serves as the clock reference source. The power and ground grids in the channels require specific design consideration, as the glue logic circuits include high-drive-strength cells with high switching activity (e.g., signal repowering buffers, state-repeating register banks).

The floorplan may include allocated routing track segments for major signal busses between blocks in the overall SoC architecture, including global repeaters. These preroutes assist with the definition of the block-level floorplan pins, area, and aspect ratio.

The development of floorplan pins is a critical facet of SoC design planning. The pin definition for each block's primary inputs and primary outputs includes the following:

- The pin width, corresponding to the interconnect wire width to use with the global signal
- The pin metal layer for the interface between global and block routing
- The pin multipatterning decomposition assignment

For advanced process nodes, depending on the metal layer, the pin definition may also need to include a multipatterning assignment that is consistent with the "color" associated with the pin's routing track. Alternative methods for pin location assignment include the following:

- **Internal pin locations**—The pin may not be assigned to the block perimeter; rather, it might be given internal coordinates. The goal of using internal pin locations would be to improve timing. As mentioned in Section 7.2, the accuracy of block-level timing closure is improved if the cells connected to block PIs/POs are placed in close proximity to the pin. An internal pin location may allow optimal placement of block netlist cells with connectivity to both global and internal signals. A high density of internal pins may have an adverse impact on global routing, however, to accommodate both over-the-block global routes and pin accessibility.
- **A flexible range of locations**—A pin may be allocated to a range of locations (e.g., a segment of a specific floorplan edge) but not assigned fixed coordinates. In this case, the block placement flow includes pin location assignment as part of cell assignment; rather than using fixed pin locations to influence cell placement, the algorithm is able to include pin placement as an optimization objective. The methodology decision to use flexible pin locations as input to the block cell placement flow introduces an interdependency between global route planning and block physical implementation.

Floorplan areas allocated to hierarchical design blocks are typically rectangular, although EDA vendor tools for physical design may support rectilinear definitions. The aspect ratio of each floorplan block is a key factor in subsequent routing and path timing closure. A high aspect ratio block has a skewed ratio of available horizontal to vertical wiring tracks, and thus it may have difficulty subsequently closing on routing.

The SoC floorplan includes blocks associated with the chip input/output pad circuits, usually located on the die perimeter. Mixed-signal IP cores are also typically associated with unique floorplan blocks, such as PLLs, data conversion functions (ADCs, DACs), and high-speed interface SerDes IP. These blocks also require unique power/ground distribution design. The I/O circuits are likely to use additional voltages different from internal cells (e.g., VDDIO, VDD_1_2, VDD_1_5). Mixed-signal cores require separate low-noise supply rails (e.g., VDDA, GNDA) that are electrically distinct from the rails for digital switching networks.

Power-gating design is reflected within each block, as represented by the power format file description (described in Section 7.6). The internal power and ground distribution to enable deep sleep behavior is not extended globally, as depicted in Figure 8.1.

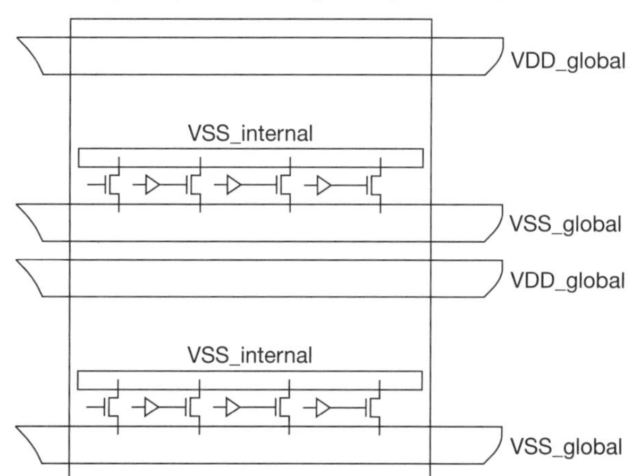


**Figure 8.1** The block internal power (or ground) distribution to support power gating is not extended globally.

## 8.2 Parasitic Interconnect Estimation

The placement flow utilizes a number of measurement criteria when selecting a candidate location for each cell or for candidate pairs of cells to swap their current coordinates. The process involves a combination of geometric and timing-driven calculations, including the following:

- Total estimated network wire length to realize all connections, using one of various net topology estimates (e.g., bounding box, star, Steiner tree) (Timing estimates from physical synthesis provide [negative slack] nets that may be given additional weighting in the total geometric wire length summation calculation.)
- Interconnect segments crossing a coarse grid overlay of the block floorplan to assess wiring track demand versus availability (to avoid congestion)
- Cell interconnect delay calculation for timing-driven placement optimization

The representation of interconnect delays during placement involves estimates of the R*C parasitics and a simplified computationally fast delay calculation algorithm (e.g., an *Elmore delay* model for the estimated net topology; see Section 11.1). The SoC methodology team needs to collaborate with the EDA vendor and the foundry to determine how to best estimate the interconnect parasitic delay during cell placement. This estimate needs to reflect the different (per unit length) R and C measures of the multiple horizontal and vertical metal routing layers available within the block. During cell placement, an average R*C delay measure across the available metal routing layers is used. An estimate for parasitic via resistances could also be included in the interconnect delay model.

In addition, the methodology team may use the physical synthesis timing data to derive "non-default" constraints for subsequent cell placement and routing:

- Preferred metal layers for routing critical nets
- Wider width segments (e.g., 1.5X or 2X width rather than 1X)

The EDA placement tool applies a different set of parasitic interconnect estimates for nets with non-default rules. Again, collaboration with the

EDA vendor and foundry is required to define how multiple wire load models for different classes of nets should be calculated for timing-driven placement optimizations.

## 8.3 Cell Placement

The SoC block designer relies on the (timing-driven) cell placement flow to provide a routable solution with minimal timing issues for a netlist with (tens of) millions of instances. Placement algorithms have evolved to provide greater netlist capacity with reasonable runtime. To help physical designers achieve improved predictability and confidence in timing closure, the EDA vendor placement tools have incorporated additional features that apply input constraints:

- Preplaced cells and hard IP macros
- Relative placement groups of cells (a set of cells with relative alignment coordinates that are placed/moved as a unit)
- Restrictive area allocation within the floorplan block for subsets of the cell netlist (see Figure 8.2)
- Guidelines for maximum local cell utilization percentage (to allow for the addition of a suitable density of decoupling capacitance cells, substrate and well contact cells, and dummy logic cells for ECOs)
- Ability to place cells with cell height that spans two rows of the placement image (see Figure 8.3)

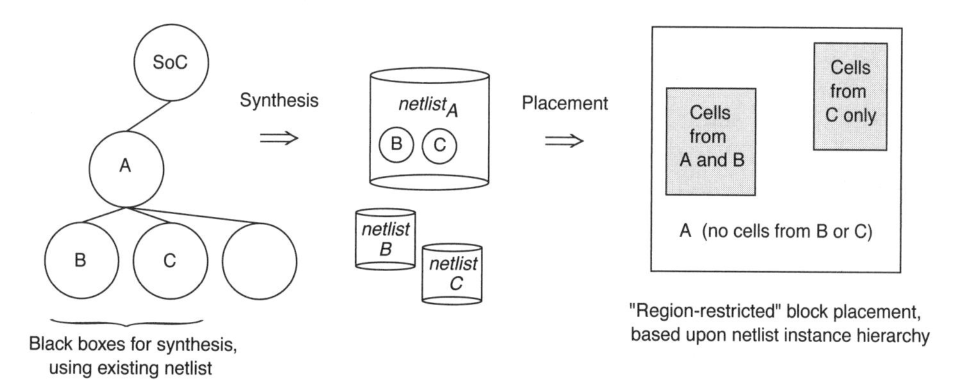


**Figure 8.2** The block placement flow may be provided with restricted areas for placement of subsets of the block netlist cells. This subset would typically be identified by a specific string in the (flattened netlist hierarchy) instance name.

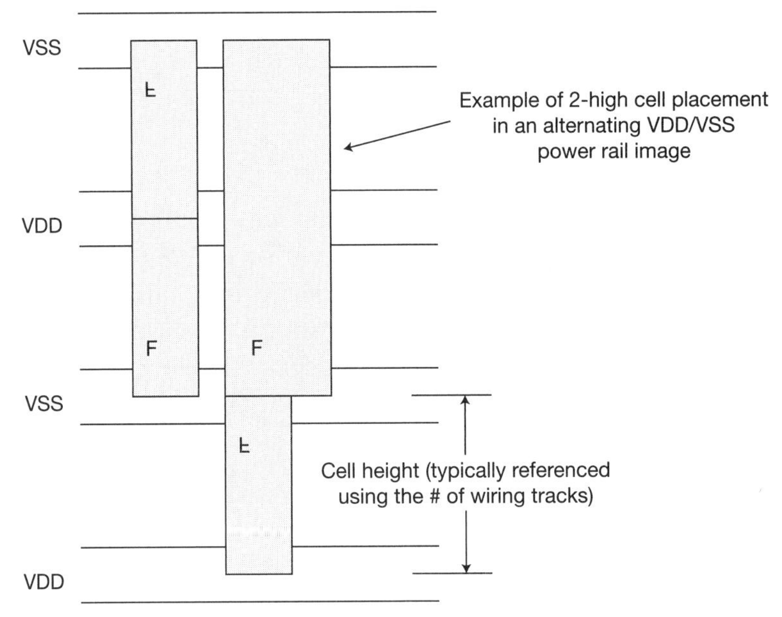


**Figure 8.3** The cell library may contain physical cells spanning two rows in height.

For current fabrication process nodes, additional cell adjacency restrictions must be observed during placement. Lithographic uniformity of (critical dimension) device gates may require the insertion of dummy gates between cells and at the ends of cell rows. The transition between cells of different $V_t$ types may also require dummy gate cells to reduce the device variation from $V_t$ mask overlay and implant dosage. Depending on the design of the cell image, the placement algorithm may also need to insert device well continuity *filler cells* in vacant locations. The methodology team needs to review the cell library techfile data and fabrication process design rules to ensure that any specific placement restrictions and/or dummy cell insertion guidelines are coded for the EDA placement tool.

Throughout the evolution of EDA placement tools, the goal has consistently been to provide a result that is ultimately routable and achieves timing targets, with runtimes that scale with the increasing block netlist instance size. Prior to the introduction of physical synthesis, placement tools consisted of *constructive* cell/macro location assignment followed by *iterative optimization* (or "successive refinement") steps. The physical synthesis methodology has resulted in a shift in EDA placement tool development emphasis to improving the iterative solutions. Numerous algorithms have been developed to select candidate cells to reposition and evaluate new proposed locations and/or to successively resolve placement overlaps from an existing assignment, with optimization objectives that address routing congestion and estimated path timing improvements.[1,2]

## 8.4 Clock Tree Local Buffer Placement

A key aspect of the placement flow is the special consideration to be given to the clock buffers in the netlist, typically added by the CTS step in the synthesis flow (see Section 7.9). The CTS algorithm attempts to balance the (estimated) loading on the branches of the clock tree in the network, whether originating from a single clock pin or connecting to a global clock grid. During cell placement, the common algorithmic approach is to select clusters of flops in close proximity and place a clock buffer in the final branch of the tree within the area spanned by the flop cluster. Once all clock tree endpoints are placed, a similar approach selects clusters of clock buffers and places a buffer from the preceding level of the tree appropriately; this process iterates recursively to the root level of the clock tree. The clock buffer placement algorithm results in output netlist updates, as the (logically equivalent) sinks at each level of the tree may be swapped during the clustering phase of the placement algorithm. The introduction of clock gating to the CTS tree implies that the cells at each level of the tree are not necessarily logically equivalent; clustering of placed sinks needs to observe gated clk_enable functionality.

For block placement with preplaced hard IP macros, the related clock buffers may also be preplaced accordingly. For relative placement groups, clock buffers may be included in the group definition. An increasing design trend is to offer multi-bit registers as an atomic cell library offering to minimize the clock routing and loading among bits. These registers are also likely to be part of relative placement groups with clock buffers (and decoupling capacitance cells).

During block routing, the attention to clock signals focuses on balancing the arrival latency at endpoints, primarily through R*C interconnect segment allocation. Performance optimization features in the routing flow may result in changes to the drive strength of logic path cells and flops; clock buffer tree cells may likewise need to receive drive strength updates in routing. For drive strength increases, any resulting cell area overlaps to the placement output locations need to be (incrementally) resolved during routing.

## 8.5 Summary

The incorporation of constructive placement algorithms in logic synthesis flows has resulted in a shift in focus for EDA vendors providing placement tools. Iterative optimization and legalization of the initial physical location cell assignment from synthesis requires judicious selection of candidate cells for re-positioning, with fast and accurate evaluation of interconnect parasitic estimates. This focus on estimation efficiency is required to support an increasing number of cell instances in a design block. In addition, tools are applying a richer set of designer input constraints to direct the resulting cell placement to a solution optimized for routability, path timing closure, and power dissipation reduction. Increasingly, physical implementation design resources for an SoC project are being re-directed from executing cell placement to addressing the complexities of interconnect routing optimizations for electrical and reliability analysis flows, such as timing, power, noise, and electromigration. Nevertheless, the quality of results for the cell placement flow is crucial to achieving subsequent design closure in routing.

## References

[1] Breuer, M., et al., "Fundamental CAD Algorithms," *IEEE Transactions on Computer Aided Design of Integrated Circuits and Systems*, Volume 19, Issue 12, December 2000, pp. 1449–1475.

[2] Pan, D., Halpin, B., and Ren, H., "Timing-Driven Placement," Chapter 21 in *Handbook of Algorithms for Physical Design Automation*, edited by C. J. Alpert, D. Mehta, and S. Sapatnekar, Auerbach Publications, 2008.

## Further Research

### Estimated Wire Length

Placement algorithms are dependent on wire length estimation calculations. Constructive placement methods often use total estimated wire length for all nets as a measurement criterion. Subsequent iterative optimization algorithms may add "weighting factors" to timing-critical and high-switching-activity nets as part of the wire length minimization objective.

Describe the various net topology alternatives commonly used for wire length estimation (e.g., Steiner tree, star, bounding box). Describe the advantages and disadvantages of the different topologies in terms of computation time and accuracy trade-offs.

### Constructive Placement and Physical Synthesis

The physical synthesis flow provides an initial placed netlist, and serial/parallel repowering cells are added during synthesis. Placement tools incorporate both constructive and iterative optimization steps and signal repowering features. As a result, a flow option could be provided to disregard the placement assignments from physical synthesis altogether and apply the constructive placement step on the complete block netlist.

Describe the trade-offs in exercising a constructive placement step on the full block netlist. Describe the sample experiments and quality-of-results criteria that could assist with this trade-off decision.

CHAPTER 9

# Routing

## 9.1 Routing Introduction

In early VLSI fabrication processes, the majority of each logic stage delay was associated with the internal circuits. The output loading model used in the cell delay calculation was a summation of the capacitances of the interconnect wires and the input pins of the fan-out cells. Routing tools used this lumped cell load model for timing-driven optimization decisions. As the scaling of VLSI processes progressed, the interconnect wire resistance (per unit length) increased, necessitating a change in the delay model. Replacing the simple stage delay model for a cell to all fan-outs, a more detailed cell plus interconnect R*C delay was introduced, resulting in a unique delay value to individual fan-outs (see Figure 9.1).

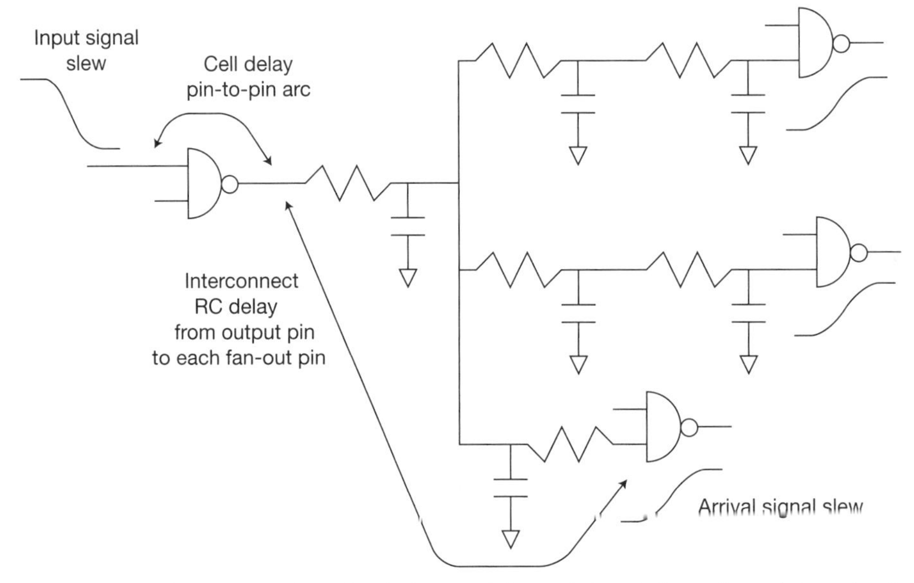


**Figure 9.1** Illustration of a timing model consisting of the cell arc delay plus the R*C interconnect delay to individual net fan-out pins.

The nature of the R*C interconnect topology required new innovations in the *model reduction* of the extensive parasitic resistive and capacitive elements, resulting in a simplified electrical representation.

### 9.1.1 C Effective Model Reduction for Delay Calculation

A perplexing issue was how to characterize the pin-to-pin delay arcs of each library cell with an unknown, complex output loading network. Library developers were providing delay (and output signal slew) tables as a function of input signal slew and total lumped capacitive load, based on characterization simulations. A key innovation that enabled the separation of cell and R*C delays was the introduction of an "effective load capacitance." As briefly described in Section 7.2, an algorithm was developed that reduced the RC tree to an equivalent pi-network and, subsequently, to an equivalent $C_{eff}$ output capacitive load for indexing into the cell delay table. Figure 9.2 illustrates the algorithmic reduction of the RC tree to $C_{eff}$.

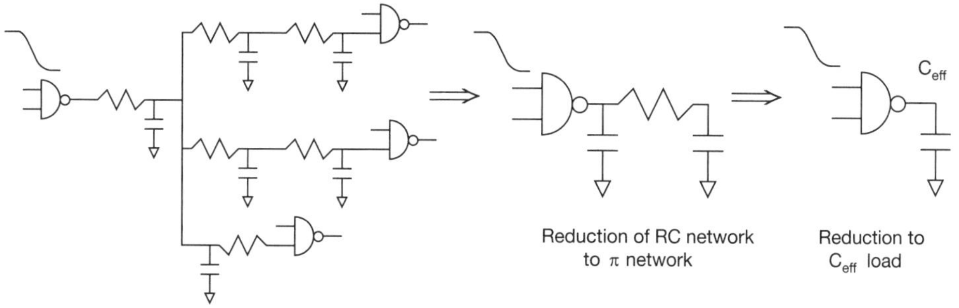

**Figure 9.2** Illustration of the reduction of an R*C interconnect tree to an equivalent pi-network.

The $C_{eff}$ algorithm reflects the nature of *resistive shielding* at the cell output, where the effective load at the output pin is less than the full interconnect and fan-out pin capacitances of the earlier timing model. The cell output signal slew from the characterization table is presented as the driving source to the RC interconnect load network to calculate the unique arrival time and arrival slew at each fan-out sink. The derivation of the signal characteristics at the fan-out sinks uses an analytical model, such as the Elmore or PRIMA calculation.[1,2] The timing optimizations in routing algorithms needed to adapt to the more detailed cell plus R*C interconnect delay-calculation method.

### 9.1.2 Sidewall Coupling Capacitance

As VLSI process lithography has continued to scale and innovative techniques for metal deposition and surface polishing have continued to evolve, the height-to-width aspect ratio of wires has increased. The contribution of the *sidewall coupling capacitance* from adjacent routes compared to the vertical parasitic capacitances has accounted for a much larger percentage of the total capacitive load. Routing algorithms needed to be enhanced to represent the coupling capacitance in RC interconnect tree model reduction and, specifically, to include coupling noise between wire segments as a constraint limiting the parallel run length of adjacent routes.

### 9.1.3 Routing and DRC Rule Complexity

Concurrently with advances in (high aspect-ratio) interconnect fabrication, the lithography design rules for metal wires became significantly more complex (e.g., line extensions past via intersections, line width versus run length and run step requirements, specific *stacked via* topologies). Figure 9.3 illustrates some of these routing rule additions.

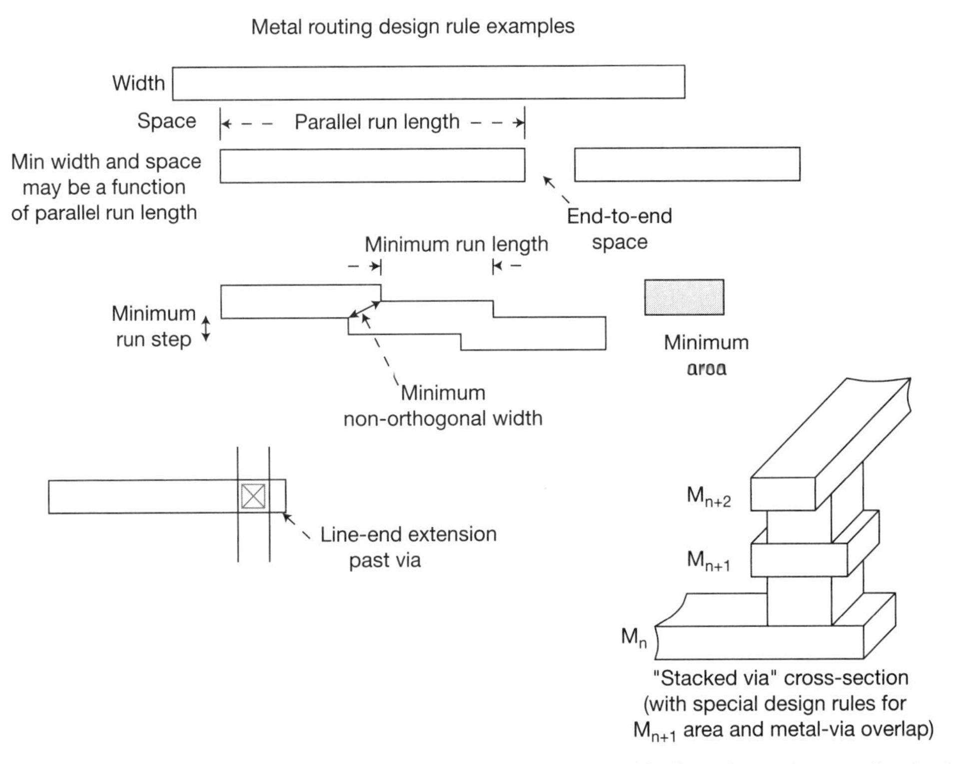


**Figure 9.3** Complex DRC rules for metal layers, with direct impact on routing tools.

These additional design rules needed to be comprehended by the routing algorithms, which began to integrate more sophisticated design rule checking capabilities. These design rule complexities were exacerbated by the increasing use of multiple wire width and spacing combinations to compensate for the increasing line resistance.

### 9.1.4 Routing with Multipatterning

Advanced process nodes have introduced the requirement for multipatterning for the densest metal layers, where the layout data of the final routed solution is decomposed into multiple masks for separate lithography and etch process steps. This introduces additional complexity to the routing tools, as *cyclic redundancy* avoidance is now required to eliminate decomposition failures, as illustrated in Figure 9.4 (also see Section 18.1).

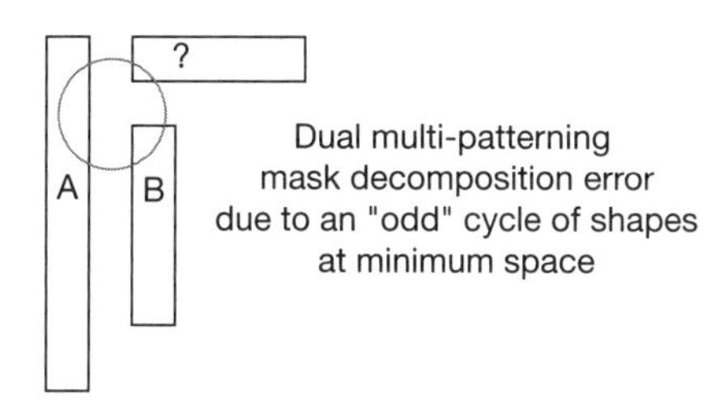


**Figure 9.4** Illustration of a multipatterning cyclic redundancy error, to be precluded by routing tools.

### 9.1.5 Via Pillars

With FinFET technology, the nature of the device channel increases the current density per unit area. The continued lithographic scaling of new process nodes has resulted in increased concerns about the reliability of high current densities through metal wires, contacts, and vias. For routing, single vias between metal layers may be insufficient for high-drive-strength signals, and via pillars may need to be assigned, with the associated wiring track blockages (see Figure 9.5).

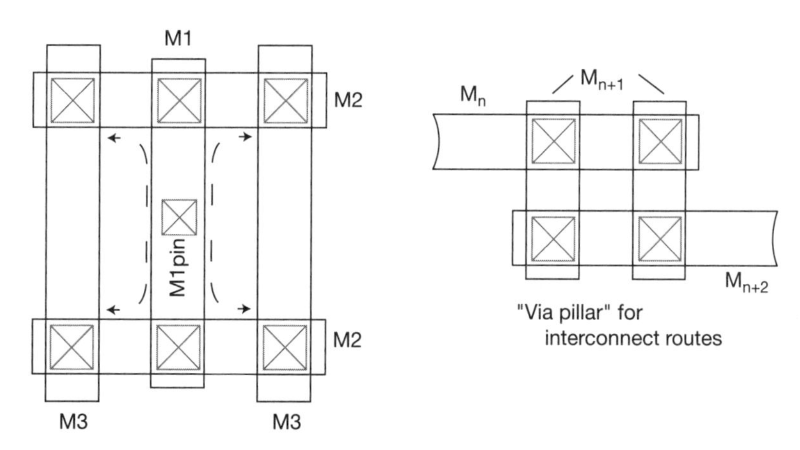


**Figure 9.5** Illustration of via pillars, required in advanced process nodes to satisfy design requirements for multiple vias for high-drive-strength signals. In addition, the design rules at advanced nodes limit the variability in allowed via sizes; large area vias are disallowed, necessitating the use of multiple vias and via pillars.

### 9.1.6 Metal Trim Mask Definition

Lithography scaling has resulted in the requirement to maintain aggressive metal line end spacing design rules, concurrent with reductions in metal line-width and space dimensions. An additional "metal trim" (or "cut metal") mask lithography step replaces the patterning of distinct metal segments at minimum line end spacing, as depicted in Figure 9.6.

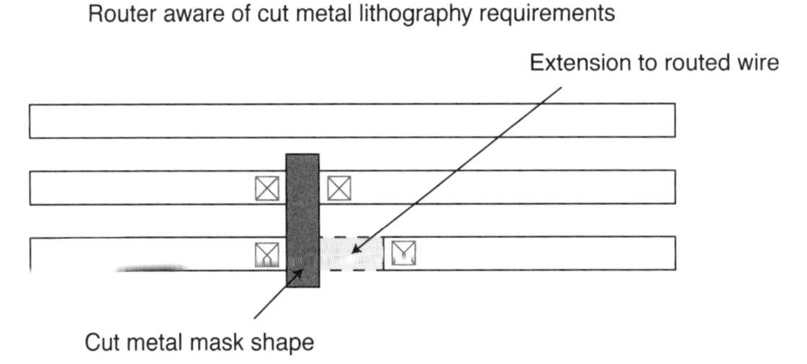


**Figure 9.6** Illustration of the cut metal mask data added to define line end spacing. Note that the metal trim shape requires router segment extensions to align adjacent line ends.

Two distinct nets are merged into a common metal line to be isolated by the new metal trim process step. The router output data needs to support this layout style; note that adjacent route segments may need to be extended to align with the (minimum area) metal trim shape.

Throughout the generations of VLSI technology scaling, the traditional adage has been "A good router will not be able to fix timing issues due to poor cell placement." While this is still accurate, the demands on routing algorithms have increased substantially: improved timing-driven model accuracy; coupled noise avoidance; adherence to complex layout design rules (including multi-patterning); and, ultimately, lifetime reliability. EDA tool vendors and foundries have collectively invested significant development resources to ensure that the routing tool features (and related technology files and design rule checking descriptions) support the requirements of each new process node.

## 9.2 Global and Detailed Routing Phases

Routing algorithms from EDA tool vendors have traditionally been divided into two distinct steps: a "global" assignment of interconnect segments to coarse grid edge crossings and a “detailed” assignment of these segments to specific metal layers and wiring tracks. (In this context, the term *global route assignment* refers to an algorithmic approach, not to the task of implementing global signals between blocks in an SoC floorplan; the global routing step applies to any design submitted to the routing tool.)

The efficiency of the global routing approach allows it to be incorporated into physical synthesis and cell placement flows. As depicted in Figure 9.7, the routing area is divided into a grid. The coordinates of the pins in each net are used to calculate the number of edge crossings, based on an initial proposed net topology (e.g., a Steiner tree). An evaluation at each grid edge of the available tracks versus the demand of the current net topologies is used to assess the potential congestion that would arise during detailed routing.

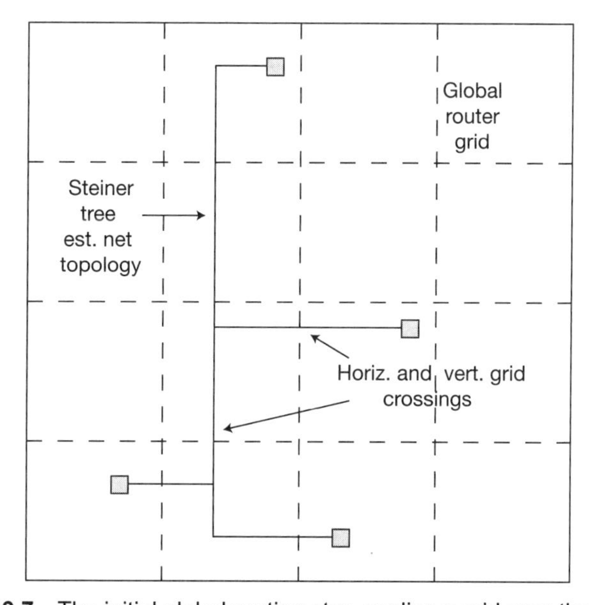


**Figure 9.7** The initial global routing step applies a grid over the block area, allowing a calculation of the wire demand (grid crossings in the estimated topology for each net) versus the available routing tracks.

An alternative net topology may need to be evaluated for specific grid edges. During global routing, designer input constraints are observed—for example, prerouted (partial or full) nets, non-default wire widths, and/or upper metal layer preferences for performance-critical nets.

There is an underlying assumption that routes will consist of orthogonal segments, with successive metal layers alternating between horizontal and vertical directionality. Process lithography design rules have evolved in this regard. Traditionally, successive layers were given preferred directions, with *wrong-way segments* allowed, albeit with more conservative design rules. Advanced process node design rules require strict unidirectional segments on (lower) metal layers.

Detailed routing provides the final topologies for each net, with specific layer, width, and track assignment for wires and corresponding vias between layers. The topology for each net in the netlist includes appropriate coverage of cell pin shapes on the proper metal layer. (The library cell abstract provides shapes data for all the pins, which are included in the route model from the placement flow.) In the case of large area pins, a cell property indicates whether full or partial pin shape coverage is required. The detailed router needs to include an algorithm that correctly constructs the pin access shape's data for the case where tapering between route width and pin width is required or a via transition from an upper-level metal is needed, as shown in Figure 9.8.

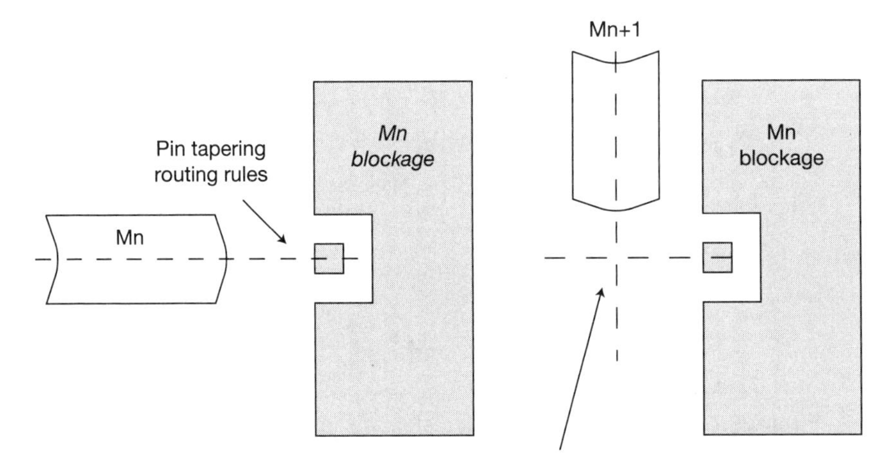


**Figure 9.8** Illustration of the transition of a route to the corresponding cell pin, highlighting the need for wire width tapering and specific via insertion.

Advanced process nodes require that the detailed router incorporate awareness of subsequent multipatterning assignment. A simple track grid for the detailed routing step would assign a multipatterning color for each track, as depicted in Figure 9.9.

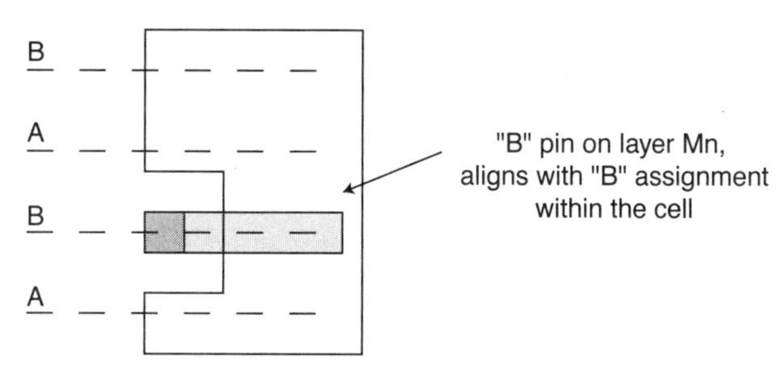


**Figure 9.9** Illustration of route track "pre-coloring" for a multipatterning metal layer.

The valid cell placement coordinates and library cell pin data would align with the multipatterning decomposition grid. The use of wire widths greater than the 1X design rule minimum (and, correspondingly, greater than minimum spacing) introduces additional complications, where the predefined track color assignment is no longer applicable; Figure 9.10 provides an example.

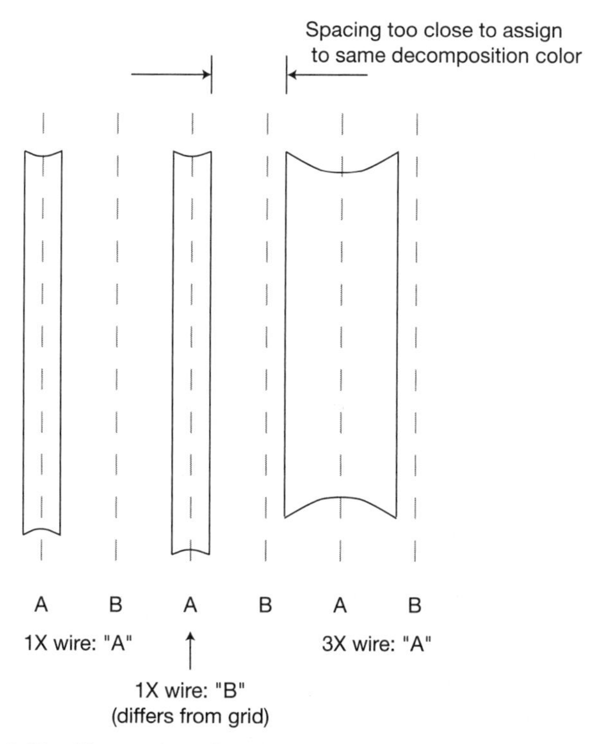


**Figure 9.10** The routing of a wide wire invalidates the pre-coloring track assignment. As a result, multipatterning decomposition checking algorithms are still required in the router.

The detailed router requires access to multipatterning layer decomposition and design rule checking algorithms to ensure that the solution is valid for mask data processing.

The output of the global routing step offers insights to the block physical designer and provides expectations for successful detailed route closure. The designer may choose to make modifications to the model and rerun the global routing flow—for example, edit the cell placement, add more preroute segments, or, most significantly, alter the block floorplan and re-execute cell placement on the full netlist.

The output of the detailed routing step generates the interconnect segments and vias to merge with the placement flow data. The detailed routing step also provides preliminary timing and noise results, although separate electrical analysis flows are used for tapeout sign-off.

The detailed router may terminate with an incomplete solution. A (hopefully small) subset of nets may not be fully routed due to any of a variety of factors—for example, congestion, unsatisfied constraints, or even "timeout" of the flow if the router is making frequent route topology changes to address congestion and constraints. In the case of an incomplete solution, the router output includes a list of *overflow nets* that are not fully implemented, perhaps with some partial connectivity provided. The EDA vendor provides a route editing tool for a layout engineer to attempt to resolve the set of overflows; this tool is typically optimized for route segment embedding and is different from a general circuit layout shapes editor. The tool reads the full netlist connectivity model and the overflow list. The selection of an overflow net highlights any partial route segments, as well as the pins that are "dangling." Detailed routes are typically displayed as lines (with a width property) rather than rectangles, to reduce the visual complexity in the editor. Route-specific edit commands are provided, as well (e.g., "shove" a selected group of route segments (and vias) to insert a new segment while maintaining valid connectivity). The support of multipatterning for routed metal layers introduces an additional feature for the route editor: The mask decomposition grid and corresponding segment design rules need to be accessible. The layout engineer may need to bypass some of the constraints originally provided to the detailed routing step to make local routing tracks available for overflow nets. The SoC physical design and electrical analysis teams collaborate with the layout engineer to ensure that timing/noise/power sign-off targets are still able to be achieved if (a small subset of) existing constraints are relaxed.

If the number of overflows is large, the SoC project management team needs to assess the likelihood of successfully (manually) completing the routing step with a reasonable schedule and resources or whether a revision to the existing placement and/or physical floorplan is warranted.

The EDA algorithms used across the entire SoC design flow have evolved throughout the process generations; recent innovations involve the integration of previously separate steps into a "predictive" methodology, enabled by a data model that spans multiple domains. Physical synthesis (see Section 7.2) is a good example, as the accuracy of logic synthesis timing calculations is improved by incorporating physical placement estimates. However, there remains a wealth of fundamental algorithm development opportunity in the areas of global and detailed routing. The process lithography design rules associated with metal and via shapes are becoming much more complex—for example, width and adjacent wire parallel run length dependencies, via overlap requirements, allowed via array combinations, and, especially, multipatterning decomposition rules for wires and vias. The number of metal layers available for block and SoC routing has increased dramatically, enabled by the wafer surface planarity realized by intermediate chemical-mechanical polishing (CMP) fabrication engineering. The number of potential metal stack combinations (of different electrical characteristics) available from the foundry has correspondingly increased, allowing greater opportunities for optimization through upper and lower metal layer route segment assignments. The scaling of physical design rules with higher aspect ratio metal cross-sections results in greater impacts of adjacent net switching noise. Increased device current density impacts the electromigration reliability of wires and vias. Design teams are relying on routing technology to address the increasing rule complexity, optimize timing on the chosen metal stack, and avoid issues during subsequent electrical analysis. Enhancements to routing algorithms remain an active field of research.

## 9.3 Back End Of Line Interconnect "Stacks"

A key facet of any SoC design is to determine the number of metal layers to be used and the specific fabrication process module selected for each layer; collectively, this choice is denoted as the *metallization stack*.

The foundry offers a variety of metal layer choices. Each layer type has a unique set of layout design rules for intra-layer segment width and spacing (W/S) dimensions, as well as via rules to layers above and below. The foundry places additional restrictions on the allowed layer type transitions. As illustrated in Figure 9.11, a set of intermediate-layer thicknesses are required as the (minimum W/S) lowest-layer metals taper up to the thicker metals used for global power/ground, clock, and signal distribution.

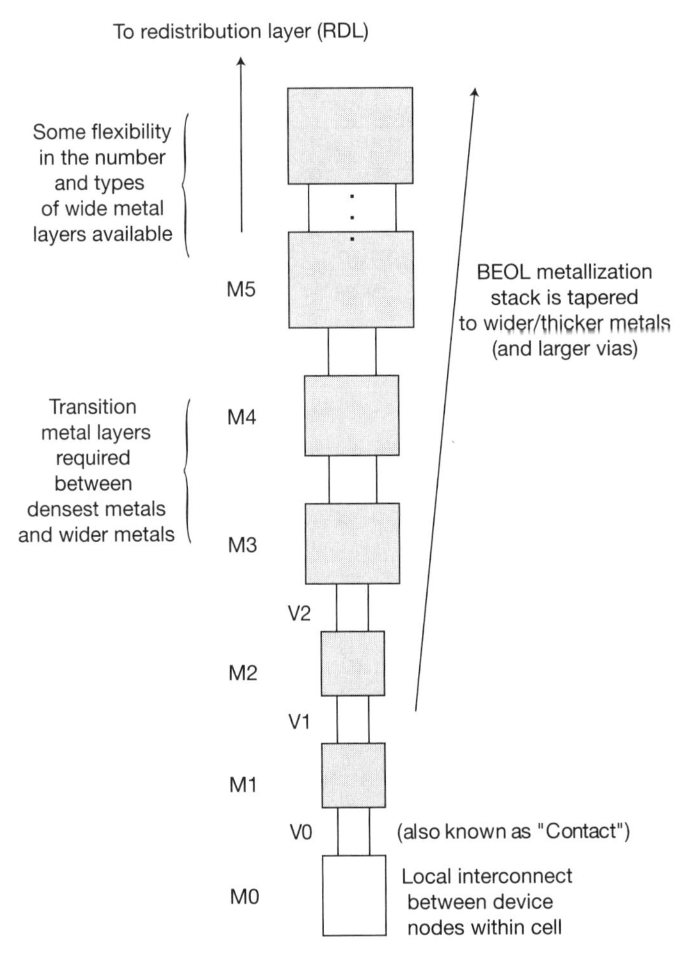


**Figure 9.11** Illustration of the BEOL interconnect stack, with specific intermediate transition layers required.

A common nomenclature to describe the chosen stack is to concatenate the process modules together in a single string. For example, a nine-level metal design example would be denoted as "1P_3Mxa_2Mxb_2My_2Mz_1Mr" to indicate a single polysilicon gate technology with four separate types of metals in the nine-layer stack and a top redistribution layer.

A crucial phase of SoC planning is to assess what stack provides the optimum balance between (global and local) routability and electrical characteristics, the latter involving the R*C parasitics per unit length and electromigration current density limits. An increasingly significant factor is the statistical variability in these electrical parameters. For example, the metal line thickness variation is a strong function of the local neighborhood density on the same layer due to the different characteristics of metal and dielectric removal during CMP. The capacitance per unit length for multipatterned layers is a strong function of the decomposed mask overlay tolerance, as depicted in Figure 9.12.

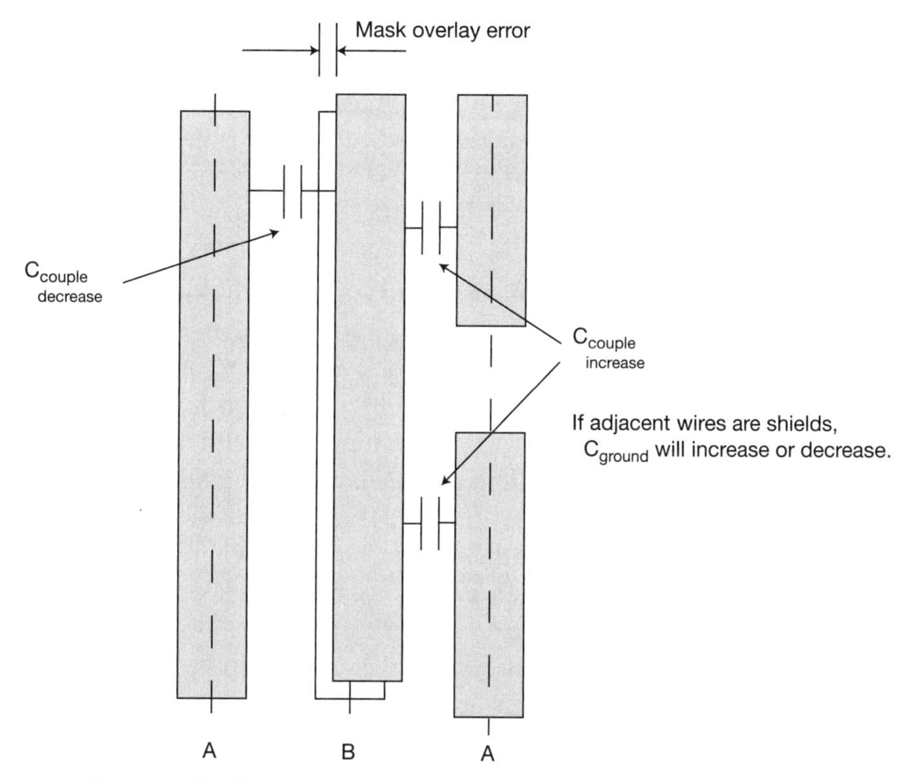


**Figure 9.12** Illustration of the multipatterning mask overlay tolerance and the impact on the adjacent wire coupling capacitance.

The SoC design team pursues a set of design experiments to assist with the metal stack decision; these experiments involve assessing both the nominal and statistical variation trade-offs in the electrical characteristics of each layer type. A high routing demand associated with dataflow logic and/or the need to distribute wide busses in the SoC design would drive the stack decision preferentially toward including a larger proportion of dense metal layers. The SoC requirement for very low clock skew and lower R*C interconnect delays to close on aggressive timing makes thicker intermediate metal layer types more attractive. The upper layers of the stack are typically among the thickest metal fabrication modules available from the foundry for global nets and the power and ground distribution grids. The very top layer is used to provide redistribution from I/O and power/ground pad locations to internal connections.

There are significant product cost implications to these metallization stack evaluations, as well. The addition of an (intermediate or global) routing layer involves additional mask and fabrication cost. The number of (lower) multipatterning layers to achieve greater route density is a major contributor to product cost due to the multiple masks and additional lithography/etch steps per layer. An additional, indirect cost adder is the need to use more advanced (expensive) equipment for aggressive lithography processing for the densest layers; the depreciation of this capital cost is assigned to the wafer lots for the SoC designs requiring this capability.

The SoC planning experiments to guide the metallization stack selection involve estimating how representative timing paths of interest in the design would be implemented. Models for cell delays and R*C interconnect delays with estimated wire lengths and layers are constructed for a selected set of critical architectural timing paths and the cumulative path delays compared to design performance targets. Different metal layer types (and non-default route widths) are explored. The vias between the layers are included in the model, as well. As process scaling has continued, the increasing via resistance has had a greater impact on path performance. Various multiple via and/or large area via design options may be available and incorporated into the path performance models. The fabrication process design rules also define what "via stacking" dimensions are supported when attempting to transition between multiple metal layers with minimal route track blockage.

The SoC foundry interface team also needs to be closely engaged in the stack selection to ensure that all the related process design kit (PDK) data are available—for example, DRC checking runsets, LVS connectivity definitions, parameterized layout generation utilities, and, especially, layout parasitic extraction techfile support (which requires detail of the full cross-section from the substrate to the top metal).

The SoC planning team also needs to participate in the metal stack decision process, as it relates to any hard IP and cell library circuitry to be integrated. Indeed, the hard IP will likely dictate the (lower levels) of the stack to be used. It may be possible to engage the IP provider to develop a modified hard IP design for a specific stack that is not currently supported. However, this development cost and the schedule will impact the SoC project, especially if additional silicon testsite fabrication and characterization are warranted. Because of the strong dependency on external IP, the number of metallization stacks adopted by SoC designs in a new process node is actually quite limited, and many foundry customers gravitate toward an existing PDK offering (certainly for lower and intermediate metal layers).

## 9.4 Routing Optimizations

The majority of SoC design closure on routability, timing, and power/ground distribution is achieved during floorplanning, physical synthesis, placement, and global routing steps. However, numerous optimizations are afforded during detailed routing, and post-route utilities add data for improved fabrication yield. Indeed, the algorithms to address coupled noise limits from aggressors to a victim net require detailed route data.

As with other flow optimizations, the key to detailed router algorithms is the integration of analysis engines that work on a shared data model. The representation of detailed routes involves the electrical characteristics of segments on each metal layer and the vias between them. Of specific importance is the estimation of route parasitics. The parasitic representation needs to be sufficiently accurate to enable optimization decisions but also sufficiently abstract so that (incremental) calculation of self- and coupling-capacitance with resistance is efficient. A key consideration is estimation of the parasitics for a route segment traversing (core or library) IP. The IP layout shapes data are typically abstracted to a gray box for parasitic capacitance calculations to the route segment, as depicted in Figure 2.17. The parasitic representation is further complicated by the variety of route segment characteristics available to the optimization algorithms, to be discussed next.

### 9.4.1 Route Segment and Via Topologies

A detailed router has the following route segment variations from which to select:

- Non-default width, such as 1.5X or 2X
- Non-default spacing

Adjacent tracks may be left unoccupied to reduce coupling capacitance, at the cost of overall routability. A “wire-spreading” algorithm exercised near the end of the detailed routing step can also reduce coupling (and minimize yield loss from *bridging faults*). However, the presence of routes “off-grid” complicates the completion of overflows and subsequent design ECOs and may introduce topologies that are problematic for advanced lithography processing.

#### *Route Shielding*

Route segments could be assigned to tracks that are adjacent to non-switching wires, such as the existing power/ground distribution or shield segments specifically added and connected to power/ground. These adjacent segments would thus never act as aggressors and thus would not contribute to a coupled static noise-level fail or coupling-induced signal transient delay. (Half-shielded wires have one adjacent non-switching segment; fully shielded wires have non-switching segments in both adjacent tracks.) Shields are a common routing constraint given for the clock net segments at the leaf levels of a clock tree. A router is likely to include a specific clock route optimization feature that assigns clock route segments to shielded tracks with high priority.

#### *“Balanced” Routing*

Another detailed routing constraint is to balance the electrical characteristics of a set of nets to their fan-outs, also applicable to clock nets. (A mixed-signal block router also incorporates this feature to apply to differential signals requiring matched electrical characteristics.)

#### *Via Optimizations*

A number of via topologies, with different resistance values, are typically available to a router (see Figure 9.13):

- Square and rectangular vias
- Via arrays (both signal and power/ground grid connections)

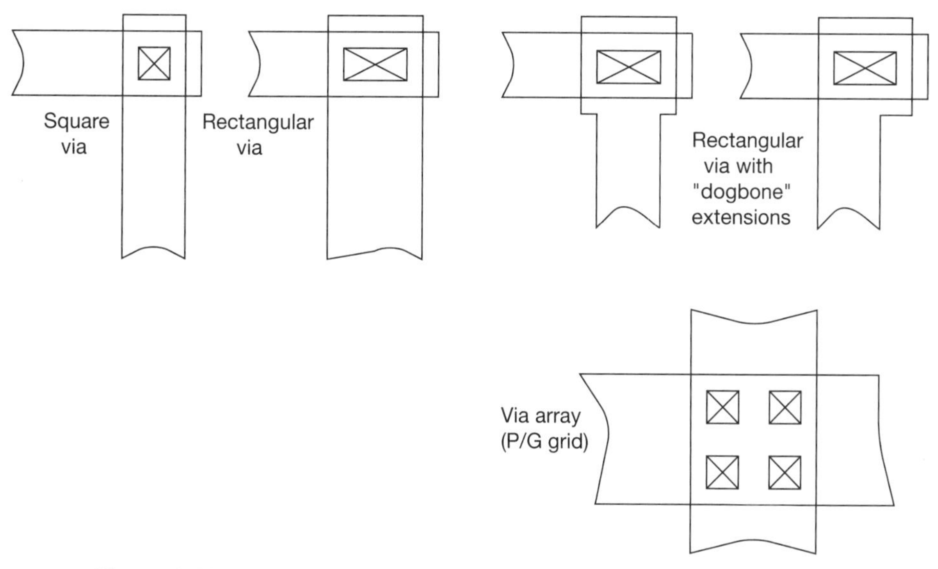


**Figure 9.13** Illustration of via topology examples. At advanced process nodes, the variability in allowed via sizes is reduced.

As with metal segment layout design rules, via lithography has also become increasingly stringent with process scaling. A single via size requirement is becoming prevalent, and multiple vias are commonly used to reduce resistance. The via array may also be subject to multipatterning decomposition in the most advanced process nodes, restricting the allowed topologies.

### *Via Pillar*

The most complex topology is a via pillar, as illustrated in Figure 9.5. This configuration may be required in advanced nodes for both the reduction in resistance and to accommodate higher current density associated with high-drive-strength circuits. The electrical model used by the router for the via pillar requires reduction of parasitic detail for efficiency.

## 9.4.2 Electrical Analysis Optimizations

A detailed router includes algorithms to address closure on multiple electrical characteristics of the net, including timing, noise, power, and electromigration reliability.

#### *Timing Optimization and Detailed Routing*

The SoC physical implementation flows prior to routing have a much higher impact than does detailed routing on path timing closure. The router is able to address critical paths identified by physical synthesis and cell placement, using a combination of approaches:

- Net priority (Nets with negative timing slack values from prior flows would be given higher priority in the assignment of route tracks to minimize any scenic traversals made to circumvent congested areas.)
- Metal layer assignment in the stack
- Route width/space/shielding

In addition to the R*C per-unit length estimates for a metal segment, there is a question on the router's electrical model for coupling capacitance to adjacent segments. As illustrated in Figure 9.14, an adjacent switching segment could contribute a range of capacitive loading to a signal transient.

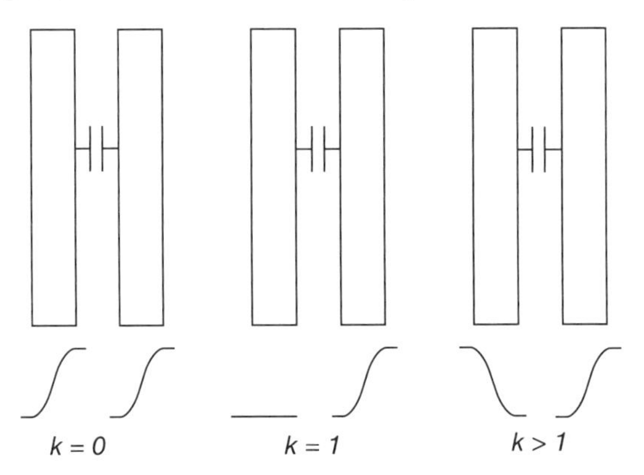


**Figure 9.14** The impact of an adjacent wire switching concurrently with a signal transient results in a widely varying coupling capacitance. A multiplicative factor for Cc is commonly applied during routing, as well as for other electrical analysis flows.

The choice of the electrical model used by the router for timing optimization could use an average coupling capacitance (~ 1 * Cc in the figure) or a pessimistic value (greater than Cc). If the latter, a track assignment that includes a (half- or full-) shield would improve the interconnect delay calculation.

### *Noise Optimization and Detailed Routing*

The detailed router is the only implementation flow that can optimize circuit stability due to switching noise, as specific information on the net track assignment is required. The sign-off noise analysis flow utilizes an intricate extracted parasitic model for calculation of the (superposition of) coupled and propagated noise, as depicted in Figure 9.15.

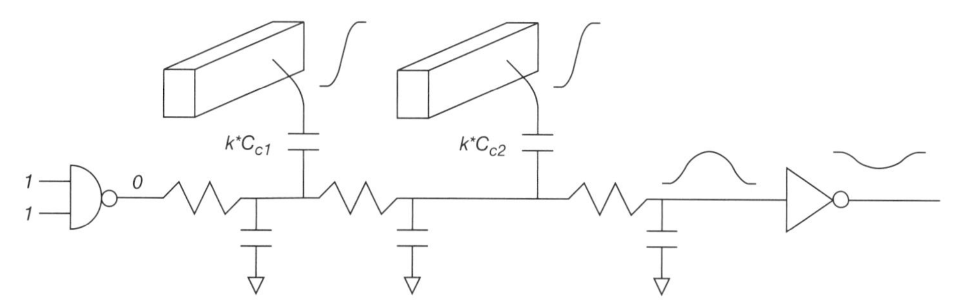


Illustration of (superposition of) coupled noise into a quiet "victim" signal — the detailed router incorporates noise avoidance algorithms.

**Figure 9.15** Illustration of the electrical model used in detailed noise analysis (see Chapter 12, "Noise Analysis"), whereas the router employs a simplified noise avoidance algorithm.

The driving circuit dissipates part of the coupled energy, and some of the noise transient arrives at the input pins of the fan-out cells. A noise transient at the outputs of the fan-out cells results, and is propagated further. The noise model embedded within the detailed router is necessarily less complex than the sign-off flow. The simplest router noise optimization method would be to apply a set of rules limiting the parallel run length of adjacent track segments, based on the source/sink cell types and drive strengths.

### *Power Optimization and Detailed Routing*

For power optimizations, multiple router algorithms can be applied, including the following:

- **Net priority (based on estimated switching activity)**—Much as with timing-driven optimizations to reduce interconnect load capacitance, a set of net priorities can be assigned based on estimated switching activity and fan-out. The SoC methodology would need to include a switching probability input file to the router flow, derived from netlist-level simulation testcases.

- **Tapering of wide wires**—A route segment may be assigned a width greater than the metal layer default to satisfy various constraints (e.g., timing optimization, alignment with output pin width). The delay of an interconnect tree is most sensitive to the resistance near the driving output pin. A power optimization to reduce the total net capacitive load, with a lesser impact on timing, involves tapering of the wide wire toward the end of the segment, as illustrated in Figure 9.16.

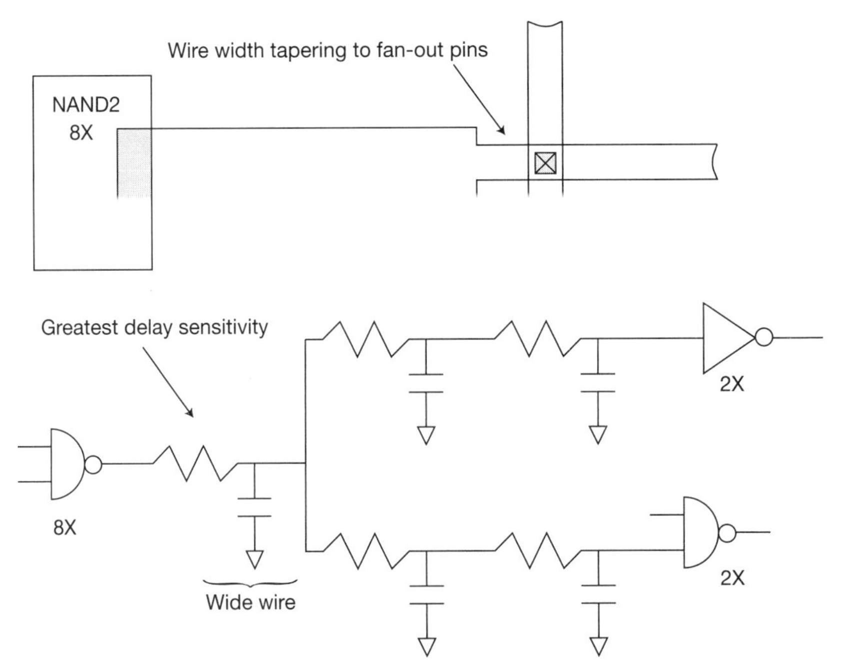


**Figure 9.16** Tapering of a wide wire at the far end to reduce $C_{total}$ and active power dissipation.

### *Electromigration and Detailed Routing*

Electromigration reliability is an increasing design issue with the device current density of advanced (FinFET) processes. Electromigration failure rate calculations are a complex function of material properties, current waveforms, temperature, switching activity, and wire length (see Chapter 15, "Electromigration (EM) Reliability Analysis"). For routing algorithms, the goal is to ensure that segment widths and vias are consistent with the drive strength and capacitive loading for the net.

A unique aspect of detailed signal routing relates to the connectivity to a cell pin. A cell designer must be cognizant of the local current distribution when a route is assigned to the pin, as the resulting current waveforms may have different (nonzero) average values through metals, contacts, and vias within the cell, based on the selected route track. Figure 9.17 illustrates a (somewhat contrived) example of an area pin; based on the tracks available for the detailed router to select for the pin, different currents in the local interconnects within the cell result. The cell designer needs to pursue additional electromigration analysis for the possible detailed router track selections.

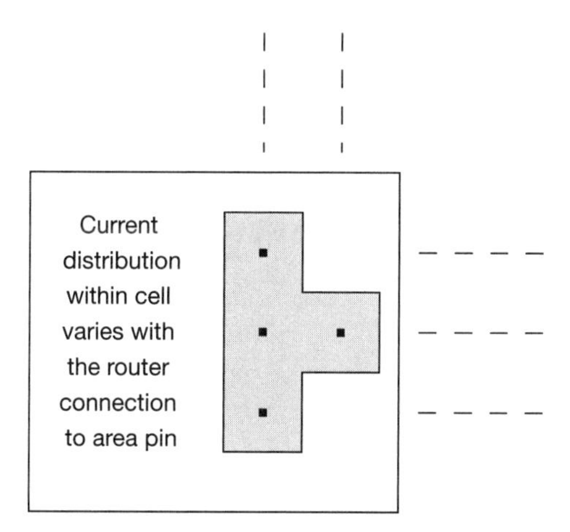


**Figure 9.17** The current distribution within a cell may differ depending on the route track selected for an area pin. Electromigration analysis of the cell needs to include any potential route connection.

Note that this discussion has assumed that a robust power/ground distribution grid has been provided with the SoC physical floorplan image. If power net routing is required to complete the connectivity to cells and IP, additional electrical analysis constraints apply for detailed routing to ensure that I*R voltage drop and electromigration current densities are within limits.

### 9.4.3 Non-Orthogonal Routes

Early VLSI process fabrication nodes supported lithography design rules that included non-orthogonal angles, specifically 45-degree metal segments. The goal was to reduce the need for "jumper" wires and vias when changing tracks for a signal route. The layout design rules required a conservative width for the angle segment to compensate for "corner rounding" after fabrication, with design rule restrictions on vias present on the angled segment. (There was also a short-lived methodology proposal and EDA router tool product release with alternating metallization stack layers dedicated to orthogonal and 45-degree routing tracks, with specific vias for an orthogonal layer to 45-degree layer transition. The goal was to achieve reduced global route lengths.) With continued lithography and fabrication process scaling, any 45-degree segments on a metal line were disallowed and replaced with orthogonal *river routes*. Figure 9.18 illustrates the layout design rules for the allowed widths and edge lengths in the track transition region.

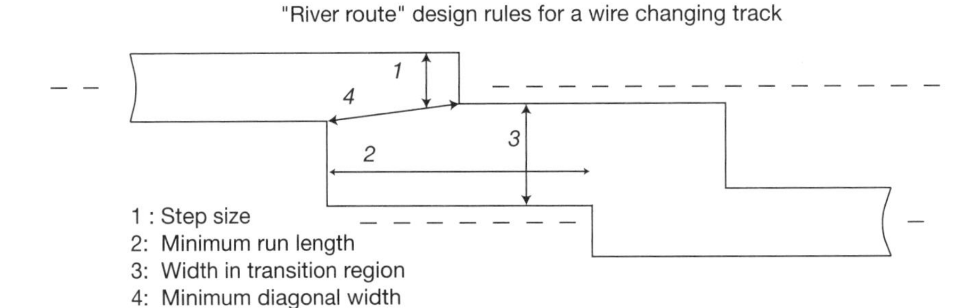


**Figure 9.18** Illustration of a "river route" configuration for a signal route changing tracks. In general, multiple, adjacent signals follow a river route topology as a group.

In advanced process nodes, only unidirectional rectangular metal shapes are allowed on the lowest layers (with multipatterning decomposition). In addition, the router needs to be cognizant of *forbidden pitch* design rules that define disallowed metal width/space combinations to avoid "wavelength grating interference" through the mask. The exposing light energy incident at the photoresist-coated wafer needs to be focused to represent the open areas of the mask. The forbidden metal pitch rules are intended to prevent an interference pattern on the mask that would result in significant light energy outside those areas. As a result, the use of non-rectangular routes—even as a post-detailed router optimization to remove track jumper segments on adjacent layers—is rapidly declining.

The exception to this orthogonality trend applies to the design rules of the redistribution layer, which connects I/O signals and power/ground from chip pads to their internal coordinates. The design rules for the redistribution layer are significantly more conservative due to the large metal layer thickness—and, in this case, 45-degree-oriented segments routed between pad locations are common. The redistribution layer (RDL) metal shapes for signals and power/ground are often designed manually or are, potentially, developed with extensive user constraints given to the detailed router.

### 9.4.4 Design-for-Yield (DFY) Optimizations and Detailed Routing

After the detailed routing phase is complete, a number of additional design optimizations are invoked prior to exporting the physical layout shapes dataset. The most pervasive design update is the insertion of metal fill to achieve a recommended metal layer density (and density gradient) over all stepping windows across the design area. The metal fill step is usually added to the SoC methodology after routing, using a separate EDA tool specifically optimized for the complex design rules, fill patterns, and density calculations associated with fill.

Other DFY optimization algorithms in addition to metal fill are pursued as part of route finishing at the end of the detailed routing step. The following sections describe them.

#### *Wire Spreading*

Wire spreading in areas of low track utilization reduces the susceptibility to manufacturing yield loss due to bridging faults between adjacent wires after fabrication. With the advanced process requirements for minimizing river route–style data, this optimization is of diminishing applicability.

#### *Line End Extensions Past Vias*

Process design rules are developed to enable aggressive route density; specifically, the metal width, via size, and (minimum) via coverage are critical layout features. A minimum metal-via overlap is preferable for density; however, the metal-to-via mask overlay tolerance results in variation in electrical resistance for the connection. A post-detailed route optimization extends the metal segment for improved characteristics where space is available, as illustrated in Figure 2.13. An efficient method to describe signal route segments is to use a "PATH" record in the layout database; this is a one-dimensional line representation with additional parameters defining the line width and the extension dimension past the segment coordinate endpoints. Thus, adding to the extension past a route endpoint is a relatively straightforward algorithm to implement, and it does not require extensive shapes data manipulations.

#### *Redundant Via Insertion*

Relatively complex wire extensions are used in combination with a *redundant via* insertion algorithm that is exercised after detailed routing. In older process nodes, these extensions were often implemented by adding a non-rectangular extension to a metal segment, as depicted in Figure 2.14. The restrictive metal shapes allowed by the layout design rules for advanced processes reduce the opportunities to exercise redundant via insertion except for wider wires.

The insertion of redundant vias is strictly motivated by the improvements in manufacturing yield related to reducing the impact of the fabricated via defect density, such as an incomplete via opening in the dielectric layer between metals. If the current density in a single via is an electromigration reliability concern, the detailed router step should have already been guided to insert the appropriate via array or via pillar structure.

#### *Antenna Diode Insertion*

Whereas line extensions and redundant via insertion are opportunistic approaches to enhance (electrical and process-limited) manufacturing yield, another post-detailed route algorithm is required for specific net topologies: diode insertion to eliminate *metal antennas* present during fabrication. Antenna diode insertion design checks are released by the foundry as part of the PDK runset. (Figure 2.11 illustrates the definition of a metal antenna.)

During fabrication, various process steps expose the wafer surface to a gas plasma, with a high concentration of ions/electrons at high energies. During these plasma processing steps, charge accumulation on metal lines connected to a device gate input without a corresponding discharge path subject the gate dielectric to a potentially damaging electric field. The antenna checks examine each route topology. If the connection between the output pin (with an inherent diode from device nodes to substrate) and the input gate pin(s) is not electrically continuous until metal layer Mn, the area of the segments connected to the input(s) on preceding layers up to Mn is measured. This charge collection antenna area is compared to the total gate input pin area, which is derived from the cell library characterization data. The foundry design rules define how the segments on each metal layer contribute to the antenna sum and the allowed ratio between antenna and gate areas. There are two main approaches to addressing antenna check violations:

- **Reroute the violating net.** Figure 9.19 illustrates how an antenna area could be reduced, although the ability to insert a jumper may be limited in highly congested areas.

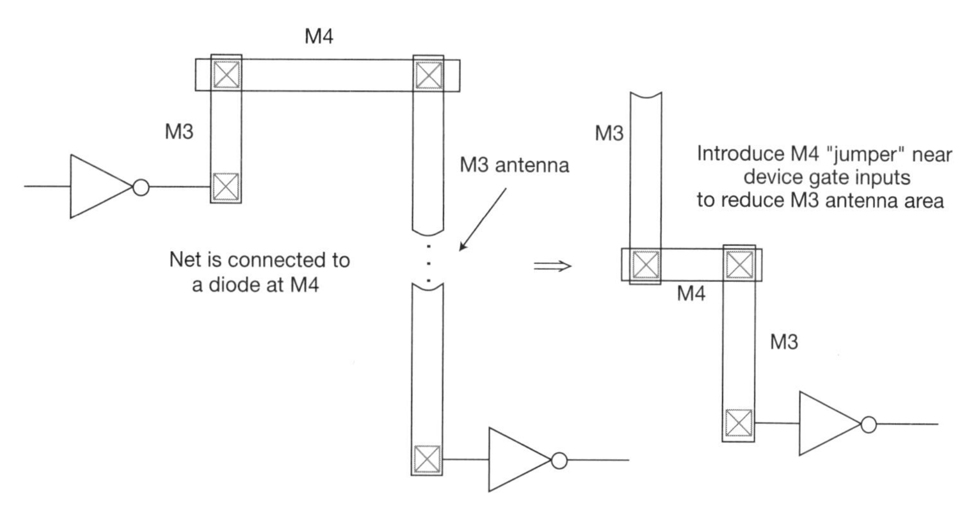


**Figure 9.19** Antenna area violations could be addressed by the router to modify the net topology accordingly.

- **Insert an antenna diode to the net.** A specific layout cell providing the antenna diode could be added to the net topology to provide a discharge path prior to completing the net connectivity to the sourcing output pin. Figure 9.20 illustrates the cell layout, which consists of a reverse-biased diode with the cathode of the diode connected to the net and the anode represented by the wafer substrate. As with the insertion of jumpers, the ability to insert the diode may be constrained in highly congested areas.

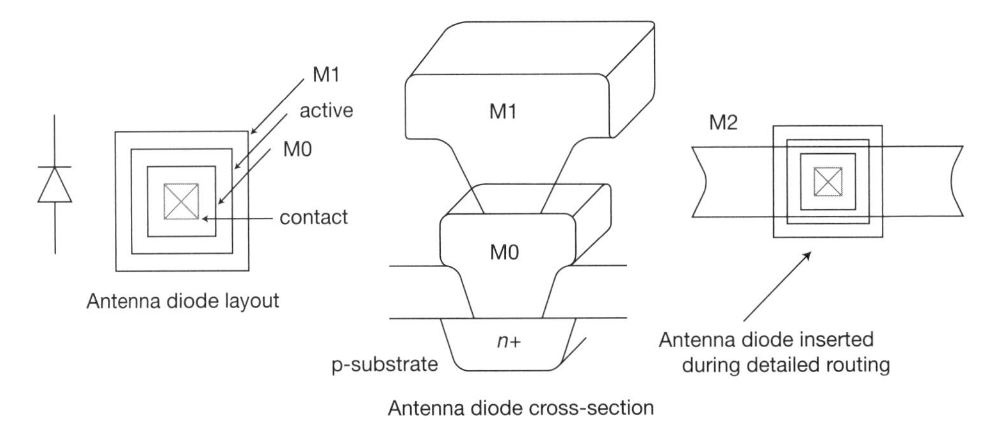


**Figure 9.20** Illustration of the antenna diode layout cell connected to the net to fix antenna area violations (either manually or automatically by the router).

A third option for (thoroughly) addressing antenna design rules is to include a diode within the IP library design itself, where all input pins are already hard-wired to the diode layout (see Figure 9.21). Although frequently used for older process node IP, this design approach has more recently been discarded in favor of router-based solutions.

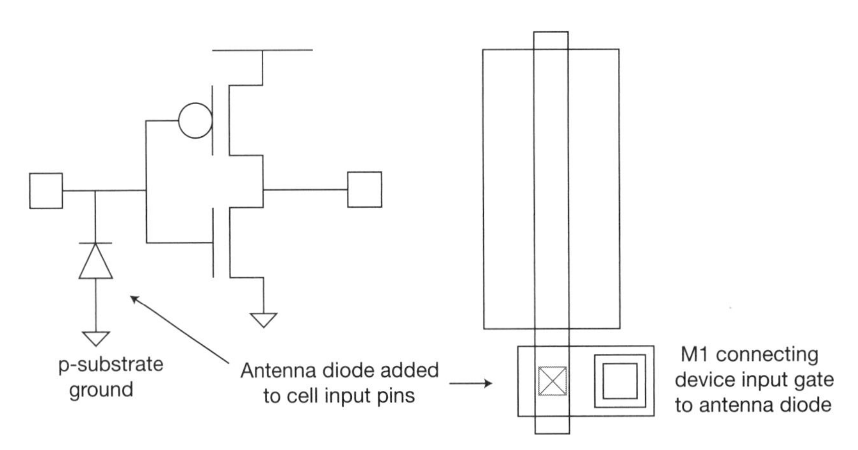


**Figure 9.21** Illustration of antenna diodes connected to library cell input pins, eliminating any antenna area violations.

Note that the EDA industry has developed the terms *design-for-manufacturability* (*DFM*) and *design-for-yield* (*DFY*) to describe the post-route optimizations. Although the distinction between DFM and DFY is a little vague, the required (and DRC-verified) nature of antenna diode insertion clearly makes it a DFM methodology.

## 9.5 Summary

The expectation for the overall routing flow is to provide a resulting physical database that is free of design rule (and electrical antenna) errors which can be submitted to final layout preparation flows for electrical analysis. The routing flow includes the addition of fill data, insertion of spare gate cells, and insertion of foundry process monitoring layout cells. However, as mentioned in Section 9.2, there may be incomplete overflow nets (and/or remaining antenna violations) after the flow is completed.

The EDA router tool includes an editor platform, where route segment modifications to the physical database can be made by a layout engineer to manually address overflows and antenna requirements. After these modifications, the router can be invoked to try to resolve remaining overflows, a step familiarly known as “rip-up and reroute.” (This EDA platform would also be the environment used to manually define any net preroute segments prior to submitting the netlist and placement model to routing.)

The optimizations discussed in this chapter assume that no netlist modifications are enabled during the routing flow (e.g., no buffer insertion, no cell drive strength or $V_t$ assignment updates). The general SoC methodology assumption is that the timing/power optimizations exercised during placement (with estimated net route models) have the most significant impact on design closure. Nevertheless, with some nets, the detailed routes differ substantially from the estimates, especially in areas of high congestion. The placement algorithm includes a metric of local route demand versus track supply, and it limits cell area utilization and pin density accordingly. Nevertheless, timing issues after detailed routing are likely to arise due to differences from placement estimates. The routing editor platform benefits from the inclusion of additional features beyond rip-up and reroute, including the following:

- Additional or modified netlist cells (e.g., inserting buffers to the netlist)
- Placement of new cells (including a local push/shove of existing cells and their routes)
- Netlist connections swapped to logically equivalent cell input pins

In short, the post-route editing environment is effectively the same EDA platform that would be used for implementing subsequent netlist ECOs to an existing design.

## References

[1] Elmore, W.C., "The Transient Analysis of Damped Linear Networks with Particular Regard to Wideband Amplifiers," *Journal of Applied Physics*, Volume 19, Issue 1, January 1948, pp. 55–63.

[2] Pileggi, L.T., et al., "PRIMA: Passive Reduced-Order Interconnect Macromodeling Algorithm," *IEEE Transactions on Computer-Aided Design of Integrated Circuits and Systems*, Volume 17, Issue 8, August 1988, pp. 645–654.

## Further Research

### Device-Level Placement and Routing

This chapter focuses on block-level routing of a cell-based netlist, with timing and area constraints associated with the SoC floorplan. Routing of global signals between blocks also needs to address constraints associated with the timing budgets between blocks. A separate area of EDA development has focused on placement and routing tools to translate a (complex) device-level schematic into a physical layout, which is a significant productivity aid to a layout designer.

There are unique tool requirements for cell and (especially) analog/mixed-signal IP layout generation (e.g., support for interdigitated devices in a centroid-style pattern, maximum alignment of common nFET and pFET device gates, support for both track-based and non-track based wiring modes, matching of differential signal wires, shielding of noise-sensitive wires). In addition, AMS schematic designs may have been simulated using (minimum and maximum) estimated device and interconnect parasitics, which are to be observed in the physical layout.

Describe the features required for device-level "custom" place-and-route tools and the types of design constraints that need to be supported for AMS circuits. Assume that device layout generation utilities are used to evaluate the (parameterized) device schematics to generate the individual layouts. Describe the features required to optimize the layout for adjacent devices of the same type that share common nodes in the schematic (leveraging the mathematical formulation known as an "Euler path"). Describe the features required of the device-level router. Note that the custom place-and-route toolset also requires support for additional structures that are not present in the schematic but that are required in the layout. Additional layout data need to be generated and correctly applied for substrate/well contacts, (floating or tied) dummy devices, guard rings, metal fill data, and so on.

Describe the physical verification and electrical analysis features required in the toolset, as well as the types of output data report feedback of interest to the library and AMS IP designer.

### Routing and Multi-Corner Optimization

Physical synthesis and cell placement flows incorporate timing and power optimization algorithms that guide cell selection and placement location. Typically, a "dominant" corner is selected to guide the optimization decisions (e.g., a "slow" corner for setup timing analysis and a "fast" corner for hold timing). These decisions are subsequently followed by analysis at multiple corners to identify issues requiring optimization iterations. Detailed routing has the additional requirement of incorporating signal-coupling noise analysis and optimization.

Describe a potential strategy for managing multiple-corner optimizations in routing, spanning timing, power dissipation, and noise.

### Hold Time Fixes in Routing

Chapter 8 discusses fixing hold time issues with the judicious insertion of delay padding cells in short paths throughout the logic network, without adversely affecting critical paths identified during setup time evaluation. An alternative to cell insertion would be to have the detailed router apply a "scenic route," where the additional R*C interconnect delay would address the hold time fail.

Describe the trade-offs between hold time fixes using delay cells, scenic routes, or a combination of both. The trade-off decisions include: the magnitude of the delay to be added, the behavior at different PVT corners, the impact on local route congestion, and subsequent application of an ECO when the most accurate static timing analysis flow results are available.

TOPIC V

# Electrical Analysis

After the design implementation of a block is complete, the layout database is presented to the electrical analysis flows to determine the suitability for sign-off to fabrication. The initial step in electrical analysis is to extract accurate interconnect parasitics, replacing the existing estimates used in physical design optimization algorithms. The new network model with the annotated parasitics is then submitted to the breadth of analysis flows, to evaluate path timing, coupled noise, power dissipation, and electromigration reliability.

The results of block-level electrical analysis are used to generate abstracts for subsequent global evaluation. An exception to the use of block abstract models for global analysis is typically made for the full-chip power and ground distribution grids, to ensure that I*R voltage drop and electromigration current density limits are observed.

A number of electrical analysis flows are applicable to specific IP block types. The electrical model for the power state transient that connects and disconnects the power/ground distribution to a block requires a specific type of circuit simulation. Memory arrays are susceptible to the disruption of a stored bit value from external radiation, necessitating an analysis of the probabilistic soft error rate. Chip I/O circuits may be subjected to an electrical stress condition applied at the package pins. An analysis of the electrostatic discharge protection circuitry connected to the I/O pad is required.

The results of the electrical analysis flows are captured in the methodology manager application, to be monitored closely by the SoC project management team. Electrical issues identified by the analysis flows may require design iterations, commonly with an engineering change (ECO) that minimally disrupts the existing implementation. If the magnitude and number of analysis flow errors are significant, a full implementation iteration loop may need to be pursued with the corresponding impact to the project schedule and resources.

CHAPTER 10

# Layout Parasitic Extraction and Electrical Modeling

## 10.1 Introduction

All electrical analysis flows are based on a methodology that incorporates a transistor or cell-based netlist with corresponding electrical parasitics from the layout interconnects annotated to the netlist to create a complete electrical model. There are layout parasitic extraction (LPE) algorithm trade-offs in terms of electrical accuracy, the number of RLC parasitic elements generated, RLC element *reduction* strategies (while maintaining a model of sufficient accuracy), and the EDA tool compute resources and runtime.

Traditionally, LPE tools from EDA vendors have used either of two methods for capacitance calculation, with distinct characteristics relative to these trade-offs:

- **A 3D field-solver algorithm**—The layout cell is translated into a three-dimensional representation of interconnects and dielectrics and presented to an algorithm that solves Maxwell's equations for an electrostatic topology. The goal is to determine the capacitance between each pair of conductors in the cell, $C_{ij} = Q/V_{ij}$. The potential difference between two wires is related to the integral of the electric field emanating from one to the other due to the electrostatic charge on the wires. In a multi-conductor model, the calculation of these electric fields from the local surface charge density on each wire results in a complex system that requires a concurrent solution for all wires. For any multi-conductor layout with strong electrostatic interactions between the conductors, an analytic solution to Maxwell's equations is intractable. There are several numeric algorithms available to calculate an empirical solution, using either the integral or differential form of Maxwell's equations. The two most prevalent are the boundary element method (BEM) and the floating random walk (FRW) approach.[1,2]

The BEM and FRW algorithms use an indirect technique. Assigning a reference conductor *i* to a potential of 1V, with all other conductors at 0V, a (numeric) solution for the total charge on each wire *j* provides the $C_{ij}$ capacitance value. Superposition theory is applied, enabling the calculation to iterate through each interconnect as a reference to complete the capacitance matrix for all wires. The BEM and FRW algorithms differ in terms of how they reach the solution for all surface charges in the layout topology; however, to achieve high extraction accuracy, either method requires substantial compute resources. As a result, 3D field-solver methods for layout parasitic extraction are commonly used only for IP library cells and smaller mixed-signal IP designs. (The FRW algorithm is much less memory intensive and extremely parallelizable, which would enable larger layout cells to be extracted, given sufficient CPU cores.)

- **Application of empirical formulas from representative layout topologies**—The other LPE method utilizes a set of layout pattern examples, analyzed using a 3D field-solver algorithm. Capacitive formulas are then derived from the field-solver results. Dimensional parameters in the layout examples are fitted to the formula coefficients. From this set of pre-characterized layouts, the LPE tool performs a pattern match to the submitted layout cell. The specific interconnect dimensions of the layout are submitted to the formulas, which provide a capacitive value by interpolation, as depicted in Figure 10.1. Although the accuracy of the capacitance calculation is reduced by the fitted interpolation, due to the high capacity and fast throughput of the layout pattern matching approach, this method is commonly used for any large block design.

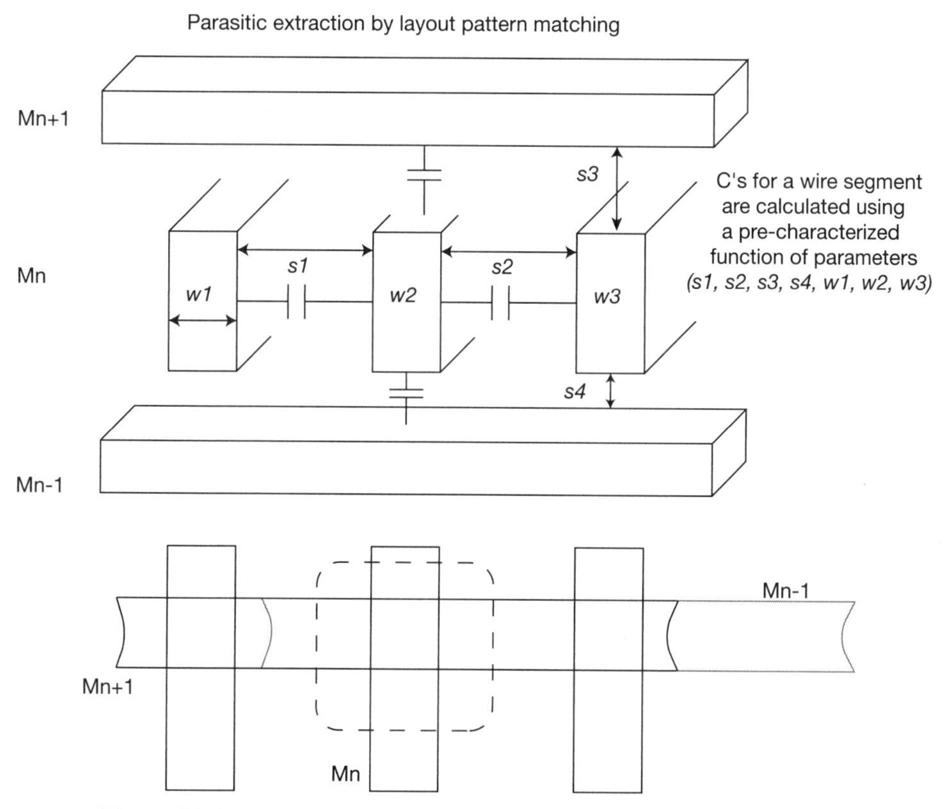


**Figure 10.1** One method for parasitic extraction uses a pattern-matching method, applying a set of formulas to the specific layout pattern dimensions.

After the capacitive matrix for the layout cell is computed (using either the field-solver or pattern-matching method), the LPE tool then incorporates resistive wire segment values to provide an RC network as the parasitic netlist output.

### 10.1.1 Inductance Extraction

Unique LPE algorithms from EDA vendors are provided for the extraction of wire inductance, whether applied to an on-chip inductor layout cell (for the self- and mutual-inductance of wire turns), the chip pad to package pin through the top layer metal redistribution, or internal power (and clock) grids. Figure 10.2 illustrates examples of the layouts for which inductance extraction is applicable.

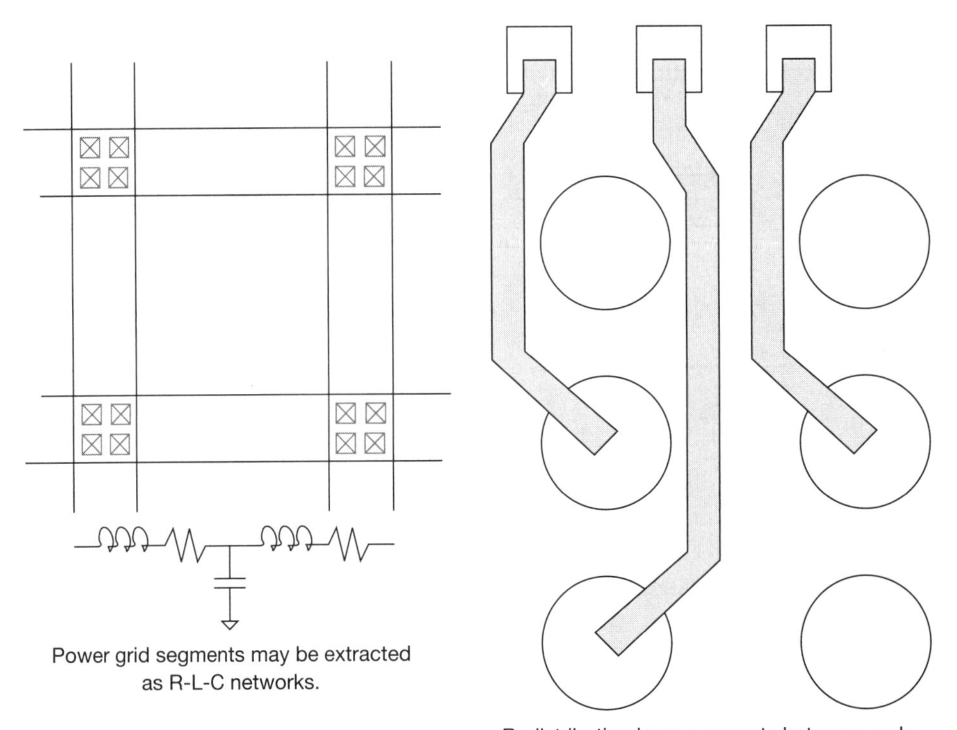


**Figure 10.2** Illustration of inductance extraction, applicable to power grids, high-speed clock grids, and top-level redistribution layer connections.

These designs require accurate L and M parasitic elements for specific simulations—for example, LC-tank circuit response in analog IP, (high-speed) off-chip driver/receiver signal integrity, or on-chip power di/dt transient analysis. The balance of this chapter focuses on RC element extraction.

### 10.1.2 Extraction Methodology Decisions

This chapter does not delve into extraction algorithms in detail. Rather, the discussion focuses on the considerations at the points where the SoC methodology team might choose one approach over another.

The sheer volume of cells, interconnects, and available metal layers in block layouts results in very large parasitic netlists. In advanced process nodes, the layout variability is reduced (e.g., track-based routing with strict width/space design rules, FinFET device/cell placements on grid). These two characteristics suggest that the LPE flow is increasingly likely to leverage a pattern library approach for extraction, with sufficient accuracy. The parasitic extraction flow for "high-sigma" circuit characterization still requires the increased accuracy of a 3D field-solver algorithm, however.

The parasitic extraction flow involves several steps to provide the annotated netlist as output. Further, the flow is divided between circuit-level extraction for library cells (see Section 10.2) and extraction of interconnect routes (see Section 10.4).

For circuit extraction, the input layout cell is evaluated using a PDK techfile, which includes the *recognition operations* necessary to do the following:

- Identify devices (and their width, length, and finger calculations)
- Identify global supply and ground connections
- Divide the layout into device versus non-device geometries
- Trace valid connectivity from device nodes through contacts and metals to other circuit nodes or cell pins
- Measure the layout dimensions of interest in the neighborhood of each device (for layout-dependent effects, as discussed shortly)

Note that the initial extraction steps are the same as for the layout-versus-schematic netlist verification flow—that is, identify and measure devices and then trace connections between devices. Indeed, the layout cell must be "LVS clean" to the corresponding schematic netlist for parasitic annotation to be successful (see Figure 10.3).

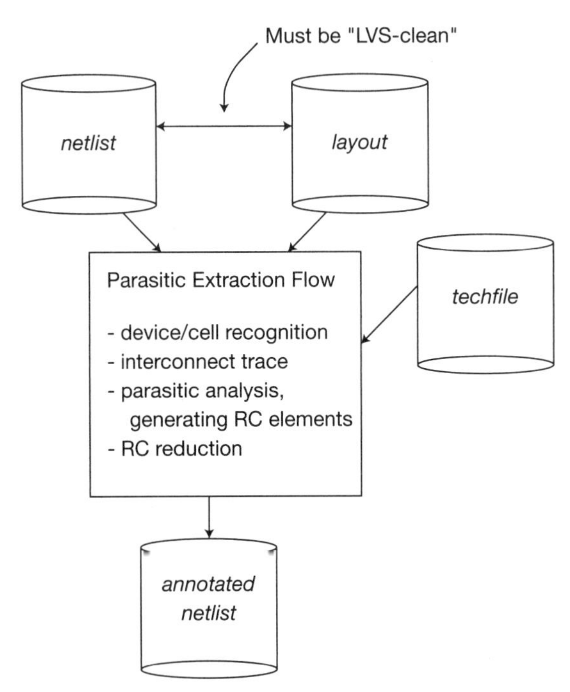


**Figure 10.3** Correct annotation of LPE parasitics to the schematic netlist requires the layout to be "LVS clean."

The layout does not need to be "DRC clean" to present to the extraction flow; annotation only requires the layout to be LVS clean. If DRC errors are present in the layout, there will be inaccuracies in the extracted resistive and capacitive elements, but annotation to a simulatable netlist would still be successful.

If the SoC design team is developing circuit-level IP, a project management decision is needed about when a layout is of sufficient quality to be presented to extraction and electrical analysis flows. If the IP layout is LVS clean but contains (minor) DRC errors, it may be prudent to submit this preliminary version layout to analysis for early identification of any major electrical issues.

### 10.1.3 Hierarchical Extraction of IP Macros

For circuit-level extraction of library cells, the annotation of parasitic elements is typically applied to a flat LVS netlist. For larger IP macros, an alternative approach would be to use a hierarchical model in which extraction and annotation are performed on (highly repetitive) instances within the schematic netlist presented to the LVS flow. To enable this efficiency, there must be a degree of consistency between the hierarchies for the schematic model and the physical layout view. The extraction flow would be provided with a list of correlated hierarchical LVS instance and layout view identifiers within the IP to extract/annotate once and reuse the results throughout the full model hierarchy. Wires connecting instances would be extracted separately, typically using a gray box visibility approach for the instances. The combination of hierarchical LVS and extraction is an effective method for highly regular IP macros, such as register files and arrays.

## 10.2 Cell- and Transistor-Level Parasitic Modeling for Cell Characterization

This section describes the extraction approaches to model custom cell layouts for analysis—specifically, the identification of the layout dimensions affecting device simulation parameters and the annotation of parasitics surrounding the devices. This section also briefly reviews the evolving nature of the characterization flow using the composite device and parasitic extraction netlist (i.e., the generation of cell electrical abstract models for release with the functional, physical, and test models as part of the IP library). The detail in these characterization models is increasing for advanced process nodes, as the EDA electrical analysis tools enhance their algorithms for increased accuracy.

### 10.2.1 Cell Extraction

The cell library layouts are presented to the LPE flow for extracting detailed netlists prior to characterization. This requires a "full custom" LPE techfile as input, with process cross-section and material properties. This techfile includes properties for the device fabrication layers, local metals and contacts, and substrate/well nodes.

A key consideration to review with the foundry is how parasitics are allocated to the device model and to separate extracted elements when the layout is measured. Figure 10.4 illustrates the capacitances between gate, source/drain, and substrate nodes for a planar FET device; both internal and external dimensions are associated with these capacitances. There is an added complication that the device model is likely to also include a capacitance calculation, given the area and perimeter measures of the source/drain nodes.

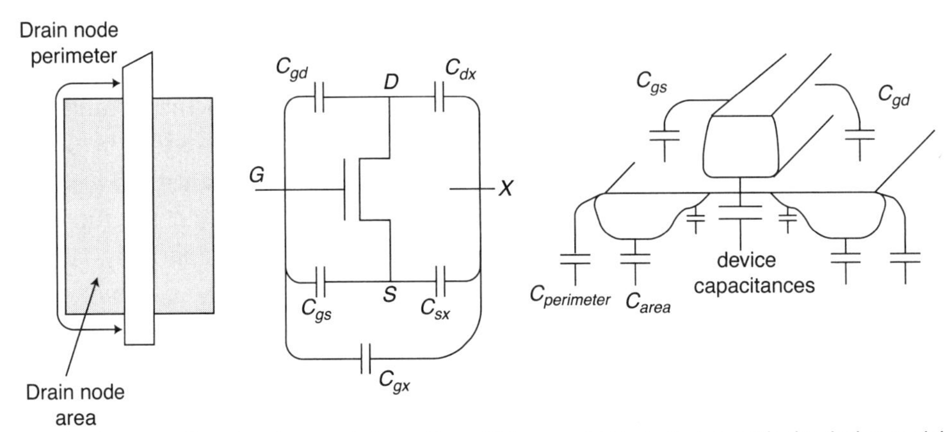


Custom parasitic extraction needs to confirm which capacitances are present in the device model, and which are to be annotated by the extraction flow from the layout.

**Figure 10.4** Device models may include the requisite parameters to include parasitic capacitances, in addition to the internal node to node capacitances, given area and perimeter measures.

Adding a local M0 interconnect layer, contacts, and metal1 wire to the layout adds to the calculation detail, as the $C_{gs}$ and $C_{gd}$ capacitances now include the vertical structures, as well (see Figure 10.5).

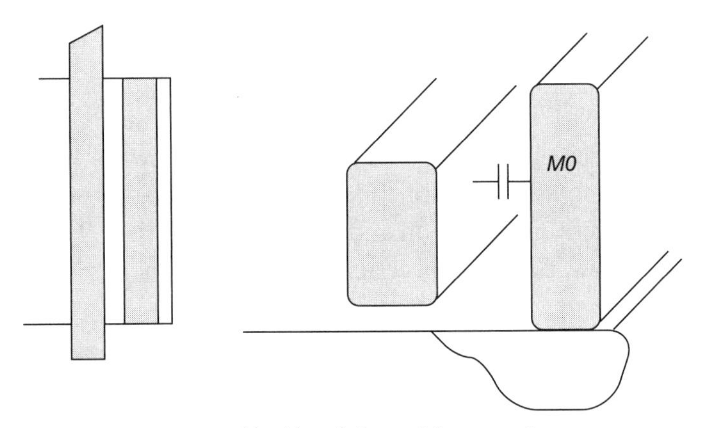


M0 local interconnect adds sidewall $C_{gs}$ and $C_{gd}$ capacitance.

**Figure 10.5** The local M0 interconnect layer adds a vertical dimension to the device parasitic capacitance extraction topologies.

The scaling of device channel length has increased the sheet resistivity of the device gate, implying that the distributed nature of the gate $R*C_{channel}$ is of greater importance, as depicted in Figure 10.6. The PDK techfile assumption for parasitic modeling of the gate also requires review with the foundry.

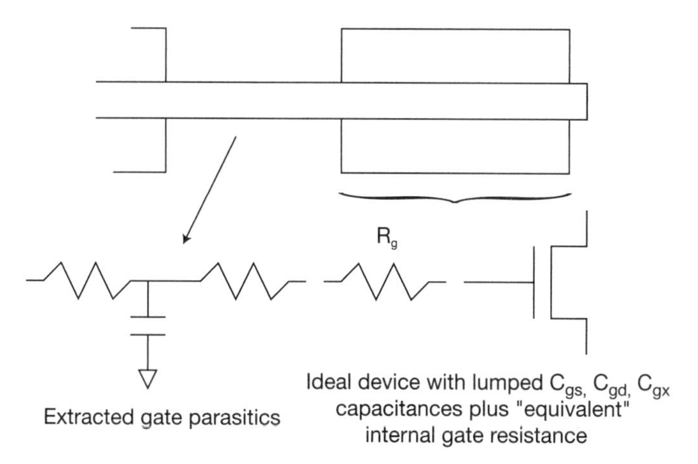


**Figure 10.6** Parasitic extraction of the gate input requires review with the foundry for the definition of both the external gate parasitics and the reduced model of the distributed gate R*C.

The device-level parasitic models are significantly more complicated for FinFET devices due to both their vertical profile and the traversal of the gate between multiple fins. The allocation of internal device model versus external $C_{gs}$ and $C_{gd}$ extracted elements is intricate, as illustrated in Figure 10.7.

A complication is present with FinFET device models that describe a single fin and represent a multi-fin device with the schematic parameter "NFIN = n". The extraction and annotation of $C_{gs}$, $C_{gd}$, and $R_g$ elements to the NFIN device model requires approximation to represent the distributed capacitance and resistance of the gate traversal between the (n − 1) fins as lumped elements in the parasitic extraction netlist.

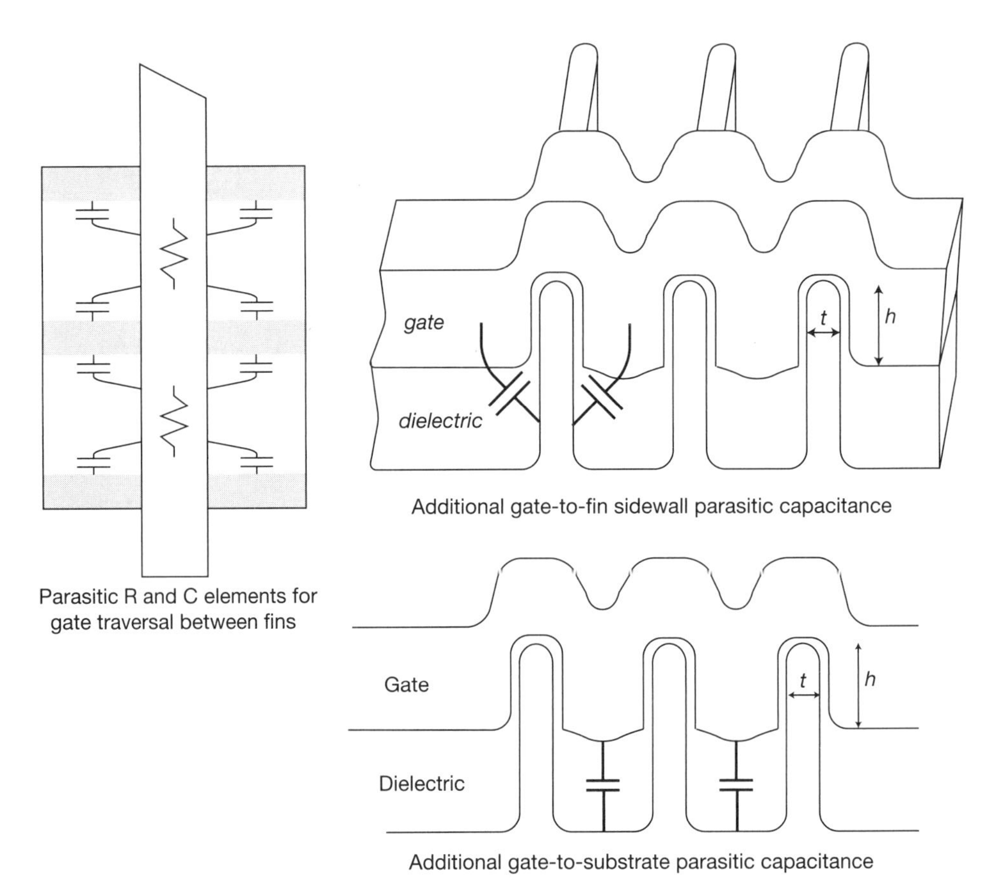


**Figure 10.7** Parasitic extraction of the external parasitic capacitances for a FinFET device requires review with the foundry for the definition of the capacitances of the gate traversing between fins.

### 10.2.2 Layout-Dependent Effects (LDEs)

For custom extraction at the cell level, there are device behavior impacts due to layout dimensions in the neighborhood of the device. Figure 10.8 illustrates some of the measures taken during extraction. *Layout-dependent effects* result in adjustments to the device channel carrier mobility and threshold voltage, using additional input parameters on the device model.

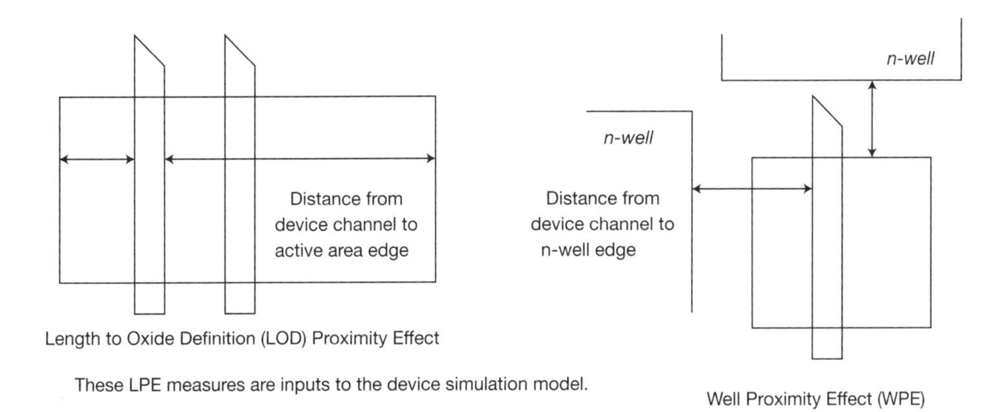


**Figure 10.8** Illustration of several layout-dependent effect measurements required for custom layout parasitic extraction. These measurements are inputs to the device model.

The LVS techfile commands incorporate additional measurements for the LDE effects. The device simulation model is enhanced to apply these parameter measurements to adjust the carrier mobility and threshold voltage. Note that the LDE measurements differ for individual device fingers, which are connected in parallel to implement a wide device. As a result, although the input schematic draws the multi-fingered device using a single symbol, the LVS output netlist expands fingers into individual device instances, with specific layout-dependent effect measures for each.

The foundry may include the impact of a new layout-dependent effect measured on fabricated devices during the bring-up of a new process, using an application programming interface (API) software layer added to the existing device model, as depicted in Figure 10.9; this allows circuit characterization to proceed prior to the release of an updated compact model standard.[3]

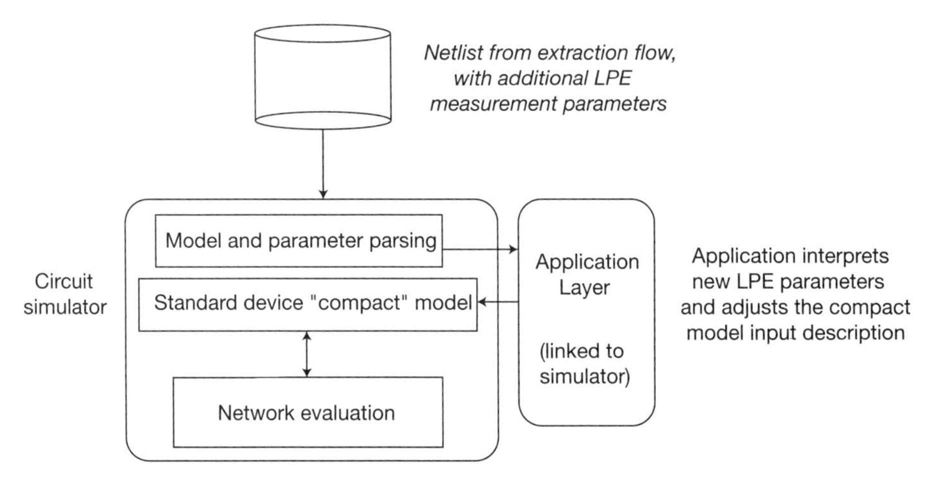


**Figure 10.9** To enable new layout dependent effects not reflected in the standard device model, an API software layer may be added. The new LDE measures are reflected in a new output netlist generated by the parasitic extraction flow. The API simulation layer then modifies the existing device model accordingly.

The layout-dependent effects apply to device extraction, not the calculation of parasitic R and C elements for device netlist annotation. As a result, the foundry PDK defines how these effects are represented. The measurement of layout-dependent effects involves a key methodology decision. When library cells are being extracted, a cell needs to be surrounded by a representative layout environment so that devices at the cell edges receive suitable proximity measures, as illustrated in Figure 10.10.

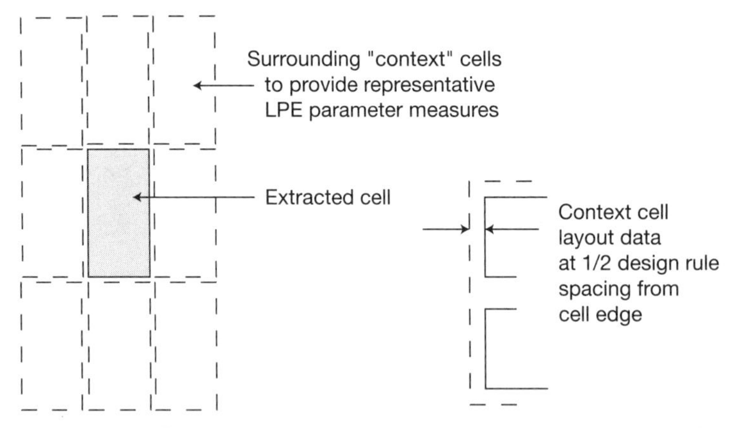


**Figure 10.10** Context cells surround the layout cell being extracted to provide representative proximity measures for perimeter devices.

The choice of the cell surround (and route overlay) data used for parasitic extraction, and thus for subsequent cell characterization, is a subjective methodology decision. The ultimate goal is to provide accurate cell pin delay arcs, consistent with the path timing margins used in the static timing analysis flow.

As an aside, the emergence of layout-dependent effects has changed the nature of library cell design engineering project planning. Traditionally, IP design engineers would capture their schematic design and define the individual device dimensions, specify the number of device fingers, and (potentially) enter layout estimates for the device node area and perimeter. This schematic representation was sufficient to submit directly to a circuit simulator. Once the optimal schematic dimensions were achieved for the PPA targets for the IP cell, the schematic was reviewed with the layout engineer for physical implementation. The engineering review might include a discussion of any specific layout assumptions made during the schematic-based simulation phase. With the introduction of layout-dependent effects, it is much more difficult to develop a suitable schematic-only model for design simulations; a representative (and iteratively refined) cell layout is required to extract the proximity measurements. The circuit and layout design engineers collaborate much more closely, with cell layout activity commencing much earlier in the library development schedule. The traditional "throw the schematic over the wall to the layout engineer" methodology is no longer adequate. In addition, there is increasing interest in the productivity benefits of automated generation of cell IP layouts (as briefly mentioned in the "Future Research" section in Chapter 9). An initial cell layout generated "semi-automatically" could provide a sufficiently accurate extracted model to capture important LDE characteristics and then be iteratively refined by the design and layout engineering team.

### 10.2.3 Extraction Corners

Each PVT characterization corner reflects a specific combination of process fabrication variations, applied voltage at the device (including supply voltage and ground distribution I*R drop margins), and device temperature to be used in subsequent electrical analysis flows. The set of process fabrication variations includes both device parameters and wire measures. The device variations are typically represented by a single set of model parameters that result in an n-sigma device current at the voltage and temperature values for the corner (see Figure 10.11).

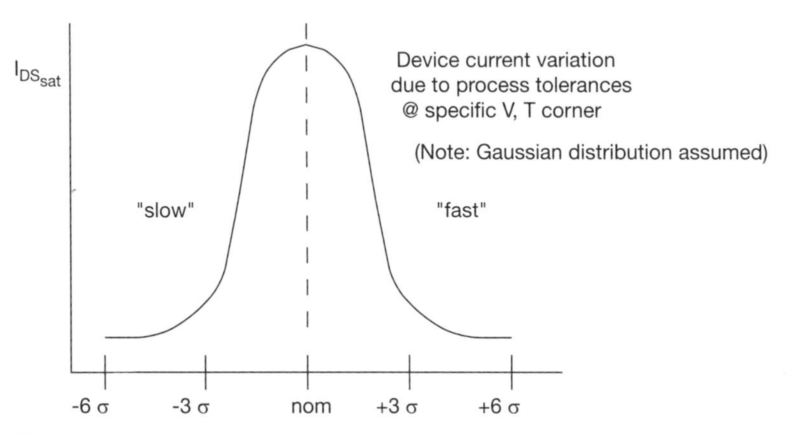


**Figure 10.11** The device model parameters selected for characterization at a particular corner represent a composite n-sigma device current. For library cells, this is typically n = 3.

The extracted elements for wires introduce unique corners, adding to the number of characterization simulations. These wire extraction settings reflect the interdependence between wire thickness, resistance, and coupling capacitance. The foundry provides PDK support for the wiring extraction corners, including the following:

- Max_$C_{total}$ (R will be low) and min_$C_{total}$ (R will be high)
- Max_RC ($C_{coupling}$ low, R*$C_{ground}$ high)
- Min_RC ($C_{coupling}$ high, R*$C_{ground}$ low)
- Nominal_RC

For multipatterned metal layers, the overlay tolerance for a decomposed mask layer introduces spacing variations between adjacent wires. This additional source of variation introduces new MP variants for existing corners. Again, the foundry assesses which of the many potential extraction settings sufficiently cover the variation space and provides the PDK techfile support. EDA tool vendors have optimized their extraction algorithms such that derivation of parasitic netlists for multiple corners requires a minor increase in runtime.

The SoC methodology team evaluates which extraction corners to annotate to the device netlist for the electrical analysis flows—timing, power, and noise.

Note that the EDA industry has proposed an extracted netlist format that would include multiple value entries for each R and C element that would represent a statistical range rather than a single element value. To date, however, this representation has not displaced the use of separate netlists for each corner.

### 10.2.4 Introduction to Cell Characterization

The extracted netlist of devices (and related layout-dependent effect parameters) with the annotated parasitic elements is presented to the cell characterization flow, which initiates a number of circuit simulations with specific input/output conditions and measurement criteria. From these measured data, a number of electrical models are derived (e.g., delay arc models, input gate load and output drive strength impedance, noise propagation from each input pin to output, cell power dissipation). The level of detail in these models has evolved substantially with process node scaling to provide greater accuracy. Correspondingly, the EDA tool algorithms have evolved to leverage this additional detail. For example, in early VLSI processes, gate delay was much larger than interconnect delay; a lumped capacitive load, rather than a $C_{eff}$ load and distributed RC interconnect network, was sufficient for delay calculation. These early delay arc models used a simple linear equation for the dependency on output capacitive load and input signal slew.

This linear model was increasingly inaccurate for submicron process nodes. The non-linear delay dependency evolved to a representation using the set of measured values from characterization simulations entered into two-dimensional tables. The Non-Linear Delay Model (NLDM) tables provided the arc delay and output pin slew as a function of capacitive load and input slew, as before (see Figure 10.12).

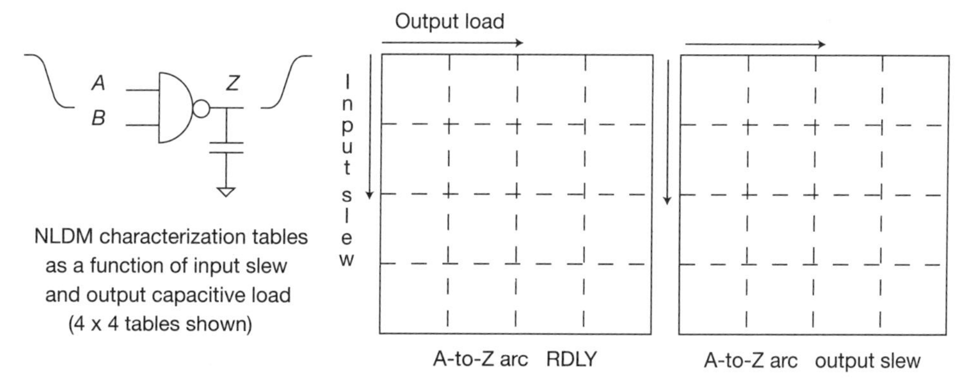


**Figure 10.12** Illustration of arc delay characterization data as a set of Non-Linear Delay Model (NLDM) tables, with output capacitive load and input signal slew as the independent variables.

Concurrently, the effective capacitive load $C_{eff}$ and separate RC interconnect network delay methodology was introduced, as described in Section 9.1. The (NxM) dimensionality of the NLDM tables for characterization of each delay arc was selected to adequately cover the $C_{load}$ and input slew ranges while limiting the (N*M) simulations required to populate the data in the tables for characterization throughput. More recently, the NLDM approach has been augmented by a more general methodology that records the output waveform in detail for each of the N*M simulations rather than using a single slew-based signal transition value.[4] Figure 10.13 illustrates one of the general modeling approaches in use to represent the output. The result of cell characterization is a non-linear output driver current source that is to be connected to the distributed RC load network.

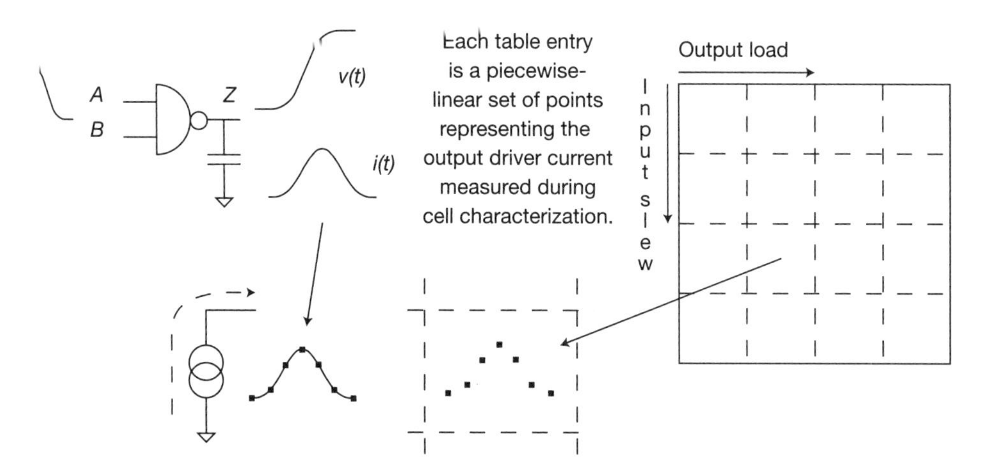


**Figure 10.13** An alternative to the single NLDM output signal slew uses a set of (value, time) data for each table entry, representing a non-linear current (or voltage) source. A new set of characterization flow measurements is required.

The waveform detail is stored using a set of time points, so the non-linear source model is actually piecewise linear. The characterization slew table no longer uses fixed NLDM value entries but rather a set of (time, value) pairs from sampling the simulation measures.

This enhanced library cell format also includes a feature to describe specific side-input pin values in the case where the measured input-to-output pin response is a strong function of the (static) values on other inputs, as shown in Figure 10.14. This *state-dependent delay* model requires significantly more characterization simulations.

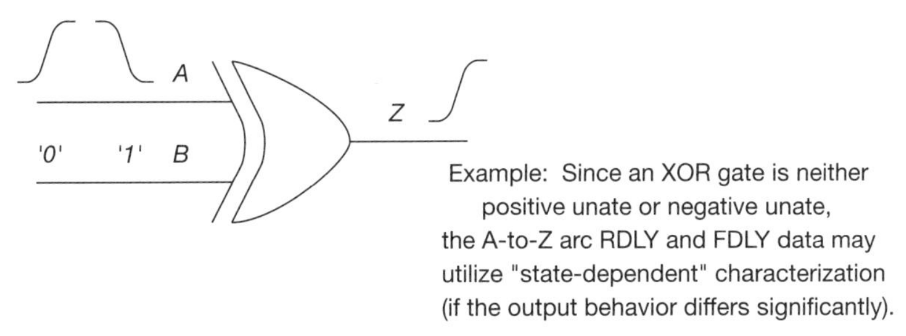


**Figure 10.14** An extension to the library cell characterization model includes support for multiple tables, based on specific (static) values at other cell inputs.

The cell's input pin capacitance model has also recently been expanded. Rather than a fixed $C_{gate}$ for the input pin devices, multiple values can be used to represent the voltage-dependent device and Miller input capacitance behavior.

With these general representations, a different delay calculation and propagation algorithm approach is used by the related EDA analysis tools. Specifically, interconnect delay and noise propagation algorithms need to solve an interconnect network model with the (piecewise-linear) driver, fan-out receiver capacitances, and extracted RC parasitic elements. The remainder of this section uses both the NLDM and general current/voltage source driver model approaches in the description of cell characterization methods.

Note that the temperature value used in characterization simulations affects both the device model and the extracted resistive elements in the RC network. The resistor model in the foundry PDK includes temperature coefficients (TC1 and TC2):

$$R(T) = R(T_{nom}) * [1 + (TC1 * (T - T_{nom})) + (TC2 * ((T - T_{nom})**2))] \qquad \text{(Eqn. 10.1)}$$

### 10.2.5 Characterization Ranges and Corner Values

The IP library provider defines the range of load capacitance and input pin slew rates over which characterization values are measured for each corner. The SoC methodology team has several engineering decisions to make after reviewing the characterization settings.

#### *Algorithm for Out-of-Range Delay Calculation*

During the delay calculation phase of static timing analysis, a specific cell instance may have an effective load capacitance or input pin arrival slew outside the characterization range. The calculation algorithm would typically attempt to extrapolate from the delay table entries. The SoC team may choose to simply accept the calculation or may request that an error be reported by the tool such that a design modification can be made. As illustrated in Figure 10.15, a large input slew or large output load implies a significant transient cross-over current and cell power dissipation, as well as delay inaccuracies associated with the extrapolation.

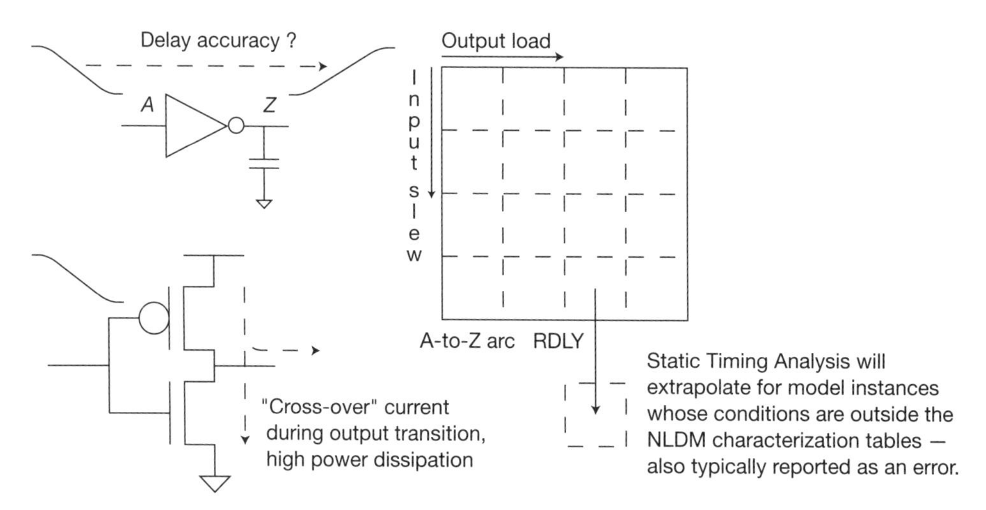


**Figure 10.15** Delay calculation requiring extrapolation from the characterization table ranges is typically reported by the timing flow, due to the high internal cell cross-over current and delay accuracy error.

#### *Algorithm for Voltage Values Differing from the Characterization Corner*

Modern SoC designs may include multiple IP voltage domains and/or dynamic voltage frequency scaling (DVFS) "boost/throttle" modes, where a voltage regulator adjusts the domain supply. Alternatively, the supply voltage regulation tolerances from nominal for the specific SoC end product application may differ from the characterization assumptions. As a result, the operating environment may include voltages that differ from the IP characterization values. Traditionally, CMOS circuit delays were adequately described as a linear function of supply voltage; for example, if characterization used a ($VDD_{nom}$ − 10%) assumption at the cell for slow timing, but the product application could ensure ($VDD_{nom}$ − 5%) was provided, a performance boost of ~5% could be assumed. However, with newer process nodes, this assumption is less accurate. The active device input *overdrive* of $|VDD — V_t|$ as a percentage of VDD is smaller because VDD has been scaled faster than the device threshold. As device dimensions have scaled, VDD has been reduced to adhere to electric field limits for reliability; conversely, to maintain suitable circuit noise rejection, $V_t$ has not been reduced correspondingly.

To support unique operating voltage conditions, the SoC methodology team needs to assess whether a single delay multiplier will be sufficiently accurate or whether additional characterization corners at specific voltage(s) are required, with project cost and schedule impact.

### 10.2.6 Multiple-Input Switching (MIS)

A fundamental cell characterization assumption for pin-to-pin delay arcs is that other input pins are at static values. However, if other inputs are also switching in a narrow time window around the pin transition, the measured cell delay and output slew may differ significantly, as depicted in Figure 10.16.

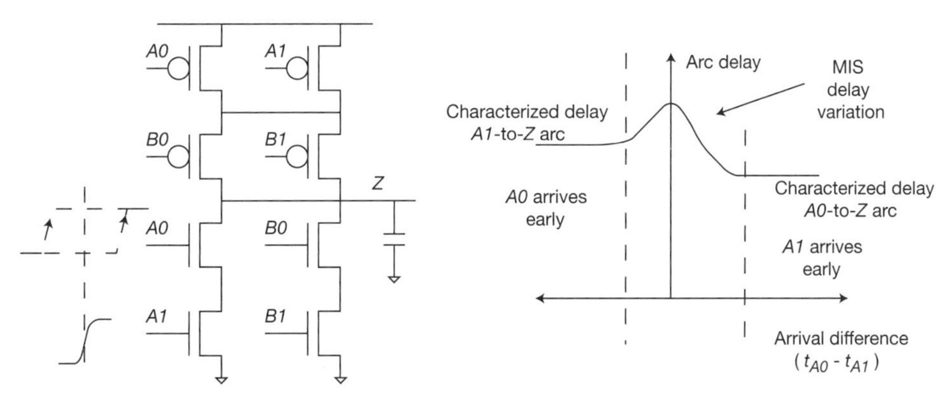


**Figure 10.16** Illustration of a multiple-input switching (MIS) event and the potential delay calculation inaccuracy associated with characterization using static side inputs (also refer to Figure 4.18).

There is no well-defined methodology for incorporating multiple-input switching (MIS) events into characterization libraries. An ad hoc approach would be to examine the critical paths reported by static timing analysis and explore the input signal arrival times on the non-critical delay arcs. If another arrival time might impact the critical delay, an additional timing margin may be warranted. There have been proposals to enhance *statistical static timing analysis* algorithms to better support MIS. Probabilistic input pin arrival times reflect cell and extracted interconnect variation. A convolution of multiple arrival time distributions during timing analysis would provide a single input distribution to use with a (statistical) gate delay model to generate an output timing distribution.[5,6,7,8]

### 10.2.7 Logically Symmetric Inputs

Figure 10.16 illustrates the impact of an MIS event on the cell delay arc. In the figure, the single input switching delay arc values differ for the two logically symmetric logic gate input pins. The library data model for each cell indicates the sets of inputs that are logically equivalent. A common physical synthesis and physical implementation timing optimization is to evaluate a swap of the nets connected to equivalent pins to move a timing-critical input arrival to a faster delay arc.

### 10.2.8 Sequential Circuit Characterization

In addition to the clock-to-output delay arc, the delay characterization of a flip-flop cell includes the measurement of the data-to-clock setup time and the clock-to-data hold time tables. The measurement criteria used during characterization by the IP provider should be reviewed by the SoC methodology team to evaluate against the delay margin assumptions used in timing analysis. Specifically, the definition of flip-flop setup time (and hold time) is typically based on the allowed increase in clock-to-output delay, as the data transition occurs closer to the clock edge, as illustrated in Figure 10.17.

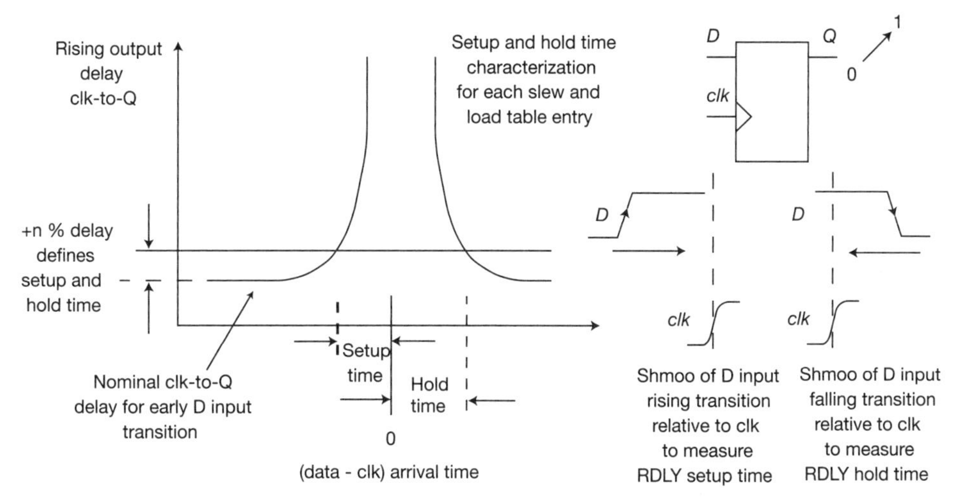


**Figure 10.17** Flip-flop setup time in cell characterization is typically defined using a clk-to-Q delay pushout criterion, measured using a simulation sweep of clock-to-data arrival transitions.

A *shmoo* of circuit simulations at each corner sweeps the data transition toward the clock edge, and the clock-to-output delay for the new data value is measured to establish the setup time; that is, the setup time equates to an n% increase in clock-to-output delay from the delay of a stable data input. Similarly, a sweep of a data transition back to the clock edge is performed, and the clock-to-output delay for the trailing data value is measured to establish the hold time.

The SoC methodology team needs to be aware of this characterization measurement. An engineering judgment may be needed to review failing paths from the timing analysis flow (especially if the project is approaching the tape-out target schedule). Referring again to Figure 10.17, there will be an increase in clock-to-output delay for data input arrivals failing the setup time. However, if the setup timing test fails by a small interval and the timing slack for the flop's clock-to-output path launch is positive by a sufficient margin, the arriving path setup test fail could potentially be waived. Any timing waiver would need to be granted judiciously; the clock-to-output delay curve in Figure 10.17 is *very* steep for data transitions not far from the selected setup time.

### 10.2.9 Input Pin Noise Characterization

A capacitive-coupled transient from aggressor nets to a victim net propagates to the input pins of fan-out cells. The fan-out cells suppress a (small) input transient, with a reduced perturbation on the output. As a result, the noise pulse presented to the next level fan-out is diminished; an example is depicted in Figure 10.18. This filtering applies to the complementary transistors of CMOS logic circuits. Other logic types are much more sensitive to input pin noise, such as precharged domino circuits or inputs to data-steering transfer gates. The circuit characterization for input noise limits involves a low-up and high-down input pin noise transient. (For completeness, a high-up and low-down transient is also being characterized: The increased electric field magnitude/duration across the device gate-to-channel for these transients would introduce a reliability concern.)

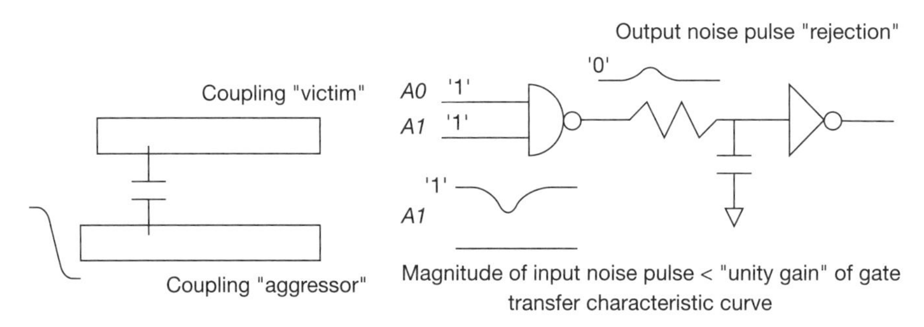


**Figure 10.18** A cell input pin noise transient is propagated to the cell output.

The pin noise characterization strategy has evolved over process node scaling as the aspect ratio of metal lines has changed and the relative contribution of coupling capacitance has increased. The most direct method would be to compare the magnitude of the input noise pulse to the DC transfer characteristic of the cell. As long as the input pin noise is well below the high-gain transition slope of the ($V_{out}$, $V_{in}$) curve, the output fully suppresses the input perturbation (see Figure 10.19).

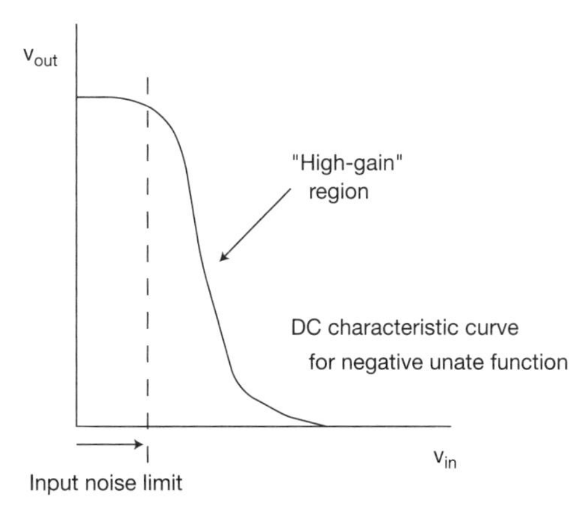


**Figure 10.19** Noise model using the ($V_{out}$ versus $V_{in}$) DC transfer characteristic curve. Suppression occurs for an input noise pulse magnitude below the high-gain region of the curve.

Cell characterization simply generates the transfer curve and selects a single noise magnitude limit. This approach is extremely conservative and does not scale well to the impact of increased coupling. Of specific consideration is that both the magnitude and duration of the input noise transient influence the output behavior (for a given load capacitance). A higher-magnitude pulse may be acceptable if the duration is limited. This behavior led to the definition of an input *noise immunity curve* (*NIC*), as illustrated in Figure 10.20.

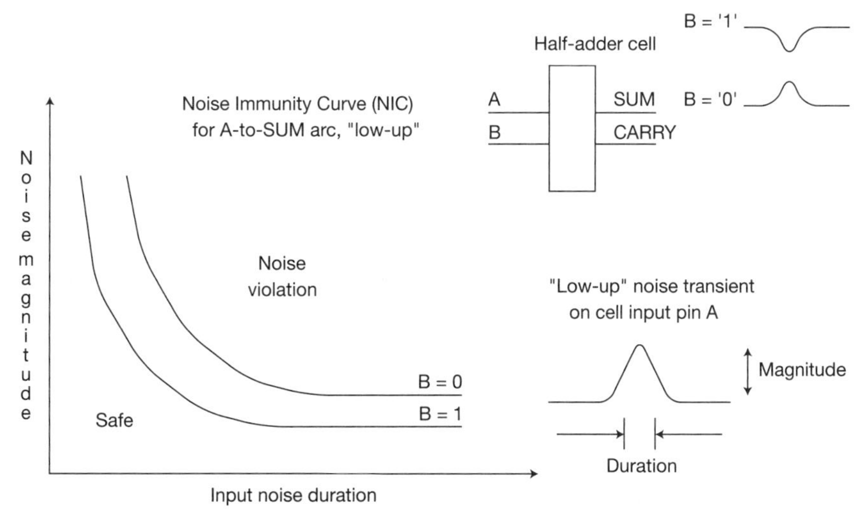


**Figure 10.20** An enhanced noise model utilizes a noise immunity curve (NIC) developed during cell characterization.

The characterization assumption is that the typical input pin noise perturbation on-chip is adequately modeled as a (smoothed) triangular ramp. The noise immunity curves in Figure 10.20 define the edge between the acceptable ("safe") and violating output response when the cell instance is being evaluated during the noise analysis flow. The figure depicts immunity curves for an arc with different static values assigned to side inputs. The IP provider is faced with the decision of how many NIC models to generate for each arc, with a commensurate increase in the number of characterization simulations. The typical approach is to release only a single NIC for each arc and use the results for the side input values providing highest sensitivity to the input pin noise event.

The subsequent noise analysis flow, like static timing analysis, would commonly be exercised without functional vectors, leading to the use of a single, conservative NIC for each arc. (The noise analysis flow would accept functional and timing exclusions to reduce the superposition of potential aggressor noise sources, as described in Section 12.2.)

The key requirement for noise analysis is whether signal transients propagate to a flop input, such that an error state value could be recorded. Rather than apply a check at each cell input pin during the noise analysis flow using the DC transfer characteristic or the NIC curve, a more general approach is to calculate the output response to an input noise event and initiate analysis of the next stage in the path, as depicted in Figure 10.21.

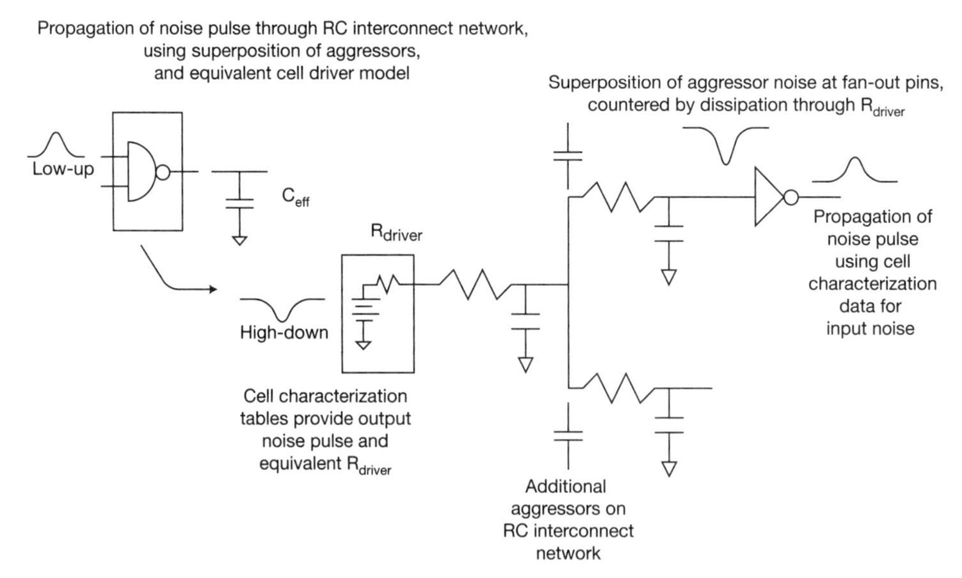


**Figure 10.21** Illustration of noise propagation to subsequent stages in a path to a flip-flop input.

Propagation involves analysis of a (linear) network, consisting of a driver voltage source and resistance model, the RC interconnects, additional aggressor sources, and the receiver capacitance. As with the general time-based output transition waveform recorded during characterization described earlier in this section, cell characterization for noise would measure and store a set of output waveform data from cell input pin noise transients. An illustration of the noise characterization propagation arc for a high-down input pulse is depicted in Figure 10.22.

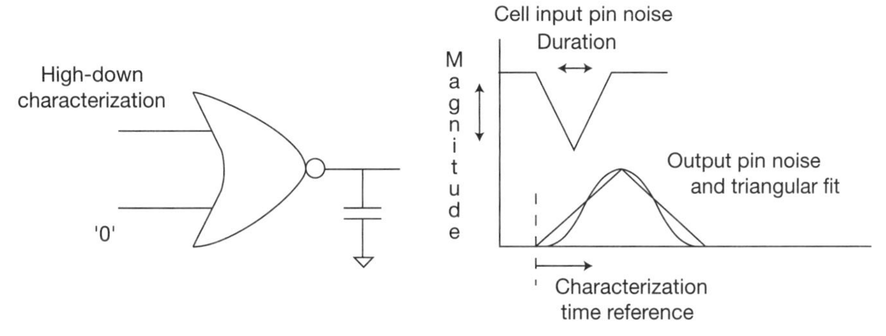


**Figure 10.22** Example of a noise propagation arc model from cell characterization.

The output pulse data would reflect the magnitude and delay of the response to the input transient; either detailed (voltage, time) points or a fitted triangular output pulse could be recorded. The characterization input pulse set should span a wide range of magnitude, duration, and $C_{load}$ values. The intent of the general analysis algorithm is to allow greater propagated noise in the network toward a flop test endpoint rather than the more restrictive individual cell limits.

This discussion of the impact of coupling noise transients has focused on the behavior of quiescent victim nets. There is also a corresponding impact on the delay of a pin-to-pin arc if the injected noise occurs during a transition on the victim net, as illustrated in Figure 10.23.

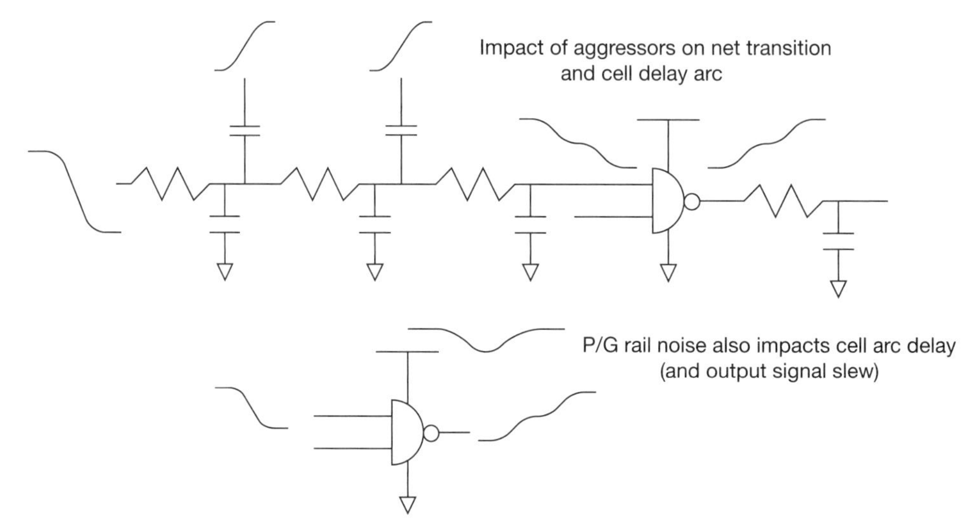


**Figure 10.23** A noise transient from aggressors to a victim net in transition impacts the arrival time at the victim fan-outs. A similar noise-delay impact arises due to voltage transients on the power and ground rails.

The presence of dynamic voltage transients on the local supply/ground rails also contributes to additional noise on the driving waveform. As a result, the traditional definition of a single input signal slew used in cell characterization does not accurately represent a noisy signal at the fan-out cell inputs. Various technical approaches have been developed to calculate an "effective slew" for the noisy input to use with existing characterization data and a cell delay adder based on the (approximate) derivative of the waveform at points in the signal transition.[9,10]

### 10.2.10 Cell Power Characterization

To support power optimization in the synthesis and physical implementation flows, a cell power model is released as part of the IP library. Characterization of this model at each corner includes both static sub-threshold leakage power and dynamic power during a switching event. The dynamic measure describes the internal power dissipation during the output transition, separate from the energy dissipated in charging/discharging the fan-out capacitive load. The magnitude of the internal power for a single-stage logic cell is related to the crossover current and thus is a strong function of the input slew and output load capacitance. The internal power is commonly represented by a table with slew and load as the input parameters, similar to the NLDM delay arc table.

As with the other electrical characterization models, the internal and leakage power dissipation for a pin-to-pin arc is dependent on the static values assigned to other cell pins. Again, the IP provider needs to assess whether the additional characterization simulations are warranted in order to provide a full state-specific pin power model. (As mentioned below, optimization flows would not have detailed simulation data that would use a state-based pin power model.)

The SoC methodology team needs to review the library power characterization data to confirm that values are provided for the following:

- **Vectorless static leakage**—For synthesis.
- **Vectorless internal power dissipation, used with an output switching activity factor measurement**—Provided to the synthesis and physical design flows.
- **Internal power dissipation for input pin-to-output pin arcs**—Used with functional simulation vectors from selected validation testcases for detailed peak/average power calculation by the power analysis flow; stateless or full state-specific pin power models could be applied.

The vectorless characterization values are appropriate for the algorithms that require fast calculations for cell selection optimization.

## 10.3 Decoupling Capacitance Calculation for Power Grid Analysis

Analysis of the voltage drop on the power and ground distribution grids is required to ensure that the cell voltage margin assumptions used during characterization are not exceeded. It is common to assume either a percentage of the supply or a fixed voltage drop as the local VDD and GND values present at the cell for characterization simulations, as shown in Figure 10.24.

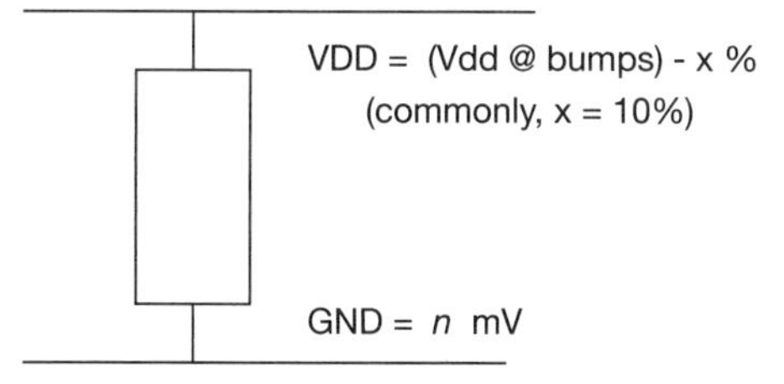


**Figure 10.24** Cell characterization uses a power and ground voltage margin for circuit simulation.

The power grid voltage drop analysis can proceed using either of the following:

- **A conservative DC static I*R drop calculation with active "on" devices drawing their saturated current**—As depicted in Figure 10.25, the placed cells are represented by current sources connected to the power and ground distribution. The current source values are equivalent to the (maximum) saturated current of the pullup and pulldown devices in the cell. The static solution assumes that all cells on the rail would be active simultaneously. The extracted model for the power and ground rails reflects the resistance of the metals, vias, and contacts in the power and ground grids.

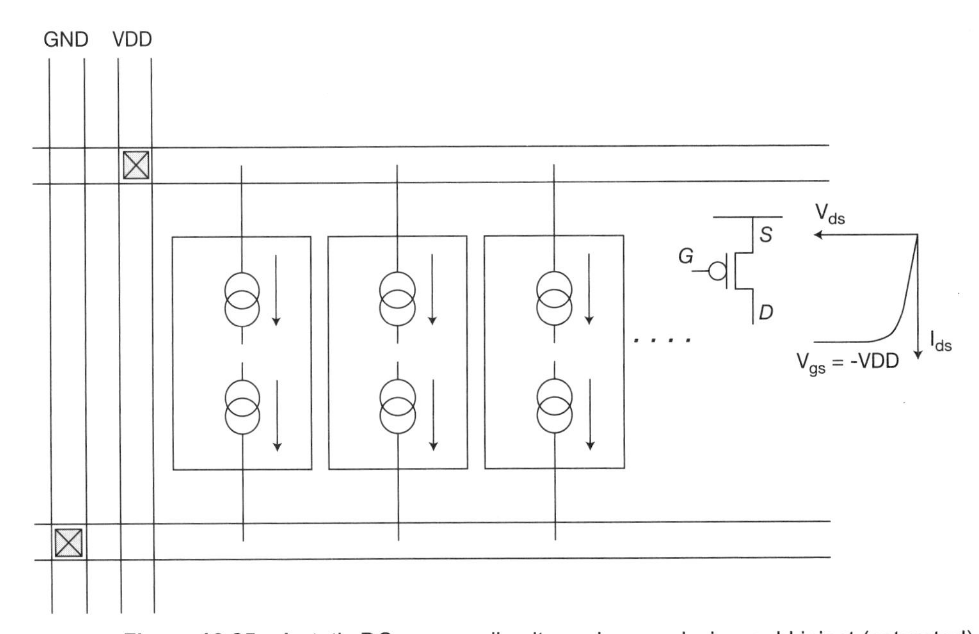


**Figure 10.25** A static DC power rail voltage drop analysis would inject (saturated) device currents from each cell, with the (conservative) assumption that all cells are active concurrently.

- **A dynamic I*R voltage drop analysis, using current pulses injected on the rails**—Functional simulation testcases exercised on the cell-based netlist are used to identify the detailed switching activity. Static timing analysis flow results provide useful information for each cell, including the following:

- Output pin driver current waveform for each delay arc for the specific $C_{load}$ and input slew in the static timing analysis model
- The cell RDLY and FLDY delay values for each arc
- The earliest and latest arrival times for each input pin for each corner

To realize the improved accuracy of dynamic I*R analysis, the extracted power/ground model needs to include both resistive and capacitive elements. The response of the local power and ground voltages to the current from multiple cells switching on a rail relies on the decoupling capacitance in close proximity to the cells. The extraction and annotation algorithms for the power/ground nets need to include the layout recognition definitions and electrical models for both internal parasitic capacitance and explicitly added decoupling capacitance, as illustrated in Figure 10.26.

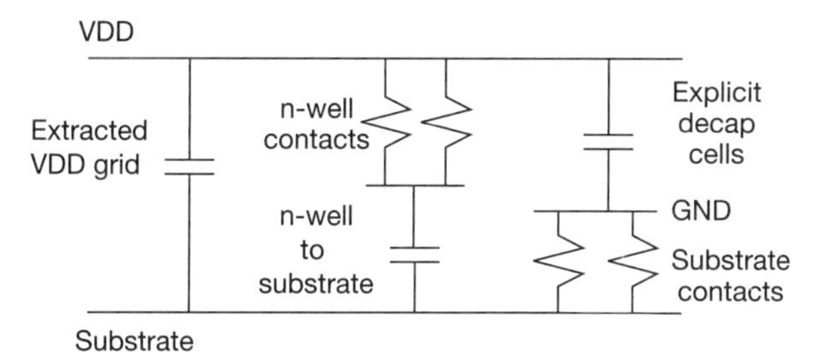


**Figure 10.26** Illustration of explicit and implicit parasitic capacitances to include with the extracted rail model for dynamic I*R analysis.

## 10.4 Interconnect Extraction

Section 10.1 describes the metal and via cross-section stack definition for the SoC and the related conductor and inter-level dielectric material properties used by the extraction algorithm. This section expands briefly on that introduction and discusses additional properties of the extracted R and C elements for the interconnect wires between cells. The most efficient method for interconnect extraction uses a library of patterns for which parameterized R and C values have been calculated. The distributed capacitance is allocated to interconnect network nodes established for the metal wire during *fracturing*, such that the sum of the discrete C elements equals the total capacitance. Resistive elements are based on the (fractured) metal segment path.

When evaluating signal interconnects, the extraction flow needs to apply a method to represent the IP layout data under the route. Specifically, the two approaches used are denoted as black box and gray box extraction, as illustrated in Figure 10.27.

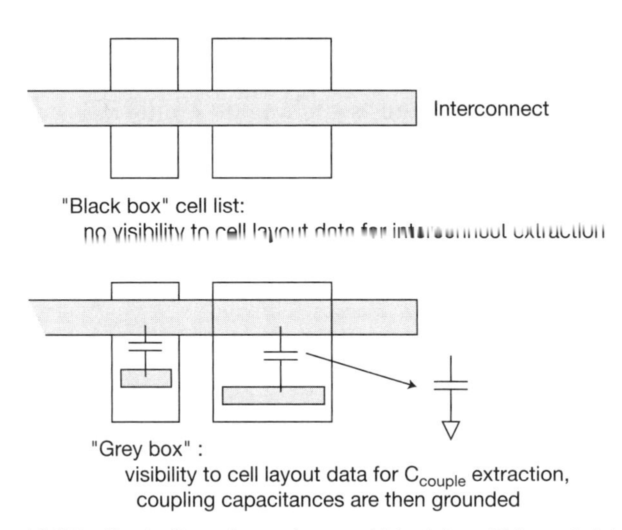


**Figure 10.27** Illustration of gray box and black box IP layout data for interconnect parasitic extraction.

If a black box cell list is provided to the interconnect extraction flow, no layout data within the cell is visible to extraction. This approach is the most efficient, which is a major consideration when extracting over a number of PVT corners. Conversely, gray box extraction exposes the detailed cell layout data with the routed interconnects. Additional coupling capacitances are generated between the route and the shapes within the cell. As there are no nodes in the route netlist for these internal cell locations, the extracted capacitive elements would be lumped to ground (coupling factor, k = 1) for parasitic annotation. The gray box method is more compute intensive, and many of the additional capacitive elements are small; however, it provides a more complete model. The SoC methodology and CAD teams need to evaluate the accuracy versus compute resource trade-offs when preparing the black box and gray box cell lists for the interconnect extraction flow.

### 10.4.1 Resistivity

The previous section highlights the fact that the resistive elements extracted for IP library cell layouts include temperature coefficients of resistivity in their models (e.g., for $R_{gate}$, $R_{drain}$, $R_{source}$, R_M0, R_M1). Similarly, the metal layers used for interconnects include TC1 and TC2 coefficients in the models provided by the foundry. The interconnect metal layers may also include a width-dependent sheet resistance calculation (see Figure 10.28).

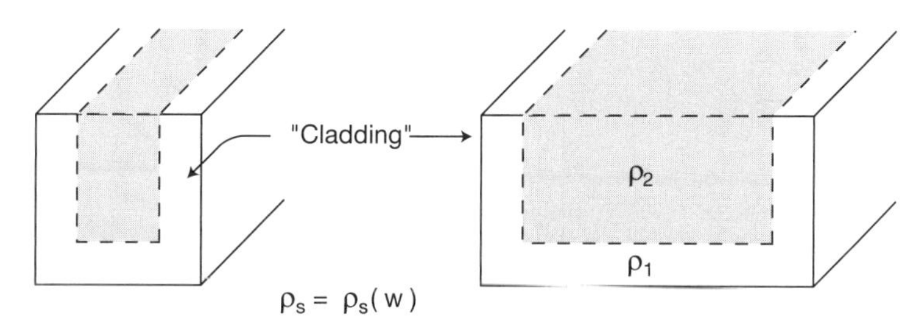


**Figure 10.28** The foundry model for parasitic interconnect extraction may include a sheet resistivity value that is a function of the route width.

The fabrication of interconnects typically involves the deposition of an initial metal "cladding" layer in the damascene trench, followed by the subsequent (predominantly Cu) metal deposition. The (fixed) thickness cladding has a different material resistivity than Cu, resulting in a sheet resistivity that is a function of the linewidth. Vias result in a resistance added to the extracted netlist, based on the interconnect overlap area. The calculation of the via (or contact) resistance is more complex if the resistivity of one of the interconnect layers is significantly higher than the other; the calculation requires identification of the current through the via/contact at the leading edge of the high-resistivity material.

### 10.4.2 Coupling Capacitances and "Multipliers"

Figure 10.1 illustrates some of the geometric topologies that contribute to the extracted capacitances for interconnect wires. The fracturing of the layout separates the different topologies based on adjacent wire spacing and wires present above and below. The parasitic netlist output from the extraction flow consolidates the coupling capacitances between signals to avoid double-counting.

However, a key SoC methodology decision is needed when submitting the annotated netlist to subsequent electrical analysis flows. When a net is analyzed, the effective coupling capacitances could be significantly different from the values determined by the geometry, as shown in Figure 10.29. A *k-factor coupling multiplier* is used by the analysis flows to represent the different aggressor and victim signal transitions in the figure to scale the extracted coupling capacitances in the RC interconnect model.

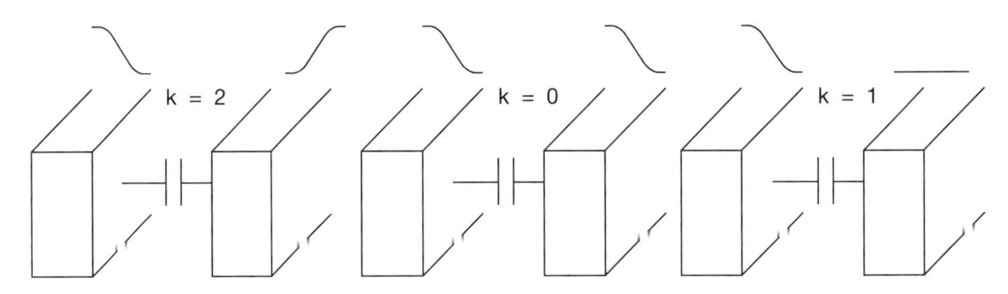


Note: The k-factors above assume that (dv/dt) is equal for both signal transitions.

**Figure 10.29** The parasitic extraction coupling capacitances between interconnects are typically subjected to a k-factor multiplier in analysis flows to reflect the "effective" coupling capacitance.

### 10.4.3 Parasitic Netlist Reduction

Layout fracturing for extraction can result in a very large number of R and C elements in the final network. (In the most detailed parasitic network output format from EDA extraction tools, layout coordinates are included with the R and C elements for reference, as an informational comment.) Several electrical analysis flows do not need this level of detail; a reduced netlist that provides a comparable electrical response for the signal's spectral frequency range of interest would be sufficient and would result in significantly improved flow runtime.

The SoC methodology team needs to establish the appropriate reduction settings in the extraction flow, such as:

- Equivalent R and C values for arrayed vias (see Figure 10.30)
- Magnitude of coupling capacitances that can be lumped to ground

A large number of coupling capacitances stresses the runtime of reduction algorithms, and converting small $C_c$ elements to grounded caps is a common network transformation, as shown in Figure 10.31.

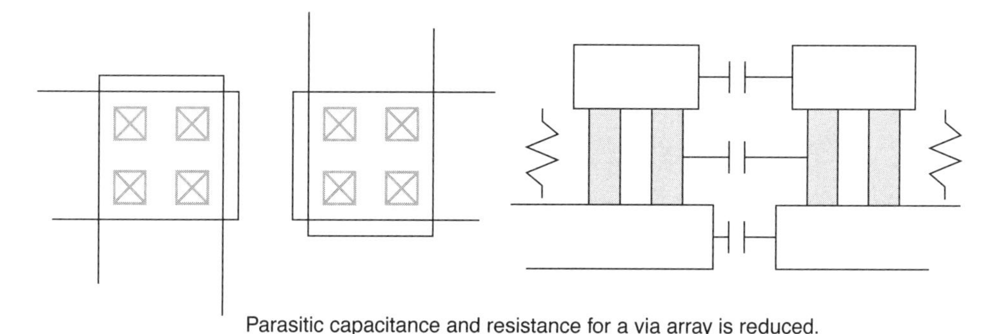


**Figure 10.30** The resistive and capacitive parasitic values of a via array are typically reduced to single R and C elements.

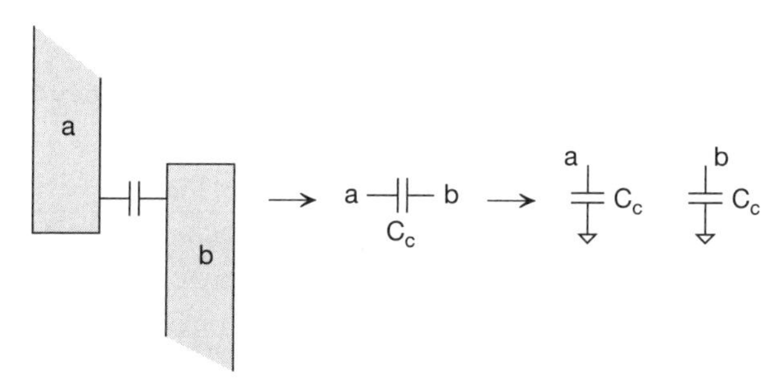


**Figure 10.31** Parasitic coupling capacitance elements below a threshold are commonly converted to grounded capacitances at the node during network reduction.

Reduced RC networks are typically suitable for signal net timing analysis, noise analysis, and power rail voltage drop analysis. However, electromigration (EM) analysis relies on a detailed calculation of the current density through all interconnects, vias, and rails. The non-reduced extracted netlist is submitted to the EM analysis flow.

## 10.5 "Selected Net" Extraction Options

Layout parasitic extraction tools from EDA vendors incorporate multiple methods that tradeoff runtime versus (non-statistical) model accuracy. The most detailed method adopts a 3D field-solver algorithm, as described in Section 10.1. However, this approach is far too compute intensive to use for all interconnect nets; the faster, albeit less accurate, pattern-matching algorithm is used instead. The EDA extraction tools offer a feature to request field solver–level accuracy on a small set of selected nets.

### 10.5.1 Clock Arrival Analysis

A unique methodology flow is commonly developed for the analysis of clock nets at each extraction corner:

1. Extract all interconnect branches of the clock repowering tree signals using the field-solver option.
2. Annotate the RC elements of the clock tree to the block/chip netlist (without reduction).
3. Excise the cell and RC instances for the full clock tree from the total netlist.
4. Submit the excised netlist to circuit simulation, measuring the arrivals at the clock tree endpoints.
5. Compare the measured arrivals against the skew targets to assess the success of the physical implementation clock-balancing steps.
6. For static timing analysis, override the delay calculation for clock buffers and assign the measured arrival from circuit simulation at the clock tree endpoints to the STA model prior to evaluating the setup and hold tests.
7. The clock tree circuit simulation testcases also provide current measures through the (non-reduced) parasitic R elements for electromigration analysis.

The justification for this additional flow complexity (i.e., replacing buffer cell and interconnect delay calculation with detailed circuit simulation for clocks) depends on the SoC clock frequency specification and skew targets. Excising the full clock model for simulation requires handling the extracted coupling capacitances to the clock nets. Commonly, the coupling capacitances are multiplied by a k-factor = 2 or a k-factor = 0, depending on the specific clock delay skews to be amplified. For very-high-frequency clocks, the physical implementation flows commonly include features to shield clock wires to the maximum extent possible to reduce the number and magnitude of coupling capacitances to be excised from the netlist for circuit simulation.

## 10.6 RLC Modeling

As mentioned briefly in Section 10.1, EDA vendors offer unique tools to extract (self- and mutual) inductance of interconnects. The common applications for extracting a parasitic network of R, C, L, and (potentially) M elements are:

- Models for the thick top metal redistribution layer (RDL) patterns from chip bumps to power/ground grids
- Models for the RDL between chip bumps and I/O pads associated with high-speed interface drivers/receivers
- Models for (very-high-frequency) clock grids

The power grid RLC models are typically merged with corresponding package models and then exercised using dynamic power current transients. Although local current transients rely on decoupling capacitance to minimize supply/ground bounce, the overall chip plus package RLC model must also be analyzed for global voltage fluctuations. The analysis of a full end-to-end driver-to-receiver model of a high-speed (SerDes or parallel DDR) interface requires circuit simulation-level detail and full RLC parasitics. The design of advanced microprocessors and SoCs pushing multi-gigahertz frequencies requires RLC modeling of the global clock grids.[11] The principal difficulty in accurate extraction of interconnect inductance is the identification of the "return current loop." The CAD team and EDA vendor need to review how the loop will be identified through the metal stack and die well/substrate.

## 10.7 Summary

This chapter briefly reviewed the extraction of parasitic elements for annotation to a netlist model for use in electrical analysis flows. For cell-level IP, extraction identifies the schematic devices in the cell layout and determines the parasitic R and C elements to connect to the devices. At advanced process nodes, device-level identification includes proximity measures associated with layout-dependent effects. The merged device and RC parasitic netlist for the cell-level design is submitted to a (large) number of characterization simulations to generate the timing, noise, and power models for the library IP release.

For block-level and global-level interconnect routes between cells, extraction generates a large number of parasitic RC elements. The methodology team needs to review the extraction algorithms used for capacitive coupling calculation and RC parasitic reduction prior to netlist annotation, to assess the resulting accuracy of interconnect delay calculation and to establish suitable margins in electrical analysis flows.

The extraction of inductive parasitic elements is increasing in applicability, for detailed electrical analysis of the chip-package power delivery network and for full transmit-to-receiver model simulation of high-speed chip I/O interfaces.

An increasing complication to any extraction methodology is the determination of the PVT corners at which extraction algorithms are to be evaluated, corresponding to the corners at which electrical analysis flows will be exercised. The methodology team needs to evaluate the trade-offs between analysis across a wide range of PVT variation and the resources/runtime required to run the flows and interpret the results.

EDA vendors are continuing to make a significant investment in extraction technology. Specifically, there is a requirement to apply field solver-based extraction algorithms to a greater set of IP designs seeking to apply models of the highest accuracy in electrical analysis flows. There is little value in pursuing high-sigma statistical simulation of device models if the accuracy of the extracted interconnects annotated to the devices is lacking. EDA vendors are focused on providing large model custom extraction with high parasitic accuracy and manageable compute resource.

## References

[1] Kao, W., et al., "Parasitic Extraction: Current State of the Art and Future Trends," *Proceedings of the IEEE*, Volume 89, Issue 5, May 2001, pp. 729–739.

[2] Iverson, R.B., and Le Coz, Y.L., "A Stochastic Algorithm for High Speed Capacitance Extraction in Integrated Circuits," *Solid-State Electronics*, Volume 35, Issue 7, July 1992, pp. 1005–1012.

[3] Compact device simulation models (for numerous fabrication technologies) are reviewed and approved by the Compact Model Coalition, a group within the Si2 Consortium: https://projects.si2.org/cmc_index.php. This compact model standard approach allows EDA vendors to qualify their circuit simulation tools against these models prior to customer release. The prevalent source for many compact model proposals is the Device Group at University of California-Berkeley: https://www-device.eecs.berkeley.edu/research.htm.

[4] Trihy, R., "Addressing Library Creation Challenges from Recent Liberty Extensions," *IEEE 45th Design Automation Conference (DAC)*, 2008, Paper 26.5, pp. 474–479.

[5] Salzmann, J., Sill, F., and Timmermann, D., "Algorithm for Fast Statistical Timing Analysis,"*2007 International Symposium on System-on-Chip*, 2007, pp. 1–4.

[6] Liou, J.J., Cheng, K.T., Kundu, S., and Krstic, A., "Fast Statistical Timing Analysis by Probabilistic Event Propagation," *IEEE 38th Design Automation Conference (DAC)*, 2001, pp. 661–666.

[7] Devgan, A., and Kashyap, C., "Block-Based Static Timing Analysis with Uncertainty," *IEEE International Conference on Computer-Aided Design (ICCAD)*, 2003, pp. 607–614.

[8] Agarwal, A., Dartu, F., and Blaauw, D., "Statistical Gate Delay Model Considering Multiple Input Switching," *IEEE 41st Design Automation Conference (DAC)*, 2004, pp. 658–663.

[9] Nazarian, S., et al., "Modeling and Propagation of Noisy Waveforms in Static Timing Analysis," *Proceedings of the Design, Automation, and Test in Europe (DATE) Conference*, 2005, pp. 776–777.

[10] Hashimoto, M., Yamada, Y., and Onodera, H. "Equivalent Waveform Propagation for Static Timing Analysis," *IEEE Transactions on Computer-Aided Design of Integrated Circuits and Systems*, Volume 23, Issue 4, 2004, pp. 498–508.

[11] Deutsch, A., et al., "On-Chip Wiring Design Challenges for Gigahertz Operation," *Proceedings of the IEEE*, Volume 89, Issue 4, 2001, pp. 529–555.

[12] Murrmann, H., and Widmann, D., "Current Crowding on Metal Contacts to Planar Devices," *IEEE Transactions on Electron Devices*, Volume ED-16, Issue 12, December 1969, pp. 1022–1024.

## Further Research

### Via/Contact Resistance

Describe the requirements for via/contact resistance modeling when the two layers connected are of significantly different resistivity (and, thus, the current density in the via/contact is non-uniform). Specifically, describe the definition of *current crowding* and its role in resistance calculation.[12]

### BEM and FRW Extraction Methods (Advanced)

The BEM and FRW methods for high-accuracy extraction differ significantly in the 3D model formulation and solution calculation.

Describe the model capacity, compute resource, runtime, and accuracy trade-offs associated with these methods (including the opportunity for algorithm parallelization).

CHAPTER 11

# Timing Analysis

## 11.1 Cell Delay Calculation

As described in Section 10.2, the library cell characterization flow provides delay models and output signal slew data for input pin-to-output pin arcs, as a function of input signal slew and output capacitive load. The arc delays are typically provided in terms of Non-Linear Delay Model (NLDM) tables. Rather than use a single output slew value, output waveform tables are a more recent IP library release format, consisting of a set of (value, time) points, recorded for each input slew and output load characterization simulation. During characterization, the local supply and ground rail voltages in the simulation testcases reflect the best-case/worst-case assumptions on chip bump, global grid, and local drop DC margins.

For cell delay calculation in timing analysis with this characterization data, it is necessary to determine an effective output load capacitance from the extracted interconnect RC network. The $C_{eff}$ calculation is used in support of the cell library characterization methodology, with separate cell and interconnect delay contributions. The $C_{eff}$ value is less than the sum of the capacitive elements in the extracted network due to the resistive shielding near the driver.

One of the first methods used to determine $C_{eff}$ is described in Reference 7.2. Initially, the extracted RC network is simplified to an equivalent pi-network, representing the *driving point admittance*. A subsequent algorithm analytically calculates $C_{eff}$ such that the average current through the pi-network and $C_{eff}$ are equal, over the waveform transition time that defines the cell delay, as illustrated in Figure 11.1.

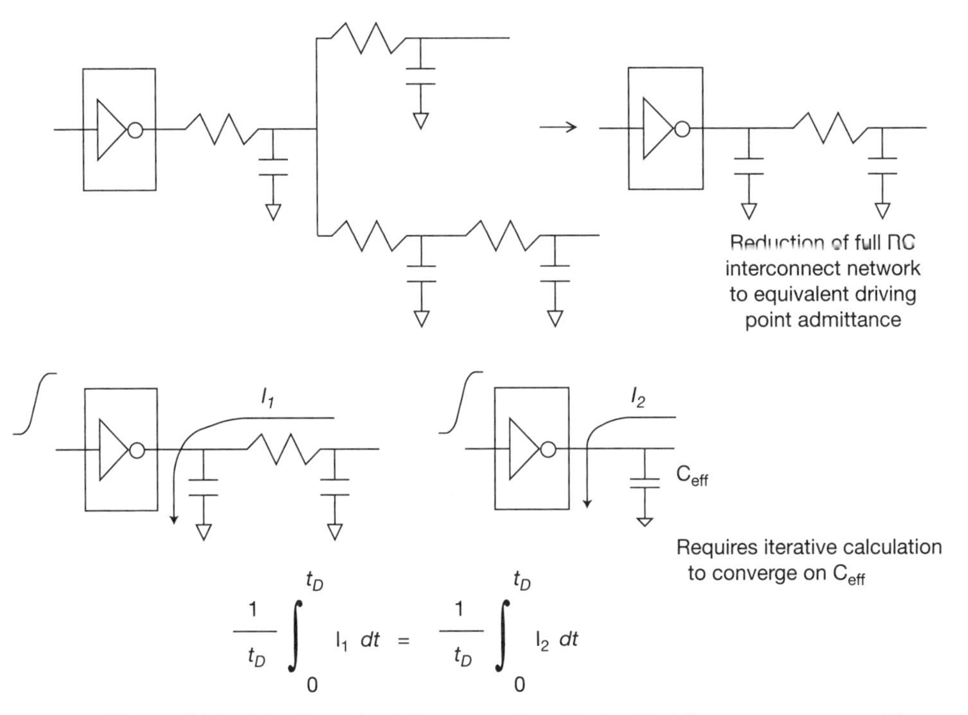


**Figure 11.1** The $C_{eff}$ calculation uses the criterion that the average current through the admittance network and $C_{eff}$ are equal.

This algorithm is iterative and involves the following steps:

1. Selecting $C_{total}$ as the starting load value
2. Indexing into the cell characterization tables to get delay and output waveform data
3. Evaluating the integral for currents $I_1$ and $I_2$ in Figure 11.1 to calculate an initial $C_{eff}$
4. Returning to step (2) to use the $C_{eff}$ value for the output load until the iterations converge

The resulting $C_{eff}$ will be between C1 of the pi-network and $C_{total}$ (= C1 + C2). If $R_{pi}$ is large compared to the cell's drive strength impedance, the resistive shielding effect will be maximal, and $C_{eff}$ will be closer to C1.

A more recent approach uses a table-based $C_{eff}$ model, replacing the integral calculation in step 3 of the $C_{eff}$ procedure above.[1] During cell characterization, additional circuit simulations are submitted over a wide range of driving point admittance pi-networks. A shmoo of lumped $C_{load}$ simulations for each pi-network example is also run to match the propagation delay and thus derive an equivalent $C_{eff}$. A characterization table of $C_{eff}$ values is released with the traditional NLDM delay arc and output slew tables. To improve the efficiency of the $C_{eff}$ data, the table is represented as a function of two (normalized) variables, as illustrated in Figure 11.2.

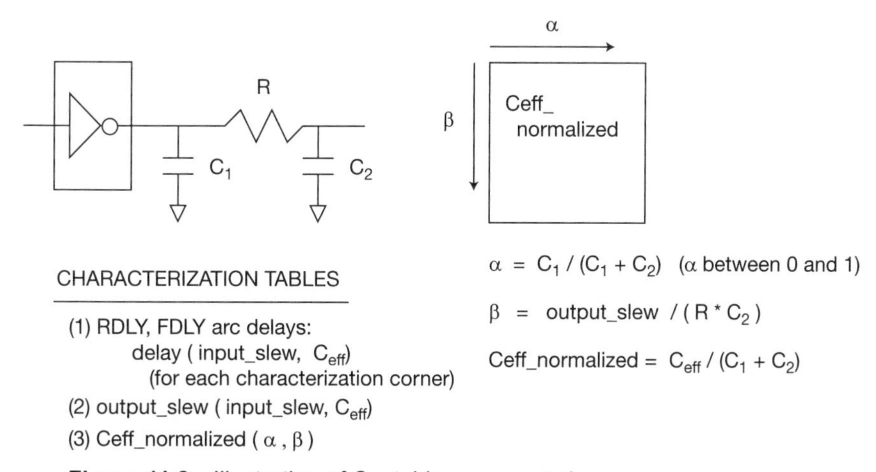


**Figure 11.2** Illustration of $C_{eff}$ table representation.

With the released $C_{eff}$ tables, an iterative procedure is still required for cell delay calculation to determine the final $C_{eff}$ and output slew values for each cell instance; the interdependence between output slew and $C_{eff}$ requires iterative convergence. Reference 11.1 indicates faster convergence with the $C_{eff}$ table lookup method compared to using an analytical model of the driver connected to the pi-network. The trade-off is the expense of the additional library characterization simulations to derive the $C_{eff}$ tables for the cell.

The (iterative) determination of $C_{eff}$ is based on matching the average current to that of the driving point admittance pi-network. However, the output driver waveform characterized with $C_{eff}$ may differ significantly from the pi-network waveform. The discussion in Reference 11.2 suggests that the output waveform using $C_{eff}$ should be adjusted for propagation through the interconnect RC network to better reflect the possibility of a "long tail." At the end of the signal transition, the driving device is not a (saturated mode) current source but a linear $R_{on}$, as depicted in Figure 11.3. If $R_{on}$ is comparable to $R_{pi}$, the output waveform will have a significantly longer time constant than represented by the $C_{eff}$ simulation.

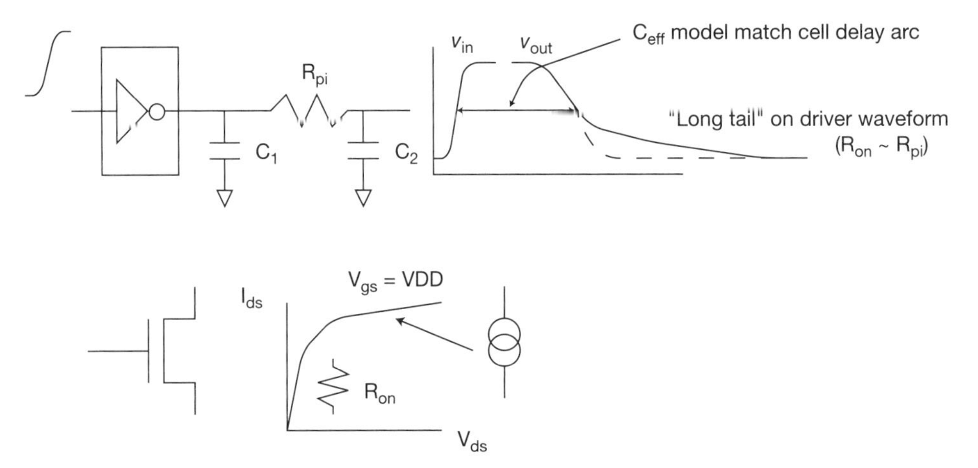


**Figure 11.3** Illustration of a long-tail output waveform, which arises when the linear driver resistance is comparable to $R_{pi}$; the $C_{eff}$ delay model will be less accurate.

## 11.2 Interconnect Delay Calculation

The timing analysis flow separates delay calculation into the cell delay arc (using $C_{eff}$) and the delay through the extracted (linear) RC interconnect network. The interconnect delay to the cell fan-outs could be simulated using characterization-based output waveforms as stimulus, although the volume of interconnect elements and paths makes this approach cumbersome for design optimization decisions. As interconnect delay calculation is an integrated analysis engine in synthesis and physical design, a computationally efficient method is required. The EDA industry has actively researched methods to approximate the interconnect delay, focused on the similarities between RC network circuit response and probability distribution functions. The origin of these approximation methods was first published in Reference 11.3 and subsequently applied to VLSI interconnect delay calculation[4]; this method is denoted as the *Elmore delay* calculation.

Consider the RC interconnect network depicted in Figure 11.4. Coupling capacitances from the original parasitic extraction flow are grounded to implement a direct RC path from the driving cell output pin to each fan-out. (Section 10.2 briefly discusses approaches to adjust the interconnect delay and signal slew at fan-outs to reflect potential coupling noise during the signal transient.) Assume that a step input is applied at the driver output at time $t = 0$—that is, $v_{in}(t) = u(t)$. The transient will propagate to the fan-outs after a delay, measured as the 50 percent crossing point at the RC tree endpoints. For a "simple" RC network (i.e., no aggressor coupling injected on the transient) the output through a linear, passive RC network will be monotonic. It is not feasible to determine the precise endpoint waveform analytically. The goal is to estimate the delay crossing point. If both the $v_{in}(t)$ step input and the voltage $v_a(t)$ at fan-out pin 'a' are differentiated, the input $v_{in}'(t)$ becomes an impulse; both this input impulse and $v_a'(t)$ are shown in Figure 11.5.

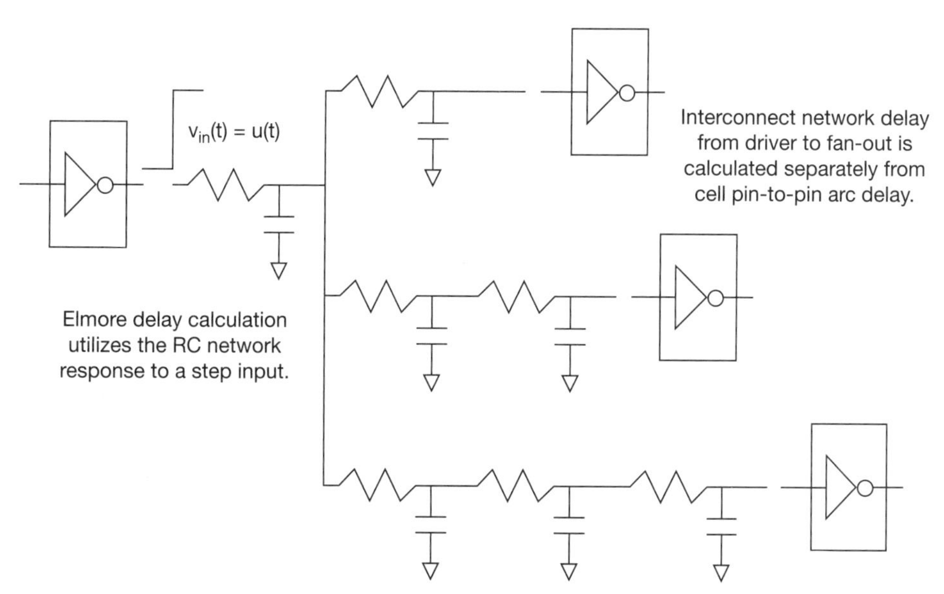


**Figure 11.4** An example of an RC interconnect tree to illustrate the Elmore delay calculation.

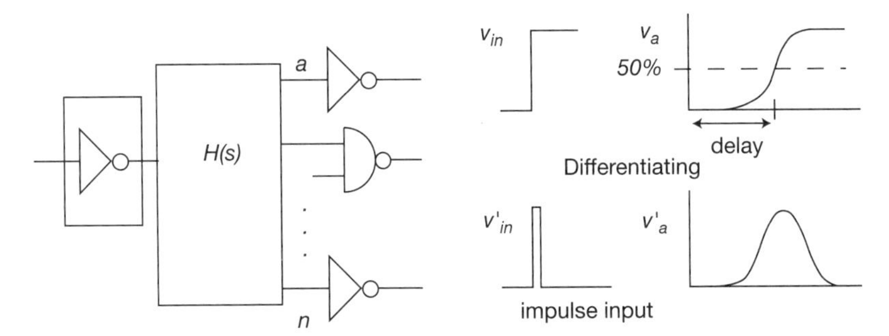


**Figure 11.5** A step input and impulse input response for the RC interconnect tree are depicted. The impulse response is the derivative of the step input response for the linear network.

Note that the circuit responses $v_a(t)$ and $v_a'(t)$ in the time domain have characteristics similar to those of a probabilistic cumulative density function (CDF) and its derivative, the probability density function (PDF); this similarity is the basis of the Elmore approximation, as will be discussed shortly.

The *transfer function* of a (linear, time-invariant) network in the frequency domain relates $v_{out}(s)$ to $v_{in}(s)$, with the expression $v_{out}(s) = H(s) * v_{in}(s)$. The transfer function is the Laplace transform of the circuit's impulse response. The transfer function can be expressed as a Taylor expansion polynomial around $s = 0$:

$$H(s) = m0 + (m1 * s) + (m2 * (s**2)) + \ldots \qquad \text{(Eqn. 1)}$$

In this format, the *m* coefficients of each term are denoted as the *moments* of the transfer function.

Now, consider the same $v_a(t)$ and $v_a'(t)$ circuit response curves as CDF and PDF probability distributions as a function of *t*. The goal would be to measure the 50 percent crossing point of the CDF as the equivalent interconnect delay; in probability terms, this would represent the "median" of the distribution. The median calculation is not straightforward for a general CDF.

The "mean" of a PDF is defined by the relationship in Figure 11.6, where the denominator normalizes the area of the PDF to one.

$$\mu = \frac{\int_0^{\infty} t * p(t)\ dt}{\int_0^{\infty} p(t)\ dt}$$

**Figure 11.6** Equation for the mean of a probability distribution function.

Comparing the probability integrals in Figure 11.6 to the Laplace transform integrals applied to the impulse circuit response, the mean of the PDF is the same as a ratio of transfer function moments (see Figure 11.7).

$$\mu = ( -m1 \ / \ m0 )$$

$$\text{where } H(s) = m0 + (m1 * s) + (m2 * s^2) + \ldots$$

**Figure 11.7** The mean of a PDF is correlated to a ratio of the moments of the transfer function.

The characteristics of the $v_a'(t)$ as a PDF curve for the impulse response of the RC network have been proven to satisfy the relationship in Figure 11.8.[5]

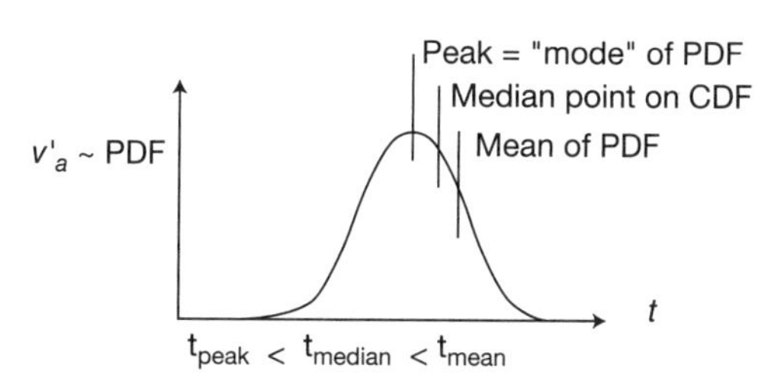


**Figure 11.8** Delay bounds associated with using the mean versus median of a PDF.

To recap, if the general response of an RC interconnect network can be expressed in terms of the *s*-domain transfer function, the 50 percent crossing point of $v_a(t)$ will be approximated by the “mean” of $v_a'(t)$, expressed as a ratio of moments from the Taylor expansion of the transfer function, $(-m1 \ / \ m0)$. The delay bounds the actual delay of the median 50 percent crossing point.

The moment *m0* of the *H*(*s*) transfer function is equal to one. The calculation of the moment m1 in an RC network is computationally straightforward:

1. Determine the (unique) path in the RC network between the driver source and the fan-out pin of interest; given the nature of the extracted RC tree, there will be no circuit loops, and the path to each fan-out pin will be unique.
2. Effectively "zero out" all resistances in side paths.
3. Follow the resistive elements along the selected path, summing all forward capacitances at the end of each resistor.
4. To find the Elmore delay, sum all the RC terms along the path between driver and fan-out pin.

The algorithm to determine the Elmore delay approximation for the moment $|m1| = \text{mean}(v_a'(t))$ is perhaps best illustrated by an example, as shown in Figure 11.9.

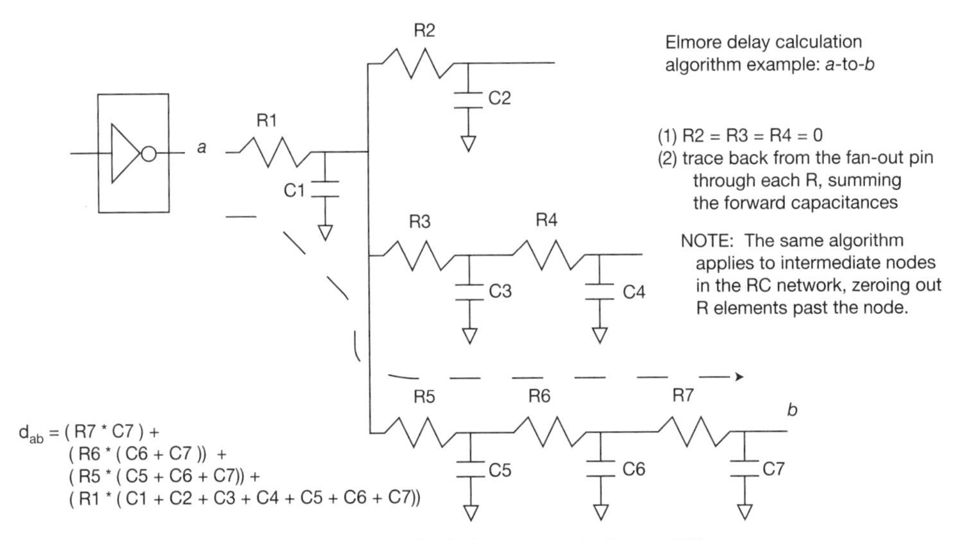


**Figure 11.9** Elmore delay calculation example for an RC tree.

There are certainly topologies in an RC interconnect network where the Elmore delay coefficient is a conservative upper bound on the interconnect delay. Consider the simple example in Figure 11.10: The Elmore delay is >30 percent larger than the actual 50 percent waveform crossing.

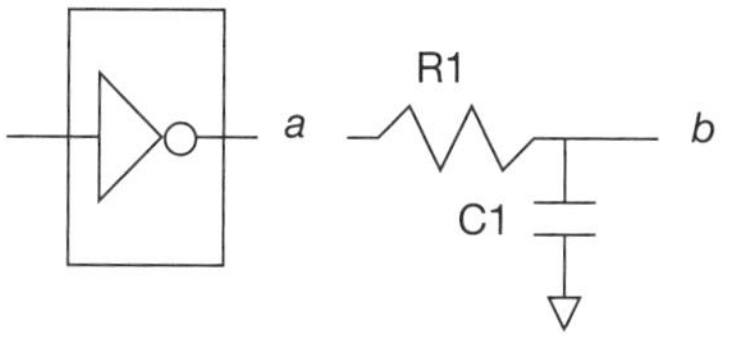

R1

a b

C1

$H(s) = 1 / (1 + s * R1 * C1)$

$m0 = H(0) = 1$

$m1 = H'(s = 0) = - R1 * C1$

$d_{ab}\_Elmore = ( - m1 / m0 ) = R1 * C1$

In response to a step input, $v_b(t) = (1 - e^{(-t / R1C1)})$

$v_b( d_{ab} ) = 0.5$ ; $d_{ab} = - \ln(0.5) * R1 * C1 = 0.7 * R1 * C1$

**Figure 11.10** Example of the conservative error in the Elmore delay calculation.

The Elmore delay calculation errors tend to be larger at the near end of the interconnect tree, especially if the resistive shielding effects are significant. The Elmore ("mean") delay calculation is efficient, conservative, and independent of the specific cell output waveform. It is suitable for basic timing-driven optimizations in physical design flows but lacks the accuracy and detail required for critical optimizations or for interconnect delay calculation in timing analysis.

EDA researchers have actively pursued improvements to the interconnect delay calculation accuracy. Specific CDF/PDF probability distributions have been proposed to approximate the typical $v_a(t)$ and $v_a'(t)$ waveforms. With a specific distribution selected, the waveform median value and slew can be calculated using additional moments of the H(s) transfer function.[6,7] Additional moments can be derived efficiently from the network RC elements, using similar path traversal calculations as the Elmore first moment.[8]

The SoC methodology team should review the interconnect delay calculation algorithm(s) used by the EDA vendor timing analysis tool to assess the (absolute and relative percentage) accuracy targets over a wide range of interconnect trees. As discussed in Section 11.4, the timing analysis flow may choose to apply *derating delay multipliers* to address anticipated pessimism (or optimism) in critical timing paths. In addition, a comparison between the interconnect delay calculation engines used in the design optimization flows and the timing analysis flow will be indicative of the potential differences between the predicted timing from physical design and the sign-off timing results and will be indicative of the resources to allocate to final timing closure prior to tapeout.

## 11.3 Electrical Design Checks

After cell and interconnect delay calculation and prior to executing the path propagation and setup/hold tests in the timing analysis flow, it is common to exercise a set of design checks on the network timing model. These checks are intended to identify potential timing issues that may result in anomalous results. The EDA tool vendor will provide a set of timing model query functions. The CAD and SoC methodology teams will code the electrical design check utilities to integrate into the timing flow. Timing model design check examples include the following:

- $C_{eff}$ load capacitance outside limits, based on the characterization table ranges (to estimate the extrapolation error based on the cell drive strength)
- Input pin signal slews outside characterization ranges
- Successful clock arrival and clock slew annotation to flop inputs (from a separate clock simulation flow, as described in Section 10.5)
- Clock skew checks
- Proper sizing of power gating cells (with a low effective $R_{on}$ resistance during the active power state for parallel power gating cells)

Electrical design check thresholds that are exceeded usually result in an interruption to the timing analysis flow so that the electrical model can be examined. The inclusion of coordinate information for extracted RC elements allows the parasitic model to be correlated to the physical layout to assist with this review.

Additional design checks have been developed to evaluate the structure of the timing model. As discussed in more detail in the next section, timing analysis is based on a synchronous model of operation; that is, clocked cells have an associated path timing test. The timing model consists of levelized gates between the clocked cells for path delay propagation. As a result, any combinational logic loops in the timing model are invalid. In addition, model inputs identified as clocks should propagate (through specific types of buffer and clock gating cells) solely to timing test endpoints; clocks should not propagate into general logic cones. The multiplexing of clocks is also likely used in the SoC design. A block may be clocked by different signals when the SoC is operating in different design modes. The presence of converged clocks at a MUX should be highlighted during timing model checks to ensure that the timing analysis flow will be provided with the necessary definitions for multiple-mode evaluation.

There is an SoC methodology decision associated with a structural check that detects PI-to-PO paths, without an intervening clocked timing test. Combinational paths through blocks are a valid timing model, with appropriate pin constraints applied to implement a timing test. However, the initial budgeting of a clock cycle fraction across more than two blocks is significantly more complicated, as is the incremental timing closure of failing tests between blocks during timing analysis. The methodology team may want to review the combinational path reports from the checks to ensure that these paths are easily managed globally.

## 11.4 Static Timing Analysis

Cell delay calculation is based on a pin-to-pin arc model. The timing analysis flow combines cell and interconnect delays to determine the set of input pin arrivals (and slews) at each cell. Each potential delay arc is then evaluated to calculate the maximum (late) or minimum (early) arrival at the cell output for both rising and falling output transitions, as illustrated in Figure 11.11.

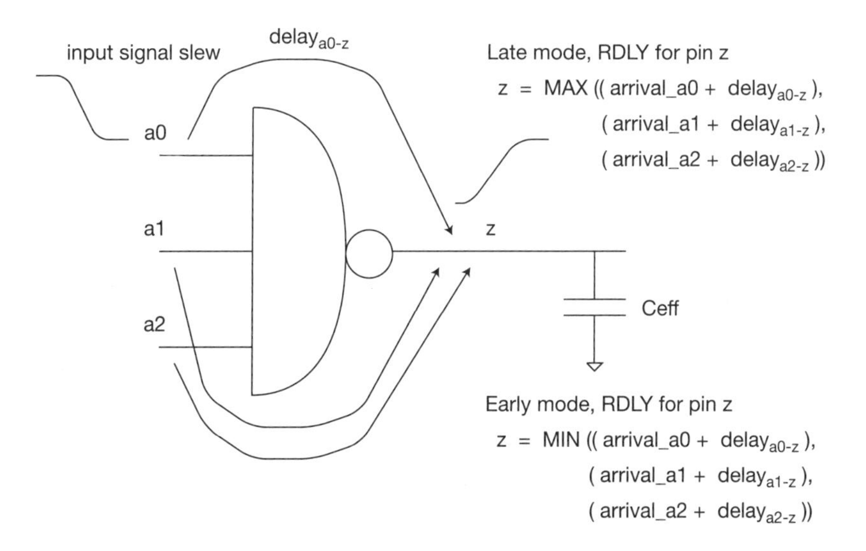


**Figure 11.11** Illustration of late and early arrival time propagation in static timing analysis.

There is a methodology question associated with this forward propagation algorithm. The results of the cell delay arc calculations are used to select the (min, max) arrival time at the output. However, each of the candidate delay arcs has an associated output slew. The propagation algorithm needs to select a slew to launch to the RC interconnect load. A conservative approach would be to use the slowest slew (in late mode) among *any* of the candidate delay arcs; the combination of the latest output pin arrival and slowest slew avoids the accuracy uncertainty of propagating a late output arrival with faster output slew versus an earlier arrival with slower slew. If the timing endpoint tests pass with this conservative calculation, no further investigation would be needed. If timing tests fail and debugging is needed after running the timing analysis flow, it is necessary to use the EDA vendor tool to select detailed paths through user-defined sets of cell pins and recalculate the specific arcs and slews.

The general delay propagation method utilizes a levelized cell network between timing testpoints (e.g., flop-to-flop, PI-to-flop, flop-to-PO). A clk_enable-to-clk timing test is included for gated clock cells. This method does not utilize any simulation testcase vectors to sensitize specific cell input-to-output (functional) transitions; as a result, this flow is denoted as *static timing analysis* (*STA*). A key advantage of this approach is the comprehensive analysis of all potential signal paths. As there is no logical evaluation of the network, a method is needed to prune timing paths that are not able to be exercised, as discussed shortly.

Although there is no logical evaluation of the levelized network, there is a logical property of the cell that is pertinent to STA. The library data for each cell input pin includes a *unateness property* that identifies whether the cell output will invert the input pin transition (e.g., NOT, NAND, NOR, AOI) or is positive unate (e.g., BUFFER, AND, OR) or is binate (e.g., XOR). This designation is needed to evaluate and propagate RDLY and FDLY arcs correctly.

### 11.4.1 Timing Slack Calculation

A signal arrival time that fails a test at a timing path endpoint is indicative of the need to make an engineering change to the physical design. However, in a large network with potentially many failing tests, the SoC designer would need additional insight into where to focus timing optimizations. The concept of *timing slack* is used to identify the criticality of all input and output cell pins in the timing model, as depicted in Figure 11.12.

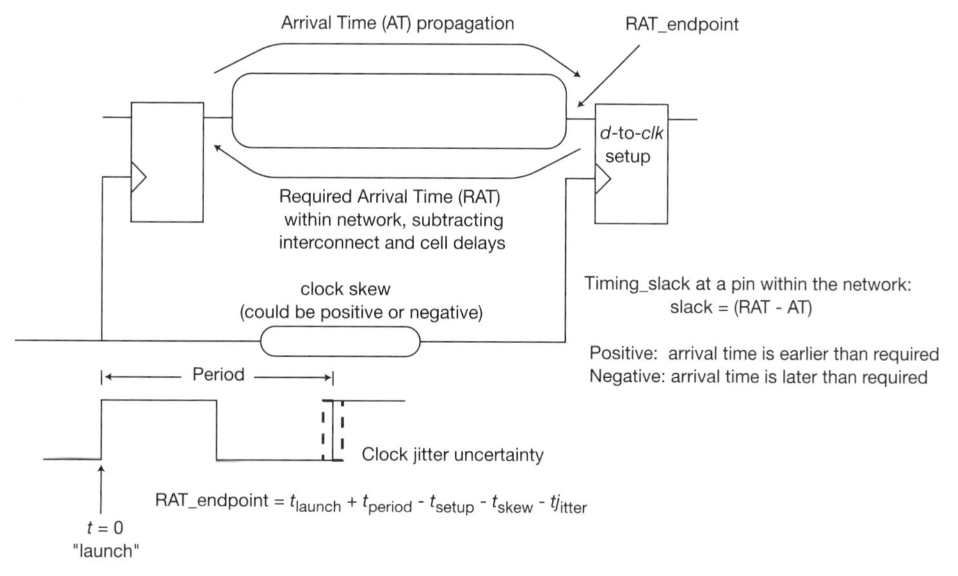


**Figure 11.12** Illustration of the timing slack calculation in the static timing analysis flow.

The forward propagation of calculated cell and interconnect delays provides the *arrival time* (AT) at each pin. The timing test involves a comparison between the AT and the *required arrival time* (RAT) at the path endpoint. In late mode, the endpoint RAT is derived from the setup time constraint, the clock period specification, the clock launch-to-capture skew, and clock jitter margin. If the network timing model were to be traversed in reverse levelized order from the endpoint RAT, subtracting the cell and interconnect delays, a specific RAT value could be assigned to each pin internal to the network.

For a cell output pin, the late mode RAT is the "earliest" required time traversing back from the fan-out pins through interconnect delays. The (RAT − AT) difference at each pin is denoted as the (late mode) timing slack. A positive slack indicates a forward propagation arrival time that precedes its required arrival time, as derived from the backward calculation from *all* timing path endpoints. A negative slack indicates an arrival time that is later than its required arrival and, thus, needs to be addressed.

In early mode, the AT propagation uses minimum path delay calculations, as depicted in Figure 11.11. The RAT at the timing endpoint is based solely on the clock skew, clock jitter, and hold time constraint of the flop cell. For early mode timing analysis, the timing slack is the (AT − RAT) difference. If the AT is earlier than the RAT, the timing slack is negative, and a hold time error is reported. A similar backward calculation through the network using the "latest" required time through fan-out interconnect delays determines the RAT and the early mode slack at all cell output pins.

The advantage of the timing slack formulation is that it quickly identifies the cells internal to the network that have a large sensitivity to timing closure (see Figure 11.13).

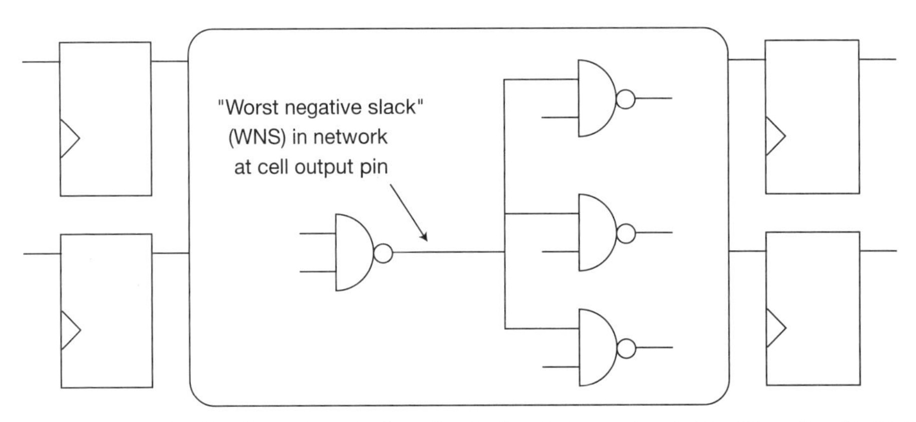


**Figure 11.13** Negative slack nodes in the timing network identify cells with high sensitivity to path timing closure.

Cells with a large negative slack are the primary candidates for (late mode) timing optimizations:

- Update the cell drive strength.
- Update the cell $V_t$ selection.
- Restructure the cell netlist to implement different fan-out repowering.
- Modify the wire layer/width/spacing for segments in the fan-out interconnect tree.

If the fan-out tree from a negative slack cell has a wide range of slacks, offloading non-critical fan-out may significantly improve the overall timing closure, as illustrated in Figure 11.14.

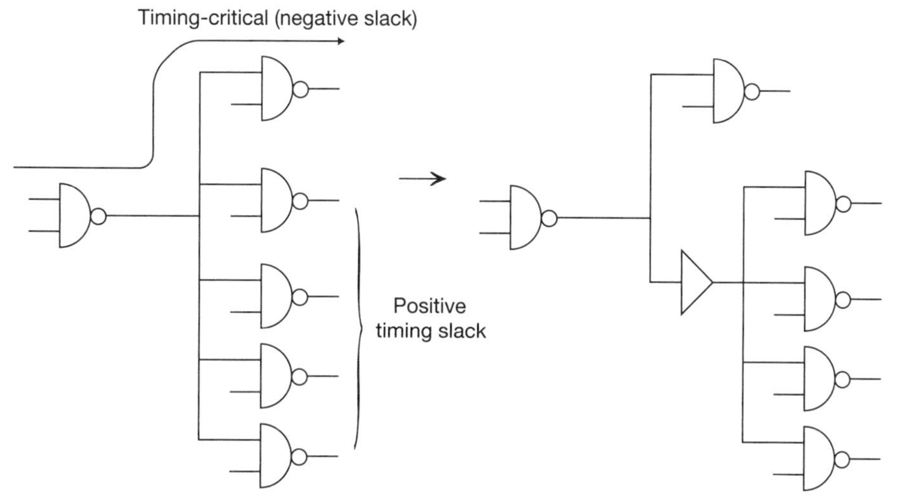


**Figure 11.14** Illustration of fan-out restructuring at a negative timing slack network node.

EDA vendor STA tools offer an interactive session feature, with access to the existing timing model and the results from the STA flow. In addition to allowing the querying of the timing results data, the interactive environment allows the designer to inject timing model changes, such as the candidate cell swaps listed above. An incremental timing feature can assess the impact of the model edits on the timing results; for fastest response, the scope of the timing recalculation is limited rather than striving for highest accuracy with a full model recalibration. The output of interactive timing debug would be to generate a set of netlist ECOs that could be applied to the physical implementation. Cell and/or fan-out repowering changes would be placed, cell overlaps resolved, and affected routes updated.

### 11.4.2 Pin Constraints for STA

The timing paths for tests involving model PIs and POs utilize constraint data similar to that provided for the synthesis flow (see Section 7.2). The distinction for STA is that the actual clock arrival data are used as the timing test reference rather than the idealized clock latency and skew targets used during synthesis.

### 11.4.3 STA "Don't Care" and "Adjust" Specifications

STA does not evaluate the logic functionality of the timing model (except the unateness) as part of the comprehensive approach to forward propagation of all potential network paths. As a result, there may be paths that are logically invalid. To exclude these paths, a directive to the STA flow indicates a timing don't care (also known as a *false path*). The STA false path constraint would indicate the connections in the levelized timing graph for which propagation is not performed (e.g., a specific path with "from-through-to" pins or a larger sub-graph with "from-to" pins). Reference 11.9 describes an algorithm for efficiently removing edges in the timing graph, given a false path specification. Note that it is necessary to update the timing graph for false paths after delay calculation; the loading of the false path pins is included in the cell and interconnect delays, but the (max, min) arrival propagation step omits the false path cells. The SoC methodology should ensure that any false paths submitted to timing analysis are independently verified to avoid exclusion of a timing path that is indeed exercisable. Formal property verification would be appropriate to prove the validity of the false path constraint.

Another input constraint to the STA flow would be the identification of any *multi-cycle paths* (i.e., paths between testpoints that are by design expected to exceed the cycle time period specification), as illustrated in Figure 11.15.

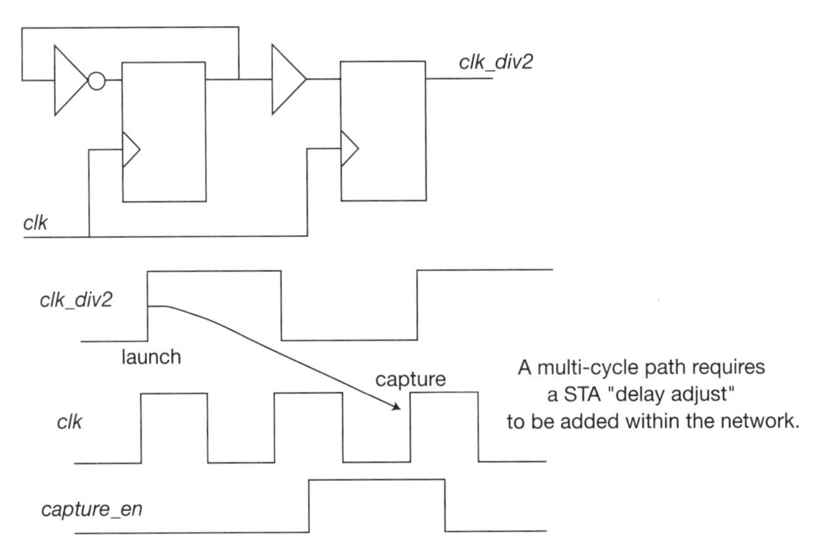


**Figure 11.15** A multi-cycle logic path requires the addition of a delay adjust constraint in the static timing analysis flow.

In the figure, correct functionality is maintained when the capture flop is updated every other clock cycle. To present this case to the propagation phase of the STA flow, a *delay adjust* constraint is given; essentially, an artificial cell is added to the timing graph in the combinational data network with a "negative propagation delay" value. To test a two-cycle path, a delay adjust equal to one clock period would be added. The location in the timing graph for the delay adjust needs to be selected judiciously, as sub-graphs in the timing model may be part of both single- and two-cycle paths. Similar to the false path constraint, any multi-cycle delay adjusts provided to the STA flow should also be independently verified.

### 11.4.4 Timing Analysis Modes

In general, STA is performed without evaluation of the logic functionality of the netlist model. However, in some cases, an operating condition for the timing model defines a specific set of timing tests to be performed. Section 7.2 first introduced the concept of multi-mode analysis, in the context of timing-driven synthesis. Similarly, the STA flow can support functional modes. Logical values are assigned to specific signals in a separate mode file, along with related clock period and PI/PO constraints. These assigned values are functionally simulated to establish specific timing model paths for propagation.

The most general cell timing models may include *state-dependent delays*, where an input-to-output pin arc has multiple representations, based on specific logic values assigned to other pins during characterization. If this modeling approach is used, a timing mode requires recalculating delays before propagation and slack calculation.

A common use of a timing mode is to evaluate the paths associated with scan-shifting during test operation, as shown in Figure 11.16. The shift clock frequency from the tester is typically much slower than the system clock, and with the relatively short scan logic paths, setup time checking should be straightforward. The emphasis in the test timing mode would be on hold time checking, using the skew and jitter for the test clock distribution to calculate the RAT at each flop endpoint.

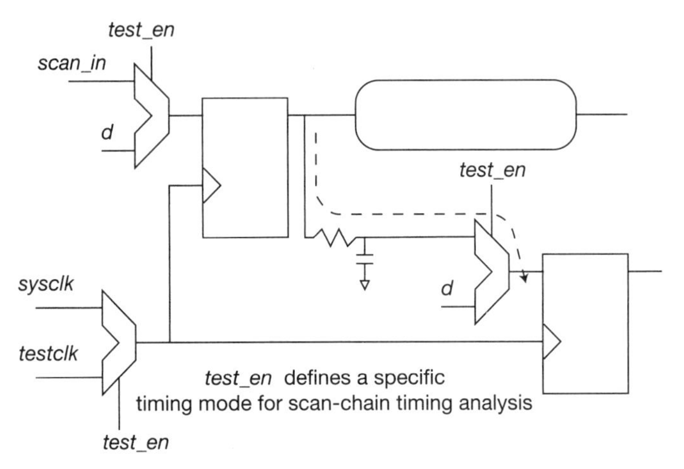


**Figure 11.16** A timing mode is included for scan shift paths.

Another common application of timing modes relates to unique timing tests associated with an embedded IP macro. The macro timing model from the IP vendor could include multiple sets of timing constraints, whose values depend upon the specific operation being performed. For example, an embedded SRAM array may have different address/data setup constraints to the array enable input for a read operation compared to a write access. The SoC timing team needs to develop the static timing analysis modes and timing model inputs to enable verification of the different operations.

### 11.4.5 Reporting

The output reports from the STA flow offer (extremely) detailed delay and slack data for all pins, for all corners and modes. The first debug step is to sanity check the results. For example, a negative slack (of significant magnitude) may be indicative of a multi-cycle adjust that may have been inadvertently omitted or an additional timing mode that needs to be defined. The next step would be to explore potential updates to the timing model, using the EDA vendor's interactive model viewer with incremental timing recalculation. From the updates to the model, an ECO netlist would be exported for physical implementation.

The design team may request a review of critical timing paths with the SoC methodology and CAD teams. There is potentially sufficient conservatism in the STA propagation method to warrant additional analysis of a failing path. A common feature of EDA vendor STA tools is to export a detailed netlist of a specified critical path—excising the cells, RC interconnect tress, side loads, and clock distribution. Note that the excised circuit simulation netlist requires attention to the k-factor applied to the coupling capacitances, which have been grounded for STA delay calculation. This netlist would be submitted for circuit simulation (at the corners of interest) to provide a more precise arrival time.

The results of the timing analysis flow are summarized in the project management scoreboard. A number of different results representations are commonly used—for example, the worst negative slack (WNS), the total negative slack summed for all failing paths (TNS), and a graphical distribution of all slack values (positive and negative) at timing test endpoints. The slack distribution is the most informative. For example, if a large number of failing paths are present, with small negative slack, the resources required to explore individual cell updates will be substantial. A review of other potential, more pervasive approaches would be warranted, such as a supply voltage increment, changes to the clock distribution to reduce the arrival skew, or modification of the design assumptions to adopt a useful skew implementation. As the tapeout schedule is likely fast approaching, the SoC and methodology teams need to be prepared to quickly address the possibility of a negative slack "timing wall" in the timing analysis path distribution curve. Finally, the review of the timing reports may result in waivers (due to assumed conservatism in timing margins), allowing the STA flow to be logged as "complete" in the SoC project management scoreboard.

### 11.4.6 Variation-Based Timing

The sources of manufacturing variation are more diverse in advanced process nodes, from interconnect RC tolerances due to CMP, to FinFET device fabrication parameter distributions, to layout-dependent effects. Parameter variations are derived from testsite data, measured across wafer lots, within each wafer, and within each die. As a result, the variations are typically associated with "global" and "local" distributions, as shown in Figure 11.17. The global and local n-sigma variation models used for library timing characterization are derived from measurements on device currents and extended to cell delay arcs. The figure depicts the statistical distribution of a cell delay arc at a specific voltage and temperature due to fabrication variations.

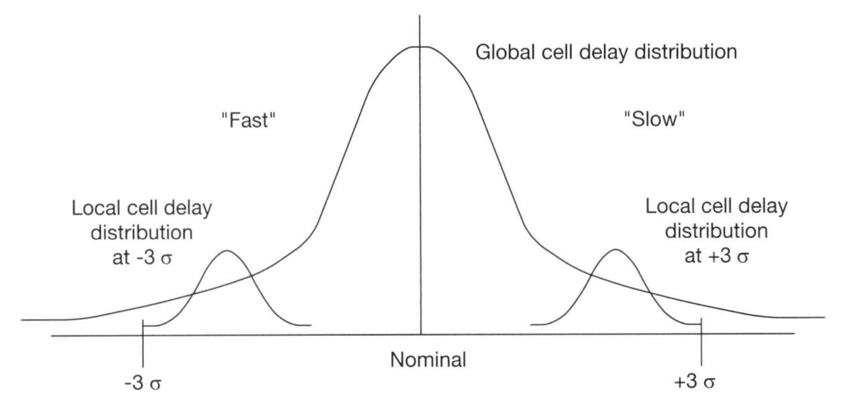


**Figure 11.17** Global and local device cell delay distributions due to fabrication variation.

Characterization is initially performed using a set of process parameters to reflect an n-sigma slow and an n-sigma fast delay from the global distribution (typically n = 3). For variation-based timing analysis, the local distributions are also incorporated. The global delay curve in the figure represents the full variation across the manufacturing process. (As an aside, the global curve drawn in the figure is Gaussian. At leading process nodes, the delay distribution is decidedly non-Gaussian; implications of timing variation with non-Gaussian distributions are discussed in Reference 11.10. Also, most characterization flows generate the nominal delay arc tables, using the typical process parameters, at the voltage and temperature associated with the corner.)

Superimposed on the global delay curve are local distributions, in which the n-sigma endpoints of the two curves are aligned. Local parameter distributions would be applicable to within-die circuits, where there would be a degree of process tracking of parameter values in close proximity. The figure illustrates the proposal that a global worst-case (or best-case) delay characterization approach applied to all cells in the timing model would be conservative. A *derating* of these global delays to represent the probabilistic nature of local delays is appropriate. An analogy for the pessimism in the use of global delay models for path timing for all cells would be to schedule a cross-country driving trip. The conservative calculation of being in "rush hour traffic" through each major city is very unlikely; the more cities on the route, the more unlikely this would be.

To enable SoC designers to compensate for timing pessimism associated with this global and local delay distribution model, the EDA vendor STA tool likely includes a feature to apply a derating multiplier. A derating factor could be applied (selectively) to any of the delays in the timing model:

- All delays
- Cell delays (to all logic cells or by library logic cell name or by logic cell delay arc)
- Interconnect delays
- Specific cell instances (differentiating clock cell derates from logic cell derates)
- Specific interconnect nets
- By levelized position in a timing path
- By relative location to other cells

Derating multipliers less than one (worst case, late mode) applied to logic cells in the timing model would be used to reduce expected pessimism in the global characterization, delay calculation, and propagation algorithms. Conversely, a multiplier greater than one would add conservatism to late mode logic paths. For early mode, a logic data path derating factor greater than one would reduce pessimism for hold time checks (although given the critical nature of satisfying hold time for functional operation, application of derates for a pessimistic hold assumption should be done judiciously).

A timing path example illustrating the use of cell derating factors is shown in Figure 11.18. The logic cells in the timing path would be given derate multipliers to reflect sampling from the local delay distribution. The cells in the clock path maybe given distinct derate values from the logic path. Also, note that levelized cells in the logic timing path would typically be assigned different derates, based on path position. For longer paths, the probabilistic sampling of variations will tend to result in an "average" cell delay closer to the local mean, for the sum of the cell delays in the path. Thus, the derate multiplier for a high-levelization-number cell is selected to provide a local delay for the cell further from the global characterization value to adjust the overall path delay accordingly.

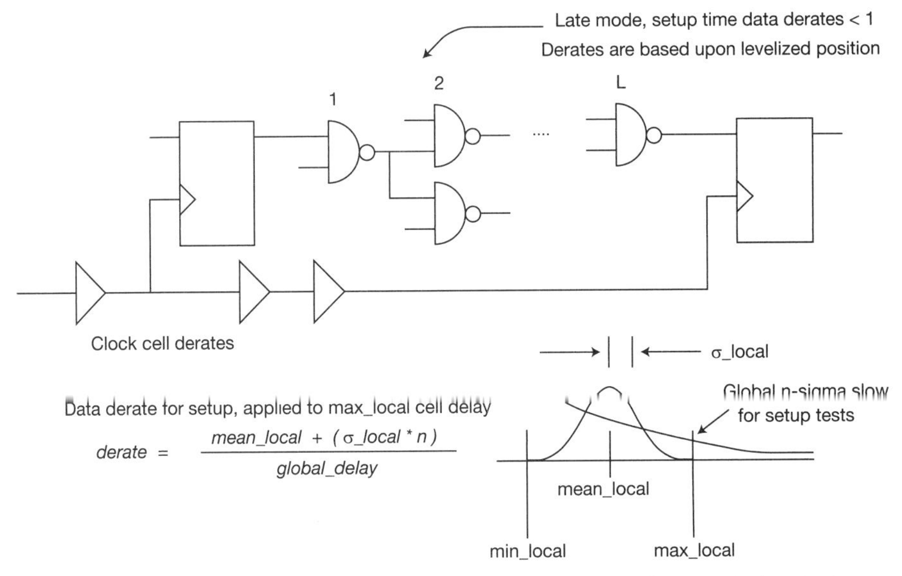


**Figure 11.18** Illustration of cell delay derating factors. A setup timing test example is depicted.

To implement a derating strategy to address pessimism in timing model delay calculation with global and local variation—or, for that matter, to address potential optimism—the SoC methodology team needs to review the statistical characterization approach used by the IP provider. The foundry releases PDK models with statistical distributions based on fabrication testsites. These distributions reflect the overall global device and interconnect parameter data, measured from lot to lot, from wafer to wafer, and across the wafer. Additional distributions represent the local variation data within die, where the $\sigma_{local}$ would be a function of the proximity of devices and interconnects on the testsite. The IP provider utilizes these statistical process distributions to establish the derate multipliers to help SoC designers address (late mode) pessimism in the STA flow. The IP provider releases *derate tables*, which are similar to NLDM tables, as illustrated in Figure 11.19.

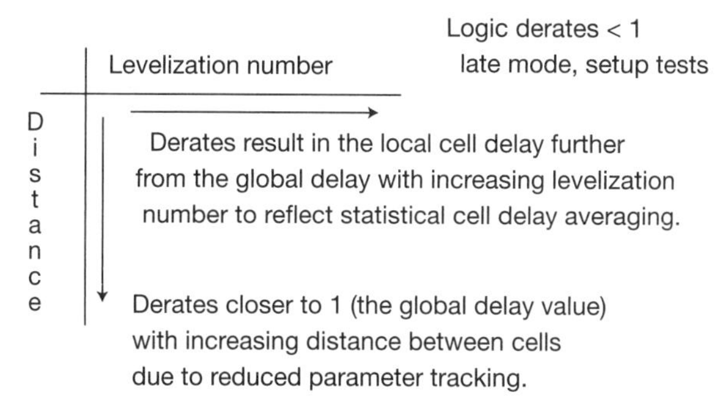


**Figure 11.19** Example of a cell derate table to address late mode timing path pessimism. Rather than all cells using the global *n*-sigma arc delay, a set of local derate multipliers are applied to the characterization table value.

In the figure, the input parameters for cell delay derate multipliers are the physical distance spanned by the timing path and the "levelization number" of the cell. For larger dimensions spanned by the path, the "tracking" of local cell delays is reduced; the derate multipliers are closer to 1, which is the global delay value. For higher levelization numbers, the cell derate multiplier results in a delay further from the global n-sigma maximum to better reflect the statistical averaging of multiple local cells in the path.

The potential number of derate tables (and additional characterization effort) is large; recall that NLDM delay and output slew tables are provided for each cell arc across a range of input slews and output loads. Derate table generation for the delay and output slew for each cell would require significant (statistical, Monte Carlo sampling) simulation resources to derive the local distributions. For efficiency, the IP provider is likely to consolidate the derate tables considerably (e.g., one derate table per cell, selected across all arcs, slews, and loads). The path length and location dependency factors in the derate calculation are also determined using Monte Carlo–sampled simulations of representative cells and interconnect layouts, applying the local distributions from the foundry for both device and interconnect variations.

Derate factor tables may also be released by the IP provider for scaling delays with respect to supply voltage and temperature environment settings that differ (slightly) from the PVT characterization corner settings.

A newer proposed variation timing approach uses a different formulation. The cell characterization data would provide an NLDM-like table with $\sigma_{local}$ values provided for delay arcs across input slew and output load ranges for each corner. A different statistical-propagation algorithm is now required within the STA tool, as opposed to the conventional max/min arrival time calculation, with scaled delays using derate multipliers. The timing test at the path endpoint also now becomes probabilistic, comparing clock and data distributions to establish a statistical confidence measure.[11]

It should be highlighted that the static timing methodology to represent increased process variation is still evolving. Timing analysis involves a complex interaction between PDK models, library cell characterization, and cell plus interconnect delay calculation. There are accuracy versus resource tradeoffs throughout:

- Circuit and interconnect extraction for various corners
- IP characterization, including the definition of the setup/hold timing tests
- Coupling capacitance modeling (e.g., the *k*-factors used in interconnect delay calculation)
- $C_{eff}$ calculation
- Cell delay calculation (nominally measured as the 50 percent-to-50 percent input-to-output waveform crossing delay, with interpolation between the NLDM table entries)
- Cell output waveform definition and interconnect delay calculation
- Interconnect delay/slew calculation at cell fan-out pins
- Local supply/ground (dynamic) voltage drop
- Local temperature differences (on device and interconnect behavior)

With all these variables, the foundry, IP library provider, and EDA tool vendor are striving to define modeling and timing analysis methods that also adequately reflect process variation, without excessive pessimism (or, worse, optimism). The SoC methodology team must assess which variation-aware timing approaches are appropriate for their resource budget and schedule, relative to the design performance targets.

### 11.4.7 Delay-Based Timing Verification

The application of STA to VLSI methodologies has significantly diminished the use of simulation testcase-based delay testing. The comprehensive nature of STA evaluating all paths in the levelized network graph has effectively covered the need for functional simulation of the cell netlist with instance-specific delays. There are two applications where delay-based simulation may offer additional insights to the SoC design team:

- The members of the test engineering team may wish to simulate specific test patterns to ensure that the chip pin input stimulus time and output pin strobe capture settings relative to the tester clock are valid.
- The functional switching activity with instance-specific delays provides more accurate data on cell-level power dissipation.

Power estimation at the RTL level uses cycle-based simulation event traces on RTL signals and registers to gauge the switching activity. These estimates are extremely useful for relative power optimization decisions prior to physical design. The use of representative simulation testcases with a cell netlist-level model (with only delta delays) offers additional insight. Even if the cell instance delays are zero, more accurate functional switching activity factors are provided compared to the RTL estimates. However, if a delay-based netlist model is used, the results also provide insights into the *glitch power* due to combinational logic toggles within a cycle, as depicted in Figure 11.20.

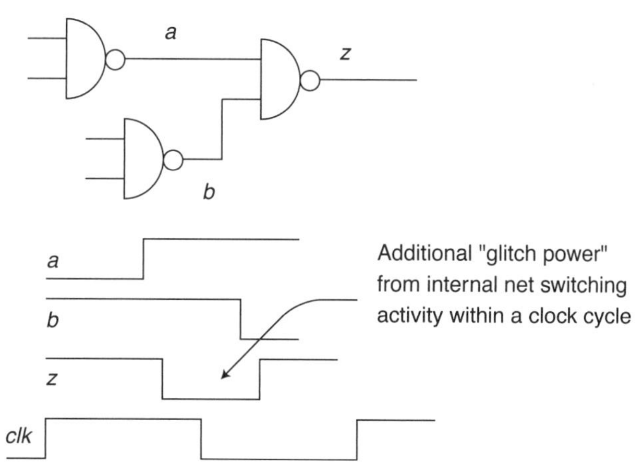


**Figure 11.20** Example of cell power dissipation due to a glitch transition within a clock cycle. This transition is not accounted for by static timing analysis, nor is it detected by (levelized) netlist simulation with zero-delay arcs.

There are several intricacies to enabling delay-based simulation, for both delay annotation and timing tests. The IP library HDL models need to incorporate the appropriate timing tests at the PVT corner corresponding to the power dissipation calculation. For complex macros and cores from IP providers, the internal event evaluation detail and checking in the functional model is likely to be quite limited. The SoC methodology team needs to review what delay-based model support is available for the IP and whether that meets the simulation testbench requirements. The STA tool needs to export the cell and interconnect delays in a format that can be annotated to the netlist instances. The EDA industry has established a de facto standard for netlist delay annotation; the Standard Delay Format (SDF) defines how timing model data is to be written by the STA tool and the interpretation of this file by an HDL simulator that supports annotation.[12] This cell delay format includes the capability to define a range of (min, typ, max) transition delay values, assigned to cell pins; the simulations need to select the delay value (or range of values) appropriate for the switching activity of interest.

### *Transport and Inertial Delay*

The SoC methodology team needs to review the HDL simulator settings to be applied when events on the simulator queue are subject to preemption. Specifically, a new event may be associated with *inertial* or *transport* signal delay, as illustrated in Figure 11.21.

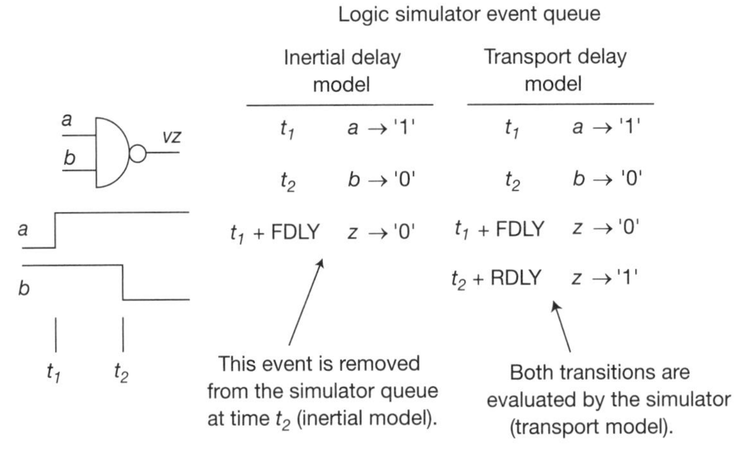


**Figure 11.21** The distinction between transport and inertial delay for an HDL simulator is depicted for delay-based simulation.

An inertial delay property implies that a pending signal update on the simulator event queue would be preempted by a subsequent event of different value; the addition of the subsequent assignment to the queue includes removal of the pending event. The "inertial" response time of the signal would preclude the interim transition. A transport delay property is equivalent to a signal with infinite bandwidth; the new signal value assignment is added to the queue after the pending event, and both transitions are evaluated. A transport delay representation offers insight into the glitch switching activity. (There is also the unlikely situation in which a subsequent signal update would occur in time *before* pending events on the queue; in this case, all future events scheduled for times later than the most recently posted update would be preempted, regardless of the inertial or transport signal property.)

Delay-based simulation for timing verification has effectively been displaced by STA and is focused on specific power and test applications. The adoption of STA modeling based on cell arc and interconnect delays still allows timing model annotation to an HDL-based netlist, using an external delay file format. Delay-based simulation is complicated by the diverse sources of IP models and the focus by IP providers on representing timing tests in the timing model abstract, rather than the functional HDL.

## 11.5 Summary

Of the three primary PPA goals associated with an SoC design project, performance is commonly the most critical. Market expectations for a new product are usually based on the relative performance compared to existing offerings. SoC architects and physical implementation engineers focus on design optimizations to achieve timing goals—yet, the final sign-off decision is based on static timing analysis flow results.

Ideally, the measured silicon performance closely correlates to the critical paths identified by the STA flow. During silicon prototype evaluation, experiments are pursued to ascertain which paths fail as the clock frequency is increased beyond the product specifications. Hopefully, the STA slack distribution provides an accurate prediction of the silicon data for the passing clock period and the failing paths when the clock period is reduced. However, the accuracy of the static timing model remains a major focus area.

The methodology for STA strives to identify steps where delay calculation and/or path arrival propagation is conservative. The impact of a conservative timing model is twofold. Engineering resources are expended to "fix" negative slack paths that are indeed adequate, and the power/area targets could likely have been further optimized for a timing model with more accurate measures. As a result, there is significant methodology development to improve STA accuracy in collaboration with the foundry, IP library providers, and EDA vendors. Cell characterization models are incorporating more accurate output waveform detail for interconnect delay calculation. Derating strategies are being more widely adopted to better reflect an n-sigma timing result encompassing a complete levelized path, rather than for each individual cell in the path. Alternatives to the propagation of the latest output arrival and slowest output slew rate from all input pin arrival data are being used to further reduce the conservatism (refer to the "path-based" propagation method highlighted in the upcoming "Future Research" section).

As briefly mentioned in this chapter, considerable STA methodology development activity is ongoing to effectively represent PVT variations as a statistical distribution of individual cell and interconnect delays for the SoC design, and efficiently calculate the statistical arrival time through the levelized network. The result of statistical STA is a confidence level in achieving the overall performance target, rather than a slack value at each timing endpoint.

The methodology for STA with the optimum trade-offs between delay model plus propagation accuracy, flow throughput, and correlation to silicon data continues to evolve.

## References

[1] Macys, R., and McCormick, S., "A New Algorithm for Computing the Effective Capacitance in Deep Sub-micron Circuits," *Proceedings of the IEEE Custom Integrated Circuits Conference (CICC)*, 1998, pp. 313–316.

[2] Qian., J., Pullela, S., and Pileggi, L.T., "Modeling the Effective Capacitance for the RC Interconnect of CMOS Gates," *IEEE Transactions on Computer-Aided Design of Integrated Circuits and Systems*, Volume 13, Issue 12, December 1994, pp. 1526–1535.

[3] Elmore, W.C., "The Transient Analysis of Damped Linear Networks with Particular Regard to Wideband Amplifiers," *Journal of Applied Physics*, Volume 19, Issue 1, 1948.

[4] Rubinstein, J., Penfield Jr., P., and Horowitz M.A., "Signal Delay in RC Tree Networks," *IEEE Transactions on Computer-Aided Design*, Volume CAD-2, Issue 3, July 1983, pp. 201–211.

[5] Gupta, R., Tutuianu, B., and Pileggi, L.T., "The Elmore Delay as a Bound for RC Trees with Generalized Input Signals," *IEEE Transactions on Computer-Aided Design of Integrated Circuits and Systems*, Volume 16, Issue 1, 1997, pp. 95–104.

[6] Alpert, C.J., et al., "Delay and Slew Metrics Using the Lognormal Distribution," *Proceedings of the 40th IEEE Design Automation Conference (DAC)*, 2003, pp. 382–385.

[7] Kar, R., et al., "Delay Estimation for On-Chip VLSI Interconnect Using Weibull Distribution Function,"*2008 IEEE Region 10 and the 3rd International IEEE Conference on Industrial and Information Systems*, 2008, pp. 1–3.

[8] Alpert, C.J., et al., "Closed-Form Delay and Slew Metrics Made Easy," *IEEE Transactions on Computer-Aided Design of Integrated Circuits and Systems*, Volume 23, Issue 12, December 2004, pp. 1661–1669.

[9] Blaauw, D., Panda, R., and Das, A., "Removing User-Specified False Paths from Timing Graphs," *Proceedings of the 37th IEEE Design Automation Conference (DAC)*, 2000, pp. 270–273.

[10] Keller, I., and Ghanta, P., "Importance of Modeling Non-Gaussianities in STA in sub-16nm Nodes," TAU Workshop, 2016. (Presentation slides available at http://www.tauworkshop.com/2016/slides/10_TAU2016_Ghanta_nonGaussian_POCV.pdf.)

[11] A brief description of the Liberty Variance Format (LVF) is provided by Bautz, B., and Lakanadham, S., "A Slew/Load-Dependent Approach to Single-variable Statistical Delay Modeling," TAU Workshop, 2014. (Presentation slides available athttp://www.tauworkshop.com/2014/Slides/Bautz_SOCV_TAU_2014.pdf.)

[12] IEEE 61523-3-2004: "IEEE Delay and Power Calculation Standards—Part 3: Standard Delay Format (SDF) for the Electronic Design Process," https://ieeexplore.ieee.org/document/7386825/.

## Further Research

### Graph-Based Versus Path-Based Delay Analysis

The discussion in this chapter highlights a conservative methodology for static timing propagation at each cell; specifically, the latest input-to-output arc arrival time and the slowest output signal slew among all arcs are selected for (late mode) propagation. This approach is referred to as "graph-based" STA.

Describe the alternative "path-based" algorithm, with either NLDM or waveform tables.

Describe the runtime versus accuracy trade-offs between graph-based and path-based analysis—specifically, when and where path-based analysis is appropriate.

### False Path Verification

Describe a methodology flow for independent verification of STA false path constraints.

### STA Results to Be Recorded in the SoC Project Scoreboard

Describe the key information from the STA flow report to record in the SoC project methodology manager. Data to consider include the following:

- Flow input timing constraints, corners, and modes
- Assigned clock arrivals and skews (e.g., from a separate clock simulation flow)
- Derate assumptions used for local n-sigma delay scaling
- TNS and WNS, for both early mode and late mode
- Slack distributions
- Slacks for PI-to-endpoint and endpoint-to-PO paths (for budgeting review; see below)
- Specific from-through-to failing paths (especially paths through hard IP macros, which may necessitate significant physical design updates)
- Paths that have been excised and submitted to circuit simulation

- Waivers (granted after a review of a path-based timing analysis for a from-through-to path or from detailed circuit simulation results)

## Block-Level Timing Constraint Budgeting

### *Budgeting Methodology*

A key methodology policy is how (and how often) to provide block-level pin timing arrival constraints, a step typically referred to as "budgeting" of the overall clock cycle time. A path spanning two blocks requires assigning driving output pin required arrival time and receiving input pin expected arrival time constraints, with an allocation of the global path delay interval. A multiple fan-out global net adds to the timing budget model detail, as global path delays may differ significantly to individual fan-out pins. In addition, timing constraint values are required for each timing corner and mode in support of MCMM static timing analysis. As individual block design teams exercise STA, local optimizations are evaluated to attempt to satisfy PI-to-endpoint and endpoint-to-PO failing paths. However, it may become evident that iterations on the budgeted timing constraints are required.

Describe a budgeting methodology for development and iteration on STA pin constraint values. Also assess when activity on block-level timing optimization with existing timing constraints should be concluded and when a project-wide iteration on timing budgets is appropriate. In addition, assess when a change to the global interconnect buffering design would be a preferred alternative.

### *Registered Input and Output Pins*

To simplify timing constraint budgeting, vendor IP designs may include "registered" inputs and outputs, without significant logic path depth between the IP pins and timing path endpoints—perhaps just a flip-flop and buffer on an output pin and flip-flops on input pins. SoC block designers may also seek to architecturally add registers at block pins.

Describe the advantages and disadvantages of adding registered block pins relative to:

- The timing budgeting methodology
- Local block timing optimizations
- The impact on global interconnect design
- Overall SoC PPA targets

### *Global Repeater Insertion*

A critical SoC design decision is whether to add a global repeater flip-flop to a signal between blocks. Modifications to global signal buffering topologies are an easier implementation decision than insertion of flop repeaters (assuming that the channel area is available). The decision to insert a repeater is especially disruptive if an issue arises later in the SoC project schedule. Ideally, paths requiring repeaters would be identified during initial global floorplanning; the area allocation, power and clock delivery, and SoC performance model would all reflect this decision. However, issues that arise during the physical implementation phase may require a design review.

Describe the budgeting methodology and STA results criteria that would necessitate a design review of global paths that were not previously identified as requiring a sequential repeater. Describe the various SoC design teams that need to participate in this review. If the decision is indeed made to add repeaters, describe the impact on SoC architectural, logical, and physical implementation models; also describe the impact on the functional validation testbenches.

### Delay-Based Functional Simulation (Advanced)

Describe the features of an SDF file used to represent the cell and interconnect delay calculation data of the STA flow (for a specific MCMM setting).

Describe the required features of the HDL model for each library cell to support annotation of the SDF information for delay-based simulation. Illustrate these HDL features using either Verilog or VHDL cell model examples.

Describe how specific timing tests at flip-flops and hard IP macros would be represented in the HDL models in support of delay-based simulation.

CHAPTER 12

# Noise Analysis

## 12.1 Introduction to Noise Analysis

With the increasing percentage of coupling capacitance to total load capacitance on nets in submicron VLSI process nodes, it is necessary to assess the potential impact of coupled transients on each interconnect to ensure that robust circuit behavior is maintained. Noise analysis is addressed in three different methodology flows:

- **Noise impact on delay, integrated with static timing analysis**—A capacitive coupling event to a net in transition alters the delay and slew at the net fan-outs.
- **Static noise analysis (a sign-off flow)**—A capacitive coupling event to a quiescent net results in current injected on the net; this current flows back to the active driver and to the net fan-outs. The result is a noise pulse on the fan-out receivers. Static noise analysis estimates the magnitude and duration of these receiver input pulses and evaluates the susceptibility of the logic network to an error.
- **Noise analysis suitable for physical implementation**—To minimize the number of noise-related issues reported by the static timing analysis and static noise analysis flows, a number of algorithms have been incorporated into

interconnect routing, ranging from simple *parallel run length* limits for net segments on adjacent tracks to a comparison of the estimated noise pulse at cell inputs against pin limits in the characterization library.

This chapter focuses on the impact of noise on delay and the tapeout sign-off static noise analysis flow. When discussing the noise analysis flow, modeling alternatives that would be suitable for incorporation into routing optimizations are highlighted.

## 12.2 Static Noise Analysis, Part I

The static noise analysis flow utilizes the extracted RC interconnect network, with suitable electrical models for the drivers and receivers on each net, as illustrated in Figure 12.1.

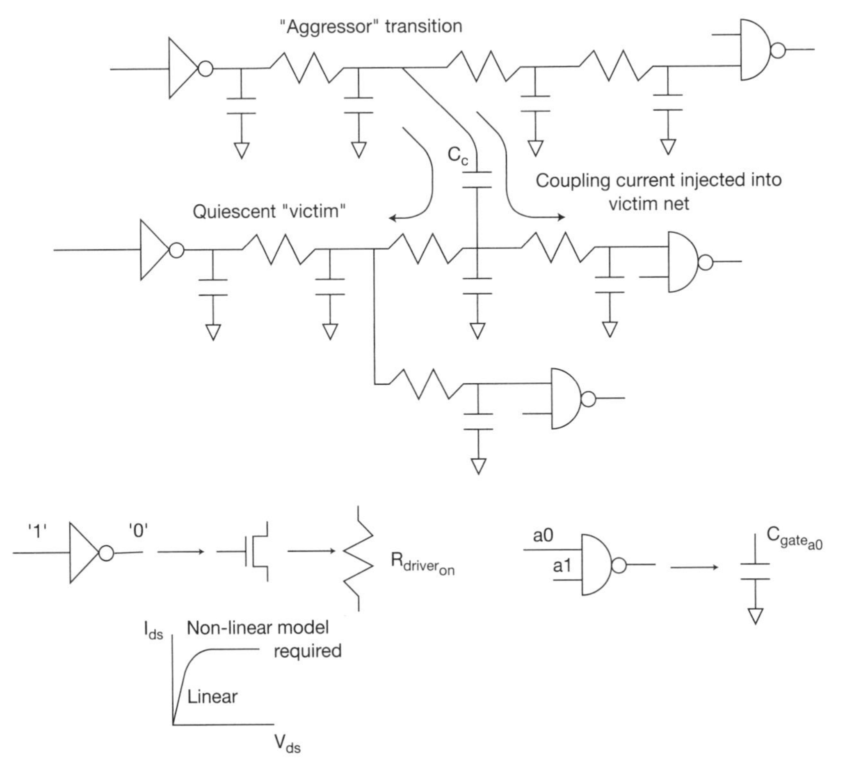


**Figure 12.1** A general RC network noise model. A single aggressor/victim coupled pair is analyzed, and superposition is applied from different aggressors to determine the coupled transient on the victim. Superposition is applicable because the (reduced) model for noise analysis is linear (see Section 12.5).

The static noise analysis flow sequences through each levelized net in the model, treating the net as the *victim*. The set of coupling capacitors on the net being analyzed is considered the *aggressors*.

Technically, a transient from an aggressor would also result in coupled current through series capacitances to other victim nets, as depicted in Figure 12.2. For simplicity, this transitive coupling current between more than two nets is not analyzed by EDA noise tools.

The total current injected to the victim is typically evaluated as the superposition of "direct" aggressor events, and the other aggressor coupling capacitances are grounded. There are certainly net topologies where the direct aggressor-only approach could underestimate the injected noise, necessitating some margining.

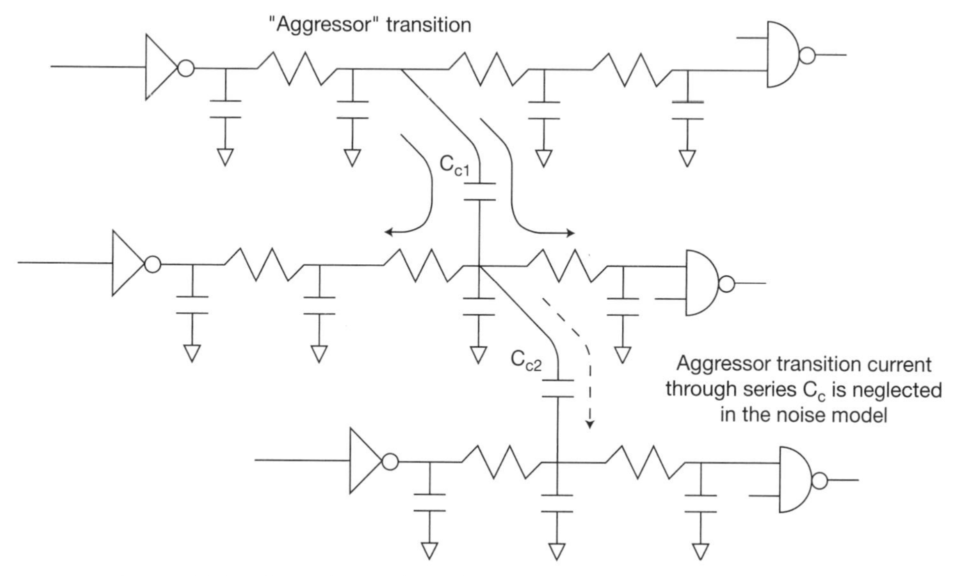


**Figure 12.2** EDA noise tools analyze "direct" aggressor coupling capacitance transients. Coupling requiring a model with more than two nets is not used, potentially necessitating noise margining.

The complexity of the net analysis is dependent upon whether a linear or non-linear model for the active driver is used; a linear model is more common, as it enables very efficient simulation of the victim noise. As illustrated in Figure 12.1, an active device connected to the quiescent victim net is operating in the linear mode. For a complex logic cell, the (highest, conservative) linear driver resistance is a composite of multiple active devices in series. One option would be to derive the maximum $R_{on}$ (for both ‘0’ and ‘1’ output values) during cell characterization by simulation with judicious selection of devices in parallel-series stacks. Another option would be to estimate $R_{on}$ during static noise analysis from the output timing waveforms in the cell characterization library data.

The static noise flow provides output data on the (estimated, worst-case) coupled noise and pass/fail results on the network stability. The detailed flow output reports list the set of aggressor nets to a victim, with a ranking of their contribution to the noise on the victim fan-outs. Such a detailed report provides insights into potential physical design changes to address noise issues:

- **Altering the netlist by increasing the drive strength of the victim net cell, inserting a buffer into the victim net, or reducing the drive strength of aggressor cells**—Note that these netlist modifications have varying degrees of influence; for example, increasing the victim cell drive strength would be relatively ineffective for strong aggressors at the far end of the net.
- **Changing the routing topology of the victim net and reducing the coupling capacitance from major aggressors**—Prior to pursuing any physical ECOs, the SoC designer should ensure that the set of aggressors is appropriate. Section 12.4 discusses the motivations and methods for filtering coupling capacitances from the noise analysis.

The initial EDA implementations of a static noise analysis tool focused on the behavior of individual fan-out cells. Input pin limits were established during cell characterization based on the ($v_{out}$, $v_{in}$) DC transfer curve, as depicted in Figure 12.3.

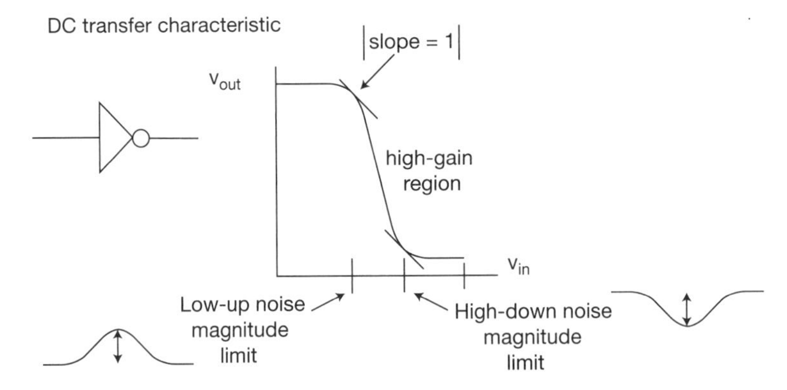


**Figure 12.3** Illustration of a cell input pin noise model based on the DC transfer curve.

If the noise pulse magnitude at the input pin exceeded the characterization limit and the logic cell entered the high-gain portion of the transfer curve, a fail was recorded. As the fraction of total net capacitance due to coupling capacitances continued to increase with process scaling, this input pin threshold limit approach was enhanced to reflect the dependency of the cell's response on both the magnitude and duration of the noise pulse input. Cell characterization was expanded to represent a dynamic *noise immunity curve* (*NIC*) for each input, as introduced in Figure 10.20. Noise inputs of magnitude greater than the DC transfer curve limits were acceptable (i.e., below the NIC limit) if the pulse was of short duration. Characterization modeled the noise input as a symmetric triangular pulse, defined by its peak magnitude and duration. With further process scaling, the number of individual cell noise violations increased, necessitating further model refinement.

The prevalent static noise analysis methodology is focused on the propagation of a noise pulse through a logic network to a flop, where the propagated pulse magnitude, duration, and arrival *timing window* are evaluated against the flop data input pin noise characterization limit and the clock arrival time, as illustrated in Figure 12.4.

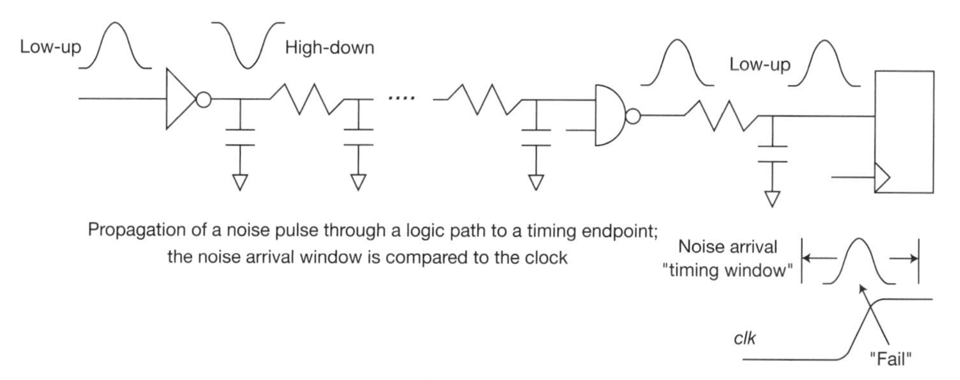


**Figure 12.4** Propagation of a noise transient to a timing path endpoint. The noise propagation includes a path delay calculation to estimate a noise arrival "timing window." If the magnitude, duration, and arrival time of the flop input noise transient may cause an incorrect value to be stored in the flop, a noise error is reported.

A propagated noise pulse of sufficient magnitude that could arrive at the flop data input within an interval associated with the capturing clock would be recorded as a static noise failure. In this case, the correct flip-flop state could be perturbed for a logically unchanged data input. The timing window of interest for the incident noise pulse at the flop input includes the clock jitter and the flop setup and hold time test intervals.

It should be mentioned that noise failures are an increased risk on circuit topologies that rely on a single transient at each logic gate input (e.g., domino circuits). This discussion assumes (fully complementary) CMOS logic circuits, with active driver devices able to restore the output voltage as the coupled event decays. Also, logic circuits that present an unbuffered device source/drain node at the cell input pin add significant complexity to the noise analysis model (see Figure 12.5); these circuit topologies are typically excluded from IP libraries. Data transfer gates are confined to internal circuit nodes with buffered inputs and very stringent noise characterization pin limits.

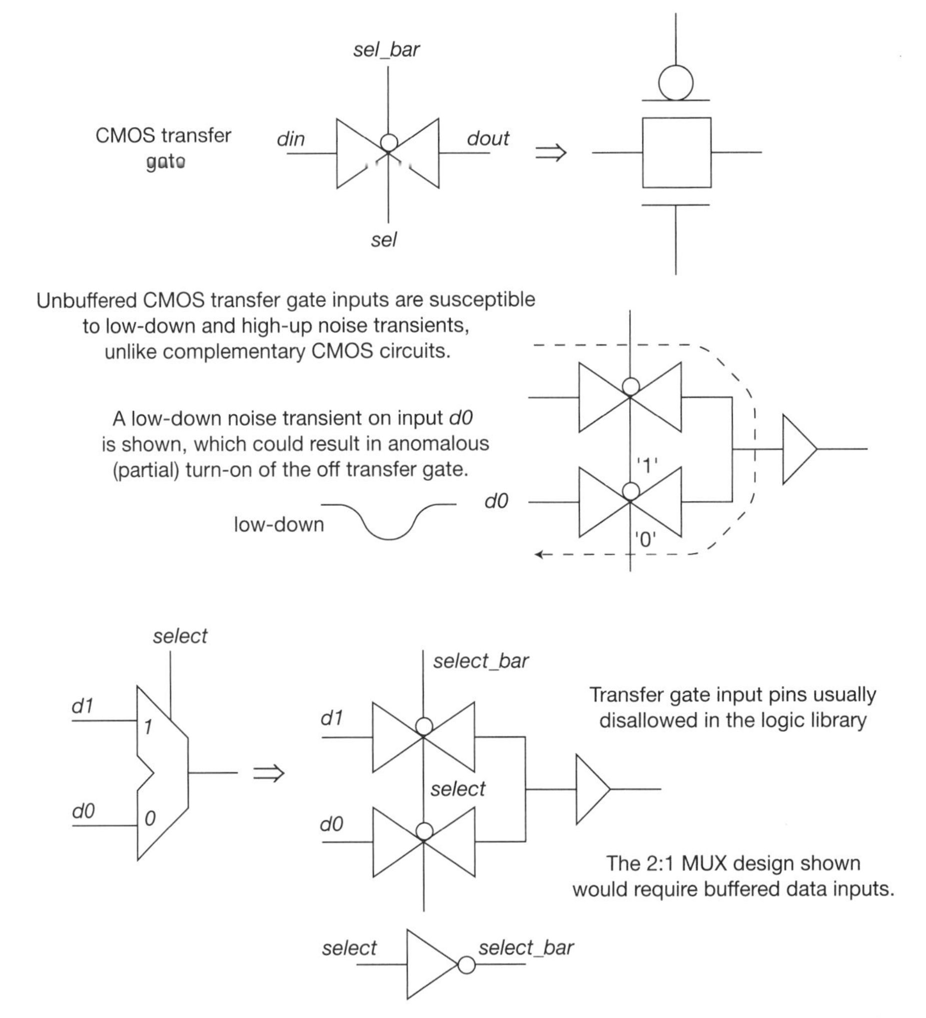


**Figure 12.5** CMOS transfer gates are not commonly provided as logic cell inputs, as their input pin noise model is not strictly capacitive and is susceptible to the low-down and high-up transients atypical of other CMOS circuits.

## 12.3 Noise Impact on Delay

A coupling event from an aggressor while the victim net is in transition between logic values results in a perturbation to the transient waveform and a change to the arrival time and waveform slew at the victim fan-outs. The impact to the delay and slew propagate further in the logic network. Figure 12.6 illustrates a simple topology.

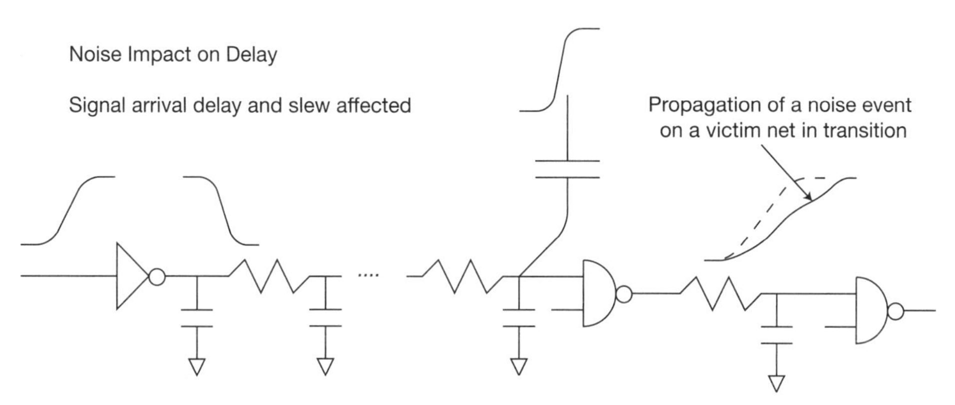


**Figure 12.6** Illustration of the impact of aggressor noise during a signal transition, propagated in the logic network.

The noise impact on delay can be modeled using an equivalent lumped capacitor replacement for the aggressor coupling to enable delay calculation with the RC tree network model described in Section 11.2.

A k-factor multiplier is applied when the coupling capacitance is split between the two nets. The k-factor is dependent on the slew rates of the overlapping victim and aggressor signals, as illustrated in Figure 12.7.

As this slew rate information is not available prior to running the STA flow, initial estimates are made on the k-factor for delay. The initial assumption would be that the aggressor signal would be active when the victim is switching. The ratio of the aggressor and victim signal slews in the k-factor equation could incorporate the relative drive strengths of the victim and aggressor circuits and their capacitive loads. The calculated k-factor value is applied for the first STA timing model evaluation. The k-factor value may be modified during STA, as described next.

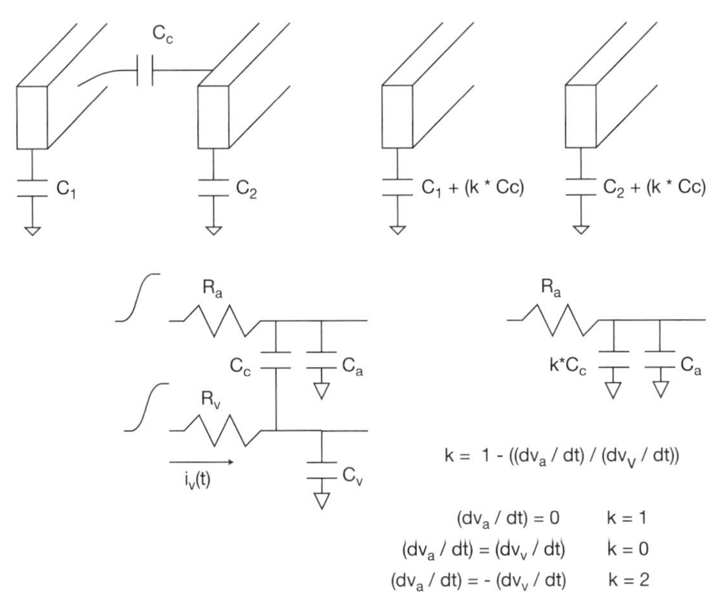


**Figure 12.7** A multiplicative k-factor is derived for the coupling capacitance as part of the analysis of the noise impact on delay. The coupling capacitor is divided between the two nets for interconnect delay calculation.

### 12.3.1 Static Timing Analysis with Overlapping Victim and Aggressor Timing Windows

The static timing analysis flow typically calculates a single arrival time at a cell output pin during the forward propagation of cell and interconnect delays. As part of that calculation, the STA tool has selected from the set of cell input arrival times. Alternatively, the static timing analysis arrival time could be represented as a *timing window* for the cell output pin and propagated to the fan-out cell input pins, as illustrated in Figure 12.8.

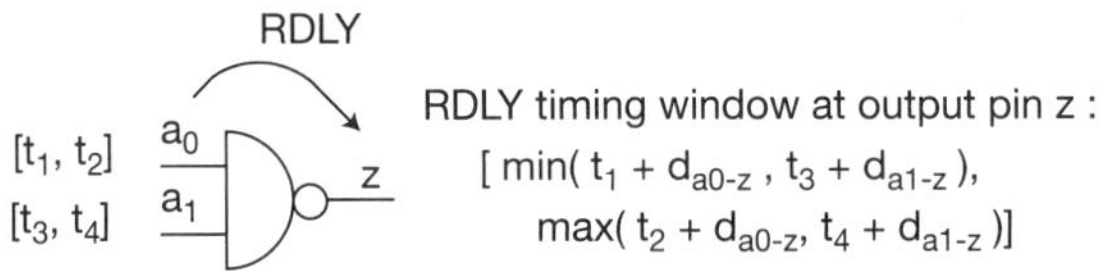

RDLY

$[t_1, t_2]$ $a_0$, $[t_3, t_4]$ $a_1$, z

RDLY timing window at output pin z :

$$[\min(t_1 + d_{a0\text{-}z},\ t_3 + d_{a1\text{-}z}),\ \max(t_2 + d_{a0\text{-}z},\ t_4 + d_{a1\text{-}z})]$$

**Figure 12.8** Static timing analysis would propagate a timing window at the cell output pin for RDLY and FDLY to determine the overlap of the signal transition to the timing windows of aggressor nets for k-factor calculation. If the aggressor could transition in the opposite direction in the same time window, a k-factor multiplier greater than 1 would be appropriate when splitting the coupling capacitance for late mode timing analysis. If the aggressor could transition in the same direction in the same time window, a k-factor less than 1 would be appropriate for early mode.

The timing window range would be maintained for both the RDLY and FDLY output transitions. The figure depicts a late arrival propagation; a similar range calculation would be used for the earliest arrival propagation for the hold timing window. The timing windows would be evaluated to modify the k-factor calculation. For non-overlapping victim and aggressor net timing windows, the k-factor would be reset to 1. If the aggressor RDLY (or FDLY) timing window overlaps the victim FDLY (or RDLY) timing window in late mode, a k-factor greater than 1 would be appropriate for the opposing transients. For early mode, if the aggressor RDLY (or FDLY) timing window overlaps the victim RDLY (or FDLY) timing window for the same signal transition, a k-factor multiplier much less than 1 would be selected to decrease the interconnect delay.

As a result, the STA flow with noise impacts on delay requires iteration. Conservative assumptions on k-factors would be applied during the first pass. The result of this initial evaluation would provide timing window and slew data, which would be used to refine the k-factor values. Subsequent timing analysis iterations would be pursued and new k-factors calculated until the resulting delay differences were below a convergence threshold.

The STA flow output timing report with noise detail would ideally be able to provide the interconnect delay difference between using k = 1 and the overlapping window k-factor calculation for insights on how to most efficiently implement physical design updates to address timing test failures.

Note that the victim signal transient waveform with aggressor coupling may no longer be well-defined by a single slew value and, thus, may no longer align closely with a specific cumulative density function (CDF) used by an STA interconnect delay algorithm (see Section 11.2). Thus, the k-factor approach introduces an additional source of error in the calculated interconnect delay. Section 12.5 discusses a circuit simulation method for static noise analysis, using a reduced RC network for the victim and aggressors. A similar approach could be used for static timing analysis with noise, replacing an analytic model approximation for the signal slew at the fan-out pin with simulation of the net; this is more computationally intensive but more accurate.

### 12.3.2 I*R Rail Voltage Drop Effect on Timing and Noise

Dynamic I*R voltage drop on supply and ground rails is another source of noise that impacts timing, as depicted in Figure 12.9. The influence of the rail noise on circuit delay is margined by characterization assumptions on VDD and GND for each cell and verified by the power I*R analysis flow (see Chapter 14, "Power Rail Voltage Drop Analysis"). Traditionally, the magnitude of the dynamic power/ground noise through active devices to the victim output is neglected for interconnect noise propagation.

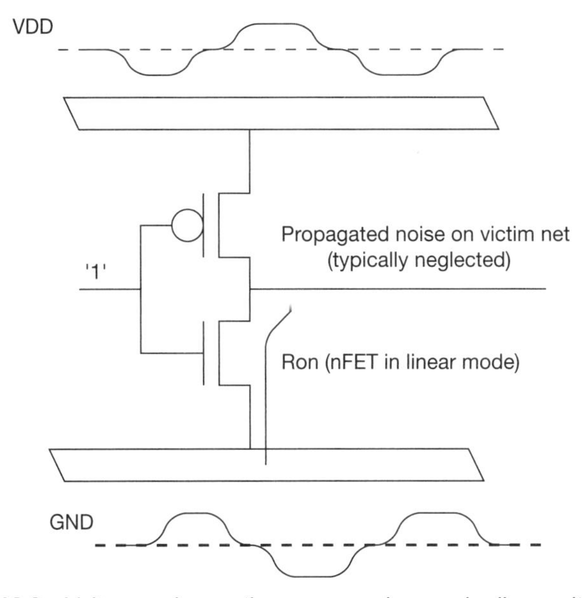


**Figure 12.9** Voltage noise on the power and ground rails results in cell circuit delay variations. This effect is reflected by margins in cell characterization.

More recently, SoC designs are aggressively scaling the supply voltage for power optimization. The typical assumptions used during characterization for the supply and ground voltages at the cell are based on a percentage of the supply as the budget for I*R voltage drop through the power and ground grids to the cell. For SoC designs operating at a low VDD voltage, closing the design to these I*R voltage drop limits is more difficult to achieve. As a result, two methodology updates are being pursued. EDA vendors are developing methods to include dynamic voltage rail noise in cell and interconnect delay calculations (using existing cell characterization data) for STA and noise analysis tools. In addition, the SoC methodology team is enabling dynamic I*R results from the power rail analysis flow that exceed the characterization assumptions to be waived if the STA and noise analysis flows do not flag errors.

## 12.4 Electrical Models for Static Noise Analysis

The victim net electrical model in Figure 12.1 is typically simplified prior to evaluating the propagated noise to the fan-outs. Optimizations to the detailed electrical model span functional, temporal, and relative magnitude considerations. Before discussing noise model optimizations, a set of *noise direction* definitions are needed.

### 12.4.1 Models for Noise Direction

As introduced in Section 10.2, there are four noise analysis victim models to evaluate for each extracted corner: (1) victim '0', coupled up; (2) victim '1', coupled down; (3) victim '0', coupled down; and (4) victim '1', coupled up. The latter two may need more sophisticated driver electrical models to include the additional restoring currents of the forward bias substrate and well junctions. If the cell IP library does not include unbuffered input pins presenting a device source/drain node, the latter two noise transients are unlikely to result in a functional failure. However, evaluating these noise directions is still important because they represent a reliability issue due to the increase in the device gate electric field above the normal PVT environment limits.

### 12.4.2 Small Aggressor Consideration

The extracted parasitics for a net usually include a limited number of dominant coupling capacitances and a potentially large number of smaller coupling capacitances. A common optimization is to filter the small (distributed) coupling capacitances to reduce the complexity of the noise model. One approach would be to compare the magnitude of the coupling capacitance to the ground capacitance at a node in the extracted interconnect network and, likely, ground small coupling elements below a threshold ratio. Another approach would be to aggregate small distributed capacitive elements into a single "virtual" coupling capacitor and then select an appropriate node to add this virtual capacitor in the RC interconnect tree.

### 12.4.3 Functional Exclusions

As with static timing analysis, static noise analysis by default does not incorporate logic functionality into the noise event introduction and propagation. In a manner similar to the false-path constraints provided to STA, a set of logical relationships can be provided to the static noise analysis flow to exclude a subset of coupling capacitances from contributing to the noise currents. For example, consider aggressor nets that belong to a set of "one-hot" signals; that is, one, and only one, of the set is a logical '1' at any time. As depicted in Figure 12.10, if the victim net is routed among this set of signals, only one coupling capacitance could contribute to the victim low-up noise analysis. The noise analysis flow would include an input constraint file providing the functional exclusivity definitions.

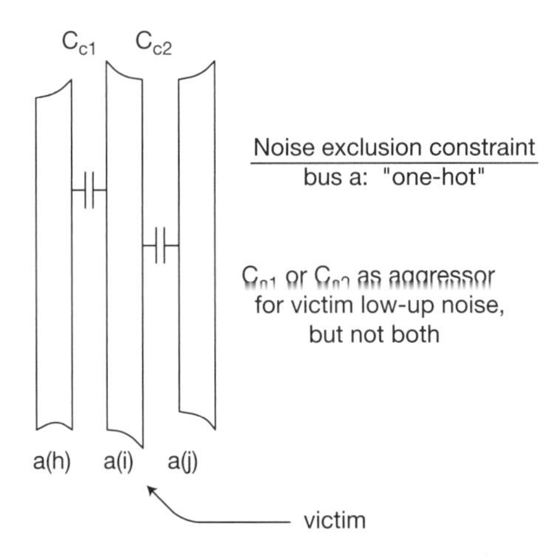


**Figure 12.10** Noise aggressors could be excluded from the superposition calculation in static noise analysis based on a functional exclusivity property provided as input to the noise flow.

Note that glitches on other signals in this set during a logic transition would invalidate this constraint; functional exclusions should be glitch free (in the time window of interest).

### 12.4.4 Timing Window Exclusions

The previous section described how static timing analysis with noise would maintain a more detailed propagation model, with a range of arrivals at each pin, for both late mode and early mode. These timing windows are used to define the aggressors that can overlap with a transient on the victim net and thus have a multiplicative (k-factor) effect on the magnitude of the coupling. Once STA with noise delay has converged, the timing window data are forwarded to the static noise analysis flow.

For noise analysis, a slightly different interpretation is applied to the timing windows, as the victim net is quiescent. The goal of the timing window model optimization for noise analysis is to identify the subset of the (dominant and virtual) coupling capacitances that have (RDLY/FDLY) arrival windows overlapping with the victim's *sensitivity window*. The superposition of these aggressor transients would maximize the injected current into the victim net for each noise direction.

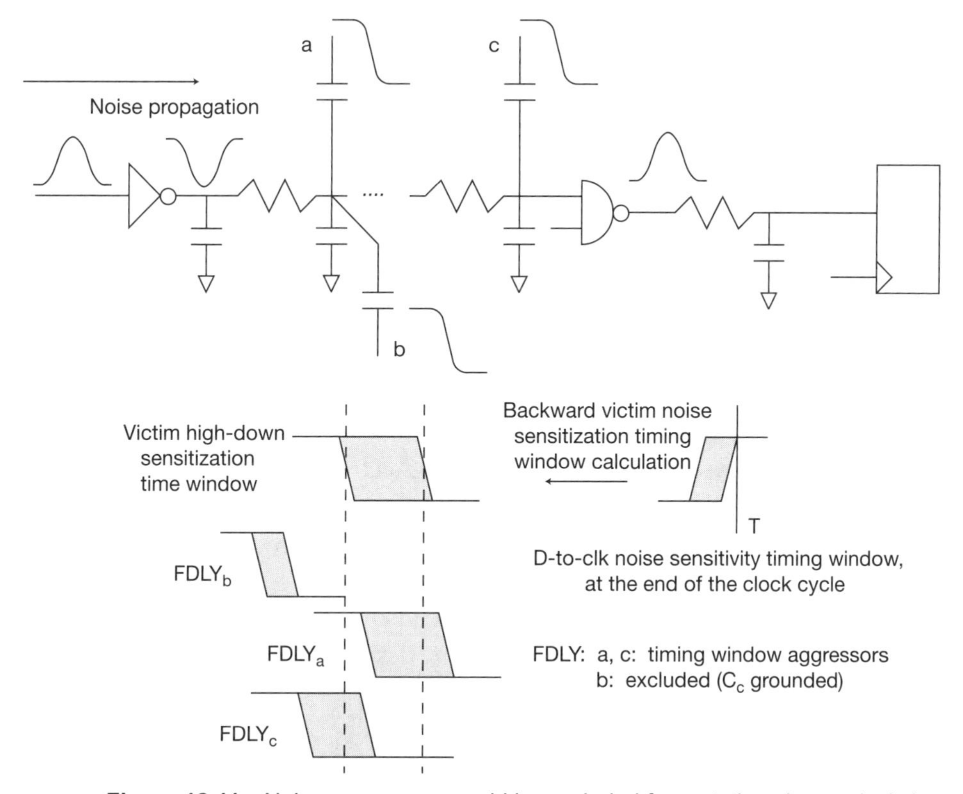


**Figure 12.11** Noise aggressors could be excluded from static noise analysis based on the aggressor timing window and the victim net timing "sensitivity window."

Consider the example depicted in Figure 12.11. For a victim high-down noise direction, the FDLY arrival time windows for the aggressor set would be reviewed. Similar to the required arrival time calculation backward from a flop endpoint used in STA to calculate the timing slack at each network node, a victim noise sensitivity window calculation would be initiated back from flops. An interval at the end of the clock period is selected, when a noise pulse on the quiescent flop data input would be a risk for invalid data capture; this interval would be centered around the data-to-clock setup time. From this endpoint timing window, a backward calculation through the levelized network is performed, subtracting cell timing arc and interconnect delays, resulting in a sensitivity window for each (victim) net. As with the RAT calculation in the STA flow, the cell unateness property would be used to manage the RDLY and FDLY noise timing window calculations for each net. The intersection of the aggressor timing windows and the victim sensitivity window would define the actual noise events of interest. The coupling capacitances for aggressors with non-overlapping timing windows would be grounded for this victim net noise direction analysis. In short, if the aggressor initiates a noise event on a victim net that would not propagate to a flop data input in the neighborhood of the clock transition, it could not possibly result in a state capture error, and thus its coupling capacitance could be grounded. (Reference 12.1 highlights that the cell propagation delay for a full-transition input pin-to-output pin delay arc differs slightly from the propagation delay through the cell for a noise pulse; when subtracting cell delay arcs during backward propagation for noise sensitivity window calculation, some delay margining would be appropriate.)

These aggressor capacitance filtering techniques (e.g., relative magnitude preprocessing, functional exclusivity, aggressor versus victim sensitivity timing window overlap) reduce the number of potential aggressors on the victim net. The "near-end" versus "far-end" position of the aggressor coupling capacitance on the victim RC interconnect tree also has a major impact on the noise pulse at the victim fan-outs. As a result, despite the reduction in the number of aggressors, the remaining topological complexity of the victim net electrical model still necessitates a simulation-based approach for calculation of the propagation of the cell output noise and injected aggressor coupling noise to fan-out pins in the victim net.

## 12.5 Static Noise Analysis, Part II

With the optimizations to the victim noise model described in the previous section, the victim noise analysis network is depicted in Figure 12.12. The driver noise pulse and active devices would be replaced with an equivalent linear model, while the aggressor inputs would be sourced with a voltage ramp (corresponding to the slew data from the STA flow).

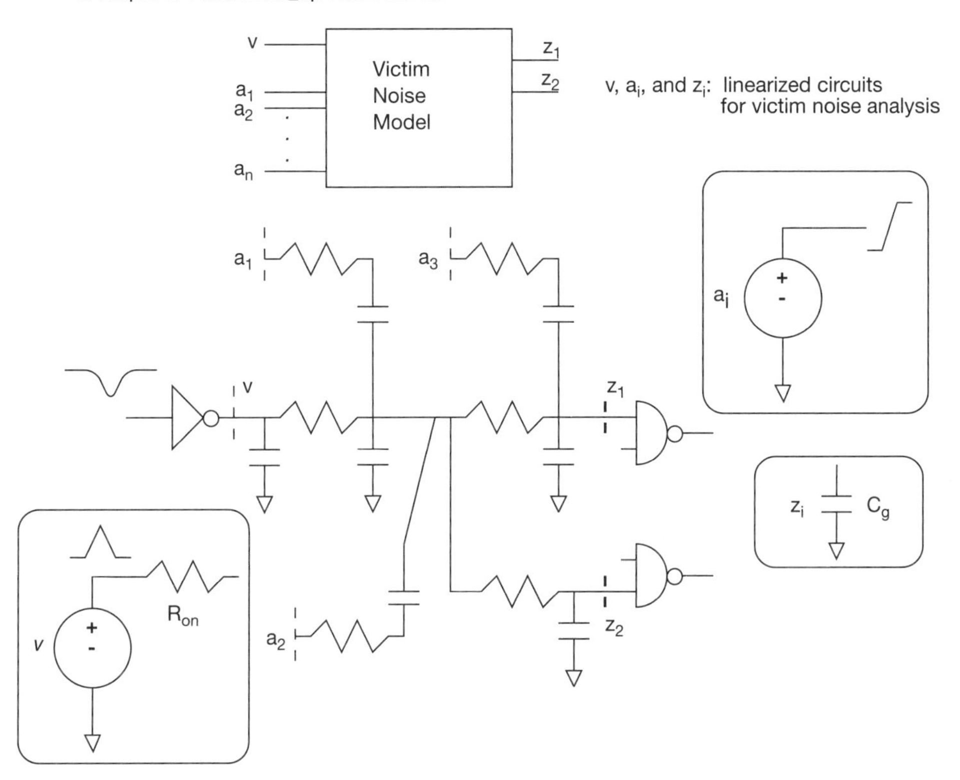


**Figure 12.12** Illustration of a reduced, linear network model for victim net noise analysis.

Receivers would be replaced with an equivalent capacitance, corresponding to the input pin device gates.

Analysis of this general network by a circuit simulator to determine the noise at the receivers would be rather inefficient. A class of model order reduction algorithms have been developed to simplify the victim net model while maintaining sufficient accuracy at the receivers; the *moment matching* algorithm for an RC interconnect tree described in Section 11.2 on interconnect delay calculation is a specific implementation of model order reduction. In general, model order reduction (MOR) algorithms represent the modified network in terms of a transfer function: $v_{out}(s) = H(s) * v_{in}(s)$, where $v_{in}$ and $v_{out}$ are vectors representing the ports of a multi-input, multi-output network. To further simplify the victim net analysis, advanced MOR algorithms provide an output netlist of passive elements from the frequency domain transfer function, which can be readily used with input port stimulus and output loads in a circuit simulation.[2] The original victim net electrical model is passive; passivity of the reduced model is a key criterion for the MOR algorithm applied to the network.

From the victim net noise simulation at each fan-out cell pin, a worst-case cell output propagated noise pulse is derived. The cell driver noise pulses are successively propagated through the levelized network to a noise pulse at the data versus clock flop endpoint, as depicted in Figure 12.4. The magnitude and duration of the noise pulse at the flip-flop data input are compared to the characterization noise immunity curve in the IP library for this cell to determine if a noise error is present.

Although the discussion has focused on a noise pulse test at a flop during a capturing clock transition, in general, noise tests are also required at hard IP block inputs. For example, memory array IP models include noise limits on data and address inputs relative to the array enable pin. In this case, the array enable timing defines the noise sensitivity window to which the data and address input noise pulses are referenced. The SoC methodology also needs to maintain appropriate propagated noise pulse data across the design hierarchy to enable block-level analysis.

The noise analysis of global signals requires special attention, as the net topologies tend to utilize long parallel segments for bus signals with significant coupling. The strategy for global net width/space constraints, and the net lengths between buffers, should be based on representative design testcases to ensure that minimal issues arise during static noise analysis. Rip-up/re-route and buffer insertion updates at the global level after (sign-off) static noise analysis are often difficult to embed as they affect the floorplan channel design.

### 12.5.1 Template Model for Fast Noise Analysis

Routing tools may integrate a noise calculation engine to avoid subsequent static noise analysis flow failures. To enable rapid evaluation during route optimizations, a vastly simplified electrical model for a net is required. One option would be to fit the driver and each net fan-out to a fixed circuit template and have a specific circuit node where aggressor coupling is introduced. Noise analysis during routing would evaluate this model for a victim net for each fan-out.[3,4] An example is depicted in Figure 12.13.

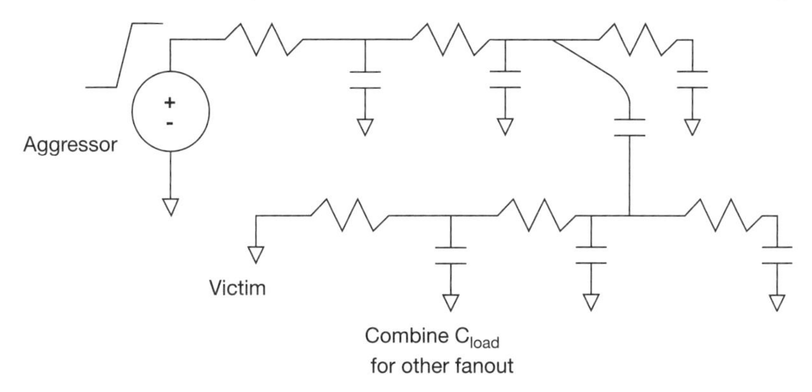


**Figure 12.13** A simplified "template-based" net model is illustrated. It is suitable for fast evaluation as part of coupled noise optimizations during detailed routing.

The router electrical model includes detailed data with interconnect RC estimates for each net that are reduced to fit the R and C elements of the noise template for the victim driver to a single fan-out. Other fan-out RC interconnect tree branches in the net electrical model are also reduced and contribute to the R and C element values in the template. Each dominant aggressor to the victim is modeled with a voltage ramp, injecting current into the intermediate node of the victim net in the template; the other aggressor coupling capacitance values are merged into the victim net model. With this fixed circuit topology and aggressor ramp, the specific R and C template values are provided to an analytic expression for the noise voltage pulse at the selected fan-out. The superposition of the victim net response for multiple dominant aggressors is used to calculate the total noise at the fan-out during route optimization and is compared to the cell pin characterization noise immunity curve.

## 12.6 Summary

This chapter discusses the analysis of coupled noise from a set of aggressors to a victim net. The aspect ratio of interconnects in advanced process nodes implies that the coupling from adjacent routes will be an increasing design closure issue.

SoC designs often include operating modes associated with dynamic voltage and frequency scaling (DVFS). Noise analysis is especially sensitive to DVFS operation, as the relative timing windows and slew rates of victim and aggressor nets scale differently with modifications to the supply voltage and clock frequency. The SoC methodology team needs to carefully evaluate the corner definitions for the static noise analysis flow if DVFS is to be adopted by the design. Also, the SoC methodology team needs to review the set of functional exclusions provided as input constraints to the static noise flow to ensure that these logical assertions are formally proven.

If the static noise flow results in victim low-up or victim high-down direction failures propagated to a test point, the SoC team needs to assess what cell sizing or route design changes are appropriate. The trailblazing blocks through physical implementation and electrical analysis provide insights into the efficacy of the noise avoidance optimizations during routing; adjustments to routing rules or electrical parameters in the routing model aid subsequent blocks. The number of noise failures identified during the block-level design phase of the SoC schedule would hopefully be addressed with a minimal set of ECO updates. The global-level noise flow failures should also be limited, as the floorplanning of major signal bus routes and buffers hopefully avoids coupling issues by design.

If the noise analysis flow results in local victim low-down or victim high-up direction pulses of significant magnitude and duration (which are not part of an endpoint test), the SoC team needs to investigate the corresponding reliability concerns with the product engineering team and/or add the nets to the correction list for ECO fixes.

## References


[1] Levy, R., et al., "ClariNet: A Noise Analysis Tool for Deep Submicron Design," *Proceedings of the 37th IEEE Design Automation Conference (DAC)*, 2000, pp. 233–238.

[2] Yang, F., et al, "RLCSYN: RLC Equivalent Circuit Synthesis for Structure-Preserved Reduced-Order Model of Interconnect," *International Symposium on Circuits and Systems (ISCAS)*, 2007, pp. 2710–2713.

[3] Cong, J., Pan, D., and Srinivas, P., "Improved Crosstalk Modeling for Noise Constrained Interconnect Optimization," *Proceedings of the 6th IEEE Asia and South Pacific Design Automation Conference (ASP-DAC)*, 2001, pp. 373–378.

**[4]** Ding, L., Blaauw, D., and Mazumder, P., "Efficient Crosstalk Noise Modeling Using Aggressor and Tree Reductions," *IEEE International Conference on Computer Aided Design (ICCAD)*, 2002, pp. 595–600.

## Further Research

### Direct Aggressor Coupling Assumption and Noise Margining

Section 12.2 highlights the noise model assumption that only the coupling from direct aggressors is analyzed and that margining may be appropriate for the "victim's victims," where indirect (series) coupling is significant (refer to Figure 12.2).

Describe a noise model development algorithm that detects victim nets where series coupling may be significant. Describe a suitable margining approach for victim nets subject to indirect coupling and demonstrate the accuracy of the proposed margining method with simulation on representative net examples.

### Victim High-Up and Low-Down Noise Transients

The static noise analysis discussion in this chapter focuses on victim low-up and high-down coupling directions.

Describe a methodology for analysis of low-down and high-up victim noise transients in which the noise immunity curve (NIC) characterization pin limits are replaced by reliability limits on the device gate electric field.

### Block-Level Noise Analysis

The noise analysis sign-off flow typically utilizes the full-chip electrical model for global paths to ensure the most accurate calculations of the victim noise sensitivity window and the propagation to an endpoint. To facilitate block-level noise analysis of PI-to-capture and launch-to-PO paths, a unique methodology approach is required.

Describe a suitable method to exercise block-level static noise analysis, including:

- Use of required output and expected input pin arrival timing constraints
- Calculation of sensitivity windows

- Initial assumptions for block PI pin noise magnitude/duration pulses and PO noise pulse limit checks (prior to detailed SoC physical integration)
- Propagation of a noise magnitude/duration pulse through a path, including global nets

### Global Net Design Planning for Noise

A critical set of design experiments is pursued prior to floorplanning to determine the "recommended" global net length between buffers. A shmoo of circuit simulations is run to determine the optimal positioning of buffers inserted in a long net to minimize the total delay. A key consideration in these experiments is to ensure that the resulting noise pulse propagation is also well managed to the path endpoint. (The buffers in a timing-critical global net are high-drive-strength cells and, thus, high-gain circuits with significant noise propagation.)

Describe the circuit simulation measures appropriate for these early global net design experiments. Describe the different variables in these buffered net simulation experiments (e.g., metal layer/width/space options, shielding options, noise pulse propagation for global clock tree signals).

Describe the methodology to integrate block-level and global noise pulse propagation results from the static noise analysis flow.

### Noise Characterization Requirements for Special IP Cells

The discussion in this chapter focuses on noise pulse propagation to flip-flop data inputs arriving in the clock transition window. However, other IP input pins are also sensitive to noise pulse events; these cases include both single pin limits and input pin tests relative to the arrival of enable signals. In these cases, the noise pulse is compared to the input pin noise immunity curve.

Describe the IP characterization data required during circuit simulation to establish the noise immunity curves for memory array inputs, clock gating cells, I/O cells (drivers, receivers, bidirectional enables), and mixed-signal data converters.

CHAPTER 13

# Power Analysis

## 13.1 Introduction to Power Analysis

The power analysis flow calculates (estimates of) the active and static leakage power dissipation of the SoC design. This electrical analysis step utilizes the detailed extraction model of the block and global SoC layouts. The active power estimates are dependent on the availability of *switching factors* for all signals in the cell netlist. Representative simulation testcases are applied to the netlist model, and the signal value change data are recorded.

The output data from the power analysis flow guide the following SoC tapeout release assessments:

- **Total SoC power specification (average and standby leakage)**—The specification for SoC power is critical for package selection and is used by end customers for thermal analysis of the product enclosure. In addition to the package technology selection, the SoC power dissipation is used to evaluate the die attach material stress due to the different thermal expansion coefficients of the die, bump/bond metallurgy, encapsulation material, and package substrate. For mobile product applications, confirmation that the power specification is satisfied is crucial to achieving active and standby operating battery life targets.

- **Peak power dissipation and peak power transient**—The peak power dissipation (along with the peak power transient) defines the limits on the package and system power and ground distribution network impedance to maintain voltages at the bumps that satisfy characterization assumptions.
- **Local thermal *hot spot* identification and analysis**—Much as a thermal resistance model from the package to the enclosure is used for end product thermal analysis, a thermal heat flow model for the die to the package is used for temperature analysis locally within the SoC. A high-power dissipation density within a small area of the die is denoted as a *hot spot*. At a hot spot, device temperatures may exceed the targeted limits used for cell characterization, invalidating timing analysis results. Temperature-dependent reliability failure mechanisms would be accelerated in these areas, as well. A hot spot may require the addition of a local *thermsense* circuit cell to signal to a central SoC power management unit that a throttling power state is required until the local temperature is reduced.

Note that several of these power analysis results criteria require detailed data about the specific location of the power dissipation on the die from the switching activity simulation data. The choice of simulation testcases is crucial to accurate calculation of average and peak power dissipation. The activity data measured during RTL simulation is a guide to selecting testcases. However, RTL validation is typically focused on coverage of unique system states and events rather than activity. An SoC project manager commonly allocates validation resources to compose power-centric testcases and allocate simulation licenses and compute resources to exercise these tests on the netlist-level model. Rather than provide SoC benchmark measurements, these *power stressmark* testcases do not necessarily reflect anticipated SoC workloads but are synthetically constructed to cover peak power dissipation.[1] (As mentioned in Section 11.4, there is a contribution to power dissipation from glitch pulse switching activity, as measured from representative testcases exercised on a netlist-level simulation model with annotated cell and interconnect delays.)

A related flow to power analysis is the power/ground I*R voltage drop analysis flow, which is discussed in Chapter 14, "Power Rail Voltage Drop Analysis."

## 13.2 Models for Switching Activity Power Dissipation

The basic model for calculation of the power dissipated by a net during a time interval is depicted in Figure 13.1.

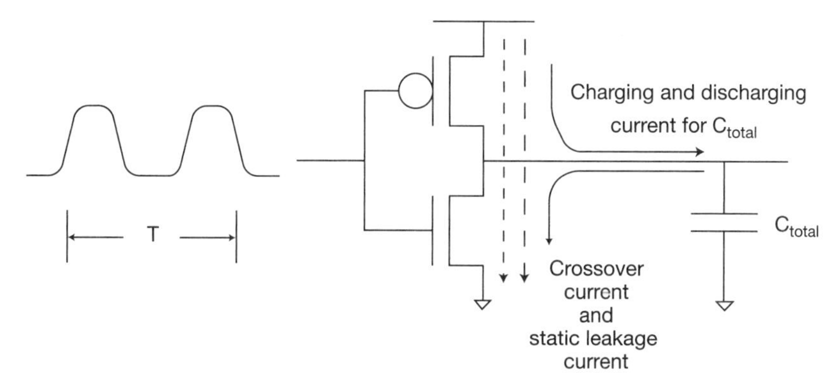


**Figure 13.1** Power dissipation model for a circuit-switching transient.

During an output rising transition, the current through the pFET device stack charges the output load capacitance, and for a duration while the input is in transition, it also sources the *cross-over current* through the nFET stack that was previously active. Conversely, during a falling output transition, the nFET stack discharges the load capacitance in combination with the cross-over current from pFET devices. Static leakage current flows in the cell through the active device stack to provide the subthreshold current of the off devices. The magnitude of the subthreshold current is strongly dependent on the input state, as depicted in Figure 13.2.

The static leakage current also includes the reverse-biased junction leakage from device source/drain nodes to the substrate (nFET) or well (pFET) potential. There is also a device gate leakage-to-channel current. With the addition of high-dielectric-constant (i.e., high-κ) materials for the device gate in modern fabrication processes, the magnitude of gate leakage currents has dropped substantially.

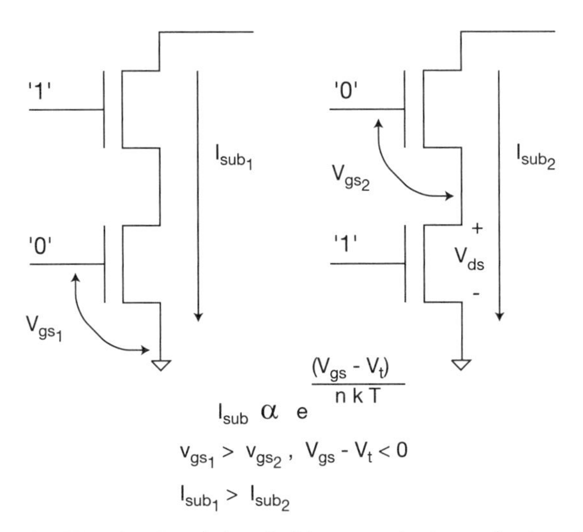


**Figure 13.2** The circuit subthreshold current is dependent on the input state.

The basic model in Figure 13.1 applies the sum of the interconnect load and fan-out gate capacitance to the cell output as $C_{total}$. For an average power calculation, a k-factor of 1 would be assigned to coupling capacitances; for peak power, a k-factor greater than 1 would normally be used. The average power dissipation associated with the current through the FET devices and the net capacitances is illustrated in Figure 13.3.

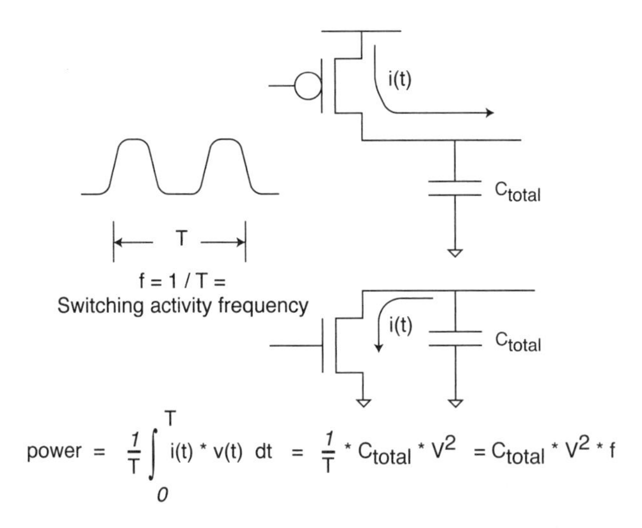


**Figure 13.3** Illustration of the average power dissipation calculation.

The cross-over current power dissipation is determined from the cell library characterization data, using the input and output slew results from the static timing analysis flow. The switching activity information from simulation testcases provided to the power analysis flow does not include the specific logical arcs for each cell that comprise the cell output signal activity factor; as a result, the power analysis flow needs to select a (conservative) cross-over power contribution, using the signal slew data from the STA flow results.

The static leakage power is also derived from the cell library data. The cell characterization flow may record the input state-specific leakage current. Again, with only signal switching activity information available, the power analysis flow needs to select a (conservative) leakage power measure from the cell model. If a specific power-state leakage current calculation is appropriate, the power analysis flow accepts a set of logic signal values and duty cycle information to access the state-specific characterization data.

The calculation of the estimated leakage power when logic cells are isolated from the power or ground distribution by sleepFETs is more complex, as shown in Figure 13.4.

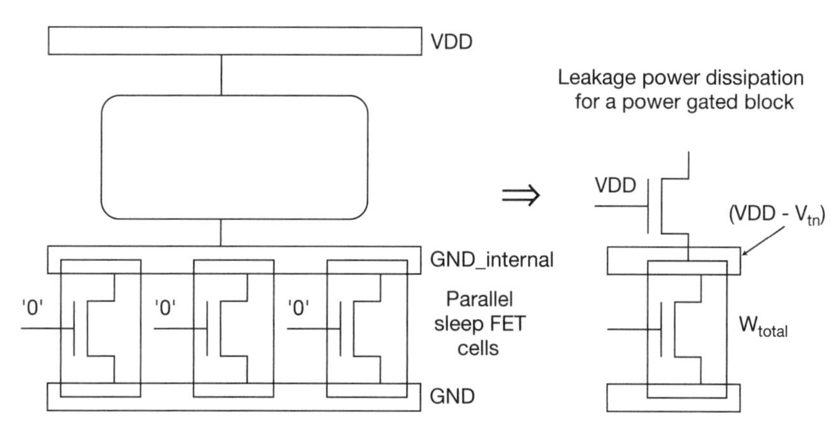


**Figure 13.4** Illustration of the leakage power dissipation calculation when the block is in the sleep state.

A conservative model would assume that the internal rail will reach a potential equal to a device threshold voltage away from the other rail during the power gated state and then calculate the device leakage current for the cumulative width of the sleepFETs at this drain-to-source voltage.

The electrical model in Figure 13.1 and the equation for average power dissipation uses the total charge delivered to the load capacitance as the integral of the current through the power-dissipating active device stack. This model neglects the additional dissipation in the distributed interconnect resistance. With advanced process node scaling, there is indeed an additional resistive interconnect dissipation contribution that is significant, especially for (global) nets driven by high-drive-strength cells, where the device $R_{on}$ is no longer the dominant resistive element in the charging/discharging current path. (Reference 13.2 describes an algorithm for estimating the power dissipation in the resistive elements of an RC interconnect tree.) The average power dissipation for each cell based on switching activity can thus be calculated as:

$$P_{avg} = P_{leakage} + P_{cross\text{-}over} + (C_{total} * (VDD **2) * f) + P_{interconnect} \quad \text{(Eqn. 13.1)}$$

### 13.2.1 Peak Power Calculation

The power analysis flow provides data on the peak power for use in thermal mapping and package analysis, as mentioned in the previous section. The characterization data for each cell include a table of power supply current waveform data, indexed by input slew and output load capacitance. As a cell may have many contacts to the power and ground rails in the layout, the characterization data consolidate the currents measured during circuit simulation. The (current, time) characterization waveform data values for each table entry reflect an average current over the interval between successive timesteps. For multi-stage logic cells, there may be multiple peaks in the current waveform data; there typically is a final dominant peak for the stage driving the cell output. As with cell delay arc characterization for static timing, multiple input switching conditions are not characterized for peak power dissipation.

For the power analysis flow, the SoC methodology team needs to review the algorithms for average and peak power measurement with the CAD team, including the following:

- The assumptions on coupling capacitance k-factor multipliers
- The summation of peak current waveforms per cycle (or within a cycle, using timing windows)
- The PVT (and extraction) corners significant to power measurement

The temperature values selected for the power measurement corners are critical to meeting the goals outlined in the Section 13.1. Product enclosure thermal analysis is required at high temperatures. Thermsense functionality is likewise focused on high-temperature operation. Package analysis is required at maximum current densities; derivation of maximum current density has become more involved in advanced processes due to *temperature inversion*. Traditionally, device current increased at lower temperatures; the increase in channel carrier mobility at lower temperatures was more significant than the reduction in input $|VDD - V_t|$ overdrive due to an increase in threshold voltage at low temperature. As the supply voltage VDD has been scaled, this mobility versus overdrive trade-off no longer ensures that the current density is maximal at the low-temperature corners. The SoC methodology team needs to collaborate with the foundry support team to establish the power analysis temperature environment appropriate for peak current calculations.

## 13.3 IP Power Models

The previous section describes power analysis applied to library cells and details signal activity factor data from netlist-level simulations. The contribution to SoC power from hard IP blocks is more difficult to model. The IP provider is not likely to provide the level of internal netlist and parasitic detail to enable a similar calculation as for a cell-based block implementation. Instead, the IP dataset includes an abstracted, execution-based model. The model defines the power dissipation associated with (multi-cycle) operations, which the power analysis flow needs to match to detailed simulation value change trace file results. For example, a memory array is characterized in terms of the power associated with a read or write operation. The IP power model describes the set of simulation pin transitions to identify each operation. The majority of the power dissipation occurs internal to the IP circuitry; to simplify power characterization, the IP power model is not likely to include any functional dependency on input pin slews or, perhaps, any power dissipation model dependency on output pin loading.

Similarly, analog mixed-signal IP includes an abstract power model. For external interface IP, the SoC driver/receiver power dissipation is a strong function of the data rate. The stressmark simulations focus on sustaining high interface IP throughput.

Global clock power dissipation is calculated separately from the cell-based power analysis flow. (The power dissipation for block-level clock buffers and clock gating cells would typically be included in the block analysis. The stressmark simulation data should be representative of clock gating enable switching activity.) The capacitance of the global clock mesh and/or H-tree is used in the C*(V**2)*f calculation. The resistive dissipation in the mesh/tree should ideally be small due to the metal level and wire width implementation. The cell location information used by the power analysis flow requires specific consideration for the global clock distribution. The clock drivers are often the source of high local power dissipation; multiple buffers with dotted outputs driving a clock mesh are often used, separated by a distance sufficient to reduce the rise in local die temperature. (The final level of buffers driving a clock mesh is typically the only case where distributed cell outputs are allowed to be dotted. The low resistivity of the mesh minimizes any cross-over current between separate buffers.)

## 13.4 Device Self-Heat Models

The die thermal map is provided for a grid superimposed over the die area. The average power dissipation for each cell location is summed over the grid cell and presented to a thermal solver with a corresponding thermal resistance model for the die and package. The interface between die and package is a critical element of the thermal model: For high-power dissipation SoCs, a specific thermal interface material (TIM) stack is applied to the (flip-chip) die backside to reduce the thermal resistance.

A specific thermal effect has emerged as a key modeling requirement in advanced process nodes. For the process technologies implementing FinFET or fully depleted SOI (FD-SOI) devices, the thermal flow is considerably different than for a bulk device, as depicted in Figure 13.5.

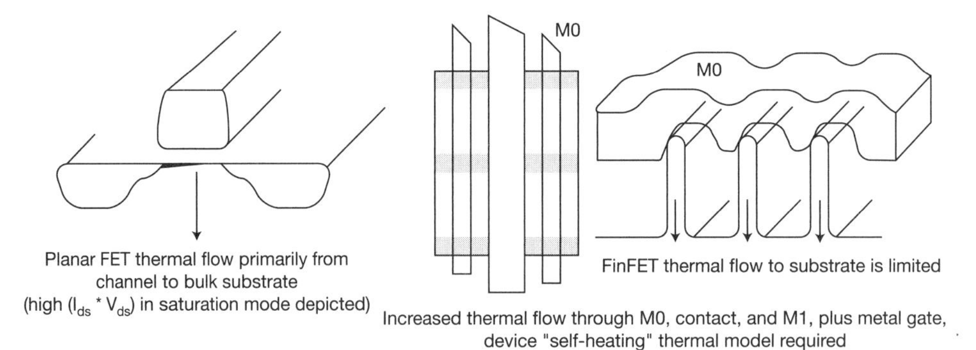


**Figure 13.5** Thermal energy flux for bulk and FinFET devices, illustrating device self-heating.

The device *self-heating effect* refers to the local temperature rise in the device channel and adjacent metal stack due to the reduced thermal flow to the bulk substrate.

For FinFET and FD-SOI technology designs, the power analysis flow still determines the active and static leakage power from cell characterization data, in the same manner as described in Section 13.2. However, the cell characterization flow is modified to reflect the device self-heating temperature increase due to the more restricted heat flux path. Given the regularity of the device and local metal layout data for library cells in these advanced nodes, thermal model assumptions allow for temperature calculations down to the device level, as depicted in Figure 13.6.[3]

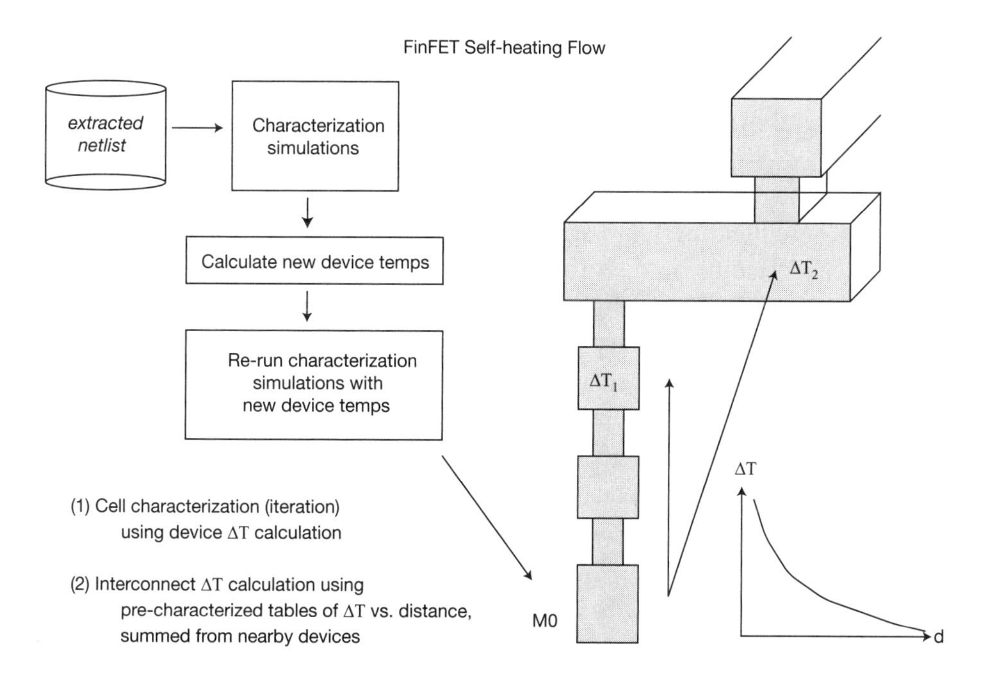


**Figure 13.6** FinFET and FD-SOI technologies require a modified characterization flow to incorporate the device self-heating temperature increase. From the device thermal calculation, the temperature increase in upper metals can also be calculated as the superposition of the heat flux from multiple active devices.

The foundry provides data for the calculation of the temperature rise in the source/drain junctions and local metals due to an active device. (For

FinFET technologies, a multiple-fin device would be represented as a single heat source.) Starting with the initial PVT corner temperature, cell characterization is performed to determine the device dissipation. The self-heating temperature increase for active devices is calculated using the foundry models. Characterization is then reevaluated using the $(T + \Delta T)$ temperature.

For upper metals, the foundry provides models for the temperature increase in each metal layer due to an active device. The temperature increase for the metal layer is a strong function of the distance between a wire and the device, as illustrated in the figure. The total temperature increase in the wire is a summation of the increase from active devices in the neighborhood of the wire. This temperature increase could be used to modify the (temperature-dependent) resistances in the RC interconnect network. Device self-heating is highlighted again in Chapter 15, "Electromigration (EM) Reliability Analysis," when discussing the strongly temperature-dependent electromigration phenomenon in metals.

## 13.5 Design-for-Power Feedback from Power Analysis

The SoC architecture team strives to optimize power dissipation. The team's high-level decisions on operational power states, allocation of functional unit resources, and logical path length versus clock frequency have the greatest impact on switching activity (and the capacitive loading of high-activity signal busses). The logic gate density afforded by process scaling while adhering to stringent power dissipation limits has led to the increased adoption of *dark silicon* (i.e., specialized functional units incorporated into the architecture that accelerate unique operations yet remain inactive for considerable intervals).

The SoC RTL team further refines the power-centric architecture optimizations, seeking to maximize opportunities for clock gating and data encoding to reduce switching activity. The SoC physical implementation team focuses on cell sizing optimizations to reduce power on positive slack timing paths and establish (local and global) routing priorities for high-activity signals. Deep sleep features may be adopted, in conjunction with (architectural) programming models.

The key is that the sign-off-level power analysis flow is typically executed rather late in the project schedule to address major power dissipation or thermal issues; the SoC team relies heavily on previous optimization efforts. The SoC program management team needs to closely track preliminary block-level power analysis results against power budgets to establish confidence in the overall specification. The stressmark simulation testcases need to be developed on a schedule that supports early block-level analysis, perhaps using RTL models for blocks that do not yet have physical implementations, with estimates for signal capacitance per unit area. Any unexpected switching activity percentages in simulation warrant further investigation to ensure that potential power optimizations have not been overlooked.

The project management team needs to engage the package engineering and product reliability groups to coordinate delivery of the SoC power analysis results with die location-based power dissipation data for thermal map calculation. The package team uses this information to analyze material stresses in the encapsulated die package model. The author is familiar with high-performance, high-power-dissipation design projects (with hot spots) where the difference in thermal expansion coefficients between the die and the package underfill materials led to partial delamination of the die attach bumps. Catastrophic failures were detected during silicon prototype bring-up. (Unfortunately, thermal stress analysis of the die-package interface wasn't pursued until after prototype tapeout.) Significant modifications to the physical floorplan were required for the next tapeout, with major engineering costs and project schedule impacts.

## 13.6 Summary

The average, peak, and transient power dissipation of the SoC design is a critical design constraint, due to the impact on package selection and the end-product enclosure thermal characteristics. With the tremendous growth in battery-powered product applications, the design architecture and physical implementation engineering teams have increased their efforts on power optimization methods. The power analysis flow provides the data to confirm that the (full-chip) power dissipation results satisfy the SoC specifications. With the fully detailed cell and extracted parasitic model, the power analysis flow relies upon the generation of representative and worst-case signal switching activity factors from selected simulation testcases. The electrical analysis team works closely with the verification team to develop the power-centric simulation testcase suite.

In addition to the full-chip power dissipation measures, the power analysis flow provides key insights into the potential for local hot spots. Thermal map data is needed to establish the design criteria for the addition of thermsense circuits, connected to a power management controller capable of (locally) introducing boost/throttle operating modes within the thermal envelope.

Recent process technologies have introduced a new facet to power analysis and thermal modeling. The self-heating characteristics of FinFET and FD-SOI devices necessitate an iterative characterization methodology using the local temperature rise from switching activity power.

Due to the design architecture and implementation focus early in the project schedule, the results of the power analysis flow should ideally identify no concerns. The flow can be exercised at the block-level as extracted parasitic netlists become available—hot spots and related thermal management design requirements from the block-level analysis can hopefully be addressed with little impact. ECO updates after the design freeze milestone to resolve (local or full-chip) power dissipation issues would be extremely disruptive to the physical implementation, with significant project schedule and resource risks.

## References

[1] Joshi, A., et al., "Automated Microprocessor Stressmark Generation," *IEEE 14th International Symposium on High Performance Computer Architecture (HPCA)*, 2008, pp. 229–239.

[2] Shin, Y., and Sakurai, T., "Estimation of Power Distribution in VLSI Interconnects," *IEEE International Symposium on Low Power Electronics and Design*, 2001, pp. 370–375.

[3] Liu, S.E., et al., "Self-Heating Effect In FinFETs and Its Impact on Devices Reliability Characterization," *IEEE International Reliability Physics Symposium*, 2014, pp. 4A.4.1–4A.4.4.

## Further Research

### Power Stressmark Testcases

Describe the characteristics of power stressmark testcases for an SoC consisting of microprocessor cores.

### PVT Corners

Describe the criteria used to select the PVT corners for power dissipation analysis for fabrication process nodes with and without temperature inversion.

### Resistive Interconnect Power Dissipation

Describe an algorithm for estimating the power dissipation in resistive signal interconnects that could be incorporated into the power analysis flow. Describe signal filtering options that would be effective in reducing the overall algorithm runtime.[2]

### Die Thermal Map

Describe the steps associated with preparing a model for deriving a die thermal map, merging the local power dissipation grid results from the power analysis flow with the thermal resistances of the active surface of the die, the die substrate, the die attach to the package (underfill and bump/bond wire connections), intermediate TIM materials, package composition, and package heatsink.

Describe options in the thermal model solution for the ambient environment—for example, the temperature within the product enclosure, air flow (forced or passive convection), and the influence of neighboring power-dissipating parts on the printed circuit board.

### Thermal Analysis Results

Describe how the die thermal map data from power analysis would be used to influence the SoC design. Considerations include:

- Integration of thermsense IP macros
- Design modifications to "throttle" power dissipation (e.g., suppress system clock pulses)
- Design opportunities to "boost" the supply voltage for increased performance
- Floorplanning for "dark silicon" functional units
- Reliability failure rate adjustments due to thermal map temperatures less than the corner temperature (see Chapter 15)

CHAPTER 14

# Power Rail Voltage Drop Analysis

## 14.1 Introduction to Power Rail Voltage Drop Analysis

The power analysis flow described in Chapter 13, "Power Analysis," provides power dissipation data used for thermal modeling of the die/package combination and for confirmation of the overall SoC power specification. The power analysis flow also provides localized power dissipation data for hot spot identification. A related analysis flow utilizes switching activity and cell characterization power supply current waveforms to calculate the voltage drop on the VDD power and ground (P/G) distribution rails.

Power rail voltage drop analysis involves applying cell current data to an extracted model for the P/G distribution network (PDN). A resistive P/G mesh model is used for a static I*R voltage drop estimate, while a more elaborate RC extracted model for the PDN is required for a dynamic voltage drop calculation. The RC network for the P/G grids would be merged with the RLCM model for the connections from die pads (through bumps or bond wires) to package pins. The self- and mutual inductances of the top metal redistribution wires from pads to the SoC internal P/G grids may be significant and would also be part of the composite die/package RLCM model.

### 14.1.1 Conservative Versus Aggressive P/G Grid Design

IP library cells and macros are characterized using an assumed limit on the supply and ground voltage drop from the package pins to the cell/macro location. The power rail analysis flow is used to confirm that these assumptions have not been exceeded.

The design of the cell layout image for the IP library includes the track allocation for the power and ground rails for the block-level and global grids. If the image design is conservative, the impedance of the P/G distribution should be low enough to satisfy the voltage drop limits when *any* SoC design incorporating this IP is submitted to the power rail analysis flow, regardless of cell placement or switching activity. In other words, the P/G image needs to be sufficiently robust so that the rail analysis should never result in a voltage drop margin error.

A more aggressive SoC project may opt to free up some of the allocated P/G rails to make more tracks available for signal routes if the voltage drop limits can be met with a reduced grid. During SoC global and block floorplanning, a physical integration engineer may opt to populate the P/G grids more sparsely than called for in the original image design. The lower-level metal rails connecting directly to cells and macros would be unaltered, but the intermediate and upper-level grids could be reduced from the IP design recommendation if the switching activity allowed.

As with the power analysis flow in Chapter 13, the SoC methodology needs to support block-level power rail voltage drop analysis for early insight into any P/G issues, especially if any modifications to the recommended image grid design (or decoupling capacitance insertion guidelines) have been pursued.

### 14.1.2 ASIC Direct Release

The origins of the semi-custom application-specific IC (ASIC) design methodology began with the availability of IP logic library cells placed and routed on a die with a pervasive cell image throughout and with *fixed* local and global P/G grids. Both standard-cell and gate-array cell library layouts were developed to align with this image. (Standard-cell and gate-array cell design styles are introduced in Section 1.1.) I/O receiver and driver cells also had a fixed placement image, connected to separate VDDIO and GNDIO rails. With this image, ASIC part numbers were offered in a range of logic gate capacities, with corresponding recommendations on cell utilization and decoupling capacitance densities.

The primary image design consideration was that *all* ASIC designs were expected to have satisfied the P/G voltage drop margins assumed during library cell characterization, regardless of the local cell logic type and drive-strength selections or switching activity factors. The goal was to facilitate the *direct release* of a placed and routed database to fabrication, without any modifications required to the P/G distribution.

Conservative design assumptions were made when designing the ASIC power and ground rails. For example, as illustrated in Figure 14.1, a high concentration of high-drive-strength cells were placed in a single row; the maximum current in the rail would be the sum of the (saturated) device current in each cell. This local model would be replicated throughout the ASIC die when analyzing the fixed P/G grids. (The figure illustrates cells placed back-to-back in adjacent rows, sharing VDD and GND rails.)

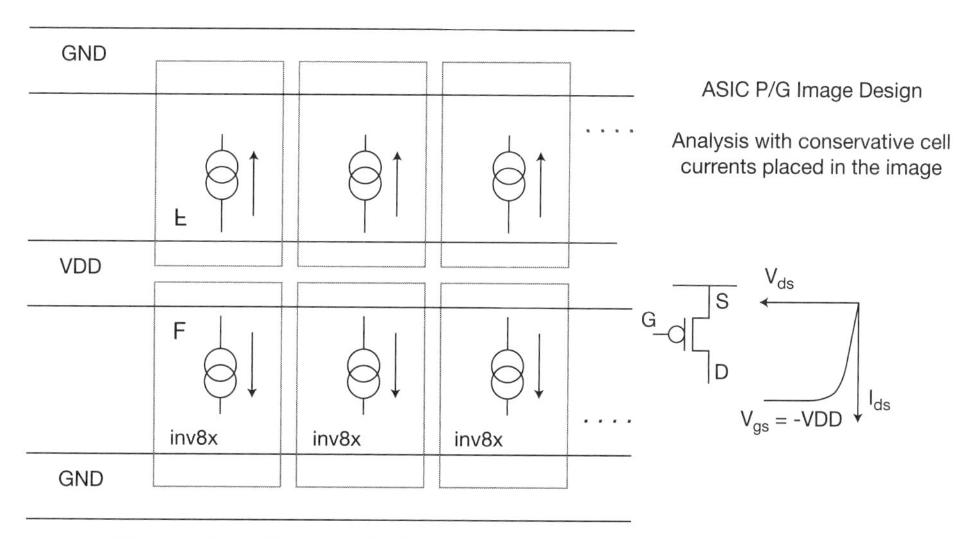


**Figure 14.1** Example of a “conservative” ASIC P/G image design assumption, suitable for direct release. It is similar to Figure 10.25 but with cell rows populated with high-drive-strength cells.

With this image released to all customers, the subsequent ASIC designs were essentially assured to be free of power rail voltage drop margin fails.

ASIC direct release was facilitated at the expense of metal resources conservatively allocated to the P/G grids.

With the emergence of SoC designs integrating a wide diversity of IP macros and cores, the direct-release, fixed-image design style is no longer feasible. As a result, there is an opportunity to pursue a more aggressive P/G design approach. Individual blocks may develop a unique internal P/G distribution. The trade-off for the aggressive PDN design is that the power rail voltage drop analysis flow is required to be exercised throughout the physical implementation project phase on both block and full-chip SoC physical integration releases.

### 14.1.3 P/G for Blocks with sleepFETs

A special methodology is required for power rail analysis of a sleepFET configuration, as depicted in Figure 13.4. The extracted network for $GND_{internal}$ and $GND_{global}$ is separated by the sleepFET cells. The P/G network model needs to be augmented prior to power rail analysis of the active block, with effective $R_{on}$ values inserted for each sleepFET device. (The turn-off and turn-on P/G currents when transitioning to/from a sleep state are outside the scope of the power rail analysis flow, as a time-dependent simulation is required to reflect the delays of the enable signal to the distributed sleepFET cells.)

## 14.2 Static I*R Rail Analysis

Static I*R voltage drop analysis uses DC currents injected into the resistive model. This calculation is much quicker than the dynamic rail analysis with a full RLCM model. Also, I*R static analysis can be pursued with a partial SoC physical design; the DC currents for blocks with an initial physical implementation can be merged with (coarse) estimates for the current of other blocks/channels and applied to the resistive network of the P/G grid.

Any excessive rail voltage drop from static I*R analysis is indicative of a design issue that needs to be addressed, without the addition of dynamic voltage analysis results. To fix static I*R flow failures, the P/G grid may need additional density, wider rails, and/or improved via arrays between metal layers.

The static I*R analysis flow is specifically applied to any distinct power domains created for analog mixed-signal IP. The DC current profile for the mixed-signal block would be derived from the IP power model or from detailed circuit simulation measures, if available. Ideally, the current profile model includes precise physical location granularity to be annotated to the nodes of the extracted analog P/G mesh.

### 14.2.1 Static I*R Matrix Solution

The algorithmic solution for the static I*R flow is represented as the *nodal analysis formulation* of the matrix equation:

$$\mathbf{G} * \mathbf{v} = \mathbf{i} \qquad \text{(Eqn. 14.1)}$$

using Kirchhoff's current law (KCL) at each node. In the equation above, **G** is the (very sparse) conductance matrix between nodes of the extracted VDD (or GND) distribution network, **i** is the vector of DC currents injected at these nodes, and **v** is the node voltage vector to solve. As illustrated in Figure 14.2, the elements of the (symmetric) conductance matrix consist of diagonal entries equal to the sum of all conductances connected to node *i* (conductance G = 1/R), and the off-diagonal values are the negative of the conductance between nodes *i* and *j* in the extracted resistive power grid network.

Matrix Formulation for Static Grid Analysis

$$\mathbf{G} = \begin{bmatrix} G_{11} & -G_{12} & \cdots & -G_{1n} \\ -G_{21} & G_{22} & \cdots & \\ & & & \\ -G_{n1} & \cdots & & G_{nn} \end{bmatrix} \qquad \mathbf{v} = \begin{bmatrix} v_1 \\ \cdot \\ \cdot \\ \cdot \\ \cdot \\ v_n \end{bmatrix} \qquad \mathbf{i} = \begin{bmatrix} i_1 \\ \cdot \\ \cdot \\ \cdot \\ \cdot \\ i_n \end{bmatrix}$$

Current sources injected into node i

**Figure 14.2** Illustration of the conductance matrix G used in the static I*R solution.

A number of approaches have been developed to solve this type of very large, very sparse matrix problem.[1]

## 14.3 Dynamic P/G Voltage Drop Analysis

A more accurate analysis of the P/G distribution is provided by the dynamic voltage drop flow, based on the supply current data from the cell characterization tables and the switching activity of (stressmark) patterns applied to the netlist model, as described in Section 13.2. The time-dependent current sources generate P/G rail voltage drops that combine the resistive losses and inductive impedance from the package and chip redistribution layer wires, offset by the (local) capacitance.

### 14.3.1 P/G Rail Capacitance

The capacitances in the dynamic RC power rail model include several contributions, as illustrated in Figure 14.3:

- Distributed capacitance of the power and ground rails
- Explicit decoupling capacitance cells added to the design
- Device node capacitance connected to the rail
- Device well capacitance to substrate

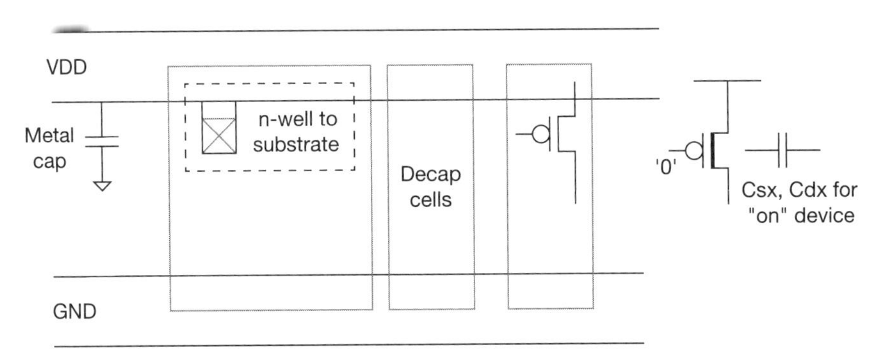


**Figure 14.3** Sources of explicit and internal capacitances on the PDN (similar to Figure 10.26).

The device capacitances that may assist in providing a low-impedance path for voltage noise depend on the operating mode of the device—whether active or off. The well-to-substrate junction capacitance includes a significant series resistance due to the high sheet resistivity of both these nodes and is thus less effective in filtering rail noise.

The parasitic extraction flow directed to the P/G net names provides only the RC parasitics of the rails. The other capacitance contributors reside inside cells and need to be specifically added to the nodes of the extracted RC network as part of the rail analysis flow. Explicit decoupling capacitance cells are readily located and added to the grid electrical model. An "averaging" algorithm is needed for active and off device node capacitance elements added to the model.

Note that P/G parasitic extraction requires special features. In general, RC reduction is not applied, to enable decoupling capacitance cells to be accurately assigned to their corresponding location; this non-reduced network is also used for power rail electromigration analysis (see Chapter 15, "Electromigration[EM] Reliability Analysis"). An exception would be applied to the reduction of metal via arrays to an equivalent resistance and capacitance to reduce the extracted model dataset size, as illustrated in Figure 14.4.

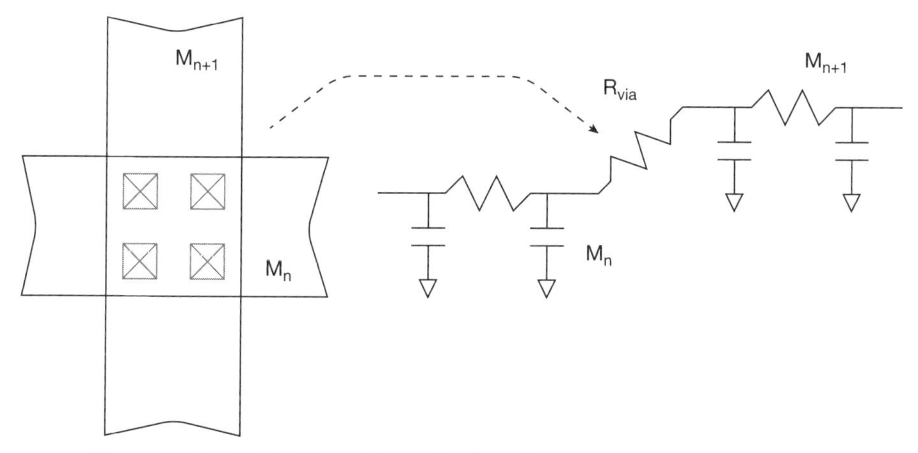


**Figure 14.4** Illustration of the reduction of via arrays in the P/G grids to an equivalent resistance and capacitance.

The package P/G pin RLCM electrical models are connected to the extracted RC (or RLCM) model of the global P/G grid, and the additional device and decoupling capacitances are included. Recall that the definition of the inductive impedance requires establishing the area of the current loop of which the wire segment is a constituent. The electrical definition of *partial inductance* allows a loop to be broken into individual segments for calculation. Once the loop current path is defined, the combination of the individual segment self-inductance and inter-segment mutual inductances allows a full electrical model to be developed.[2] The SoC design team should closely evaluate the loop model assumptions used in preparing the individual inductive elements in the RLCM network used for dynamic P/G voltage drop analysis. Note that the package (and die RDL) current loop model should include mutual inductances between VDD and GND package pins and metal traces, necessitating a *combined* dynamic analysis of both grids.

### 14.3.2 Matrix Solution

Circuit simulation of the dynamic P/G model presents extreme challenges due to the size of the extracted RC network, augmented with the addition of the decoupling and parasitic capacitances, the time-dependent current sources, and the package model. Note that the dynamic power grid analysis model is relatively constrained, consisting solely of RLCM elements, local current sources, and supply voltage sources applied at package pins; no non-linear device models are present. Rather than attempt traditional simulation, a matrix formulation and incremental timestep solution is typically pursued. The previous section briefly describes the conductance matrix used to represent the network for static I*R voltage drop analysis. Similarly, dynamic analysis can be represented by an "admittance matrix":[3]

$$\mathbf{G} * \mathbf{x}(t) + \mathbf{Y} * \mathbf{x}'(t) = \mathbf{s}(t) \qquad \text{(Eqn. 14.2)}$$

where **G** is the conductance matrix, **Y** is the time derivative matrix, and **s** is the vector of applied current sources. Whereas the solution vector in the static I*R analysis is the set of network node voltages, the vector **x**(t) is expanded in the admittance model to include both (time-varying) node voltages plus inductor branch currents. The dynamic analysis proceeds by solving the matrix equation at successive timesteps. Unlike traditional circuit simulation, with adaptive timestep advance algorithms, a fixed timestep interval is used for dynamic P/G analysis. A backward Euler approximation for the time derivative of the solution vector, **x**'(t), is used in Reference 14.3. Rather than expressing the solution at time increment (n+1) in terms of the value and the derivative at time increment n, the backward Euler method relates the solution at time $t_n$ in terms of the value and derivative at $t_{n+1}$:

$$\mathbf{x}(t_n) = \mathbf{x}(t_{n+1} - h) = \mathbf{x}(t_{n+1}) - h * (\mathbf{x}'(t_{n+1})) + \text{error } O(h^2) \qquad \text{(Eqn. 14.3)}$$

where h is the timestep increment, and the first term of the Taylor expansion for **x** is used. Figure 14.5 illustrates the backward Euler relation (for a one-dimensional curve).

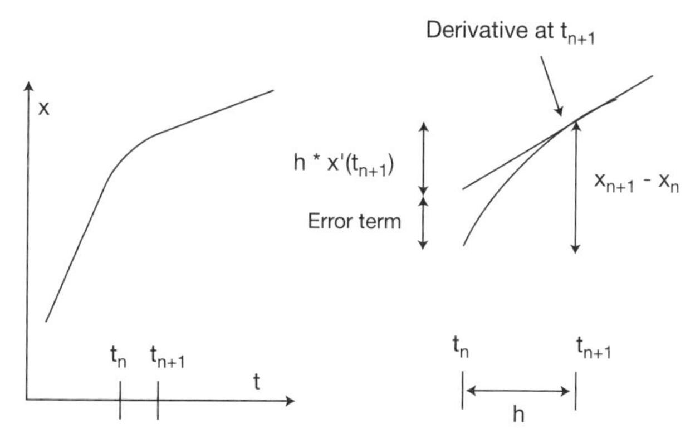


**Figure 14.5** Illustration of the backward Euler approach for dynamic P/G analysis timestep solutions. A simple one-dimensional curve (and a close-up of that curve between $t_n$ and $t_{n+1}$) is shown.

To simplify the notation, let:

$$\mathbf{x}'(t) = \mathbf{f}(\mathbf{x}, t),\ \mathbf{x}(t_n) \rightarrow \mathbf{x}_n,\ \mathbf{x}(t_{n+1}) \rightarrow \mathbf{x}_{n+1}$$

$$\mathbf{x}_{n+1} = \mathbf{x}_n + (h * \mathbf{f}(\mathbf{x}_{n+1}, t_{n+1})) \qquad \text{(Eqn. 14.4)}$$

Equation 14.4 is an approximation of Equation14.3, neglecting the $O(h^2)$ error term. Transforming the general admittance matrix equation into the backward Euler time increment format of Equation 14.4 results in:

$$\mathbf{G} * \mathbf{x}(t) + \mathbf{Y} * \mathbf{x}'(t) = \mathbf{s}(t)$$

$$\mathbf{Y} * \mathbf{f} = \mathbf{s} - (\mathbf{G} * \mathbf{x})$$

$$\mathbf{Y} * \mathbf{x}_{n+1} = \mathbf{Y} * \mathbf{x}_n + h * (\mathbf{s} - (\mathbf{G} * \mathbf{x}_{n+1}))$$

$$(\mathbf{G} + (\mathbf{Y} / h)) * \mathbf{x}_{n+1} = (\mathbf{Y} / h) * \mathbf{x}_n + \mathbf{s}(t) \qquad \text{(Eqn. 14.5)}$$

The solution vector $\mathbf{x}(t)$ is solved at each successive timestep $t_{n+1}$, using this matrix equation and the solution at time $t_n$. The time-varying current sources in the power grid, $\mathbf{s}(t)$, are recalculated at each time increment.

The matrix analysis introduced in the previous section using KCL to solve for node voltages needs to be modified to represent the current through the inductive elements, with:

$$v_{L1} = L1 * (di_{L1} / dt) + M12 * (di_{L2} / dt) \qquad \text{(Eqn. 14.6)}$$

An example of the matrix formulation is given Figure 14.6 for a simple RLCM network, from Reference 14.4.

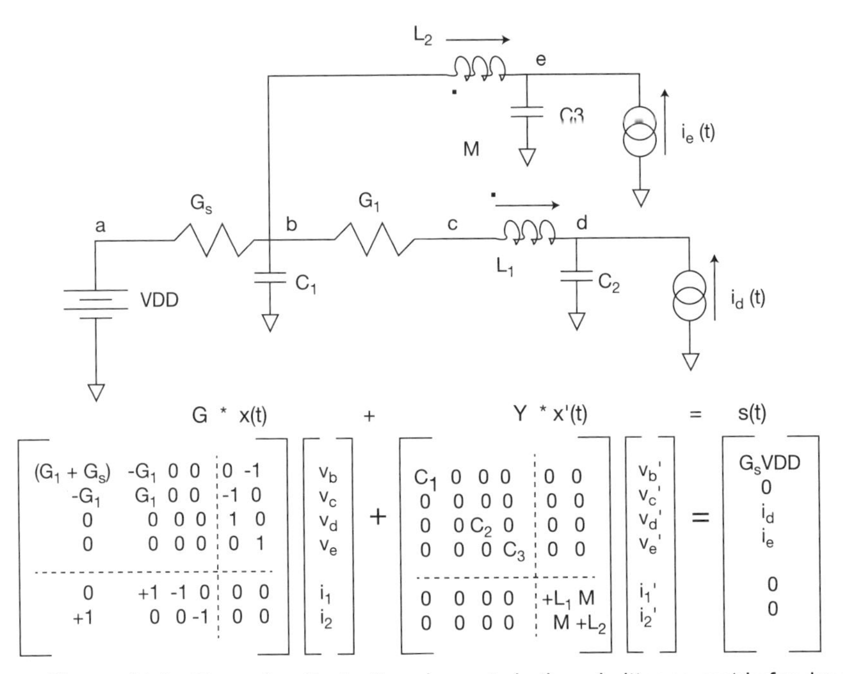


**Figure 14.6** Example of inductive elements in the admittance matrix for dynamic P/G rail analysis. The G and Y matrix entries reflect the network R, L, C, and M element values connected to nodes b through e.

The inductor branch currents in the solution vector **x** are represented by the additional rows and columns in the conductance matrix **G**, with 0, 1, and −1 entries, and by the addition of a sub-matrix to the time-derivative matrix **Y**, with the self- and mutual inductance values.

A unique optimization to the dynamic matrix model leverages the common topology in which a series resistor is present with the inductor. The time derivative of the inductor current can be equally represented as the time derivative of the resistor voltage, allowing the inductor branch current to be optimized out of the solution vector, as depicted in Figure 14.7.

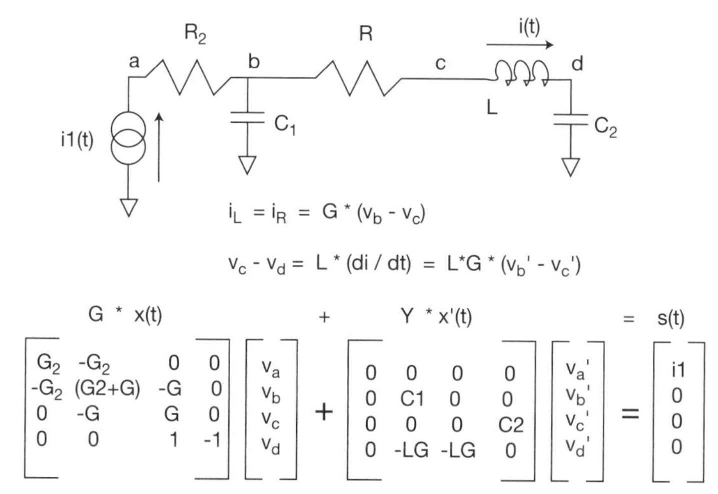


**Figure 14.7** Matrix optimization for a series R and L connection.

To review, the matrices **G** and **Y** are populated with the extracted P/G grid and package model network values. The cell source current waveform data are pulled from the characterization tables, using the output load and slew measures determined from static timing analysis. The time-varying source current vector, **s**, represents these cell currents assigned to P/G network nodes, at the switching times recorded during the power stressmark simulations. (The power supply reference voltage inputs at package pins are also part of the current source vector.) The P/G network voltages at each time step are solved using Equation 14.5. The results of the dynamic simulation would be compared to the P/G rail margin assumptions used for library IP characterization.

### 14.3.3 Hierarchical Analysis with Abstracted Partition and Global Models

The dynamic power rail network model described above is still *very* large and is therefore unsuitable for many matrix solver algorithms. Reference 14.3 describes a method for dividing the full-chip P/G analysis into a single global and multiple partition-level models to solve. For global analysis, the partitions are abstracted with their ports attached to nodes in the global grid, as illustrated in Figure 14.8.

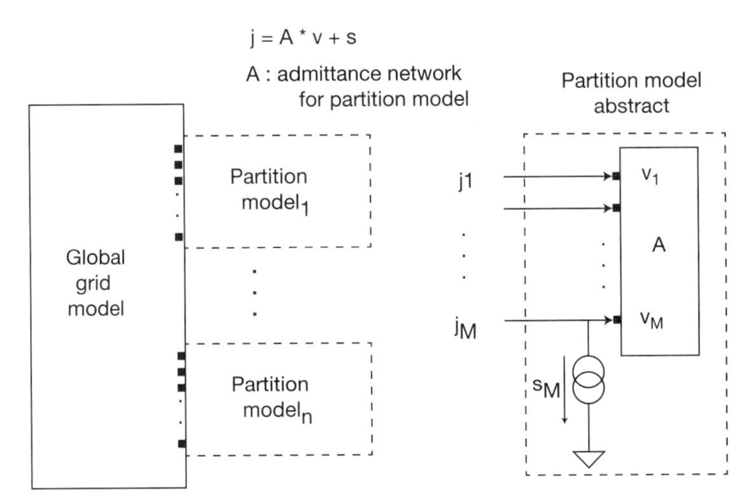


**Figure 14.8** Illustration of global and partition-level models for dynamic I*R analysis.

The partitioning step typically utilizes the physical floorplan block hierarchy. Each partition-level P/G model is composed solely of RC elements with the annotated local cell current sources. Inductive elements are likely to be present only in the global model.

Consider the global grid model, with abstracts for the partitions. In the figure, the current into each partition model port from the global grid is denoted as $j_i$. The vector **j** is of length **M**, where **M** is the number of partition ports. The goal of the dynamic rail analysis at the global level is two-fold:

- Measure the time-varying node voltages for the global RLCM model, using the extracted parasitics for the PDN network merged with the package model and the partition abstracts
- Calculate the current into each partition model port to subsequently perform dynamic rail analysis for each partition

Referring to the expanded view of the partition abstract in Figure 14.8, the current vector, **j**, is equal to the sum of the time-varying internal partition currents abstracted to the partition ports, **s**, plus the product of the admittance matrix for the partition times the partition port voltages:

$$\mathbf{j} = \mathbf{A} * \mathbf{v} + \mathbf{s} \qquad \text{(Eqn. 14.7)}$$

The admittance matrix **A** in the partition abstract is of dimensionality M x M, and the current vector **s** and voltage vector **v** are of length M. Reference 14.3 describes an algorithm for efficiently deriving the matrix **A** and vector **s** from the internal RC network and time-varying current sources within the partition. The partition abstracts are then attached to the global grid and become part of the global matrix solution, using the formulation of Equation 14.5.

During the dynamic rail analysis of the global model, the port voltages for each partition, **v**(t), are recorded. These values are used to calculate the current vector, **j**, using Equation 14.7. With the current sources from the global grid into each partition now available, the partition abstract is replaced by the detailed PDN network, and each partition is solved individually (again using Equation 14.5) to determine the local grid voltages over time.

The admittance matrix **A** has a unique characteristic. An entry $a_{ij}$ in this matrix is nonzero if there is a (direct or indirect) conducting path between the two partition ports. Given the grid-based topology of the PDN, with many potential conducting paths likely between partition ports, the matrix **A** is therefore dense. For improved throughput, the reference includes an algorithm to zero out small $a_{ij}$ admittance matrix values while maintaining an error bound in the overall voltage solution.

The partitioning algorithm seeks to maximize the inclusion of P/G nodes linked only to other internal nodes, while concurrently constraining the number of ports, M. The partition abstract is effective in global analysis if the M**2 entries in matrix **A** are much less than the number of P/G nodes internal to the partition. As mentioned earlier, the partitioning algorithm would typically follow the SoC physical hierarchy, as the physical block design should have far fewer connections to the global P/G grids than the large size of the extracted RC parasitic PDN model within the block.

This hierarchical approach enables a project schedule for dynamic power rail analysis where early blocks in physical design could exercise the flow. The blocks would be abstracted to connect to the global grid, with (coarse) estimates for the global grid current for other, as-yet-unfinished blocks.

### 14.3.4 Analysis Results

The results of the dynamic P/G analysis flow identify any nodes where the calculated rail voltage drop exceeds characterization limits. Typically, the EDA tool provides a die voltage drop image that is similar to a die thermal map. The SoC methodology team is faced with a myriad of design choices, based on the location, breadth, and magnitude of the errors:

Local:

- Refine the block-level P/G grid (e.g., widening rails, adding more grid segments, adding more port connections to the global grid)
- Add more explicit decoupling capacitance cells
- Examine the stressmark switching current profiles and assess whether changes to existing cell placements are warranted to distribute high-activity-factor, high-load cells (which may also alleviate thermal hot spot issues)

Global:

The first step in addressing dynamic rail voltage drop issues on the global grid is always to review the loop current modeling assumptions used in calculating the self- and mutual inductance (L and M) elements present in the global grid network. If the loop area and (partial) inductance calculations are valid, modifications to the global grid to address rail drop issues are required:

- Refine the global P/G grid (e.g., widening rails, adding more grid segments)
- Add more decoupling capacitance, which is typically done for the global grid as metal-insulator-metal (MIM) parallel plate capacitors (as an additional foundry process option)
- Evaluate packaging technology options to reduce pin inductance and/or add (surface-mount technology[SMT]) capacitors to the package

Any significant physical design changes to the global grid late in the SoC project schedule can be extremely disruptive. For example, adding more segments to the P/G grid impacts global net routing track availability. The SoC project management team is likely to require that the power rail analysis flow be exercised repeatedly during the physical implementation phase, using successively improved models for block abstracts as the design progresses.

Also, note that the power rail analysis results provide insight into the current associated with each die bump or bond wire. The foundry includes maximum current limits associated with each die pad connection as part of the PDK release. A separate check in the power rail analysis flow (using the global solution data) is needed to ensure that these current limits have not been exceeded.

### 14.3.5 Global Power Delivery Frequency Response

An informative early analysis of the global PDN can be performed in the frequency domain, where the impedance of the (simplified) die and package global model is plotted versus frequency. Figure 14.9 illustrates a model used to represent the PDN impedance, from the voltage regulator through a PCB, a package, and a die.[4] A preliminary impedance target for the PDN is set, based on the allocated rail voltage swing from the anticipated peak current transient: $Z_{max} = \Delta V_{limit} / \Delta I_{max}$.

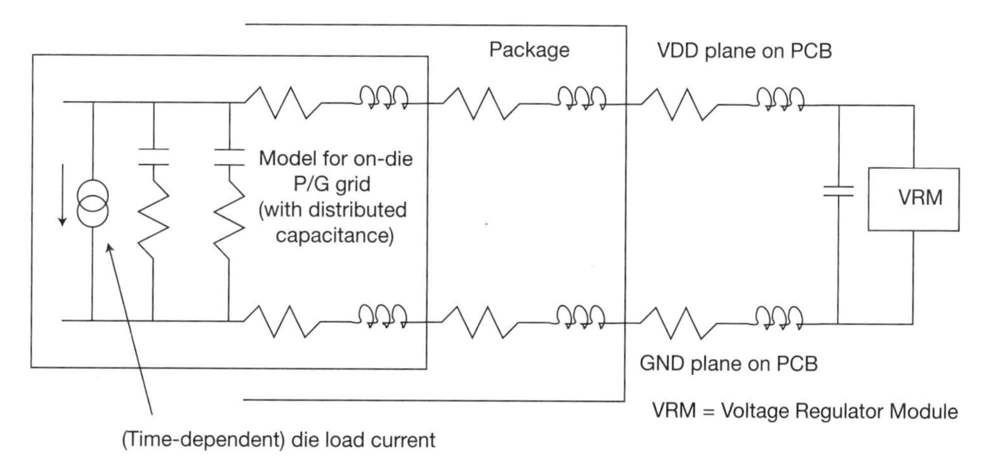


**Figure 14.9** Example of a frequency domain model for the die, package, printed circuit board, and voltage regulator power distribution network.

The SoC project management and product engineering teams undertake a preliminary impedance analysis of the overall P/G model, using estimates of the transient currents drawn by the SoC, both in functional operation and sleep state transitions.

### 14.3.6 SSO Analysis

An additional analysis task utilizes the global die and package model to estimate the VDDIO and GNDIO rail voltage behavior when subjected to the transient current profile of *simultaneous switching output* (*SSO*) drivers. Figure 14.10 depicts a (fraction of) the SSO circuit model, with off-chip driver and receiver cells connected to pads around the die perimeter, powered by specific I/O supply, ground pads, and rails.

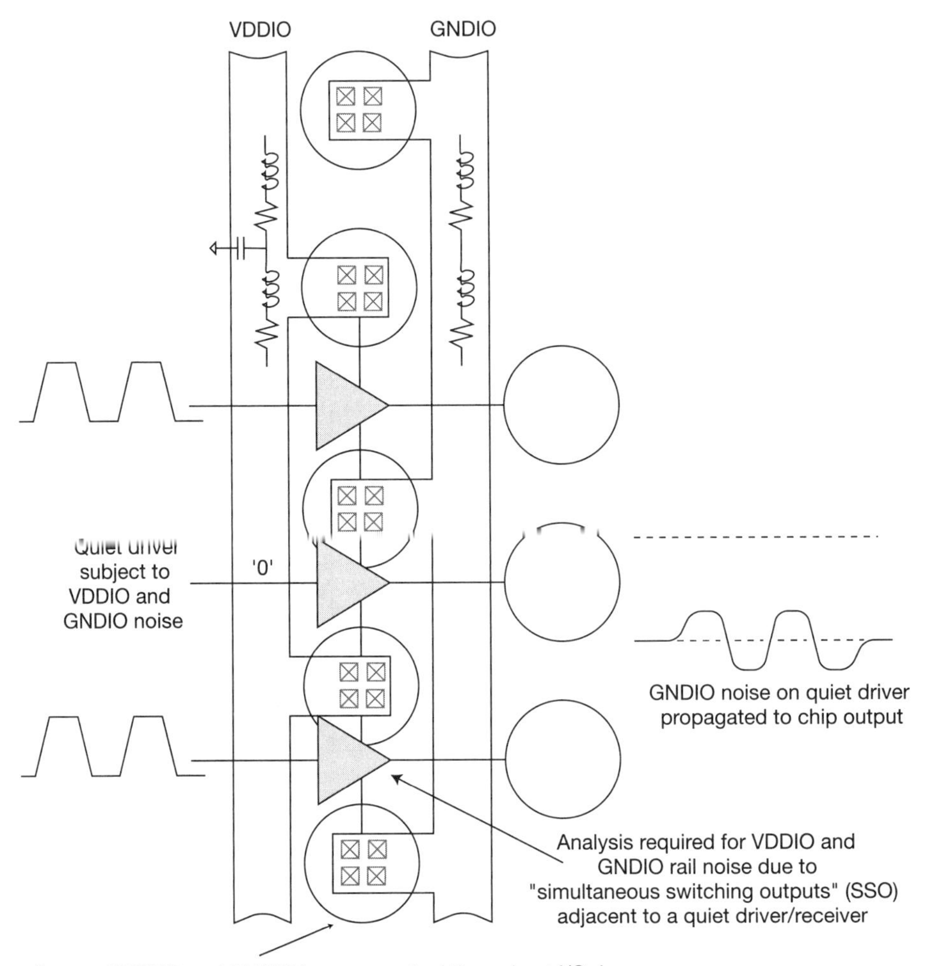


**Figure 14.10** Illustration of a simultaneous switching output (SSO) simulation model.

A set of (single-ended, full-swing) drivers switching in the same direction in a common time window creates a current transient that results in noise on the I/O rails. This P/G noise is propagated through quiet I/O cells. SSO analysis is required to ensure that the propagated logic level noise on SoC I/O signals does not exceed margins.

SSO analysis needs to be undertaken early in the SoC project schedule, as the results are integral to confirming the following:

- The signal pad and package pin assignments
- The corresponding pattern of VDDIO and GNDIO pads/bumps interspersed with the signal pads

- The VDDIO and GNDIO package trace distribution
- The VDDIO and GNDIO rail design through the driver and receiver cells

Unlike the conductance and admittance matrix formulation for internal power rail analysis, with only passive RLCM elements and (time-varying) current sources, SSO analysis includes the non-linear device models for the pad drivers and receivers; thus, a circuit simulation flow is required. Fortunately, the full-chip SSO model can be partitioned into relatively small circuit networks for simulation. By necessity, the density of VDDIO and GNDIO pads among the drivers must be high to keep the inductive current loop as small as possible. As a result, the switching currents are quite localized, and the full SSO model is divisible into die subsets for simulation.

A common feature of the I/O cells is adaptive control, as depicted in Figure 14.11. Over the full operating range of PVT corners, the driver current could vary widely, which would be unacceptable for applications seeking a narrow range of active output impedance.

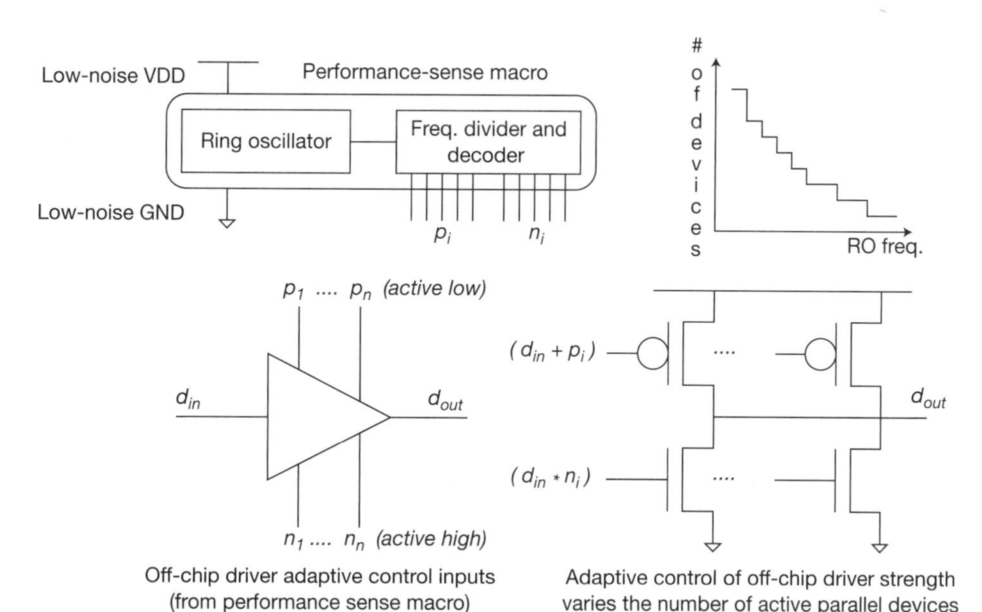


**Figure 14.11** Adaptive compensation for controlling off-chip driver output current over a range of process variation and operating conditions.

To provide driver current compensation, a separate *performance-sensing macro* provides signals to drivers to control the number of parallel devices currently active. The SSO analysis simulation model and testbench includes the performance-sensing macro with a shmoo across the operating corners and output data vectors to measure the SSO noise, driver impedance, and driver power. The additional load on the package pins is selected from the SoC interface signal specification.

## 14.4 Summary

This chapter reviews the model formulation and measurement criteria for the P/G voltage rail drop from (internal and I/O) die currents. The complexity of SoC designs, with greater IP diversity and increasing power management features, makes the design of the global and local P/G distribution extremely challenging. Static analysis, followed by dynamic voltage drop analysis, needs to be exercised repeatedly throughout the SoC project schedule. Whereas other full-chip electrical analysis issues uncovered late in the SoC project schedule are typically resolved with cell or signal route ECOs of limited scope, addressing issues reported by power rail analysis may impact the entire physical implementation, from package technology to global floorplanning to block-level routing.

Section 14.1 reviews both conservative and aggressive approaches to power distribution network design. If an aggressive distribution strategy in the SoC floorplan has been pursued, power rail analysis needs to be exercised often to (incrementally) address any voltage margin failures as additional block physical design detail is available. However, given the pervasive impact of PDN modifications, as the preparation for tapeout sign-off is approaching, the goal of SoC power rail analysis is to have no voltage drop margin fails at all.

## References


[1] Sriram, M., "A Fast Approximation Technique for Power Grid Analysis," *Proceedings of the 16th IEEE Asia and South Pacific Design Automation Conference (ASP-DAC)*, 2011, pp. 171–175.

[2] Shepard, K., and Tian, Z., "Return-Limited Inductances: A Practical Approach to On-Chip Inductance Extraction," *IEEE Transactions on Computer-Aided Design of Integrated Circuits and Systems*, Volume 19, Issue 4, April 2000, pp. 425–436.

[3] Zhao, M., Panda, R., Sapatnekar, S., and Blaauw, D., "Hierarchical Analysis of Power Distribution Networks," *IEEE Transactions on Computer-Aided Design of Integrated Circuits and Systems*, Volume 21, Issue 2, February 2002, pp. 159–168.

[4] Smith, L., Anderson, R., and Roy, T., "Chip-Package Resonance in Core Power Supply Structures for a High Power Microprocessor," *Proceedings of the ASME International Electronic Packaging Technical Conference (IPACK)*, 2001.

## Further Research

### Performance-Sensing Macros and SSO

Describe the features of a performance-sensing macro, as it is integrated into a bank of off-chip drivers for simultaneous switching (and impedance matching) control.

Describe the design of the driver devices that connect to the performance-sensing macro outputs.

### Current Limit Specifications for Die-to-Package Attach Metallurgy

Describe the current limit specifications for wire bond, (lead-free) solder bump, and copper pillar attach metallurgy.

CHAPTER 15

# Electromigration (EM) Reliability Analysis

## 15.1 Introduction to EM Reliability Analysis

The electrical analysis flows described in previous chapters pertain to the correct functionality of the (tapeout) design database. An additional analysis requirement is to ensure that the design will satisfy product lifetime reliability requirements. The product lifetime is typically denoted as "N power-on hours per year for Y years." Often, the package engineering team adds the consideration "M power cycles per year" for its analysis of the mechanical stress on the die/package attach technology due to thermal cycling.

### 15.1.1 Design Robustness and Reliability

There are two classes of SoC design specifications to address for product lifetime operation: (1) robust design, in the presence of circuit changes due to parameter drift or noise sources, and (2) reliability, due to failure mechanisms.

*Robust design* refers to the steps taken to maintain functionality during the operational lifetime. For example, the SoC is subject to high-energy incident particles, either from extraterrestrial sources or radioactive decay of (trace) materials in electronic packages. (Since the initial identification of radioactive decay alpha particles as a major source of soft errors in DRAM modules, package and die attach materials have been modified, reducing this particle flux substantially.) These particles traversing through the die may result in a "collision" and the generation of free electrons and holes. A sufficient concentration of particle collision-related free charge near a sensitive circuit node may disrupt a stored logic value. The soft error rate (SER) in the SoC circuitry can be minimized through robust circuit and layout design so that the magnitude of free charge collection does not cause a flip in stored value (see Section 16.3).

Section 4.2 highlights device-level operational effects that result in parameter drift and device current degradation (e.g., hot carrier injection, bias temperature instability). The SoC methodology team needs to evaluate these device model impacts and establish an appropriate design strategy to maintain functionality. In addition, analog mixed-signal IP needs to ensure proper behavior in response to device noise mechanisms (e.g., thermal, shot, and flicker noise).

Design robustness addresses maintaining the correct circuit response to electrical disruption during operation. SoC reliability relates to *wearout mechanisms*, typically associated with material stresses that result in (significant) changes in material properties or mechanical fractures. As mentioned earlier, thermal cycling exerts stress on the die and package attach materials due to coefficient of thermal expansion differences; over time, a material fracture may result in a (catastrophic) failure. The major reliability issue within the die relates to electromigration (EM). This chapter briefly reviews EM as a failure mechanism and techniques to evaluate the related die failure rate.

### 15.1.2 MTTF and FIT Rate Specification

The end customer market for the SoC design defines the acceptable chip failure rate specification over the product lifetime. The SoC methodology and package engineering teams need to demonstrate that these specifications will be met, whether for consumer, automotive, medical, or aerospace applications. There are two common metrics to represent SoC reliability: the mean time to failure (MTTF; this acronym is also used for *median time to failure*) and the FIT rate.

### *Mean Time to Failure*

Given a population of parts, the MTTF is the expected (mean) time until a part no longer meets the functional specifications. Specifically, if the failure probability density function is denoted as f(t), the expression for the MTTF is given in Figure 15.1.

$$\text{Mean Time To Failure} = \int_0^\infty t * f(t)\ dt$$

**Figure 15.1** MTTF equation, in terms of the failure probability density function, f(t).

The failure probability function is related to several other metrics:

- **The reliability function, R(t)**—R(t) is also referred to as the *survival function*, as it is the probability of the part surviving until time t:

$$R(t) = 1 - F(t) \qquad \text{(Eqn. 15.1)}$$

  where F(t) is in the integral of the failure probability—namely, the cumulative density function (CDF) of the probability density function, f(t).

- **The hazard rate, h(t)**—The hazard rate at time t is the (conditional) probability of failure among the remaining part population that has survived to time t. Mathematically, the hazard rate is represented by:

$$h(t) = f(t) / (1 - F(t)) = f(t) / R(t) \qquad \text{(Eqn. 15.2)}$$

- **The failure rate**—In general, the failure rate is the probability that a product will fail at time t among the remaining product population, and the broadest definition includes repaired products that remain in the population. As a failed SoC is not repairable, the hazard rate and failure rate are equivalent. The term *failure rate* is used in the subsequent discussion, denoted as h(t); the notation f(t) refers to the failure probability density function as opposed to the failure rate.

Failure rate data are measured from a sample and thus represent an estimate of the total population. For reliability calculations due to electromigration, the sample data are derived from fabricated testsites that allow specific failure mechanisms to be isolated. This reliability data is included as part of the foundry PDK release.

The mean time to failure measure is often replaced by *median time to failure* (also, confusingly, denoted MTTF), which is the solution to the integral in Figure 15.2.

$$0.5 = \int_{0}^{MTTF} f(t)\, dt$$

**Figure 15.2** Definition of the median time to failure (also denoted as MTTF). When discussing reliability failure calculations, it is important to clarify which interpretation of MTTF is being referenced.

The median time to failure is perhaps a more meaningful metric, given the very asymmetric nature of the typical failure rate function for microelectronics. Figure 15.3 depicts a *bathtub curve* to illustrate the SoC population failures over time. (Note that *infant fails* are ideally screened and removed by burn-in stress testing, as discussed in Section 21.1.)

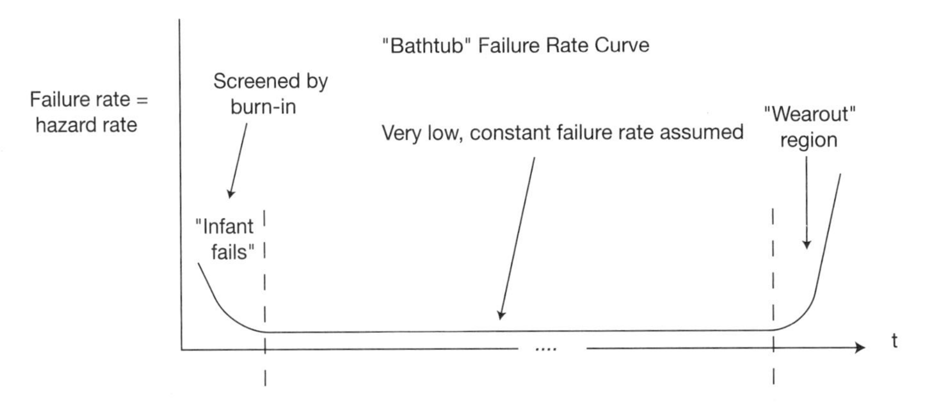


**Figure 15.3** Illustration of a typical "bathtub curve" failure rate function, h(t). Assuming that infant fails are screened, the part population failure rate is relatively constant until the wearout region.

The probability density function f(t) associated with the bathtub curve failure rate h(t) is skewed. A composite Weibull distribution is often used, with shape = 1 for the part of the population representing the constant failure rate and shape > 1 for the *wearout region*. The mean value of the distribution is significantly larger than the median due to the wearout region. As a result, the median time to failure (assuming a constant failure rate during the useful product lifetime) is more informative than the mean time to failure. Due to the nature of statistical sampling, there is also a confidence level typically associated with the median time to failure specification.

### *FIT Rate*

The *failure in time* (*FIT*) *rate* is the reliability metric that is more commonly used, as it is numerically straightforward to apply it to a large SoC product volume with the power-on-hours lifetime specification. The FIT value is an estimate of the number of failing parts in 10**9 hours of operation accumulated across the entire part population. The 10**9 hours is a de facto standard for reference. Mathematically, the FIT value and MTTF are related, as follows.

Assume a *constant* failure rate, h(t), over the useful life of the population, neglecting the wearout region. The FIT rate is equal to ((10**9) * h(t)). For example, if the estimated SoC failure rate is h(t) = 3.5 * (10** − 8) units/hour, the FIT rate for the part is 35. (Note that a single constant failure rate assumes that the part is consistently operating with a device junction temperature equal to $T_{max}$; if the temperature profile over the operating lifetime is known to differ from $T_{max}$, a composite calculation using multiple failure rates may be more appropriate.)

For a constant failure rate, the surviving function R(t) is a very slowly decaying exponential. The R(t) and MTTF are illustrated in Figure 15.4.

R(t), F(t), and f(t) for a constant failure rate

$$R(t) = e^{-(t/\tau)}$$ ($\tau$ is a large value, R is a slowly decaying exponential)

$$F(t) = (1 - R(t))$$

$$f(t) = dF/dt = \left(\frac{1}{\tau}\right) * e^{-(t/\tau)} \quad \text{and} \quad 0.5 = \int_0^{MTTF} f(t)\, dt$$

**Figure 15.4** Illustration of the surviving function, R(t), and the failure probability density function, f(t), for a microelectronic part with a constant failure rate during the operational lifetime.

The motivation for the discussion of both failure rate and MTTF is that the key electromigration mechanism is described in terms of MTTF, whereas SoC specifications commonly use the FIT value. Product developers add the FIT rates of individual components to calculate the overall expected failures.

Note that there are additional reliability metrics used when referring to a (repairable) end product outside the field of microelectronics. The mean time between failures (MTBF) is a probabilistic estimate of the time between successive failures of a product. The mean time to repair (MTTR) is an estimate of the resources required to return a product to functional use. These metrics are evaluated by product developers when estimating service costs and maintenance (downtime) schedules, including (possibly preventive) replacement strategies for field-repairable units (e.g., swapping an existing printed circuit board with a new replacement). The (non-repairable) SoC MTTF or FIT specifications are incorporated into the overall product MTBF calculation.

These SoC reliability projections are inherently statistical approximations, as the corresponding failure mechanisms are probabilistic in nature; this is certainly the case for electromigration. As discussed in the remainder of this chapter, the goal is to establish current density limits for metal interconnects, vias, and contacts such that the total failure rate for the entire SoC due to electromigration does not exceed the reliability target. SoC electrical data from other flows are analyzed to compare actual current densities against these limits.

### 15.1.3 Sum of Failure Rates

The typical reliability model used for a large electronic system is the *sum of failure rates*.[1] This model is based on the following assumptions:

- Each failure rate is independent (at least up until the first failure occurrence).
- The first failure among the independent mechanisms represents the failure of the system.

The failure rate data measured experimentally by the foundry are based on the testsite population, using elevated current density and temperature to accelerate the electromigration mechanisms. Using a key relationship (i.e., Black's equation), the measured failures can be scaled to calculate the operational MTTF and FIT with corresponding current density limits. Note that the definition of "a failure" in testsite measurement data subsequently needs to be applied to the analysis of the electrical model of the SoC; this is discussed further in the next section.

## 15.2 Fundamentals of Electromigration

The metals used in the fabrication of wires, contacts, and vias are subject to several forces during operation:

- **Electric field**—The electric field across a metal segment results in electron motion and current; however, there is no significant local net charge concentration, and thus there is no electric force on metal atoms.
- **Electron momentum transfer**—The electric current in the metal gives momentum to electrons. An electron collision with a metal atom/ion imparts a force on the atom, which may result in a permanent displacement of the atom, potentially moving it to a lattice *vacancy*. Metal atom vibrations in their lattice position increase with temperature; thus, the collision probability is strongly dependent on temperature. The net flux of metal atoms due to the electron collision force over the lifetime of the SoC is denoted as *electromigration*.
- **Thermal gradients**—The current in metal wires results in resistive power dissipation. The Joule self-heating energy due to the ((I**2)*R) dissipation results in a temperature rise in the metal and thermal gradients throughout the local neighborhood. The Joule heating temperature increase (over the reference die substrate) has a multi-faceted impact:
  - Electron-collision migration is accelerated for the wire.
  - Thermal self-diffusion (*thermo-migration*) moves metal atoms to vacancies.

    Although the thermo-migration flux is relatively small, it may result in additional paths for electron collision-driven atom displacements, accelerating electromigration.

  - The temperature of neighboring structures increases, based on the relative thermal resistances between adjacent wires and the substrate.
  - The temperature coefficient of resistance results in higher R, adding to interconnect delays.

The deposition and patterning of metal wires has characteristics that enable migration. The metal volume is composed of *grains* with abutting grain boundaries and contains lattice vacancies, as mentioned earlier. As depicted in Figure 15.5, the electron momentum imparted to a metal atom is in the direction opposite the current flow. Metal displacement results in accumulation (or *hillocks*) toward the anode, and the cathode end of the wire is subject to the *nucleation* and subsequent growth of *voids*. The paths for metal migration include vacancies, grain boundaries, and the area along the surrounding metal/dielectric material interface.

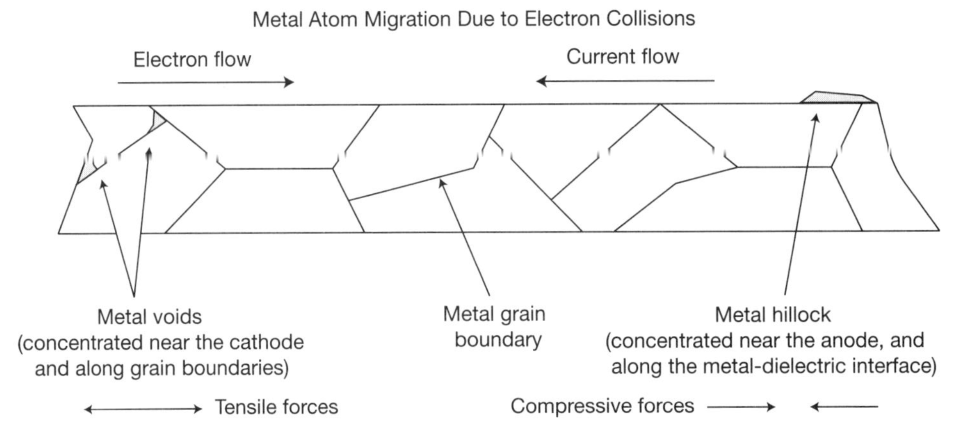


**Figure 15.5** Metal atom migration due to current flow. The atom flux is in the opposite direction of the current flow as a result of electron momentum collisions. Examples of metal grains, hillocks, and voids are depicted.

Fabrication process development has focused on reducing the diffusivity of metal atoms along these paths:

- Wire and via metallurgies have evolved, from Al to AlCu to Cu.
- Wire patterning using damascene technology includes thin trench metal barrier and seed growth layers, plus capping layers, with lower interface diffusivity.
- Local device-level and contact metallurgies have evolved, and refractory metals are increasingly being used.
- Metal grain structures and size have received considerable process development focus.

If the grain size and orientation can effectively span the cross-section of the wire, the overall cathode-to-anode diffusivity is reduced.

### 15.2.1 Black's Equation and Blech Length

The study of electromigration in integrated circuits began in the 1960s. Subsequent research has expanded the understanding of the metal migration mechanisms, the influence of bidirectional versus unidirectional current, and the influence of circuit-level device self-heating on the thermal profiles of local metals. The original EM investigations resulted in a model known as Black's equation for the median time to failure due to electron collisions that has continued to be used in the subsequent decades (see Figure 15.6).

Black's Equation for Median Time to Failure in a Wire

$$MTTF = t_{50_{DC}} = \frac{A}{j^n} * e^{(E_A / kT)}$$

A: constant, $E_A$: activation energy, j: current density in the wire;
n: fitting exponent; k: Boltzmann's constant; T: temperature

**Figure 15.6** Black's equation for the median time to failure due to electromigration.

In the equation, the factor A is an empirical constant, j is the current density in the wire, and $E_A$ is the *activation energy* for metal ion displacement. The exponent n depends on the relative contribution of void nucleation ($n \sim 2$) followed by void growth ($n \sim 1$). For AlCu wires, the migration kinetics typically result in using $n = 2$ as the best fit to Black's equation; for Cu wires, the kinetics favor $n = 1$. The key characteristics are the strong dependency of measured failures on current density and temperature. As mentioned in the previous section, this equation enables the fitting of accelerated test measurement data, which can then be applied to operating MTTF targets, as shown in Figure 15.7.

Scaling measured failure data at accelerated conditions using Black's equation

$$j_{limit} = j_{accelerated} * \left( \frac{MTTF_{accelerated}}{MTTF_{target}} \right)^{\frac{1}{n}} e^{\left\{ \frac{E_A}{k} \left( \frac{1}{T_{target}} - \frac{1}{T_{accelerated}} \right) \right\}}$$

**Figure 15.7** Measured failures on silicon testsites utilize accelerated stress conditions. The MTTF data is subsequently scaled back to the SoC operating environment using Black's equation.

The general modeling assumption is that equation coefficients A and n are independent of temperature; the ratio of the activation energy to kT represents the temperature dependence.

Subsequent experimentation demonstrated that there is a substantial reversal of the metal atom displacement flux due to electron collisions when the wire is subject to a (high-frequency) bidirectional current, as would be the case for signal interconnections with moderate-to-high switching activity. A typical model for the MTTF for wires with bidirectional currents is shown in Figure 15.8.

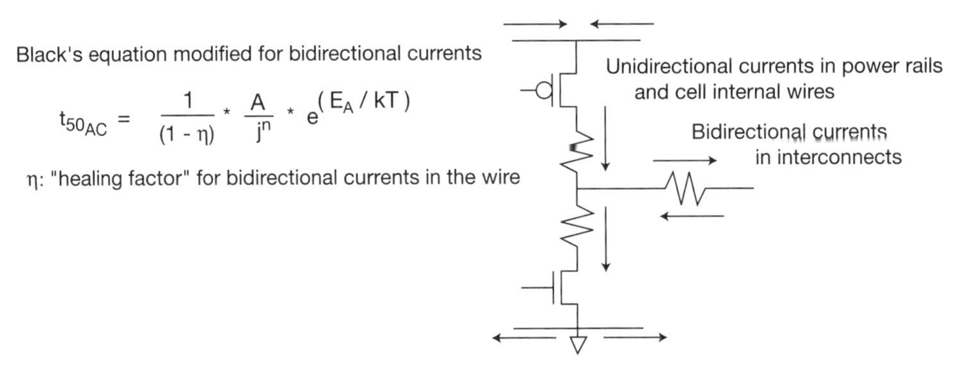


**Figure 15.8** MTTF model incorporating a healing factor to reflect the reversal of atom displacement for bidirectional currents in the metal.

In this revised model, the *healing factor* increases the reliability current density for bidirectional currents. As a result, signal interconnects between cells are analyzed using different criteria than are used for unidirectional (power rail and cell internal) metals.

Another experimental observation on EM failures noted a wire-length dependence. For test structure wires below a certain length, the failure rate dropped dramatically. As metal atoms are displaced, compressive stress in the metal is present at the anode, and tensile stress exists at the cathode. For short-length wires, these additional material stresses result in a force on metal atoms that counteracts the electron collision flow. The *Blech length* relationship is used to define metal segments for which significantly higher current density limits are applied. (The *definition* of an EM fail is discussed shortly.) The Blech length expression is ($j * L_{Blech}$<process_limit), specific to each metal layer. The dielectric materials (and damascene barrier metal layers) influence the Blech length, as these interfaces also contribute to the metal stresses.

### 15.2.2 $J_{DC}$, $J_{AC}$, $J_{RMS}$, and $J_{PEAK}$

The electromigration reliability of wires, contacts, and vias is analyzed using four factors:

- Unidirectional DC (average) current density, $j_{DC}$ or $j_{AVG}$
- Bidirectional current density (with healing), $j_{AC}$
- The wire temperature increases due to resistive Joule heating, represented for EM analysis by $j_{RMS}$
- Peak current density, $j_{PEAK}$

#### *Unidirectional DC (Average) Current Density and Bidirectional Current Density (with Healing)*

Using Black's equation, the calculated $j_{DC}$ and $j_{AC}$ current density for all SoC wires, vias, and contacts are compared to the foundry PDK current limit to scale the PDK failure rate. The PDK provides the following:

- $j_{DC}$—the reference PDK current limits published for all metal and via layers
  - Blech length current multiplier for metal layers, using a B(Lwire, Wwire) relationship
  - Via current limit multiplier, based on the Blech length of connected $metal_n$ and $metal_{n+1}$
- $j_{AC}$—healing factor per layer applied to $j_{DC}$ limits (see Figure 15.9)
- FIT value for the reference $j_{DC}$ current limit

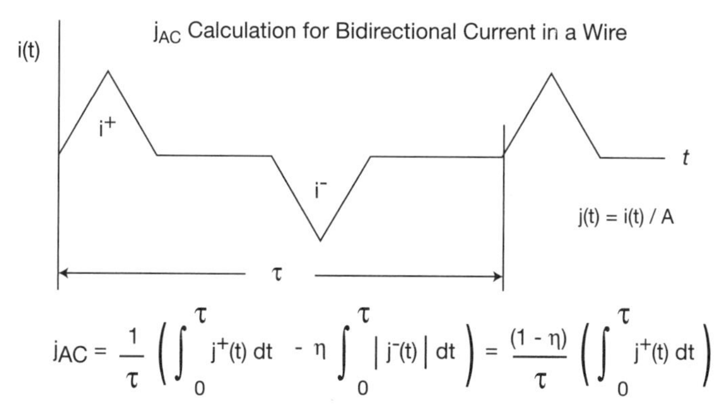


$$j_{AC} = \frac{1}{\tau}\left(\int_0^{\tau} j^+(t)\,dt - \eta\int_0^{\tau} |j^-(t)|\,dt\right) = \frac{(1-\eta)}{\tau}\left(\int_0^{\tau} j^+(t)\,dt\right)$$

**Figure 15.9** Example of the calculation of $j_{AC}$ for a bidirectional wire current.

Each SoC metal segment has a FIT value. The current density in the segment is compared to the PDK reference to scale the published FIT using Black's equation. The $j_{AC}$ calculated for the bidirectional current density in the metal segment is submitted to the $j_{DC}$ failure rate calculation; the healing factor is assumed to be a constant over the typical range of signal switching frequencies. As there is an assumption of complete signal net charge/discharge for each transient, the integrals of the two currents shown in Figure 15.9 are (by definition) equal.

#### *Wire Temperature Increase Due to Resistive Joule Heating*

The development of a self-consistent model for the wire $j_{RMS}$, the local temperature profile, and EM reliability is difficult due to the interdependencies listed earlier, at the beginning of Section 15.2. A typical approach is to:

1. Estimate the $j_{RMS}$ current in the wire (as discussed in subsequent sections).
2. Utilize a (pre-characterized) model from the foundry for the wire temperature increase due to $j_{RMS}$ self-heating for the wire metal layer.
3. Add the thermal contribution from local device power dissipation to the wire temperature.

The pre-characterized foundry model for the temperature increase due to $j_{RMS}$ is specific to each BEOL metallization stack, with the corresponding metal/dielectric layer thicknesses and thermal resistances.

For FinFET and FD-SOI device technologies, the thermal flow to the substrate is reduced over planar devices; an increasing fraction of the active device thermal energy flows into the metallization stack. The foundry provides the models for the (three-dimensional) temperature increase to neighboring wires due to device dissipation. The EDA vendor providing the EM analysis tool links the circuit-level data from the power dissipation analysis flow to these $\Delta T$ tables in the PDK to calculate the wire temperature adder due to device self-heating.

The "steady-state" wire temperature is the sum of the $j_{RMS}$ Joule self-heating $\Delta T$, the $\Delta T$ from device self-heating, and the substrate temperature (e.g., $T_{substrate} = T_{junction\_max}$). If a more accurate die thermal map is available from the power analysis flow, the values from that map could be used instead of a single substrate temperature when calculating the wire temperature.

The calculated wire temperature would be used to further scale the wire failure rate from the foundry PDK data. This method is depicted in Figure 15.10. The calculated wire temperature combining the self-heating, device dissipation, and die substrate temperature factors would be compared to the foundry PDK reference temperature, and a failure rate multiplier would be determined for the wire using Black's equation for the wire metal layer.

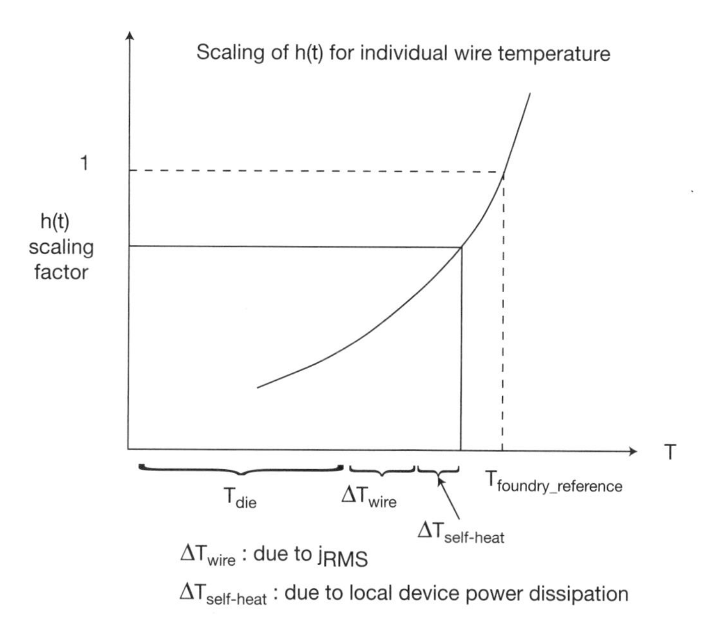


**Figure 15.10** Illustration of the scaling multiplier for the failure rate of a specific wire, based upon the calculated wire temperature.

Note that the $j_{RMS}$ temperature increase calculation applies to metal layers. The Joule heating contribution in vias is small (assuming that the layout includes sufficient vias to satisfy the $j_{DC}$ limit).

### *Peak Current Density*

The average, bidirectional, and RMS calculations integrate current waveforms over a longer interval. There is also an EM current limit applicable to a short duration current pulse. Each wire segment has a peak current density limit, $j_{PEAK}$. The definition of the peak current measurement is illustrated in Figure 15.11.

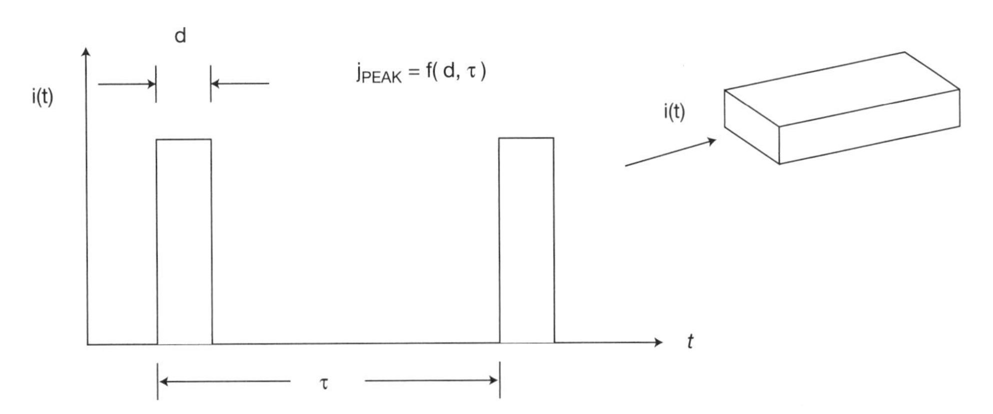


**Figure 15.11** Illustration of the $j_{PEAK}$ calculation in a wire or via.

This current density limit is intended to avoid excessive Joule heating, which could lead to metal deformation. Exceeding the $j_{PEAK}$PDK limit values for a wire or via should be regarded as an *absolute* failure, which requires design changes, rather than as a contribution to the SoC operating lifetime failure rate.

Also, it should be noted that different interconnect layout topologies may have significant *current crowding* at the transition between a wire and via or in wire jogs. However, EM analysis uses the extracted parasitic electrical network from the layout and thus is not able to accurately reflect non-uniform current densities in the cross-section of a resistive element. Separate EM-avoidance layout design guidelines are typically added to the PDK design ruleset, especially for via arrays between wide metals.

### 15.2.3 Definition of an EM Failure

The foundry's process development and PDK design enablement teams fabricate a set of testsites to measure EM failure rates, consistent with Black's equation (with AC healing) and the Blech length expression. The published $j_{DC}$, $j_{AC}$, $j_{RMS}\Delta T$, device power $\Delta T$ contribution, and $j_{PEAK}$ data for each layer are referenced to a specific failure rate and temperature. For example, (except for $j_{PEAK}$, which is absolute) the current limits for each wire, via, and contact could be defined as follows:

h(t) = (1 * 10 ** − 9) at 120 degrees C (e.g., 1 FIT)

The CAD team collaborates with the EDA tool vendor on the software utilities required to submit the wire current density and temperature results from the electromigration analysis flow to Black's equation. This calculation with the actual current and temperature scales the published PDK FIT rate for each wire, via, and contact. The full-chip SoC reliability evaluation applies the sum-of-failure rates assumption to the entire population of wires, vias, and contacts to provide an overall FIT estimate.

However, it is also crucial for the SoC methodology team to review the testsite measurement criteria that denote failure. Metal atom flux results in the formation of voids, originating from the nucleation of metal lattice vacancies. The focus of accelerated testing at the foundry is the *resistance increase* in the wire or via due to the growth of voids. A "failure" is recorded when the starting resistance of a testsite structure increases by N%—typically N = 10%, or $\Delta R = (R_{stress} / R) = 1.1$. A scattergram of ΔR versus current density is plotted for the resistance data measured after stress testing for each structure on the testsite wafers. This plot is analyzed to select the current density limit for each metal, via, and contact layer, corresponding to the reference FIT for that layer published in the PDK.

Although a wire segment resistance increase is reported as an EM failure, the SoC design may be sufficiently robust to maintain functionality. For example, the grid topology of the power/ground distribution may still be able to maintain an adequate voltage drop in the presence of (multiple) EM failures. The contribution of a "failing" power grid segment to the FIT calculation may be overstated. The EM analysis flow targeting the power grids integrates the voltage drop analysis algorithm described in Chapter 14, "Power Rail Voltage Drop Analysis," with a modified conductance matrix due to high-resistance segment(s)—perhaps even removing the conductance of the failing EM segment altogether. If the power grid is still adequate, the FIT value merits correction.

The methodology for EM analysis for a signal wire in the presence of current density failures is more complex. The robustness of a static timing or noise path may remain in the presence of wire segment resistance increases of N%. Conversely, a critical path may fail static timing analysis if the resistive element(s) in the interconnect model are increased by N%.

The SoC methodology team is faced with multiple options in response to the results from the sigEM flow:

- First, any $j_{PEAK}$ limit failures *must* be resolved.
- Design ECOs may be required to address any "high FIT" wires/vias. If a signal wire has a high FIT due to its current density, this may be indicative of a physical layout mismatch between cell drive strength, wire widths, and capacitive load. Min/max signal slew limit checks applied to the static timing analysis results would have previously identified appropriate design changes and hopefully reduce the mismatch resulting in the high current density. Reference 15.2 compares the current density from a "matched" driver and interconnect to typical process technology $j_{AC}$ EM limits. If a signal via has a high FIT contribution, multiple vias (on wider wires) are warranted. If the $\Delta T$ temperature adder is a significant factor in the FIT value, more significant layout changes are required to increase the distance from power dissipation sources.
- Additional performance margin for EM resistance increases over the SoC lifetime may be appropriate, in addition to BTI and HCI device parameter drift mechanisms. This is an unattractive option, to be sure.
- The static timing analysis slack results for the paths corresponding to signal wires with high FIT are evaluated and an assessment is made to determine whether a potential increase in interconnect delay has sufficient positive slack to warrant reducing the FIT value.

Reference 15.3 presents an algorithm to correlate the wire EM failure probability density function, f(t), to a probability distribution for the increase in wire resistance. Then Monte Carlo-sampled circuit simulations are used to evaluate probabilistic path delay increases over time (e.g., 5 to 20 years). The SoC and CAD teams may opt to deploy a similar flow to assess the sensitivity of timing paths with high-FIT wires to the resistance increase as a function of operating lifetime.

### 15.2.4 Extraction Model for Electromigration

To apply the sum-of-failure-rate algorithm for the SoC reliability calculation, it is necessary to evaluate the current densities in a detailed parasitic extraction network without reduction. In addition, the parasitic extraction flow needs to be exercised in verbose mode, where resistive elements include full layout detail to enable current density calculation (see Figure 15.12). The resistor example in the figure includes a thickness parameter. The foundry may provide an analytical model for the chemical-mechanical polishing metal planarization process module. The layout extraction tool from the EDA vendor may include a corresponding CMP analysis algorithm to derive a (non-default) wire thickness estimate. (CMP analysis is also applied to yield prediction models, as any non-planarity of the wafer surface affects lithographic exposure accuracy due to depth-of-focus limits.)

```
Rn node1 node2 value
    $W = ... $L = ... $layer = ... $thickness = ...
    $X1 = ... $Y1 = ... $X2 = ... $Y2 = ... ;
```

Additional dimensional and coordinate parasitic extraction data for EM analysis

**Figure 15.12** Illustration of resistance extraction in "verbose" mode, with layout detail to enable current density calculation.

Electromigration models and the related failure rate calculations are complex probabilistic relationships that utilize an empirical fit of testsite data to the model of Black's equation. As mentioned earlier in this section, the kinetics for atom displacement are unique for unidirectional and bidirectional currents. As a result, EM analysis is divided into power and ground rail (*powerEM*) and switching interconnect (*sigEM*) flows, discussed next.

## 15.3 Power Rail Electromigration Analysis: powerEM

The static I*R voltage drop analysis flow described in Section 14.2 applies the (average) circuit currents to the extracted resistive model of the VDD and GND grids. The matrix equation to solve for static I*R was presented as $\mathbf{G} * \mathbf{v} = \mathbf{i}$, where **G** is the conductance matrix between grid nodes, **i** is the vector of circuit currents, and **v** is the vector of node voltages to solve. The powerEM analysis flow leverages the I*R voltage drop data. The $j_{DC}$ through each conductance element in the matrix is calculated using the equation in Figure 15.13.

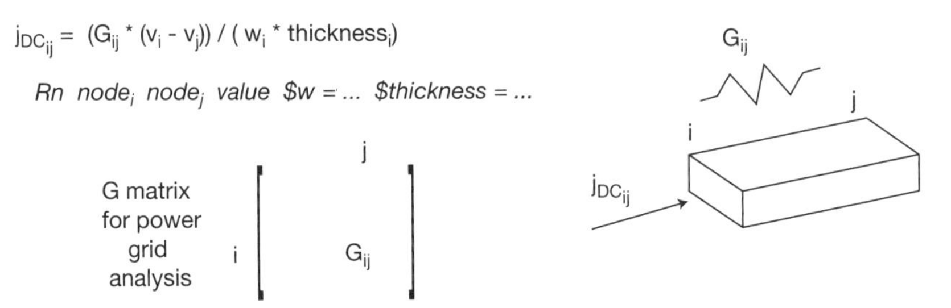


**Figure 15.13** Calculation of $j_{DC}$ in a power grid wire for the powerEM flow.

The assumption is made that the power and ground grids are adequately described by the static I*R rail voltage drop data, using average cell currents. The $j_{DC}$ failure probability spans years, as opposed to the time scale for the dynamic I*R rail voltage analysis flow results. For a DC current in the power and ground grids, $j_{RMS} = j_{DC}$.

Using the $j_{DC}$ value derived from static I*R analysis, the powerEM flow calculates the failure rate measure for each resistor, using the parameters provided in the foundry PDK. The FIT is scaled from the PDK data by the calculated $j_{DC}$ and wire temperature. The metal line length parameter from the parasitic extraction detail multiplied by the $j_{DC}$ current density defines whether Blech length kinetics apply to the failure rate calculation.

The powerEM flow then provides an output report with the cumulative FIT value and a listing of the individual high-FIT wires. Using the (x,y) coordinate information included with the parasitic resistive element, the powerEM flow also provides a physical view that highlights the high-FIT segments to allow the layout engineer to more easily visualize the wires and vias of concern.

The voltage solution from the dynamic I*R rail voltage drop flow enables a time-based current for conductance elements to be calculated (see Section 14.3):

$$\mathbf{G}^*\mathbf{x}(t) + \mathbf{Y}^*\mathbf{x}'(t) = \mathbf{s}(t) \qquad \text{(Eqn. 15.1)}$$

The solution $\mathbf{x}(t)$ from the dynamic I*R rail analysis provides the voltage at each node in the PDN network and is used in the calculation of the time-varying current in the metal segment. The $j_{PEAK}$ limit check applies to this dynamic conductance current. As mentioned in the previous section, the peak current density limit is based on the Joule heating rise:

$$j_{PEAK}\_limit = f\,(j_{PEAK\_DC}, pulse\_duty\_cycle) \quad \text{(Eqn.15.2)}$$

For the power and ground grids, the current profile differs from the short-duration pulse of a switching signal, typically with small fluctuations about the average rail current. Applying the maximum current from dynamic I*R analysis for each conductance element (with duty_cycle ~ 1) would give an appropriate approximation in the $j_{PEAK}$_limit comparison. Any powerEM $j_{PEAK}$ failures would require physical design modifications; these are hard violations rather than contributions to the SoC FIT calculation.

The SoC methodology team reviews the powerEM results and proceeds with (potentially) multiple recommendations to address issues:

- **Modify the global power and ground grids, if there are regular high FIT "weak spots."** Global grid layout data are often created using a script developed by the CAD team. PowerEM flow results may suggest a pervasive change to the grid—for example, wider wires (on specific layers), more vias between segments on different layers, or a larger stacked via area. The impact on global routing track density would need to be assessed. Note that the powerEM flow could be exercised right after (early) static I*R analysis to identify this impact as soon as possible in the overall project schedule.
- **Locally modify the P/G grid.** A unique version of the P/G grid layout cell could be manually generated in response to specific high-FIT segments. (A review of the I*R voltage drop and $\Delta T$ results in this local area would be appropriate to determine the root cause of the high FIT segment calculation.)
- **Iterate on static I*R and powerEM flows by removing the high-FIT segments from the grid.** As mentioned in the previous section, the definition of an EM segment failure is based on a (probabilistic) percentage increase in the segment resistance. The power and ground grids may be able to continue to provide an adequate voltage drop if (a small number

of) grid segments are higher resistance. The CAD team provides a utility to remove selected high-FIT segments from the parasitic network to allow iteration on the static I*R drop and powerEM flows. (Although the EM failure criterion is a percentage increase in R, a conservative approach would be to remove the conductance segment in the revised static I*R calculation.) The overall flow is depicted in Figure 15.14.

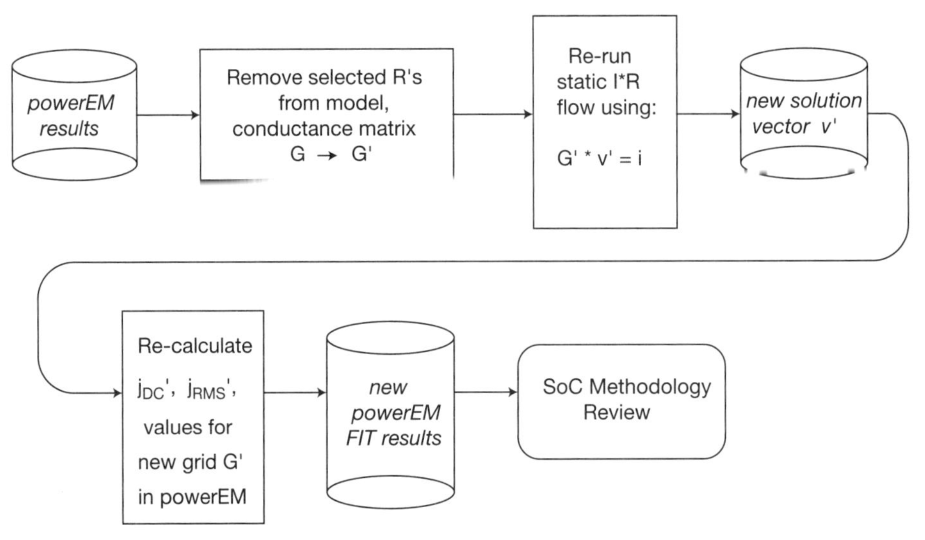


**Figure 15.14** Illustration of the (iterative) use of the powerEM and static I*R analysis flows to determine the impact of a power grid segment EM "failure"; the $g_{ij}$ element in the conductance matrix is modified.

If the new power and ground voltage drop solution is still adequate with the segment removals in the conductance matrix, the high-FIT grid segments in the initial powerEM results could potentially be waived if the subsequent powerEM results do not demonstrate a significantly higher failure rate.

## 15.4 Signal Interconnect Electromigration Analysis: sigEM

As described in Section 15.2, the EM reliability analysis of interconnects between cells applies a different $j_{AC}$ current density calculation due to the different metal atom migration kinetics with healing for bidirectional currents. As a result, a specific sigEM flow is developed for the SoC methodology, different from the powerEM flow. These two flows may be exercised at different times in the project schedule and at different levels of the SoC design hierarchy. For example, the powerEM flow may be exercised early to evaluate the global P/G grid design, using average current density estimates for individual blocks with a "constant current per unit area" measure at the grid conductance nodes. Conversely, the sigEM flow would be exercised (repeatedly) during block-level physical design iterations to ensure proper route and via construction for each cell driver (using a block-level thermal map). A budget for the total FIT value would be assigned to each block (and global signals and clocks, as well) to enable block-level analysis to proceed independently.

Before discussing the sigEM flow, there are two topics to highlight: cell-level EM and EM analysis for clocks.

### 15.4.1 Cell-Level EM Analysis

The development of a cell IP library for release to SoC designers requires analysis of the FIT rates associated with the wires, vias, and contacts within the cell layout. The cell contains a mix of unidirectional and bidirectional wire currents, necessitating calculation of both $j_{AVG}$ and $j_{AC}$ current densities during the cell characterization circuit simulations.

The library developer needs to ensure a very low FIT rate for each cell, across a wide range of characterization and usage conditions. The calculation of $j_{PEAK}$, $j_{AVG}$, $j_{RMS}$, and $j_{AC}$ for each metal segment in the extracted model requires a conservative (high) estimate for the assumed circuit-switching activity. The current density calculations need to use the maximum values measured during characterization for any of the pin-to-pin delay arcs and the range of input pin slews and output loads.

There are several electromigration concerns specific to cell-based analysis, as illustrated in Figure 15.15:

- The proximity to device self-heating results in a significant $\Delta T$ contribution, in addition to the $j_{RMS}$ wire self-heating.
- The availability of local-interconnect metal layers in a process technology results in both horizontal and vertical current density in the metal, with different PDK EM limits for the M0 layout for the two current directions.

- The introduction of area pins enables the router to select among multiple wiring tracks; the specific pin location selected modifies the unidirectional and bidirectional current profiles; a worst-case pin selection EM analysis is required.

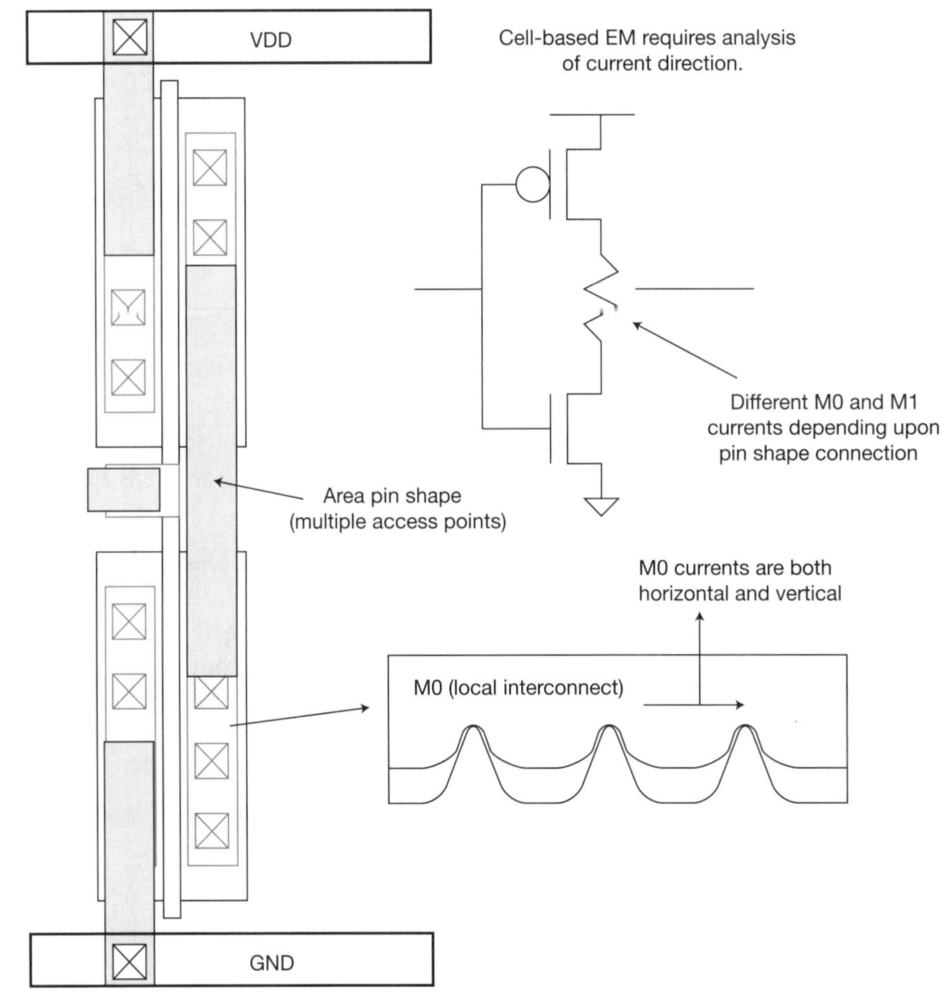


**Figure 15.15** The cell-level EM analysis requires interpretation of the current flow in metal segments. The M0 layer local interconnect has both horizontal and vertical currents. The specific connection to an area pin results in different current directions for metal segments under the pin.

These same EM analysis considerations apply to the reliability estimates for larger IP macros, as well. The SoC methodology team needs to review the algorithms used by the IP vendor when evaluating the library IP FIT rate specification.

### 15.4.2 EM Analysis for Clocks

The design of clock repowering grids and H-trees requires focus on optimal sizing of clock buffers and interconnects. Early EM analysis of clock $j_{AC}$ and $j_{RMS}$ is encouraged, as soon as the (global) clock distribution is available, to confirm the buffer and wire design assumptions.

Like the device-level BTI and HCI parameter drift mechanisms, the evolution of resistance increase in clock wires contributes to arrival skew variation at clock distribution endpoints. The previous section references an algorithm to correlate interconnect electromigration (void growth) data with a probability distribution model for resistance increase; this model would be applicable in clock simulations to develop a consolidated BTI/HCI/EM lifetime skew margin to adopt in static timing analysis.

### 15.4.3 SigEM Current Density Calculation

The sigEM analysis flow for IP library cells measures the current densities of interest using the circuit simulation data from cell characterization. For (block and global) signals between cells, a different method for current density calculations for sigEM is required.

The static timing analysis flow provides a model for the driving current source on each cell output for both RDLY and FDLY transitions (for each STA mode/corner). Although there are different current source profiles for different input pin-to-output pin arcs, STA propagates a single arc for path delay analysis. Assuming that this arc represents the greatest sensitivity to an increase in the interconnect resistance, using the STA-propagated driver current profile for sigEM analysis is appropriate.

The circuit simulation of each interconnect tree to determine $j_{AC}$, $j_{RMS}$, and $j_{PEAK}$ would be computationally expensive. Reference 15.4 describes an alternative approach, which involves solving a matrix-based formulation to calculate these currents. The translation of the interconnect model into a linear matrix is based on evaluating the total charge delivered to the network from the driving current at the cell output, as depicted in Figure 15.16.

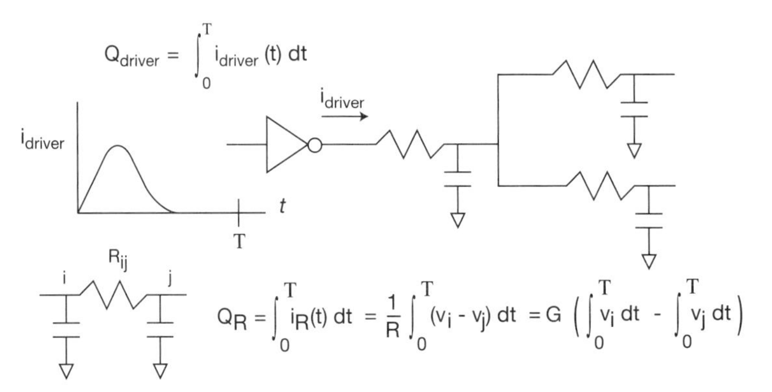


Matrix formulation to calculate total charge through each resistor due to a driver transient.

**Figure 15.16** Illustration of an algorithm used to compute the current through each resistive element in an RC interconnect tree from a transition on the driving output pin.

The matrix equations for the signal interconnect RC trees are similar to those developed for the dynamic power voltage drop flow using Kirchhoff's current law, as described in Section 14.3:

$$\mathbf{G} * \mathbf{v}(t) + \mathbf{C} * \mathbf{v}'(t) = \mathbf{s}(t) \qquad \text{(Eqn. 15.3)}$$

In Equation 15.3, **G** is the interconnect conductance matrix of dimensionality (N x N), where there are N nodes in the interconnect tree (including a ground reference node), and $g_{ij}$ is the conductance between nodes i and j. For the case of an extracted signal interconnect, this conductance matrix is very sparse (and there is no conductance between any extracted node and ground). **C** is the network capacitance matrix, and **v** is the vector of interconnect node voltages. The vector s describes the current sources into the network. For sigEM analysis, it is assumed that the only current source is from the driver node; no other side currents are injected into the interconnect network during the cell output transient (thus neglecting aggressor noise-coupled transients).

Integration of both sides of Equation 15.3 from t = 0 to t = T provides the following equivalent relationship:

$$\mathbf{G} * \mathbf{w} = \mathbf{Q}_D - \mathbf{C} * (\mathbf{v}_T - \mathbf{v}_0) \qquad \text{(Eqn. 15.4)}$$

The integral of the vector **v**' results in two simple vectors for the node voltages at t = 0 and t = T. For the case of a signal interconnect, assume that the initial and final values of all (capacitive storage) node voltages are known (e.g., $\mathbf{v}_0$ = [ 0 ] and $\mathbf{v}_T$ = [ VDD ] for a RDLY transition). The integral of the current source vector **s** provides a vector representing the total charge delivered by the driving current source, $\mathbf{Q}_D$. This vector is nonzero only at the cell output node. The total driver charge can be easily calculated: It is equal to the charge delivered to all the network capacitances during the transition, ($C_{total}$ * VDD). The vector **w** is the voltage integral at each node as a result of the driver transient, and it is the vector to be solved in the matrix equation.

Solving for the vector **w** also provides the total charge through each interconnect resistor as a result of the driver transient: $q_{ij} = g_{ij} * |w_j - w_i|$, where $g_{ij}$ is the conductance (1/R) between nodes i and j. The $j_{AC}$ for each resistive element in the signal interconnect can thus be determined from the total charge calculation, as shown in Figure 15.17.

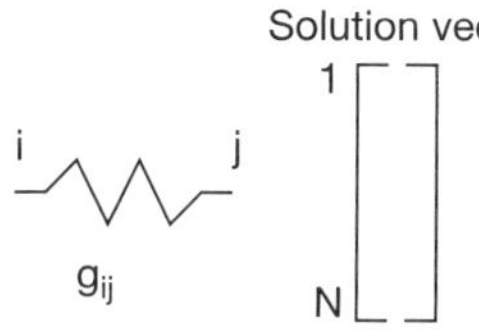

Solution vector w (1 … N)

$$q_{ij} = g_{ij} * (w_j - w_i)$$

$$j_{AC} = \frac{1}{(W * \text{thickness})} * \frac{1}{T} * (1 - \eta) * \left(\int_0^T i^+(t)\,dt\right)$$

$$j_{AC_{Rij}} = \frac{1}{(W * \text{thickness})} * s * q_{ij} * (1 - \eta)$$

$q_{ij}$ is the total charge through $R_{ij}$ from the matrix solution
s is the switching activity factor for the signal
N is the number of nodes in the signal interconnect tree

**Figure 15.17** Calculation of $j_{AC}$ for a segment in the RC interconnect tree, using the solution to the matrix formulation.

The calculation of $j_{RMS}$ and $j_{PEAK}$ for a signal interconnect element requires waveform detail. To avoid the computational demand of circuit simulation, Reference 15.4 also proposes an approximation for these waveforms, leveraging the earlier matrix solution for the total charge through each resistive element during a transition. Referring to Figure 15.18, assume that the current waveform through each resistive element is (roughly) triangular. The total charge delivered for the transition, combined with a (minimum) transition time for the signal from static timing analysis results, defines the approximate current waveform and the subsequent $j_{RMS}$ and $j_{PEAK}$ values for the sigEM FIT calculation.

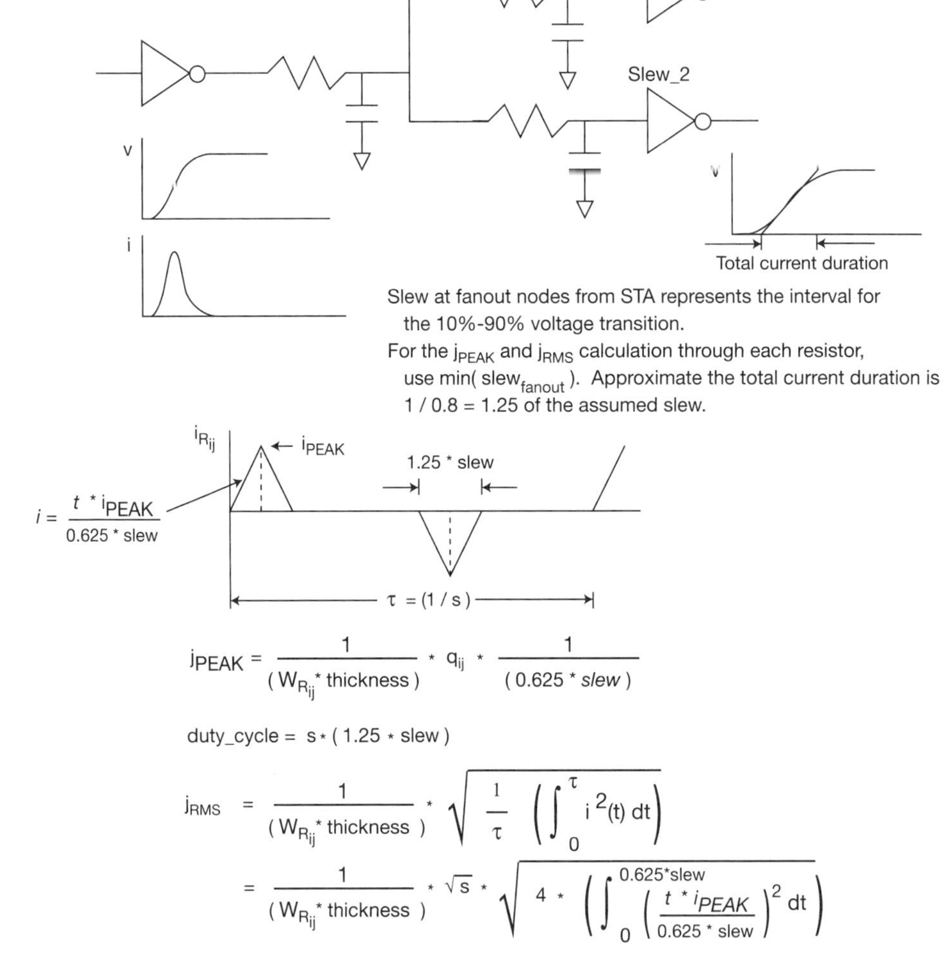


**Figure 15.18** Illustration of the approximation used for the current waveform through each resistive segment to calculate $j_{RMS}$ and $j_{PEAK}$, from the total charge through the resistor during the transient.

With the estimated values for $j_{AC}$, $j_{RMS}$, $j_{PEAK}$, device self-heating $\Delta T$, and duty cycle for each resistive element, the foundry PDK limits and FIT scaling factors can be calculated.

The sigEM FIT results at block, core, and full-chip hierarchy levels require review by the SoC methodology team to determine whether significant FIT contributions by resistive elements require physical design updates or further simulation analysis to assess the sensitivity of the (probabilistic) resistance increase to overall functionality.

## 15.5 Summary

This chapter reviews what is likely to become a limiting factor to the pace of VLSI process technology scaling: the SoC reliability impact of electromigration in metal wires, vias, and contacts. The transition to device technologies that are thermally distant from the die substrate will exacerbate the issue due to the local temperature increase from device self-heating. The SoC markets for extremely high-reliability products (e.g., medical, aerospace, automotive) are growing rapidly. The challenging temperature environments of aerospace and automotive applications require increasing focus on EM reliability analysis and sufficient design margins to continue to function despite the probability of a wire “failure,” as reflected in the resistance increase.

The sheer magnitude of the number of SoC signals multiplied by the number of resistive elements in the extracted interconnect tree for each signal presents an interesting question: Is the sum-of-failure-rate assumption still the best reliability model? Individual resistive elements may have an infinitesimally small failure rate, but their quantity is large, potentially resulting in a significant FIT estimate. The validity of the summation of a large number of very small failure rates warrants further research. With the increasing demand for high-reliability product applications, and with continued process scaling providing even greater numbers of signals (and parasitic elements), this research will be crucial for future SoC designs.

## References

[1] White, M., and Bernstein, J., *Microelectronics Reliability: Physics-of-Failure Based Modeling and Lifetime Evaluation*, National Aeronautics and Space Administration (NASA) Jet Propulsion Laboratory, JPL Publication 08-5, 2008.

[2] Banerjee, K., and Mehrotra, A., "Coupled Analysis of Electromigration Reliability and Performance in ULSI Signal Nets," *Proceedings of the 2001 IEEE International Conference on Computer-Aided Design (ICCAD)*, 2001, pp. 158–164.

[3] Mishra, V., and Sapatnekar, S., "Circuit Delay Variability Due to Wire Resistance Evolution Under AC Electromigration," *IEEE International Reliability Physics Symposium (IRPS)*, 2015, pp. 3D.3.1–3D.3.7.

[4] Oh, C., et al., "Static Electromigration Analysis for Signal Interconnects," *Proceedings of the Fourth International Symposium on Quality Electronic Design (ISQED)*, 2003, pp. 377–382.

## Further Research

### Flowcharts for EM Analysis with Data from Other Flows

This chapter highlights algorithms used for powerEM and sigEM analysis of the FIT rate and the data required from other flows to enable this analysis; however, the chapter discussion lacks visual descriptions. Develop comprehensive flowcharts for EM reliability analysis, including:

- powerEM
  - The (non-reduced) model from parasitic extraction
  - The required data from the power rail voltage drop analysis flow
  - $j_{AVG}$, $j_{RMS}$, and $j_{PEAK}$ calculations
  - Scaling of the foundry FIT data for the calculated current density of each power and ground resistive segment
  - Power rail voltage drop analysis with modified $g_{ij}$ elements for high-FIT segments
- sigEM
  - The model from parasitic extraction
  - The required data from the STA flow for the output driver current for each signal net
  - The switching activity data from (stressmark) simulation testcases
  - The $j_{AC}$, $j_{RMS}$, and $j_{PEAK}$ calculations (with healing factor)
  - Scaling of the foundry FIT data for the calculated current density of each RC interconnect segment
  - Additional path timing analysis options for high-FIT interconnect segments

### Electromigration and Wire/Via/Contact Metallurgy (Advanced)

Process development engineers are continually evaluating new metallurgies for wires and contacts/vias. These engineering teams assess the resistivity and electromigration parameters versus the difficulty in material deposition and patterning.

Research and describe the (temperature-dependent) resistivity and electromigration activation (Black's equation) for metals used in SoC fabrication (e.g., Al, Cu, Co, W, Ti).

Describe the process development techniques used to increase grain size and/or orient grain boundaries to minimize atom displacement.

### Electromigration "Resistive" Fails

The discussion in this chapter proposes addressing EM failures with additional analysis using increased resistance values in the conductance model of the power grid or the RC interconnect model for a timing path delay. The intent is to justify the specification of a lower overall FIT rate for SoC reliability. However, the nucleation and growth rate of voids in a metal wire leading to a resistance increase is a non-linear function, increasing over time.

Research and describe the rate of resistance increase in a metal wire subject to high current densities and void growth for various metallurgies. Describe the rate of change in the resistivity after an increase of N~10%.

Describe the risks associated with the justification of reducing the contribution from high-FIT segments, based on subsequent power rail voltage drop and path timing analysis with modified (R + 10%) elements.

### Blech Length (Advanced)

For efficiency, the SoC methodology team may opt to skip the current density and FIT rate calculation altogether for wire segments less than the Blech length.

Research and describe the Blech length and MTTF correction factor for metals (and surrounding dielectric materials) in advanced process nodes. Determine a strategy for optimizing EM flow runtime for segments less than the Blech length.

CHAPTER 16

# Miscellaneous Electrical Analysis Requirements

## 16.1 SleepFET Power Rail Analysis

### 16.1.1 I*R Voltage Drop During Active Power State

The discussion of power rail I*R voltage drop analysis in Chapter 14, "Power Rail Voltage Drop Analysis," briefly highlights the need to model both rails associated with the design of power domains, as illustrated in Figure 16.1. The SoC methodology and CAD teams review how to insert the $R_{on}$ and C elements for the sleepFET cells into the matrices used for static and dynamic I*R analysis.

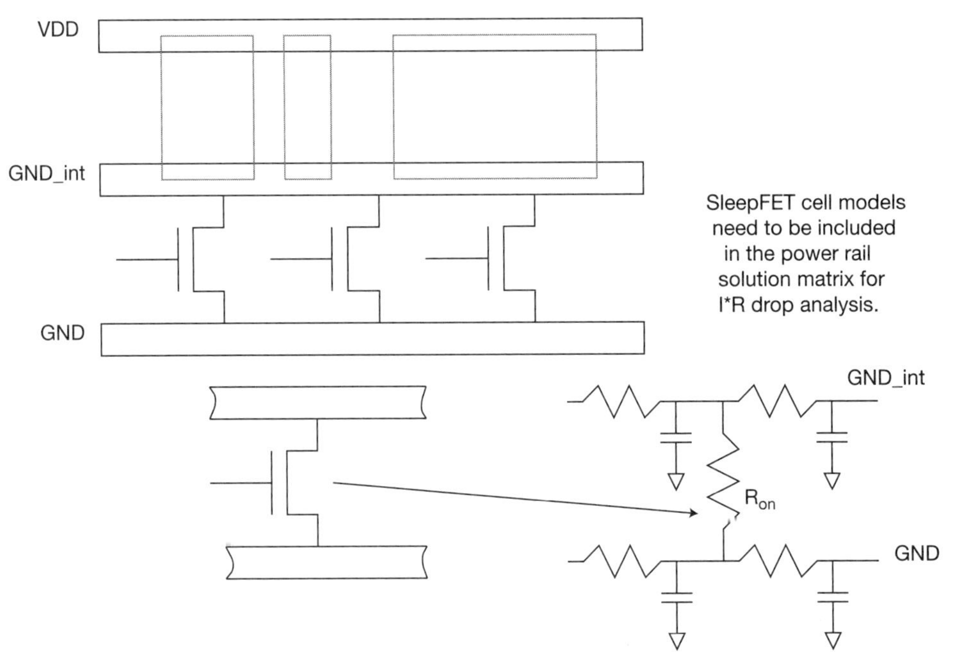


**Figure 16.1** Model for sleep (internal) and global rails for voltage drop analysis.

Leakage currents from state retention cells are connected to the non-gated rail, as shown in Figure 16.2.

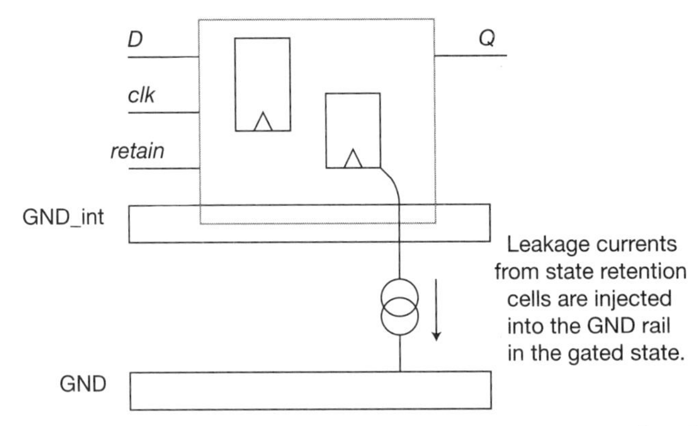


**Figure 16.2** State retention cells provide leakage current to the global rail in the power gated mode.

### 16.1.2 Sleep-to-Active State Transition

A special flow is required to evaluate the transient when the power domain transitions from the sleep state to the active state (see Figure 16.3). In this case, a full circuit simulation is appropriate to reflect the device characteristics as successive sleepFETs are enabled by the delay of the enable input. The (simplified) model for the cells in the power gated block should reflect the unknown voltages that may be present on the internal node capacitances from leakage currents during the sleep state. Figure 16.3 depicts the capacitances in the block starting at VDD for nFET sleep devices.

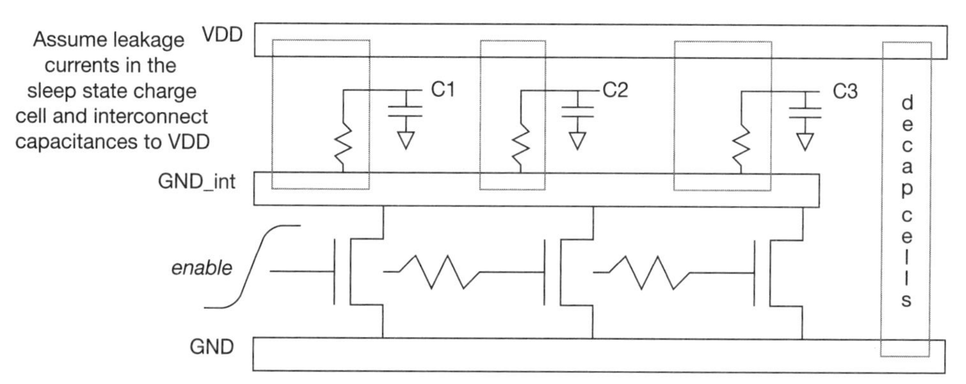


**Figure 16.3** Simplified circuit simulation model for the gated-to-active transition to measure the power and ground rail behavior.

The purposes of this sleepFET transition flow are to:

- Confirm the recovery time specification from the enable transition to the resumption of functional clocking
- Measure the $j_{PEAK}$ current density and duration for all the rail resistance elements and check against the PDK electromigration limits

Chapter 15 describes the powerEM electromigration reliability analysis flow for power rails. The sleep state transition has the potential to result in large peak currents during the sleep-to-active transition; the block capacitances are being discharged through a number of parallel devices, providing a low $R_{on}$ resistance. The purpose of introducing an RC delay to turn on successive sleepFET devices is to control the magnitude of this current in order to prevent a $j_{PEAK}$ failure. The risk of high $j_{PEAK}$ current densities is exacerbated by the goal of minimizing the area impact of introducing the internal gated power rail; the width of this rail is designed aggressively. These current surges are assumed to be sufficiently infrequent to not significantly impact the $j_{DC}$ ($\approx j_{RMS}$) calculations of the powerEM flow.

## 16.2 Substrate Noise Injection and Latchup Analysis

Circuit switching transients result in capacitive currents to the device substrate/well, as illustrated in Figure 16.4. The injected free carriers have several paths—for example, to the closest contact, in recombination with opposite free carriers within the substrate/well or to the depletion region electric field between the well and the substrate.

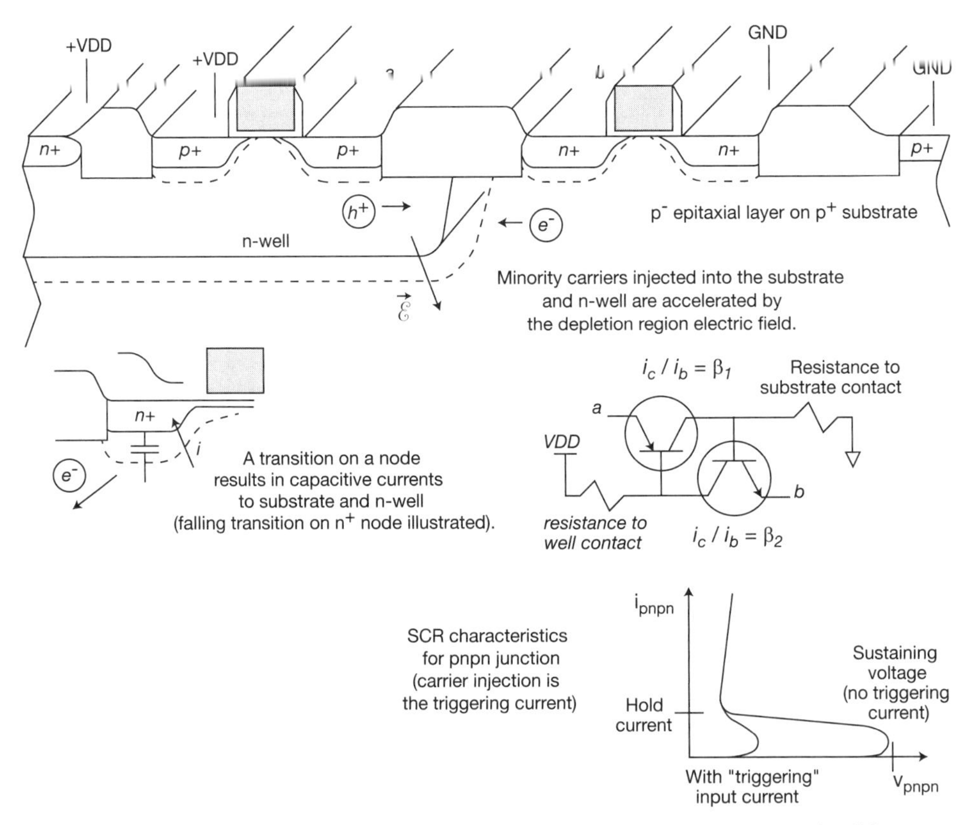


**Figure 16.4** Illustration of the injection of current into the substrate/well from a switching transient. A pair of (lateral) n-p-n and p-n-p bipolar transistors is highlighted.

The figure depicts the substrate and well structure as a p-n-p-n junction. If the product of the two lateral bipolar p-n-p and n-p-n transistor current gains, $\beta_1$ and $\beta_2$, is greater than 1, a very high current flow (and anomalous node voltages) could result from the injected current, in a phenomenon known as *latchup*. The injected current and resulting bipolar bias must be active long enough for both devices to turn on. The result of this high current flow may be a (non-permanent) data upset or a permanent catastrophic failure. Latchup currents are sustained until the supply (holding current) energy source is removed; this action is equivalent to that of a silicon-controlled rectifier (SCR).

As illustrated in Figure 16.4, the effective resistance to substrate/well contacts has a significant role in suppressing latchup. Fabrication process engineering steps are taken to increase the well/substrate free carrier recombination rates and reduce the resistive path below the devices to the well/substrate contacts.

A distinct latchup analysis flow is typically not developed; rather, the foundry PDK includes layout design rules to be verified physically by the DRC flow. Design rules ensure that well/substrate contacts are included frequently (e.g., a maximum distance between any device node and a well/substrate contact).

Note that FD-SOI device cross-sections are not subject to latchup due to the absence of a contiguous p-n-p-n junction structure.

### 16.2.1 Substrate Noise and Analog IP

Analog IP blocks are typically provided with unique AVDD and AGND distribution from specific SoC pads to minimize the exposure to power/ground noise from digital switching. Nevertheless, substrate noise propagated to analog IP can be problematic. Fabrication processes may offer a *triple-well* option, in which an nFET resides in a p-well that is electrically isolated from the bulk p-type substrate by the addition of a "deep n-well." Analog circuits may apply a unique p-well "substrate voltage" to these triple-well isolated nFETs for device $V_t$ control and/or to reduce the substrate noise. The injected substrate currents from digital circuit switching would normally disturb adjacent analog IP; an isolated substrate triple-well is extremely effective in suppressing such injected noise.

An alternative design option to reduce the fluctuations in substrate potential from injected carriers is to add surrounding *guard rings* to provide additional high-efficiency free-carrier-collection contacts away from sensitive circuitry.

EDA vendors extend their dynamic I*R voltage drop tools to include an RC mesh for the bulk substrate and injected current sources for analysis of the potential local fluctuations in substrate voltage. Analog IP designers can review these substrate noise analysis flow results to subsequently incorporate into their circuit simulations. The EDA vendor and foundry collaborate on appropriate electrical models for the substrate mesh and the guard rings. The SoC methodology team should review the analog IP power distribution and guard ring design to ensure that the substrate mesh modeling assumptions are satisfied by the physical floorplan.

### 16.2.2 Latchup and I/O Pad Circuit Design

The injected substrate/well capacitive current from an internal cell-switching transient is relatively limited. The PDK design rules relative to devices and substrate/well contacts reflect the foundry's analysis of the latchup sensitivity, incorporating the (optimized) substrate resistance impurity concentrations. However, the off-chip driver circuitry connected to I/O pads may be subjected to much larger energy transients due to the possibility of signal ringing. Impedance mismatches between the SoC driver and the package/PCB trace load topology result in reflected energy returning to the driver. This reflection may exceed IOVDD ("overshoot") or IOGND ("undershoot") and inject significant current into the well/substrate.

Note that significant overshoot on SoC power supply pins due to insufficient regulation may also trigger latchup, in a manner similar to exceeding the *sustaining voltage* of a silicon-controlled rectifier without an injected triggering current.

To reduce the latchup risk, the foundry PDK includes additional physical design rules relative to guard ring design surrounding I/O driver devices and the minimum distance to internal cells. Figure 16.5 illustrates the introduction of additional layout contact straps (a "partial" ring) between driver nFET and pFET devices to effectively create separate n-p-n and p-n-p devices rather than the composite p-n-p-n structure. The figure also illustrates a contact strap design in an advanced process node, where gate lithography uniformity is an additional constraint. pFET devices in the p-substrate with shorted source/drain nodes are depicted.

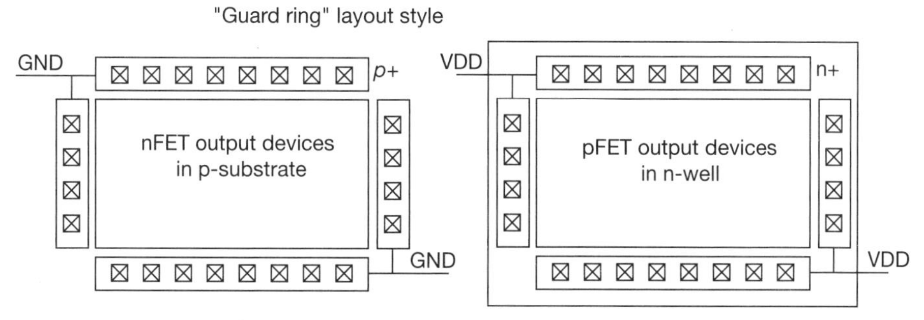


Substrate and well contacts surround an output node that may be subject to "ringing" (and charge injection).

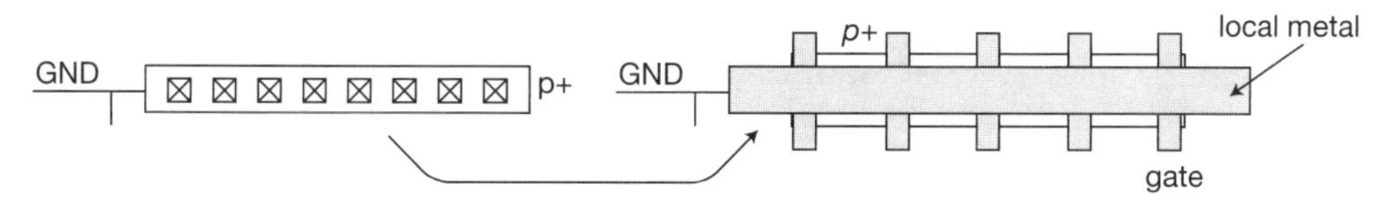


Advanced process nodes require gate layout uniformity — guard ring layout design is modified to include devices (with shorted nodes).

**Figure 16.5** Illustration of collection nodes for injected carriers added to off-chip driver device layouts.

### 16.2.3 Latchup Qualification

The foundry PDK latchup design rules for internal circuits address carrier injection and bipolar device current gain reduction with the requirement for frequent well/substrate contacts. The PDK latchup design rules for I/O circuits require a different analysis method due to the external source of (reflected) energy. An industry standard, JEDEC 78, specifies how latchup rejection is to be defined and experimentally verified on silicon. Figure 16.6 depicts how JEDEC 78 injected currents are applied on any I/O signal pin connected to a device source/drain (carrier injection) node.

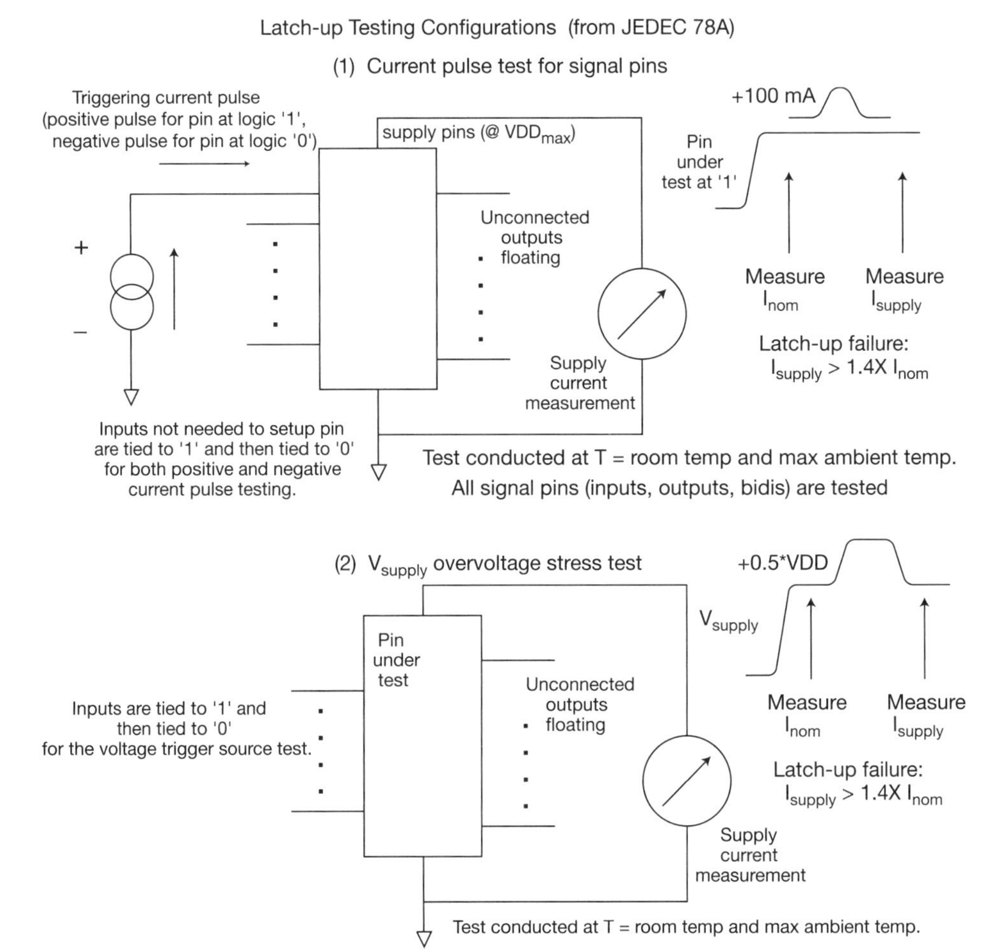


**Figure 16.6** Standard test fixture for JEDEC 78 latchup qualification.

After applying a static logical '1' (or '0') voltage value, a "nominal" current is measured. Then a positive (for '1') or negative (for '0') current pulse is superimposed on the applied signal voltage. (The JEDEC 78 standard requires ~30 microseconds duration for the current pulse.) After a brief waiting period, the change in current over $I_{nom}$ is measured. An increase in supply current that exceeds the JEDEC limit is used to indicate that a latchup event has occurred. The magnitude of the pulse is successively increased until a latchup event is indeed observed. The final magnitude prior to the latchup event defines the latchup specification limit in the SoC product documentation. (The baseline JEDEC 78 specification for I/O pins is a pulse magnitude of 100mA.)

For supply pins, an overvoltage stress pulse is applied to represent a voltage transient on the pin originating from an external source. An increase in supply current that exceeds the JEDEC spec limit after the overvoltage pulse indicates that latchup has been triggered.

The substrate/well resistances in Figure 16.4 increase with temperature: At higher temperatures, an injected node current increases the bipolar base-emitter junction bias, triggering latchup more readily. Latchup qualification testing would commonly be performed at multiple temperatures, with specific attention to $T_{junction} = T_{max}$.

An SoC project manager needs to oversee preparation of the JEDEC 78-based latchup test procedure as part of the tapeout release dataset. The test procedure documentation requires identification of:

- Signal pins that may be carrier injectors
- Supply pins
- Any test setup patterns to enable signal pins to be asserted to '1' and '0' values
- Signal current and supply overshoot pulse definitions
- Signal $I_{nom}$ and IOVDD supply current overshoot measurement specifications

An appropriate sample size of SoC parts should be selected for signal and power pin latchup testing to provide meaningful results. The JEDEC 78 specification requires testing of a minimum of six modules.

The latchup qualification data are included in the SoC release documentation to provide end customers with constraints on allowable signal reflection and supply transients.

## 16.3 Electrostatic Discharge (ESD) Checking

Sections 15.2 and 15.3 describe the calculation of $j_{PEAK}$ in a resistive wire and the application of the powerEM and sigEM flows to verify Joule heating reliability. In that model, the $j_{PEAK}$ current density represents a periodic event, in which the duration and duty cycle of the current pulse are part of the reliability evaluation. There is also a risk of a catastrophic current pulse with energy sufficient to both deform the metallization and permanently result in dielectric breakdown of connected devices. This *electrostatic discharge* (*ESD*) event is specifically applicable to circuitry connected to the I/O pads of the SoC. The risk of an ESD event arises from two mechanisms:

- **An external source of electrostatic charge contacting an SoC pin**—A human-body model (HBM) has been developed to reflect accumulated static charge on a human touching a package pin. A machine model (MM) is also occasionally used to represent charge on equipment contacting the package module; the HBM model typically covers this case.
- **Accumulated charge on the die substrate subsequently discharging through a package pin**—A charged-device model (CDM) is used to represent this event. The common example given for CDM is the motion of parts in (plastic) carriers during the PCB assembly process: The die substrate may pick up electrostatic charge, which is subsequently released through a pin.

Figure 16.7 depicts an HBM test model used to qualify the SoC package pins for this event with the stored capacitive energy that may be delivered. In the simplest form, a capacitance is precharged to the HBM voltage, and then it is discharged through a 1.5kΩ resistor to the package pin.

A more common HBM test procedure uses the transmission line pulse (TLP) method, also shown in Figure 16.7. A transmission line is precharged to the HBM voltage, and then it is floated, and finally it is discharged through the SoC pin. When using testsite silicon to characterize the HBM robustness of an I/O pad cell design, the applied HBM voltage is increased through a TLP sequence, and the pin leakage current is measured to identify a breakdown event. Production testing would apply a TLP sequence at the HBM spec voltage to each pin on a sample of parts to confirm that no breakdown has occurred.

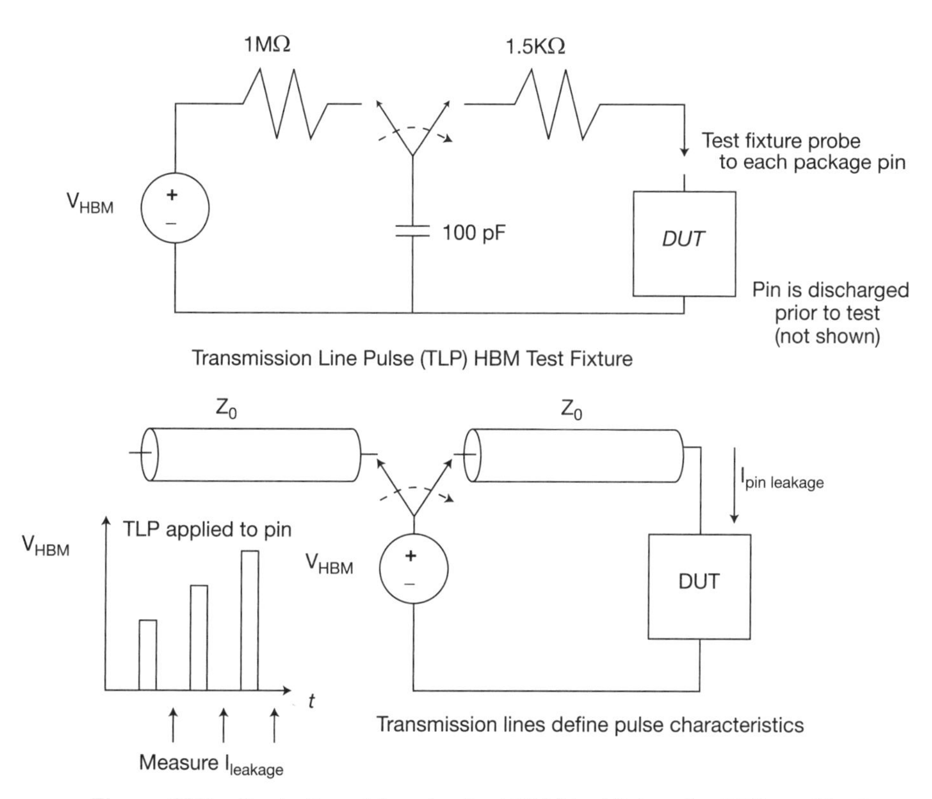


**Figure 16.7** Illustration of the standard HBM test fixture for ESD qualification.

The CDM test method is illustrated in Figure 16.8. The device is placed pins (or bumps) up on a fixture consisting of a field plate and dielectric. The field plate is charged to the CDM voltage through a high resistance.

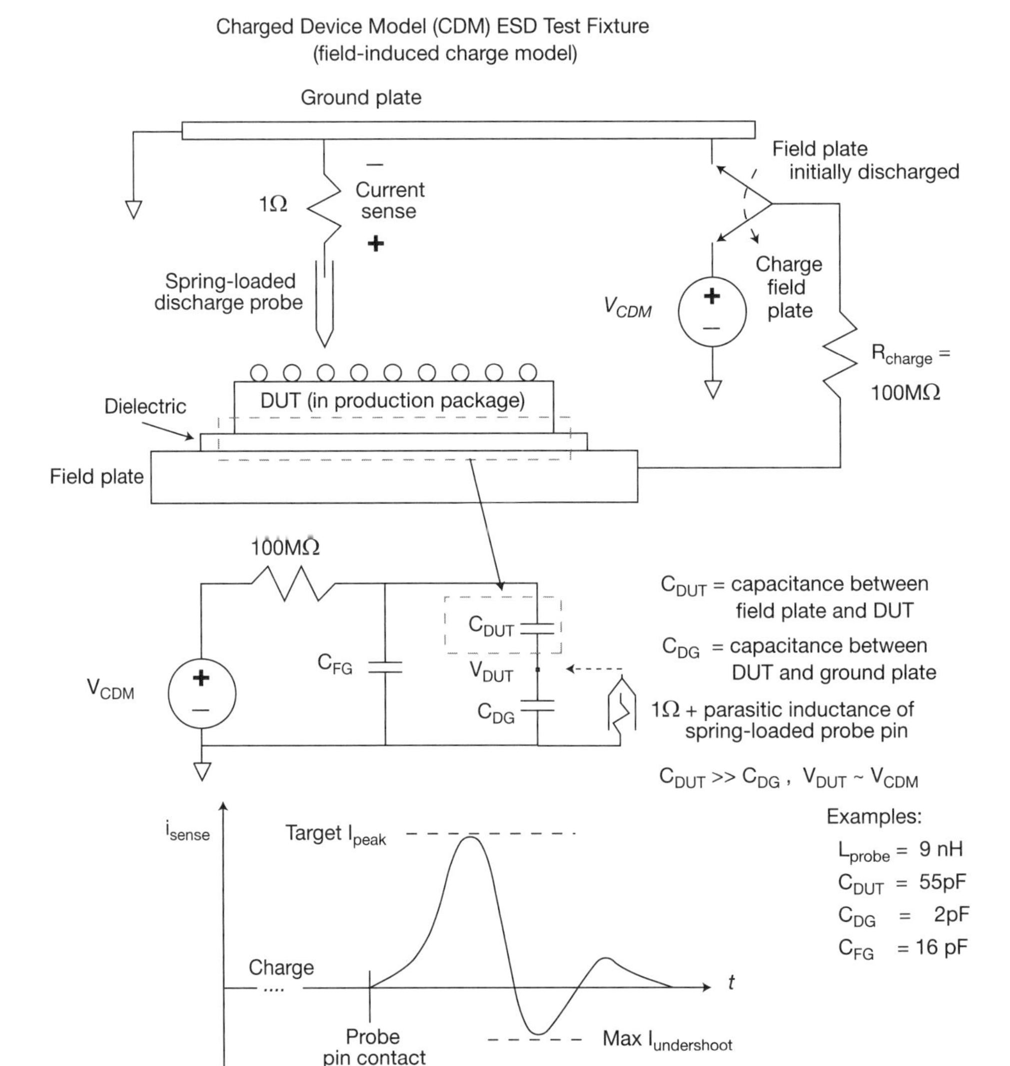


The target $I_{peak}$ and max $I_{undershoot}$ values in the JEDEC spec vary with the CDM voltage testing target (e.g., $V_{CDM}$ = 125V, 250V, 500V, ...). The JEDEC spec also includes current pulse width and rise time targets.

Note: The CDM test fixture is calibrated with a metal disc, before the DUT is applied.

**Figure 16.8** Illustration of the standard CDM test fixture for ESD qualification, part 1.

The figure also includes an effective circuit model, indicating how the device under test (DUT) is part of a capacitive divider, accumulating charge during the voltage ramp to the field plate. After charging, a (spring-loaded) probe is placed in contact with a package pin, discharging the die to ground through the pin and a very low resistance (i.e., 1Ω). The current waveform through the low resistance is measured, and the profile is compared to the JEDEC CDM standard to demonstrate that sufficient energy has been discharged through the pin. As shown in Figure 16.9, after the positive discharge event, the probe is lifted, and the field plate voltage is then switched. The capacitive divider results in negative charge on the die; when the grounding probe is then re-applied, a negative current stress pulse flows through the pin. Each package pin is subjected to multiple positive and negative pulses at the target JEDEC currents associated with the CDM voltage specification.

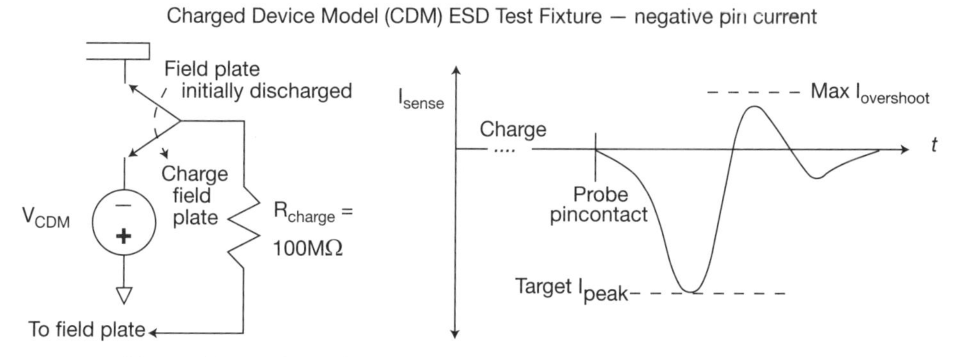


**Figure 16.9** Illustration of the standard CDM test fixture for ESD qualification, part 2.

After the CDM stress procedure is applied to a sample of parts, the modules are retested using the production parametric and functional test patterns. Any test failures are indicative of ESD damage and failure to meet the CDM qualification. Successful retesting implies that the CDM qualification has passed.

### 16.3.1 ESD Protect Circuits

ESD protection circuitry is added to the SoC I/O cells to provide an alternative discharge path for an HBM or CDM event and protect functional gate dielectrics from high-current-density breakdown. Figure 16.10 illustrates some of the typical diode and device structures that comprise the ESD protect circuit.

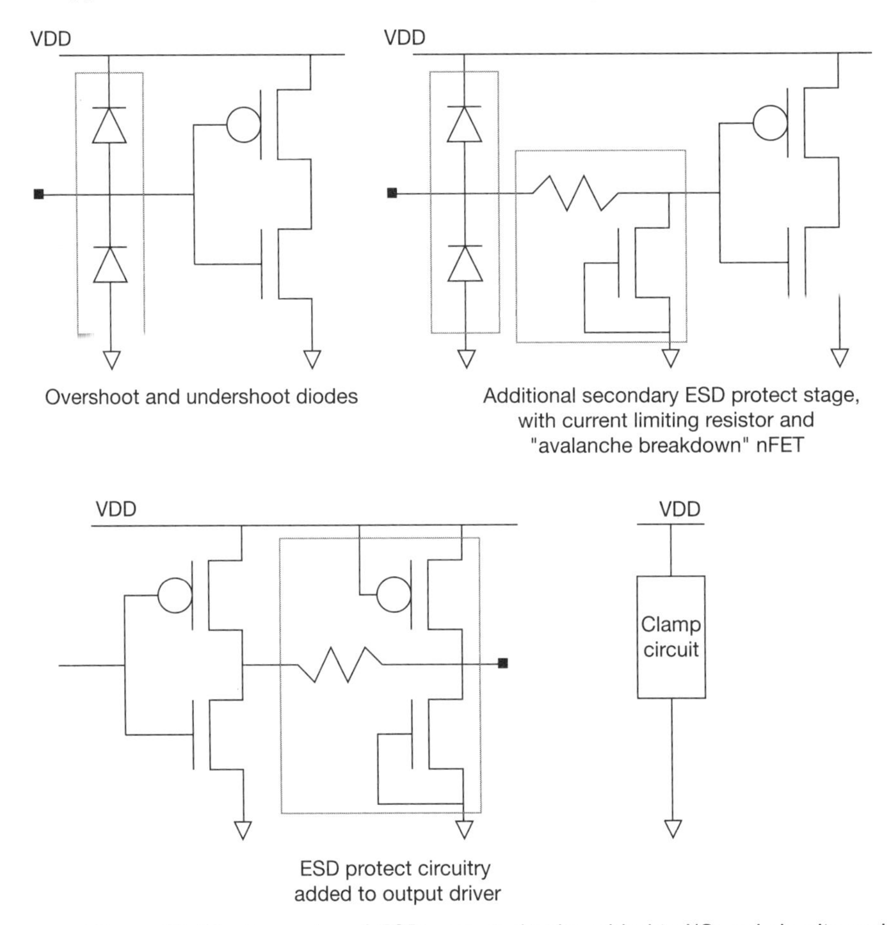


**Figure 16.10** Examples of ESD protect circuits added to I/O pad circuits and to the I/O power rails.

There are several characteristics of note in the ESD protect circuit examples in the figure:

- ESD protect circuits are present on all pins—input receivers, output drivers, bidirectional I/Os, and power/ground pins.

- Inputs typically receive a primary and secondary set of structures, sized to discharge the ESD energy through two paths; output drivers include an ESD structure (in addition to the internal diodes within the driver).
- Power and ground pins receive a unique *clamp circuit.*

For inputs, a resistor is typically included between the two structures to limit the current for the low-impedance CDM ESD event. The diodes in an ESD structure provide an alternative current path from the pad for an applied pulse that overshoots VDD or undershoots VSS. (The diode layout includes guard rings or straps to suppress latchup.) The devices in the secondary discharge path have gate and source connected in a normally OFF topology. For ESD protection, the FET device enters a mode in which a high voltage across the drain-to-substrate junction results in an *avalanche current* due to additional hole/electron carrier generation in the depletion region from high electric field impact ionization. As illustrated in Figure 16.11, the impact ionization holes serve as a triggering base-emitter current for the (parasitic) bipolar device below the nFET channel. The ESD energy is discharged by the resulting parasitic bipolar collector current, as shown in the IDS versus VDS curve in the figure (for VDS well beyond the normal device operating range).

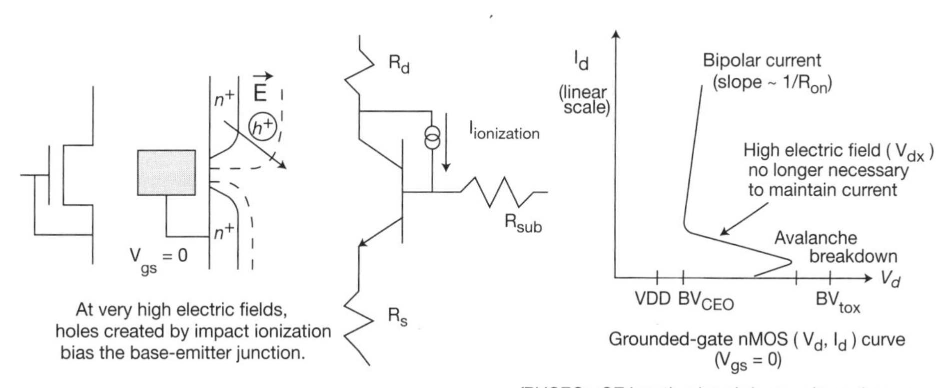


**Figure 16.11** Illustration of the impact ionization "avalanche" current discharge path for an ESD event.

This device is added in the secondary ESD stage to limit the voltage presented to the gate input of a functional receiver cell. A thick-oxide ESD FET protects a thick-oxide receiver gate (used where the $v_{in}$ interface voltage level exceeds the SoC internal VDD supply), whereas a thin-oxide ESD FET protects a thin-oxide receiver gate. Once the bipolar device current is triggered, the FET enters a *snap-back* mode with reduced VDS voltage, protecting the functional input devices.[1]

The foundry pursues process engineering optimizations to ensure the avalanche breakdown voltage, bipolar current profile, and local self-heating temperature increase are non-destructive to the snapback device when sized to accommodate the secondary ESD current. The foundry releases ESD FET design rules relative to the snapback device, diodes, and the connectivity to the SoC pads.

In addition to overshoot/undershoot diodes and snapback devices, the ESD protect circuitry in Figure 16.10 includes a supply *clamp circuit* path between VDDIO and GND near the signal pad. An example of a supply clamp is illustrated in Figure 16.12.

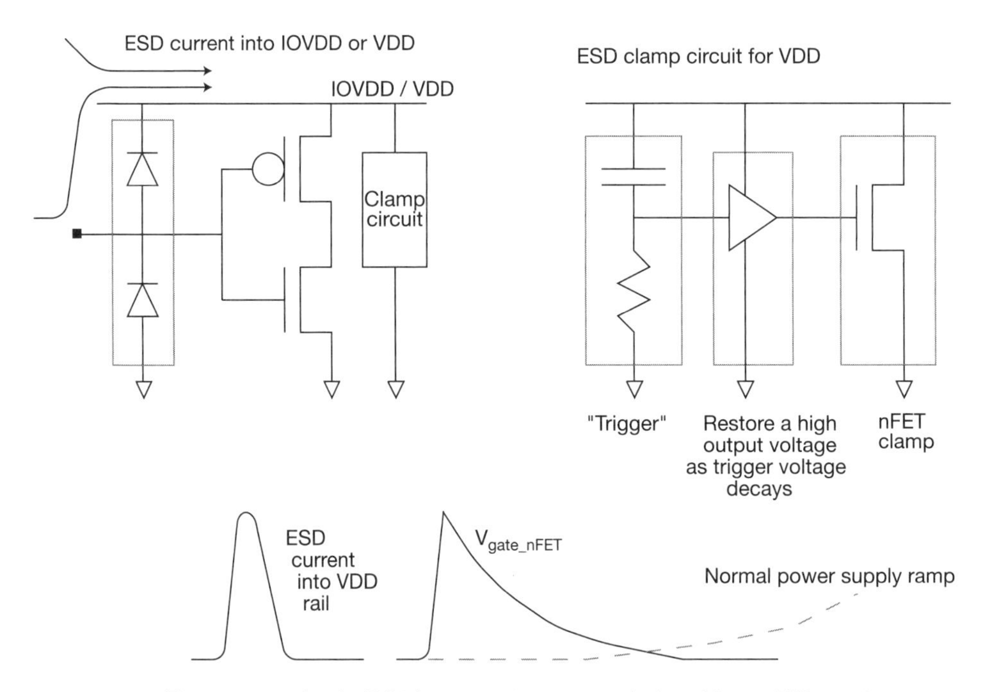


**Figure 16.12** A supply clamp between VDDIO and GND is included in the ESD protect circuit design.

A normally OFF nFET is present between the supply rails, and the gate input is connected to a trigger circuit that detects an ESD event and (temporarily) turns on the nFET to provide a low-impedance ESD discharge path. A very simple trigger circuit is shown in the figure: An ESD transient on the VDDIO rail is capacitive-coupled to the clamp nFET gate input, turning on the device. The R*C time constant of the trigger defines the duration when the clamp is active. The assumption is that this time constant is very short, as appropriate for the HBM or CDM discharge model. The rise time of a normal supply voltage ramp during power-on is much slower; the trigger circuit should not activate the clamp during a conventional power-on transient.

The size of the protect diodes and devices, the supply clamp trigger circuit, and the resistances define the temporal response to an ESD event. A key layout consideration for the ESD protect cell is to ensure that the impedances to each element connected in parallel are matched to avoid *current crowding* (e.g., to provide symmetric layouts for the parallel fingers of a multi-finger device).

The foundry PDK provides design rules related to the connection between the SoC pad, the ESD protect circuitry, and the functional I/O cell. ESD design rules are unique in that they include both physical *and* electrical rules:

- Minimum metal wire and via dimensions for each layer in the protect layout cell to carry the primary and secondary ESD currents
- Minimum primary and secondary diode sizes (and guard rings) and power clamp device size
- Minimum resistance between primary and secondary discharge paths
- Maximum parasitic resistance from protect structures to the rails

EDA vendors have provided additional DRC tool support for the PDK ESD electrical rule checks, adding algorithms to perform the necessary net tracing and resistance calculations.

The SoC project management team needs to oversee the merging of an ESD protect layout cell with the I/O cell IP to ensure that both physical and electrical design rules are satisfied. A specific consideration is to ensure that the PDK design rules and SoC qualification testing should support the HBM and CDM voltage levels corresponding to the end customer requirements for the part.

The ESD protect structures illustrated in Figure 16.10 add parasitic capacitance to the functional I/O pad driver/receiver circuits. The frequency response of high-speed interface circuitry is adversely affected by the connection to ESD protect cells, which is especially challenging for SerDes serial interfaces. The IP vendor providing an I/O macro might want to reduce the ESD robustness (HBM and CDM voltage levels) to satisfy the target data rates, incorporating a specific ESD protect cell with the interface IP. The SoC project management team may therefore need to assess the subsequent end customer concern associated with a range of I/O datasheet ESD specifications.

### 16.3.2 Electrical Overstress (EOS)

In addition to the ESD events to which the SoC may be subjected prior to and during assembly, there is also a potential for *electrical overstress* (*EOS*) during end customer system operation (e.g., signal ringing due to printed circuit board transmission line impedance mismatch, incorrect power sequencing to parts on the board). These phenomena apply a much lower voltage than an ESD event, but the total energy to be discharged is typically much larger. Whereas the JEDEC model standards apply to HBM and CDM events, there is not a clear definition of the EOS qualification requirements for an SoC. A project manager needs to ensure that the part documentation thoroughly describes the supply sequencing requirements, the I/O pad impedance specifications, and maximum (periodic) EOS current waveforms. The SoC design team may be asked to release circuit models for the pad circuits (merged with package pin parasitics) to enable the end customer to perform EOS simulations. The industry-standard *I/O Buffer Information Specification* (*IBIS*) format is typically used for abstracted I/O circuit model exchange. EDA vendors provide utilities to define circuit simulation testbenches for I/O cells and extract IBIS models from the simulation results.

## 16.4 Soft Error Rate (SER) Analysis

Chapter 15 introduces the concept of FIT rate to provide a reliability measure for the SoC part population. In the context of the electromigration mechanism, the FIT metric represented a "hard failure," implying that the erroneous functional behavior is permanent and unrecoverable. Another class of failures contributes to the FIT estimate; in this case, *soft errors* refer to an unintended change in functional state induced by an external, non-electrical stimulus.

There is no physical disruption to circuit or device characteristics; a subsequent (write) clock operation will complete successfully.

Over the evolution of VLSI process technologies, failure diagnosis and experimental research have identified two principal sources of soft error disruptions: alpha particles and cosmic rays.

### 16.4.1 Alpha Particles Incident on Circuit Nodes

The decay of (trace amounts of) radioactive elements in package and package attach materials releases a charged alpha particle composed of two protons and two neutrons ($^4He^{2+}$). The typical energy of the incident particle is ~1–8MeV; although the emitted alpha particle has a discrete energy from the radioactive source, there is energy loss as it traverses to the silicon surface, broadening the range of incident energy.[2] The trajectory of this particle may enter the die, as depicted in Figure 16.13.

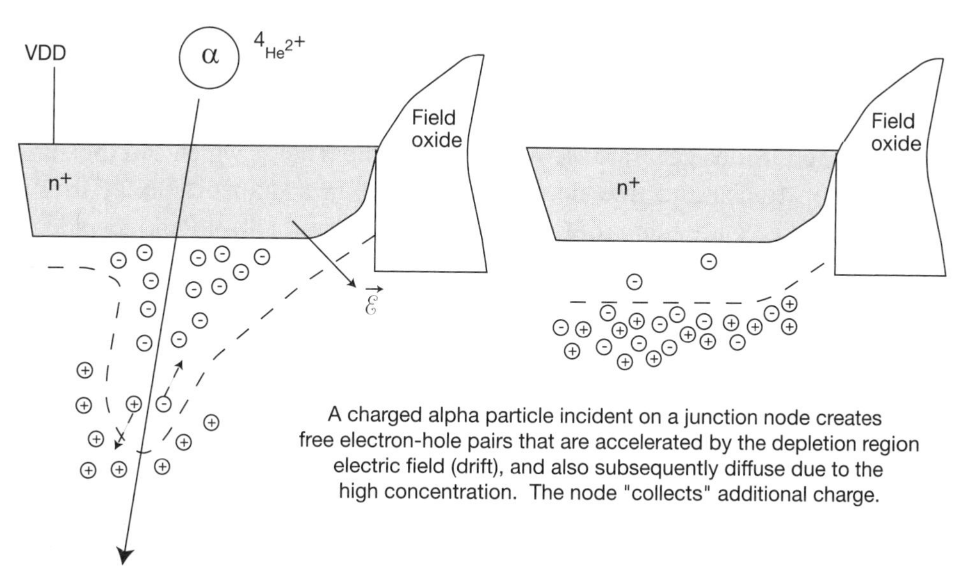


**Figure 16.13** Illustration of the trajectory of a charged alpha particle entering the die substrate.

The interaction between the charged alpha particle and the electrons in the silicon lattice results in the generation of free electron-hole pairs and a loss of particle energy. As the particle energy is reduced, more local free electron-hole pairs are produced. The free carriers generated by the alpha particle result in a current to device nodes from two mechanisms:

- Carriers near the junction depletion region electric field are rapidly accelerated, comprising a *drift current*.
- The local free carrier concentration gradient outside the electric field force results in a *diffusion current* (assuming that the electron-hole recombination rate is slow).

Note that the additional free carriers distort the local potential and electric field at the p-n junction, extending the force on the carrier into a "funnel-like" volume.

### 16.4.2 Cosmic Ray–Generated High-Energy Neutrons Colliding with the Silicon Lattice

The neutron flux incident on the SoC die may result in elastic (neutron-stable) and inelastic collisions with the silicon lattice. Whereas the charged alpha particle interacts directly with the electrons in the lattice, the mechanism of interest for the uncharged neutron is a collision with the silicon nucleus. The inelastic collisions generate secondary charged particles, which can then create ionization tracks and free electron-hole pairs, in a manner similar to alpha particles. There are numerous, complex events possible from an inelastic neutron collision (e.g., starting energy ~200MeV). The charged secondary particles include protons, $^{2}H^{+}$, and alpha particles (along with other heavier ions), with energies from around 4MeV to 12MeV. Whereas the energy (and silicon range) of external alpha particles is limited, cosmic neutrons are incident to the internal lattice with a much higher energy, and the secondary particles have varied origins and direction vectors. After an inelastic collision, the number of potential charged particle tracks and junction node interactions is larger, which typically results in greater total charge perturbation to the circuit node.

Note that the neutron flux is a strong function of altitude: Aerospace applications are subject to a greatly increased strike event probability. For reference, the high-energy (>10MeV) neutron flux at New York City is ~20/cm**2/hr, which increases ~100X at an altitude of 40,000 feet. The vast majority of the terrestrial neutrons traverse a silicon die without an inelastic collision, making estimation of the SER extremely challenging.

### 16.4.3 SER Diagnosis

The seminal work on SER diagnosis was performed by Texas Instruments when investigating high failure rates in commodity dynamic RAM (DRAM) modules.[3] As a result of this discovery, significant focus has been placed on transitioning to low-α-particle-emission materials in IC assembly and packaging.

Subsequently, a dielectric material used in IC fabrication—borophosphosilicate glass (BPSG)—was found to be a source of low-energy alpha particles, as a result of an inelastic collision between $^{10}B$ and a low-energy neutron. Process development has transitioned away from using this material as an inter-level dielectric. Presently, the dominant source of soft errors is due to high-energy neutrons originating from cosmic radiation.

The common model used for the current into a circuit node affected by the generation of free electron-hole pairs is given in Figure 16.14.

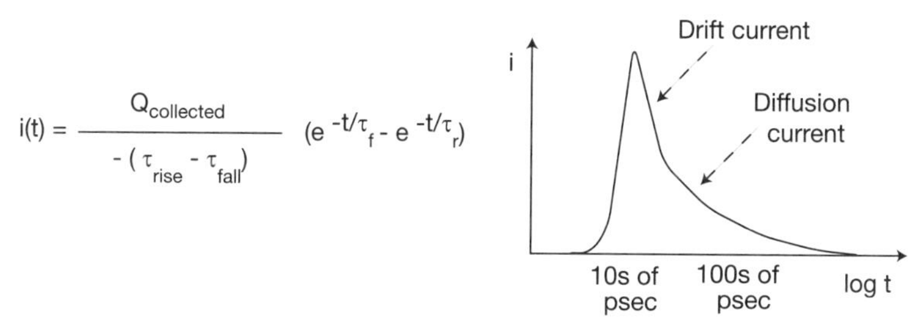


**Figure 16.14** Model for the current into a circuit node due to an SER event.

In the figure, $\tau_{rise}$ represents the rapid current increase due to the electron-hole pair generation along the particle track and depletion region drift (e.g., 10s of psec). The term $\tau_{fall}$ reflects the free carrier diffusion current (e.g., 100s of psec). Initially, the excess free electron and hole carrier concentrations diffuse together in all directions; a fraction of these free carriers diffuse to and are "captured" by the electric field surrounding the sensitive device node. The integral of the current expressed in Figure 16.14 is the total "collected charge" at the device node. The circuit layout, process cross-section, impurity profiles, and applied voltages help define the *collection volume* for the generated electron-hole pairs. Figure 16.13 illustrates the collection of electrons at an $n^+$ node at VDD (logical '1'); the other free carrier current of interest is the collection of holes at a $p^+$ node at logical '0'. Because the drift current mobility of electrons is much higher than holes, the circuit nodes most susceptible to disruption by a particle strike event are associated with n-channel source/drain areas.

Note that there is also a potential for bipolar current action from the particle strike, in addition to the junction drift and diffusion current. The increase in charge at the sensitized device is typically represented using only the drift/diffusion model. The process optimization (and circuit layout) steps taken for latchup avoidance reduce the bipolar amplification for the injected charge from a particle strike. A planar bulk FET device is depicted in Figure 16.13, and the collection volume is defined by the device source/drain node area. The susceptibility to soft errors is significantly improved for FinFET devices, as the source/drain node charge collection volume is reduced due to the fin topology. The susceptibility for FD-SOI devices is dramatically reduced due to the dielectric isolation between the die substrate and device source/drain nodes.

### 16.4.4 Qcrit

The collected charge from a strike may disrupt the (temporal) voltage at a circuit node. Consider the six-transistor SRAM bitcell illustrated in Figure 16.15. In this figure, the nodes most susceptible to a soft error upset due to collected charge are highlighted.

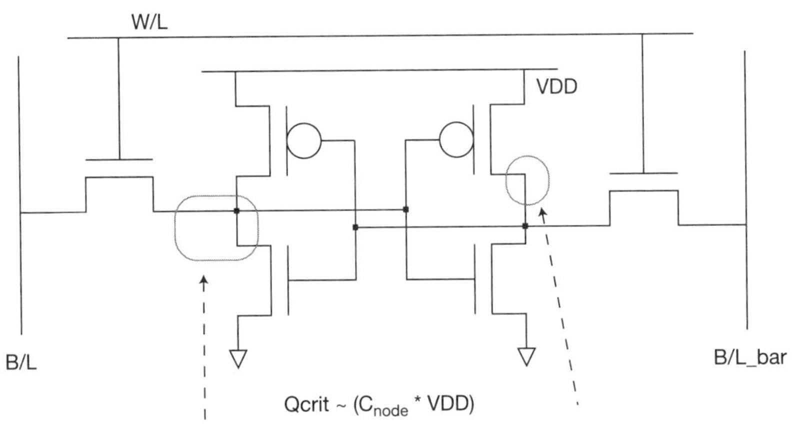


**Figure 16.15** A (six-transistor) SRAM bitcell is shown, with the sensitive nodes to an SER event highlighted. The bitcell stored value would be disrupted by charge collection of the magnitude comparable to Qcrit.

Consider the nFET devices in the SRAM bitcell highlighted in the figure. The bitcell is storing a logical '1', and the nFET pulldown device highlighted is off. The charge on this circuit node is approximately equal to:

$$Qcrit = (C_{node} * VDD) \quad \text{(Eqn. 16.1)}$$

If the collected free electron charge from a strike approaches this "critical charge," the voltage reduction on the node could be sufficient to initiate the cross-coupled inverter currents and flip the bit stored value, resulting in the soft error. A more accurate model for Qcrit would include the additional charge required to compensate for the pFET pullup device current that flows as the node voltage is in transition while the inverters flip state—for example, adding a term to Equation 16.1 related to ($Ids_p$ * flip_time). For reference, in sub-micron process technologies, the Qcrit of a 6T SRAM cell is 10s of femtoCoulombs (fC). In more advanced (FinFET) process node technologies, Qcrit is ~1fC.

### 16.4.5 Linear Energy Transfer (LET) and Range

As the incident charged particle interacts with the electrons in the silicon, it imparts sufficient energy to create free electron-hole pairs. The ionization potential for silicon is 3.6eV, although not all interactions generate free pairs. As a result, the average energy loss by the incident particle for each pair generated is a factor greater than 3.6eV. The rate of energy transfer is a strong function of the particle energy; it is maximum when the particle velocity is comparable to the electron velocity. Figure 16.16 illustrates the energy loss for alpha particles in silicon for different particle energies. Note the peak in this curve at ~1MeV. (The graph shape is similar for other secondary charged particles in silicon originating from a neutron collision.)

The vertical axis of the graph in the figure is the *linear energy transfer* metric. It represents the particle energy loss per distance traveled, normalized by the mass density of the material:

$$LET = (dE / dx) / \rho, \text{ with units } (MeV / cm) / (gm / (cm ** 3)) \quad \text{(Eqn. 16.2)}$$

Specifically for silicon, a more common term is the *stopping power*:

$$S = LET * \rho, \text{ with units } MeV / cm; \rho \text{ for silicon is } 2.33 \text{ g / cm**3} \quad \text{(Eqn. 16.3)}$$

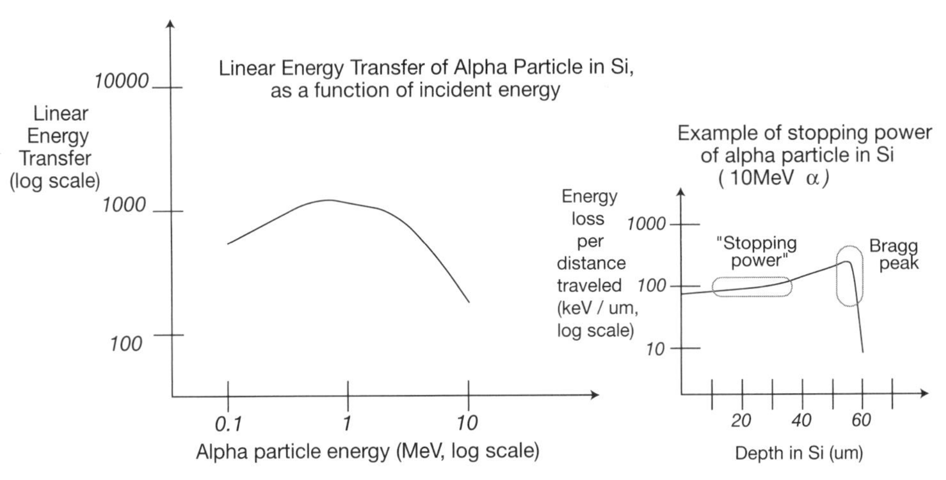


**Figure 16.16** Energy loss for alpha particles in silicon as a function of incident particle energy.

The use of a single value for the stopping power of a (high-energy) charged particle in silicon is not precisely accurate. As the particle traverses the lattice, the energy loss per distance traveled increases due to the increased interaction time. Near the end of the traversal, the rate of energy loss increases dramatically—to what is known as the "Bragg peak"—before the particle comes to rest.

The (average) distance traversed from the initial energy of the particle is denoted as the *range*. There is a statistical distribution (or *straggle*) in the actual distance for each incident particle energy value. The range for proton and alpha particles in silicon is depicted in Figure 16.17. For example, the range of a 6MeV alpha particle is ~30μm, and the range for a 1MeV particle is ~4μm.

For circuit sensitivity analysis, the key measures are:

- The (estimated) charged particle type, flux, and energy distribution
- The free carriers generated by the particle
- The collection efficiency at a sensitive node

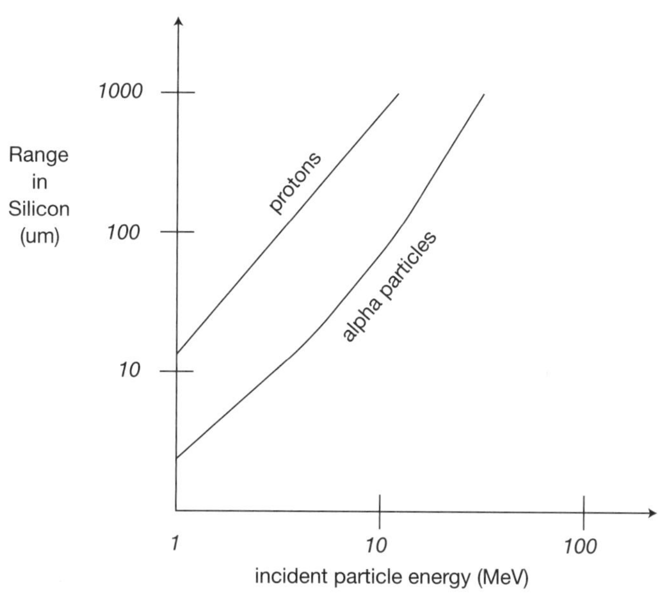


**Figure 16.17** Illustration of the range of proton and alpha particles in silicon.

The collected charge is then compared to Qcrit of the circuit node to calculate the soft error FIT contribution. A number of simulation tools are available to assist with this calculation; because the particle interaction with the die is highly probabilistic, Monte Carlo simulation techniques are applied.[4,5] As an example, using the data from Figures 16.16 and 16.17, a 1MeV alpha particle would generate ~10fC/μm and a total of ~40fC over its 4μm range (assuming a constant stopping power). The key factor in related process simulations would be the collection efficiency of this generated charge. The collection volume is reduced with each scaled process node. As a result, an increasing percentage of the generated free carriers simply recombine in the bulk substrate. Nevertheless, as the magnitude of the total generated charge is much greater than the Qcrit of SRAM nodes in leading process technologies, the risk of a soft error from a particle strike is significant. If the SRAM array is being licensed from an external IP provider, the methodology team needs to review the simulation strategy used by the IP vendor to determine the soft error rate specification.

### 16.4.6 Soft Errors for Flops and Combinational Logic

The SRAM circuit in Figure 16.15 highlights the sensitive nodes for a potential flip in the static stored bit value. The storage nodes in library flip-flop circuit layouts are becoming an increasingly important soft error risk. The flip-flop Qcrit for current processes is of a comparable magnitude to that of SRAM bitcells. Thus, the soft error contribution from flip-flops needs to be included in the FIT calculation. (The large percentage of SoC die area typically associated with SRAM array IP implies that the incident particle flux is higher for the arrays; nevertheless, the flip-flop soft error FIT contribution should be included.)

A particle strike at a combinational logic cell also results in collected charge and a perturbation to the circuit node voltages. The soft error risk from this event is dependent on various *masking factors*:

- **Cell logic inputs**—The specific sensitized node in the logic cell may not impact the output, based on the current logic input values.
- **Cell electrical response**—The collected charge on the sensitized node may not be sufficient to result in an electrical transient on the output, depending on the circuit topology and circuit drive strength to restore the logic value.
- **Propagation of output transient to a capturing flop**—If a particle strike does result in a logic output waveform transient of sufficient magnitude, that pulse needs to be sensitized through a logic path and arrive at a flop input in a temporal window when the flop is active (e.g., $t_{setup} + t_{hold}$). (This would be comparable to the propagation of aggressor noise transients in the static noise analysis flow, described in Chapter 12, "Noise Analysis.") References 16.6 and 16.7 present a methodology for analysis of the soft error risk in logic networks, derated by these masking factors.

Note that a particle strike on other sensitive nodes may result in anomalous behavior (e.g., clock buffers, analog IP circuits), which may warrant unique soft error identification and FIT calculations.

### 16.4.7 Multi-Bit Errors

The discussion so far has assumed that the free carrier collection volume is defined by a single circuit node. The free charged particles generated from a high-energy inelastic neutron collision may have sufficient energy and spatial separation to sensitize multiple nodes (e.g., from adjacent SRAM bitcells). Process scaling increases the multiple-bit upset rate, which is mitigated somewhat by the reduction in collection volume.

### 16.4.8 Soft Error Rate Mitigation: Process, Circuits, and Systems

Various optimizations are available to reduce the soft error rate, as described in the following sections.

#### *Process Optimizations for SER Improvement*

The process development team may pursue the following process optimizations for SER improvement:

- **Addition of a p-well impurity profile for nFETs (a *twin-well process*)**—A buried p-layer in a (low-impurity-concentration) p-type substrate limits the extent of the depletion region and limits the free carrier diffusion, reducing the collection efficiency.
- **A shallow n-well impurity depth for pFETs**—A shallow n-well for pFET devices also results in a reduced collection efficiency for a sensitive node, as an increasing percentage of the free carriers are generated and recombine in the p-substrate.
- **Dielectric *trench isolation* between devices**—Electrical isolation between adjacent devices is improved by the introduction of a recessed oxide dielectric (and additional surface impurity concentration at the oxide/silicon interface). The extension of this fabrication technique in more recent process nodes creates a trench oxide region, improving both the electrical isolation and the soft error sensitivity.
- **Adoption of FinFET or SOI technology**—The vertical profile of the FinFET device channel reduces the collection volume for free carriers generated from a particle strike in the substrate. The thickness of the fin defines the lateral extent of the depletion region between the device channel and substrate. The thin silicon layer for an SOI device minimizes the collection volume, resulting in a substantial reduction in the SER.

#### *Circuit Optimizations for SER Improvement*

The circuit-level modifications to enhance soft error robustness have an adverse impact on area, performance, and power; the assessment between soft error FIT rate and PPA targets is a difficult trade-off. Potential circuit design changes to reduce the SER include:

- **Increasing Qcrit**—The most straightforward approach would be to modify the circuit layout to increase Qcrit on sensitive nodes. An increase in the device node area increases $C_{node}$ but is counterproductive in that the collection volume also increases. Adding metal capacitance to the node would be a better alternative.
- **"Filtering" the collection current**—A modified SRAM bitcell circuit is shown in Figure 16.18. The additional resistances between the sensitive nodes reduce the likelihood of inducing the regenerative cross-coupled inverter currents after a particle strike. There would be a corresponding bitcell area increase and an extension to the write cycle.

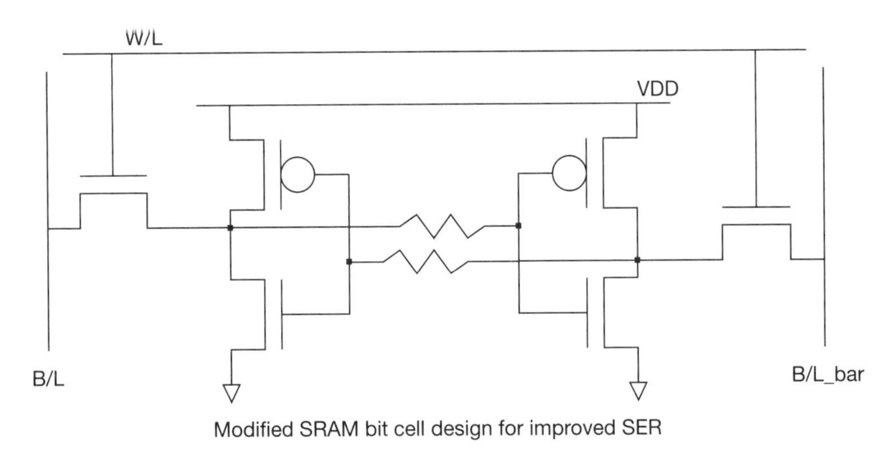


**Figure 16.18** SRAM bitcell redesign with improved SER robustness.

- **Modifying logic flip-flop circuits to improve soft error sensitivity**—A number of approaches have been proposed to modify circuit topologies to harden the circuit to the charge collection transient. These methods commonly involve propagation of redundant signals and a unique circuit response when a difference is detected. Figure 16.19 illustrates a two-input, two-output "inverter" for use in constructing latch and flop circuits.

If a particle strike causes the input values to differ, the outputs enter a high-impedance condition. The (dynamic, capacitive) output voltages retain the existing logic value until the transient on the input ends and active current drive is restored. Reference 16.8 illustrates the use of this inverter in a flip-flop circuit design, where the dynamic method is used when a strike hits a sensitive storage node.

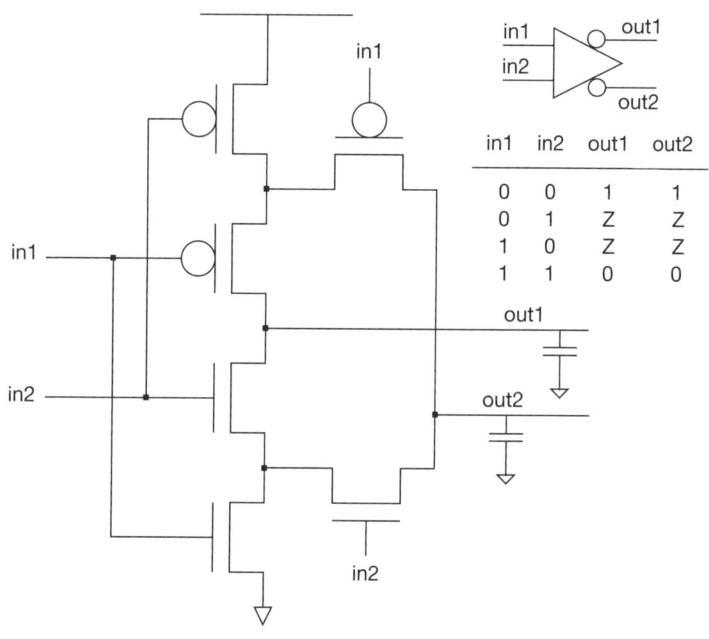


**Figure 16.19** Unique two-input, two-output inverter circuit for SER detection.

### *System Design Optimizations for SER Improvement*

From the system design perspective, an SoC architect needs to address the acceptable soft error FIT rate for target customer applications against the estimated FIT value for the IP blocks to be integrated on the SoC. The requirement to include additional functionality to detect and potentially correct data storage errors incurs significant area and performance overhead. The simplest approach would be to add a parity bit to an architected bus, with the overhead of parity generation on a storage write and parity error checking on a read cycle. If a single-bit soft error occurred, the parity checker would raise a condition so that a corrective step could be taken. A common method on wide busses used with larger on-chip arrays is to include additional encoded bits with the data word to provide single-bit error correction and double-bit error detection, known as an error correcting code (ECC) memory architecture.

The most elaborate architectural decision would be to implement significant redundancy throughout logic networks. For the most demanding nonstop applications, a *triple modular redundancy* (*TMR*) method uses three copies of a subnetwork with a “majority voting” decision gate added to propagate the “correct” value. Note that the TMR architecture is based on a single-bit logic soft error, likely a suitable assumption for (non-array) logic networks. There is always the risk of a strike at the voting circuit at a critical time; soft errors can be mitigated but are always a (probabilistic) risk. An alternative to the spatial redundancy of a TMR logic implementation would be to employ temporal redundancy—an area-versus-performance trade-off. A set of three flops could be triggered by clocks sufficiently staggered in time; these three signals would be input to a majority gate, whose output would be clocked into a fourth flop to represent the final “correct” value. These TMR options are depicted in Figure 16.20.

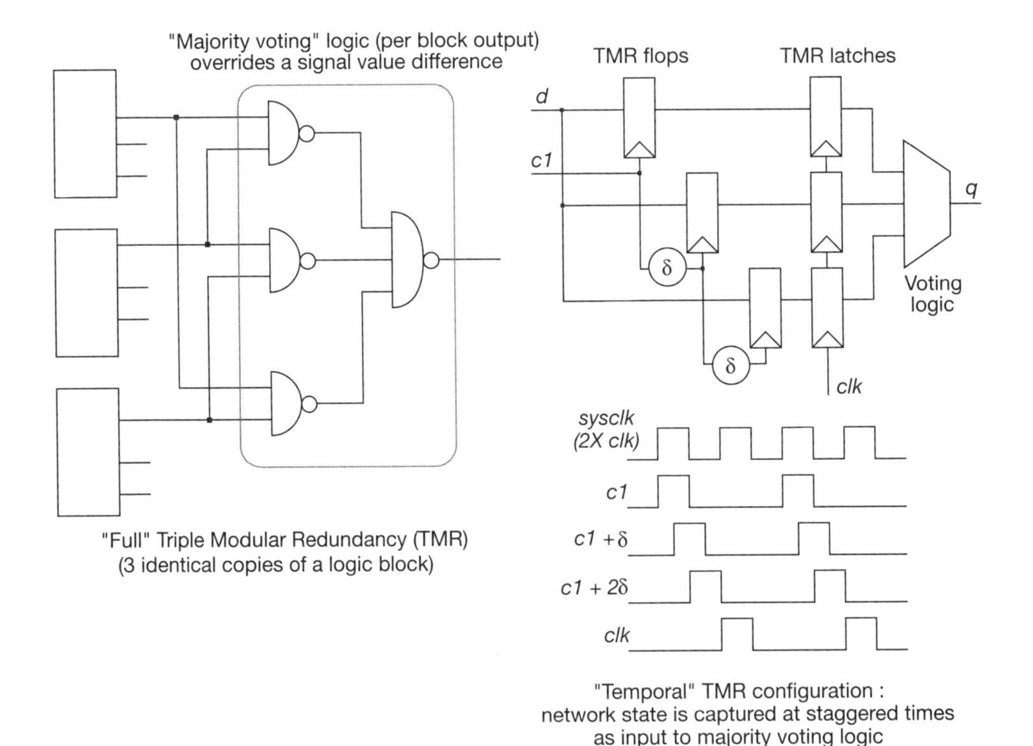


**Figure 16.20** Illustration of triple modular redundancy (TMR) design techniques.

An SoC project manager seeks to provide a soft error FIT calculation to accompany the hard fail reliability estimates to guide the architects on the error detection and correction functionality required and to guide IP designers on circuit-hardening preferences. Assumptions on the incident alpha particle and cosmic neutron flux and energy distribution can be made upon consideration of the (trace radioactive percentages in the) materials used and the customer application environment. The modeling and simulation of the strike event is complex, requiring detailed knowledge of the fabrication process, layout cross-section, and applied voltages to calculate the charge collection efficiency. The particle strike analysis typically integrates a conventional circuit simulator for the remainder of the circuit outside the sensitized node to evaluate the behavior during the charge collection interval.

The foundry reliability engineering team may pursue (empirical) testing of the collection efficiency as part of process technology development, using unique laboratories able to provide high-energy particle sources. The foundry may be able to provide models useful for the collection current, which could readily be added to the extracted circuit netlist in a conventional (statistical) characterization simulation.

For vendor IP to be integrated on the SoC, the project manager needs to collaborate with the vendor on the soft error FIT calculations to ensure consistency with the end customer application (e.g., medical, automotive, aerospace, consumer). For large IP arrays, the availability of an optimized ECC implementation may be a requirement.

Due to the random nature and complexity of the (primary and secondary) particle strike event and subsequent charge collection, an SoC project manager should expect significant "error bars" on the soft error FIT calculation. Whereas the FIT reliability for hard errors translates directly to product lifetime and serviceability costs, the soft error rate pertains more to the continuity of functionality and data integrity. Even if the soft error FIT estimate has a large uncertainty, the architectural and circuit-hardening trade-off decisions are likely unaffected if high-reliability applications are to be supported.

## 16.5 Summary

This chapter focuses on the mechanisms that affect the reliability of the production SoC, as represented by the overall FIT metric. Failures can be due to any of the following:

- The risk of electrostatic discharge to the SoC during assembly
- Device parameter drift during operation (i.e., bias temperature instability, hot carrier injection into gate dielectrics)
- Material changes that occur during operation (i.e., metal electromigration)
- Soft errors in storage values due to collected charge generated during a particle strike

The SoC project manager needs to maintain focus on evaluation of the FIT estimate relative to the requirements of the customer application(s). The breadth of factors influencing the FIT calculation is great, spanning process selection, IP selection, package and assembly procedures, and the optimization of circuit and wiring design during physical implementation. Each of these factors also has a strong impact on the resulting PPA of the SoC; the corresponding reliability-versus-PPA trade-off assessments are extremely intricate.

## References

[1] Gao, X.F., et al., "Implementation of a Comprehensive and Robust MOSFET Model in Cadence SPICE for ESD Applications," *IEEE Transactions on Computer-Aided Design of Integrated Circuits and Systems*, Volume 21, Issue 12, December 2002, pp. 1497–1502.

[2] Kumar, S., Agarwal, S., and Jung, J.P., "Soft Error Issue and Importance of Low Alpha Solders for Microelectronics Packaging," *Reviews on Advanced Materials Science*, Volume 34, 2013, pp. 185–202.

[3] May, T.C., and Woods, M.H., "Alpha-Particle-Induced Soft Errors in Dynamic Memories," *IEEE Transactions on Electron Devices*, Volume 26, Issue 1, January 1979, pp. 2–9.

[4] Murley, P.C., and Srinivasan, G.R., "Soft-Error Monte Carlo Modeling Program, SEMM," *IBM Journal of Research and Development*, Volume 40, Issue 1, January 1996, pp. 109–118.

[5] Zhang, M., and Shanbhag, N.R., "Soft-Error-Rate-Analysis (SERA) Methodology," *IEEE Transactions on Computer-Aided Design of Integrated Circuits and Systems*, Volume 25, Issue 10, October 2006, pp. 2140–2155.

[6] Zhou, Q., and Mohanram, K., "Cost-Effective Radiation Hardening Technique for Combinational Logic," *IEEE International Conference on Computer Aided Design (ICCAD)*, 2004, pp. 100–106.

[7] Joshi, V., et al., "Logic SER Reduction Through Flipflop Redesign," *Proceedings of the 7th IEEE International Symposium on Quality Electronic Design (ISQED)*, 2006, pp. 611–616.

[8] Dutertre, J.M., et al., "Integration of Robustness in the Design of a Cell," *Proceedings of the 11th IEEE International Conference on Very Large Scale Integration of Systems-on Chip (VLSI-SOC)*, 2001, pp. 229–239.

## Further Research

### SleepFET Rail Analysis

Describe the steps associated with a sleepFET rail analysis methodology flow for the gated-to-active transition, including the following:

1. Model build
2. (Multi-corner) simulation testcases
3. Restoration of network state from retention flops
4. Measurement of the state transition recovery time and rail peak currents
5. Results to be recorded in the SoC project scoreboard, with design sign-off by the functional validation team

### IBIS Format for Pad Circuits

Describe the I/O Buffer Information Specification (IBIS) format for the electrical model of driver and receiver pad circuits, suitable for simulation of the (transmission line) behavior of the chip-package-board connectivity.

Describe the EDA vendor characterization techniques used to derive the IBIS model for I/O cells.

TOPIC VI

# Preparation for Manufacturing Release and Bring-Up

The initial chapters in this topic describe the steps for the preparation of the full-chip SoC design database to release to the foundry for fabrication. Chapter 17 discusses the application of an engineering design change (ECO) to the chip database, a change request that arises after the database has completed the design freeze project milestone. The characteristics of the ECO methodology flows are unique, as a primary goal is to implement the change with a minimal disruption to the existing physical design database. Chapter 18 reviews the physical verification of the database to the foundry PDK checking runsets to ensure manufacturability and yield. Chapter 19 discusses various approaches to the generation of the manufacturing test patterns to be applied by the foundry and the package assembly service provider. In all these cases, specific EDA tool algorithms have been developed to efficiently manage the data volume of the full-chip design.

Chapter 20 covers the project management tasks associated with the tapeout release to the foundry. This critical milestone requires stringent review of the information recorded in the methodology manager application.

The project management team needs to assess the quality of the design as reflected by the results of functional validation, test pattern generation, and the electrical analysis flows.

Chapter 21 describes the activities undertaken by the engineering "bring-up" team. This team exercises the prototype silicon received from the foundry, pursuing performance measurements and verifying circuit behavior at extended environmental conditions. These electrical and thermal stress experiments are intended to identify any issues with design robustness and product lifetime. Successful completion of these experiments qualifies the design to proceed to production fabrication and product volume ramp.

CHAPTER 17

# ECOs

## 17.1 Application of an Engineering Change

An *engineering change order* (*ECO*), as introduced in Section 1.1, reflects a netlist-level modification to an existing design. It is applied at a point in the SoC project schedule after the HDL model, the logically equivalent netlist, and the physical implementation have exited the first “design freeze” milestone, in preparation for the initial SoC tapeout. It may also be applied between tapeouts to represent the design changes identified during prototype silicon bring-up. The key methodology impact to the integration of an ECO is that the changes are not a result of the “top-down” RTL-to-netlist-to-physical design flows used during the main project design phase. The ECO design changes originate for reasons other than an RTL version update. As a result, a specific set of ECO flows is applied to realize the requested changes; EDA vendor tools have a unique ECO mode to integrate the design changes.

The ECO changes typically utilize a different syntax to represent the design updates. The ECO netlist (or “commands”) would include additions and/or deletions of cell instances and signal nets, generated manually as part of the post-freeze debug. To apply the netlist changes, the physical design tool platform in ECO mode follows a unique set of steps:

1. Read the ECO netlist/commands.
2. Remove deleted cells and signal nets, adding fanout cell input logical tie-up and tie-down connections, as specified.
3. Insert added cells (resolving any cell placement overlaps).
4. Reroute new signal connections (incorporating potentially new ECO timing constraints while maintaining an awareness of power, noise, and EM rules).

The netlist changes could reflect functional modifications to the design, such as to fix a bug discovered in validation after the design freeze. Or the cell updates could be applied to optimize an electrical analysis flow result (e.g., implementing a different cell sizing or signal repowering strategy).

The goal of the EDA tools operating in ECO mode is to minimally perturb the existing physical design during the clean-up resolution step, after the cell and signal updates are applied. Ideally, the ECO does not introduce a significant number of errors in the subsequent tapeout-release electrical analysis flows.

The SoC project management team needs to exercise judgment about the design freeze milestone criteria and evaluate the risk of undetected functional bugs (using validation coverage measures) and the magnitude of remaining analysis flow errors that require manual investigation and optimization. The frequency of ECO releases and the number of edits in each should be minimal; otherwise, the likelihood of unsuccessful resolution during cell insertion and rerouting is high. The time required to implement the changes in ECO mode needs to be optimal, as well: The project schedule between design freeze and tapeout is always aggressive. The need to apply multiple ECOs of large scope may result in reverting the affected design block to an "unfrozen" full re-implementation status, with adverse impact on the project schedule and resources.

As an aside, in an attempt to minimize the time needed to generate a functional ECO netlist, EDA vendors have attempted to automatically generate the netlist from (minor) edits made directly to the frozen RTL model. Rather than attempt to manually implement a functional bug fix at the cell level, the RTL changes in the version update are localized to specific logic cones, new cells are synthesized, and then these cells are correlated to the existing physical database. The result is an automatically generated ECO netlist. The RTL ECO flow offers significant productivity advantages over a manually generated ECO netlist, and this flow remains an active area for EDA tool research.

The application of an ECO netlist that involves significant changes to the number of flops/registers is more involved. The block exiting the design freeze milestone has a (local) clock distribution tree that is tuned to existing loads. The length and connectivity of DFT scan chains has also been optimized. The introduction of new state elements involves clock optimization in ECO mode, which is an especially intricate algorithm. Although not preferable, a significant ECO update to registers in the design block may necessitate removal of the existing clock distribution and rerunning of a full clock optimization step.

### 17.1.1 Physical Design Updates

The physical database exiting the design freeze milestone may subsequently require updates without requiring the insertion or deletion of netlist cells. Although technically not considered an ECO, edits to the physical model share some of the same characteristics as the ECO flows. In addition to an interactive edit mode invoked by a layout engineer, the physical design tool platform may provide a unique command syntax that can perform physical data updates. The sequence of commands could potentially be generated algorithmically rather than applied manually in an interactive session. (An interactive session would also log the layout engineer's actions/commits.) The goal would be to have a detailed record of the version updates to the post-freeze physical design. For example, a revision to IP library physical cell views that are released after the design freeze may result in the decision to update the physical database, necessitating a new design version release. Although not a full ECO flow, per se, post-freeze physical updates typically follow similar strict version management policies. As with the application of an ECO netlist, physical database updates trigger the need to re-execute electrical analysis flows, with updated flow results recorded in the project status tracking data.

### 17.1.2 Metal Fill Data and ECO Flows

The design freeze database includes metal fill data that are added to satisfy the metal density and density gradient design rules associated with chemical-mechanical polishing and advanced photolithography. An ECO is likely to involve re-routing of new signals, potentially including local displacement of existing route segments as well. The presence of metal fill data makes successful resolution of the ECO much more difficult.

One flow option would be to completely remove the metal fill layout cell(s) from the block, apply the ECO, and fully re-insert the metal fill data. However, as the fill data is an integral part of the extracted parasitic model, the perturbation to electrical analysis flow results could be substantial. Alternatively, the physical design tool platform in ECO mode could include algorithms to "trim" existing metal fill shapes data to accommodate additional routing tracks when applying the ECO netlist commands.

The methodology and CAD teams need to ensure that the metal fill cells are properly identifiable by the ECO routing algorithms. Also, the teams need to collaborate with the EDA vendor to understand what trim algorithms are used, confirm that they satisfy the foundry PDK design rules, and record what modified metal fill cell instances will be present in the updated physical output database.

### 17.1.3 ECOs and the HDL Model Hierarchy

A common scenario for the generation of an ECO netlist is that a late functional bug is uncovered, and the micro-architects and physical design engineers determine the required logical fix and identify the corresponding netlist-level changes. If a logical synthesis flow was used to generate the original cell netlist from an HDL model, the correlation required between the HDL and netlist to make the same logical changes in both models can be difficult. Section 7.5 describes methods used by synthesis algorithms to derive signal names and cell instance names in the netlist to provide additional correspondence.

The functional bug fix developed at the netlist level may involve adding/deleting signals throughout the context of the frozen physical netlist. The ECO netlist syntax supports identifying existing cell instances and signals with a hierarchical reference path from the top level of the physical model. The RTL hierarchy needs to be maintained due to the extensive validation and test infrastructure that references registers and signals within the hierarchical RTL model. As a result, once the ECO netlist fix is identified, the micro-architecture team reflects an equivalent logical update in the RTL model; this may involve adding and deleting module ports to the existing RTL hierarchy. (The HDL language semantics may support assignment statements at a higher level of model hierarchy that directly reference signals in descendant models, but they add significant complexity to model validation, and this coding style is generally discouraged.)

## 17.2 ECOs and Equivalency Verification

A netlist-based ECO does not originate from an RTL model update. If the ECO includes a functional update, the RTL model also needs to be modified to reflect the corresponding change. Regardless of whether the ECO represents a functional or non-functional revision, the ECO methodology flow requires equivalency verification to be completed as soon as possible so that both the physical implementation and RTL functional validation flows can proceed. Figure 17.1 depicts the ECO equivalency checking project milestone.

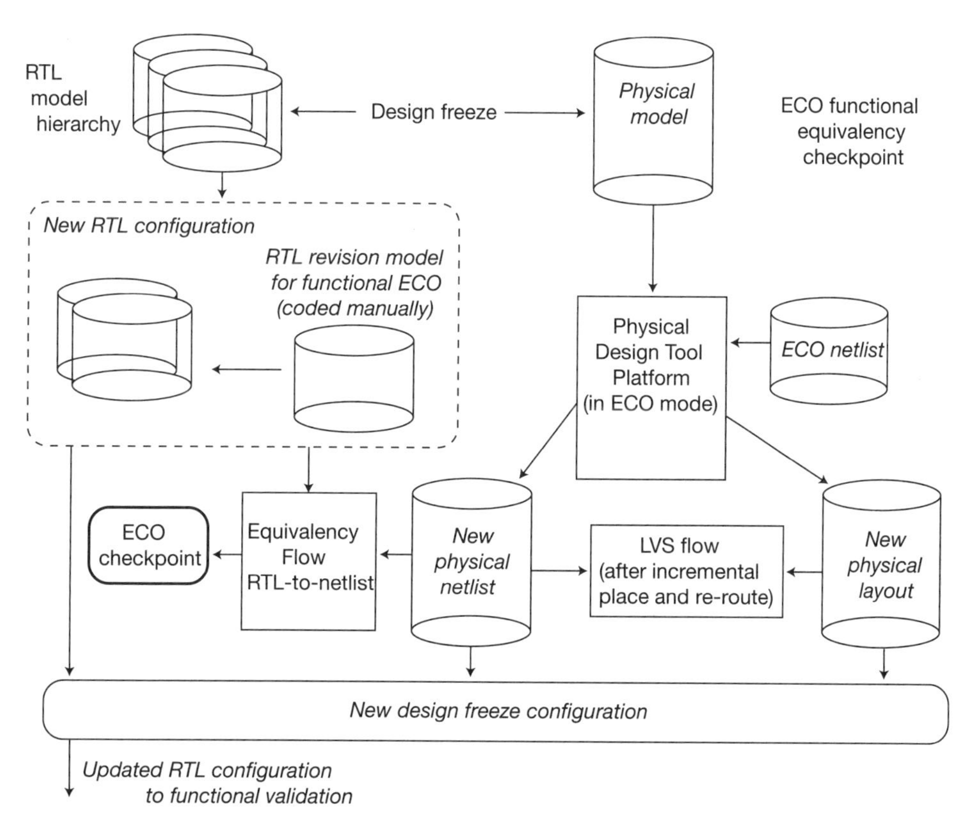


**Figure 17.1** Illustration of the project milestone in ECO mode confirming equivalency between the updated physical netlist and RTL model.

The first priority after exiting equivalency is to exercise any RTL validation testcases that originally uncovered the functional model error to confirm that the ECO successfully addressed the bug. (If the validation methodology utilizes a simulation acceleration toolset that accepts a netlist-level model, the new netlist may be directed to the accelerator immediately to exercise the testcases highlighting the bug, and RTL equivalency can be pursued concurrently.) In Figure 17.1, note that the physical tool platform generates a new netlist model after ECO netlist updates have been applied. This netlist is in the "conventional" syntax—not the unique ECO format—to submit to other flows.

The expectation is that the physical design platform in ECO mode can also successfully maintain proper connectivity throughout application of the incremental ECO updates. Nevertheless, exercising the layout-versus-schematic (LVS) flow with the generated netlist after completing the ECO is also a high priority, before resuming with the tapeout signoff-level analysis of the revised design.

It should be highlighted that functional ECOs also necessitate a high-priority investigation into the impact on the achievable test pattern coverage. In preparation for tapeout, the test engineering team compiles patterns for the various fault models of the netlist. An ECO that may adversely impact internal signal controllability or observability within the design model needs to be reviewed by the test team before committing to design implementation.

## 17.3 Use of Post-Silicon Cells for ECOs

In addition to ECO design updates identified after design freeze before an initial SoC tapeout, there may be revisions originating from the product bring-up testing of prototype silicon. Or post-tapeout revisions may be developed to address "known" (but not critical) functional defects in the original design whose fixes were deferred to avoid affecting the tapeout schedule. A common strategy is to include "spare cells" in the original tapeout database, as introduced in Section 1.1. The inputs to the uncommitted cells are tied to an inactive level during tapeout preparations but could be inserted as part of applying the ECO netlist. The goal would be to implement the post-silicon ECO with modifications to as few mask layers as possible; using existing spare cells would imply that no new device lithography masks would be needed. (Ideally, the ECO would also minimize the number of metal and via layers above the devices that are modified to embed new signal routes.) Utilization of spare cells for ECOs has two major benefits:

- The cost of making new masks for the subsequent SoC tapeout is reduced.
- The turnaround time (TAT) to receive updated silicon is reduced significantly.

The foundry (prototype lot) fabrication facility provides customers with an estimate of the tapeout-to-packaged hardware turnaround time, typically based on a "days per mask layer" target metric. If a set of wafers from an initial prototype lot are held after device fabrication, an ECO tapeout revision that only modifies BEOL metal/via masks would use these wafers, realizing a much shorter TAT (roughly half). Two approaches are commonly used for the design of spare cells:

- **All spare cell logic implementations utilize a common device/contact layout pattern, and the specific logic gate is defined by lower-level metal and contact *personalization.*** This "gate array" approach involves the design and characterization of a separate cell library, whose layouts combine the device/contact pattern with specific personalization for each logic function. The advantage of this approach is the simplicity with which spares can be inserted.
- **Specific logic cells from an existing library (with varying device patterns) are judiciously selected and "sprinkled" as spares throughout the physical model.** This approach requires less library development resource and offers a richer logic library with better individual gate delays. However, the selection and placement of specific logic spares increase the risk that a (local) solution to the application of an ECO netlist will be difficult to achieve.

In either case, the diversity of logic gate $V_t$ and drive strength options available for the ECO will be limited.

For either spare cell approach, the ECO methodology flows in metal-only revision mode need additional features:

- The insertion of new logic gates in the ECO netlist must correctly identify and only use the spare cell instances and locations.
- The embedding of ECO signal routes should adhere to a specific range of metal layer constraints.

For either spare cell library approach, the population of uncommitted cells into a block physical design requires collaboration among the CAD team, library IP team, and SoC analysis team. The library team assumes a routable utilization cell density when developing the overall routing and power/ground track image; although it will vary by design block size, a utilization target of ~80 percent is common. The analysis team needs to ensure that decoupling capacitance cells of a suitable density are inserted, likely occupying most of the vacant locations. This leaves a small percentage of the block area for spare cells (e.g., less than 1 percent). The CAD team needs to ensure that the placement algorithm applies a suitable trade-off between logic "clustering" for performance and maintaining an appropriate density of vacant locations for insertion of decoupling capacitance and spare cells.

## 17.4 ECOs and Design Version Management

During the initial SoC project design phase, the focus is on RTL model definition and functional validation. Design version releases are somewhat informally defined; a mainline release increment is usually associated with preparations for an upcoming functional validation milestone. For the physical implementation in this early stage, the global floorplan data and individual block design releases are commonly aligned to a major RTL release once logical-to-physical equivalence has been demonstrated. This version release policy becomes extremely strict when the design freeze milestone has been met and an ECO-only mode is enforced, as depicted in Figure 17.2.

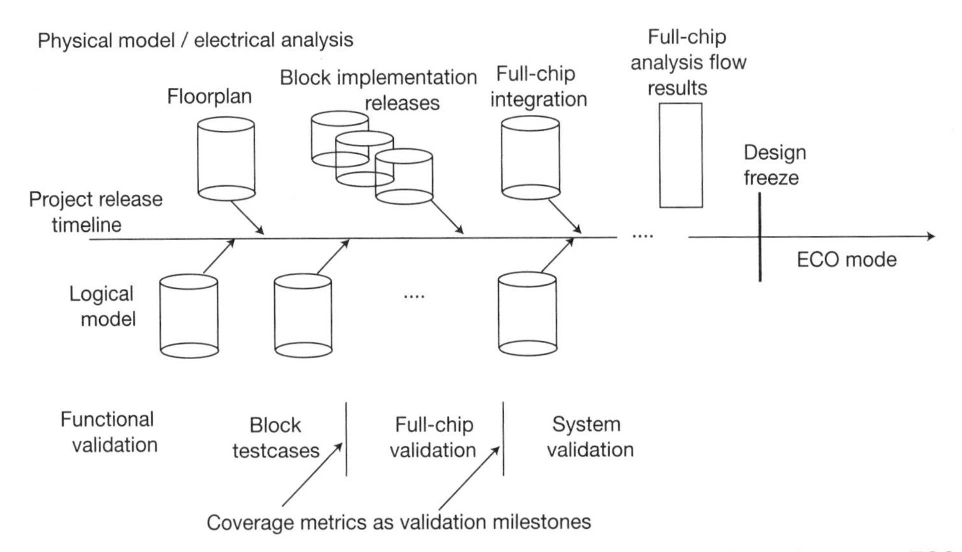


**Figure 17.2** Illustration of the version release policy when the project enters ECO mode.

Every proposed ECO netlist is reviewed by the SoC project management team before approval to apply the ECO is granted. If the ECO reflects a functional bug fix, the ECO approval includes a resource commitment from the micro-architecture team to make corresponding updates to the HDL model. The purpose of the ECO revision is documented as part of the ECO release dataset. The project management team assesses the following:

- What is the scope of the ECO? What time and resources are needed to implement and verify fixes?
- Is the functional bug or electrical analysis issue to be addressed by the ECO of critical impact? What significance would deferring the proposed update have on the silicon prototype bring-up phase?
- What is the potential impact to the planned tapeout date? What are the ramifications of a schedule slip if accepting the proposed ECO (and subsequent tapeout preparations) will push out the tapeout date?
- Is the ECO easily implemented using spare cells?

Note that prototype silicon bring-up is not especially focused on production yields or long-term product reliability. Addressing analysis flow issues that pertain to design-for-yield or reliability could potentially be deferred to a subsequent tapeout. Planned functional features (of low priority) could potentially be deferred if it becomes evident that frequent ECO requests are arising from related validation bug discovery.

Each accepted ECO triggers a new SoC model version release. This version update resets the corresponding block and full chip project methodology status. Also, it may become necessary to revert to a previous version, backing out an accepted ECO, if the impact of the update proves to be greater than expected. The most disruptive example of rescinding an ECO would be if the planned use of spare cells for a metal-only tapeout revision proves to be unsuccessful. The decision to adopt an all-layer tapeout release for an ECO instead of a planned metal-only update necessitates extensive review of the impact on development cost and resources and, ultimately, the SoC qualification and release schedule.

### 17.4.1 SoC Tapeout Project Plan

The specific ECO implementation approach to be adopted for an SoC is an integral part of the initial project planning cycle. Ideally, the SoC design would require only a single-pass tapeout to achieve all functionality and PPA targets. Realistically, the project plan would assume a multi-pass iteration to achieve a design that would be ready for volume production. The ECO methodology is used to help establish the project timeline.

A common nomenclature used in SoC planning is to use an integer version number for an all-layer tapeout release and a decimal increment for a metal-only tapeout. For example, as illustrated in Figure 17.3, a project plan may be developed assuming two all-layer releases and an intermediate metal-only ECO revision.

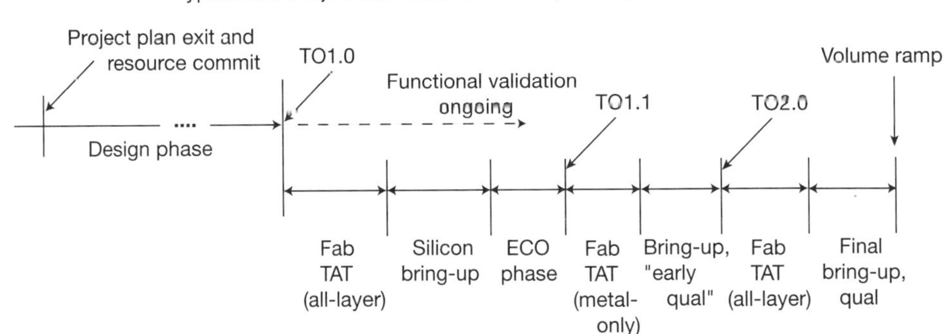


**Figure 17.3** Example of the nomenclature used in project tapeout planning.

As illustrated in the figure, functional validation continues after TO1.0, with additional (system-level) testcases. Any functional bugs uncovered during the TO1.0 fabrication turnaround time are added to the set of deferred bugs for ECO review. If a metal-only alternative is not available, the project schedule in Figure 17.3 is then modified to eliminate the TO1.1 option. As a result, the scope (and duration) of the TO1.0 silicon bring-up testing would be expanded, as TO2.0 represents the only opportunity to embed all approved ECOs.

To be sure, the most successful SoC projects ramp to volume production with the TO1.0 (or TO1.1) revision, which is also a positive reflection on the design methodology employed.

## 17.5 Summary

This brief chapter describes the unique nature of the application of an engineering change order (ECO) to a design implementation that has previously exited the design freeze project milestone and/or a design database associated with prototype silicon. The ECO may involve a functional change, typically due to the closure of an outstanding bug identified during functional validation. Alternatively, the ECO may represent a non-functional change to the physical implementation, typically to address issues found during full-chip sign-off analysis flows. In either case, cells may be added or deleted, signal wires may require rip-up and re-route steps, and in the worst case, clock distributions may need to be re-optimized.

The ECO is described using a specific netlist syntax, to efficiently represent the cell and signal modifications. As the ECO is commonly generated manually, the ECO semantics needs to be straightforward. EDA vendors have adapted their implementation tools in support of an "ECO mode," with specific features that use the ECO netlist and attempt to minimize the perturbations to the existing physical design. The incorporation of backfill cells in the physical block specifically for ECO cell additions is a special methodology consideration for ECOs to a prototype silicon design, to minimize the number of mask layers to be changed and expedite the fab turnaround time for the revised design.

The application of a functional ECO requires separate updates to the RTL description. The adoption of both functional and non-functional ECOs necessitate exercising the full-chip functional equivalency flow.

Project management of the post-freeze silicon database applies a more stringent set of criteria for a proposed ECO. A review team with broad representation across the engineering teams considers each proposal thoroughly. Their assessment considers the ECO resource and cost impacts versus the priority of the intended design fix. ECOs are inevitable, as issues almost always arise after design freeze and/or during prototype silicon evaluation. The most successful projects implement a well-defined, comprehensive ECO methodology.

## References

[1] Jayalakshmi, A., "Functional ECO Automation Challenges and Solutions," *IEEE 2nd Asia Symposium on Quality Electronic Design (ASQED)*, 2010, pp. 126–129.

[2] Krishnaswamy, S., et al., "DeltaSyn: An Efficient Logic Difference Optimizer for ECO Synthesis," *IEEE International Conference on Computer-Aided Design (ICCAD)*, 2009, pp. 789–796.

[3] Noice, D.C., et al., "Method and System for Implementing Metal Fill," US Patent No. US7661078B1.

## Further Research

### Automated ECO Netlist Generation from RTL Changes

Describe the algorithms used by EDA vendors to automatically generate ECO netlists directly from RTL revisions. (References 17.1 and 17.2 provide good background to start the research.)

### Metal Fill Trimming in ECO Mode

Describe the features of an algorithm to "trim" existing metal fill data in ECO mode after the execution of ECO command updates. (A good place to start the research is provided in Reference 17.3.)

CHAPTER 18

# Physical Design Verification

## 18.1 Design Rule Checking (DRC)

As introduced in Section 2.2, the DRC tool applies a set of layout operations and measures to the physical design data and reports any rule violations. The sequence of operations and measures, called the *runset*, is released by the foundry as part of the PDK. Currently, there is no (de facto) industry standard for the runset command syntax (e.g., operations such as INTERSECT and EXPAND/SHRINK and measure checks such as SPACE, OVERLAP, and AREA). An SoC project manager needs to review which EDA vendor DRC tools are supported by the foundry and coordinate with the CAD team for installation and flow support of the PDK runset corresponding to the EDA software licenses available. With the introduction of advanced process nodes, the number and complexity of design rules has grown, necessitating the addition of new operations to the EDA tool features, including:

- Measures based on parallel run length of adjacent shapes
- Measures applied after *binning* of shapes by dimensions (to support forbidden pitch rules)

- Layer density and density gradient measures (using a repeated, overlapping *stepping window*)
- Measures specific to a "color" property assigned to shapes on multipatterning layers

As a result, the foundry may support a diminishing number of EDA tools and/or stagger tool support in subsequent PDK releases, as the EDA vendors add the requisite features.

The EDA toolset chosen for the DRC flow is also influenced by the layout design platform used. The platform is likely to include an "interactive DRC" mode, in which the layout designer and/or chip integration engineer is provided with immediate rule checking feedback (on a reduced set of runset commands). Thus, the PDK runset for this EDA layout platform is also required. There are license cost implications involved with selecting a different (batch) DRC tool for sign-off than for the interactive design platform. The EDA toolset used for cell netlist placement and routing also embeds a DRC algorithm, which requires foundry PDK support.

To maintain throughput for (chip-level) DRC evaluation, EDA vendors have augmented their tools to leverage the multiple layout cell instances in the model hierarchy and to partition and distribute a large model across multiple processors.

The layout data operations in the DRC tool have also been expanded to add shapes data to the physical model as part of the fill uniformity requirements. The tool strategy has evolved to support *intelligent fill* to reduce the impact of fill on the database size. Multiple instances of common, repetitive fill area pattern cells are inserted hierarchically to the physical model.

Another DRC operation used in both PDK runset checking and other layout utilities is the support for *pattern matching* of the layout design compared to a set of input patterns. (For efficiency, a pattern may be described by a range of width/space dimensions among the shapes.) In addition to being included in the description of a lithography design rule, pattern matching is useful in identifying the number of instances of known yield-sensitive structures (see Section 18.4).

The (batch) DRC flow is applied throughout the SoC physical implementation phase. The release and promotion of a block to global chip integration typically requires that the block be “DRC clean.” An additional layout utility is invoked to generate a block abstract, reducing the overall data volume to layer blockage maps and pin shapes. Although final DRC verification prior to tapeout could potentially use abstracts with global layout data, the full chip flow is commonly exercised with all shapes data visible. For example, a layout density or density gradient rule check at full chip may be applied in a window across block boundaries and would not be confirmed with abstracts.

The DRC flow needs to record results in the SoC project tracking database. The EDA tool also provides a related debugging environment, in which the error log from exercising a DRC runset is cross-referenced to the layout cell. Ideally, the error log is linked directly to the layout editing platform, as well, so that layout modifications can be made and (interactively) checked. DRC errors related to unsuccessful multipatterning decomposition can be particularly difficult to address due to the (potentially) long-range shape interactions, as shown in Figure 18.1.

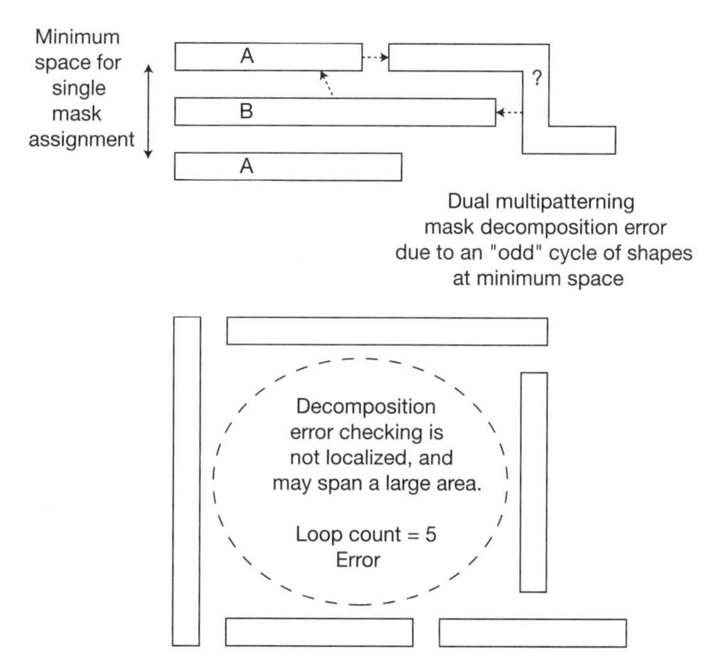


**Figure 18.1** Examples of a multipatterning cyclic assignment layout error. Multipatterning decomposition checking differs from the traditional distance-based measurement DRC operations.

EDA vendors are enhancing DRC tools to provide guidance on shape modifications to alleviate the interactions that are preventing decomposition color assignment.

## 18.2 Layout-Versus-Schematic (LVS) Verification

As applied to the physical verification of a library cell or an IP macro, the LVS flow checks that the device dimensions and connectivity for a cell layout match the corresponding schematic view. When applied to block physical verification, the LVS flow confirms the equivalence of the inter-cell route connectivity in the layout view to the corresponding netlist.

In block verification mode, the cell detail views are regarded as black boxes. The algorithms for traversal of the block layout and netlist hierarchy are provided with a *stop list* consisting of the library cell names, as these will have been previously proven to be LVS clean.

The final verification that the full SoC functional RTL model is equivalent to the (tapeout) physical database is a transitive proof, consisting of:

1. LVS for library and IP layout and schematic views
2. Formal equivalency of library and IP schematics to their RTL models
3. LVS of chip/block netlist to layout (with cell IP stop list)
4. RTL-to-netlist formal equivalency

This overall application of the LVS flow is depicted in Figure 18.2.

The LVS runset consists of many commands shared with the DRC runset (e.g., layout operations and measures to identify device types and their dimensions). In addition, the LVS runset consists of a set of connectivity relationships that enable the layout traversal algorithms to identify electrical continuity, as illustrated in Figure 18.3.

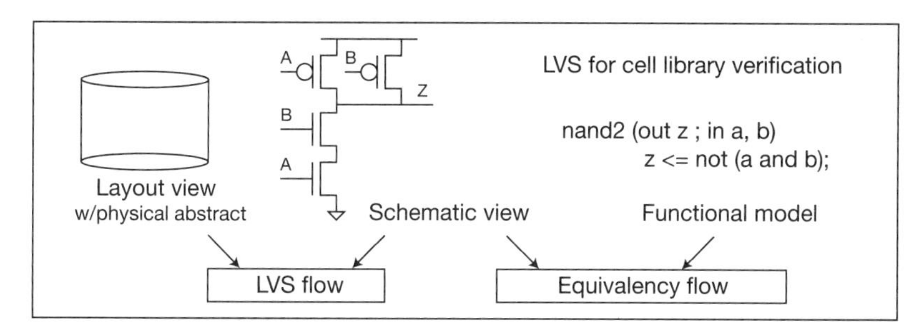


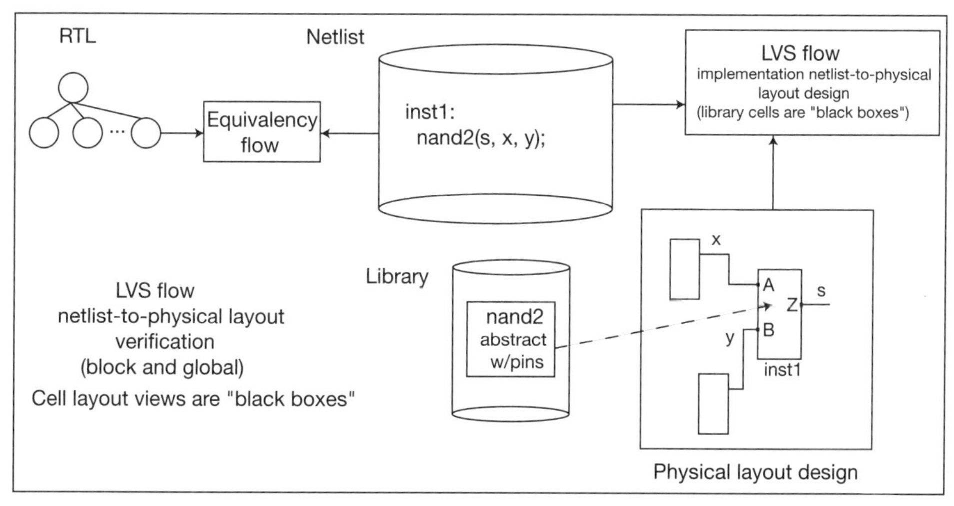


**Figure 18.2** Illustration of the LVS flow, initially for library IP and then for the full-chip SoC model.

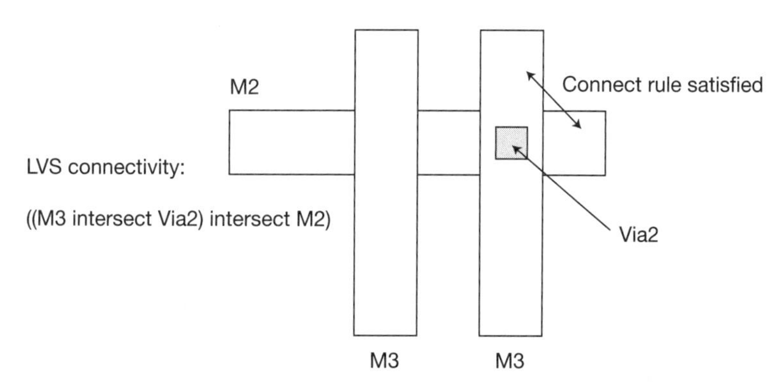


**Figure 18.3** Illustration of the LVS connectivity properties for tracing signals in the layout.

Note that connectivity is established by the intersection of shapes on the continuity definition. These intersections may not be DRC clean, in that the LVS flow is separate from design rule checking. Indeed, it is often productive to forward a layout to subsequent electrical analysis flows while the remaining design rule errors are being addressed; the assumption is that the current layout parasitics will be extremely close to those of the final layout release. Forwarding the layout would provide an early assessment of any electrical issues. Although the layout may have DRC errors, it will be necessary to be LVS clean so that extracted parasitic RC elements can be correctly annotated to the netlist model.

The layout traversal algorithm also needs to identify the pin shapes and pin name properties of the top cell and hierarchical cells in the stop list for correlation to the schematic or netlist.

When performing the correspondence checking, there are LVS flow options that the SoC methodology and library teams need to evaluate. The following sections discuss these options.

### 18.2.1 Representation of Device Width

The schematic and layout both have the option of representing a wide device as consisting of multiple, parallel fingers, as shown in Figure 18.4.

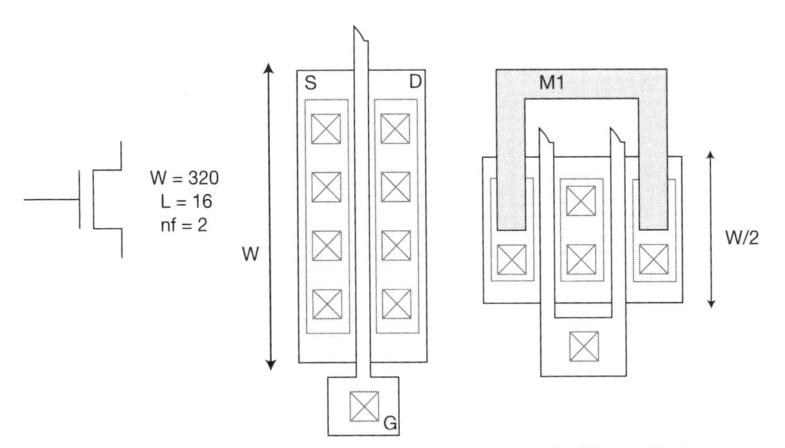


**Figure 18.4** A flow setting for cell-level LVS determines whether the total layout device width is compared to the schematic or whether the specific number and size of individual device fingers is compared. Due to the influence of layout-dependent effects on device models, the latter comparison is the norm.

The LVS flow could simply seek to verify that the total device width in both models is identical (within some small allowed error tolerance perhaps). Or the flow could be directed to verify a specific parallel device topology in the two models. Note that the device source/drain node parasitic estimates used in circuit simulation for the pre-layout schematic will differ, depending on the "number of fingers" value property on the schematic device instance. In current process nodes, there are also a number of layout-dependent effects that modify the device threshold voltage, carrier mobility, and statistical variation tolerances. The LDE dimensions used in pre-layout simulation would be estimated and would ideally be representative of the final layout. As a result, the common LVS flow setting in use is to explicitly match the *W*, *L*, and *nf* parameters between the schematic instance and the physical layout. The LVS output netlist would expand the individual fingers, with the specific LDE measurement dimensions from the layout added to the device parameter list for each finger instance.

### 18.2.2 Equivalence of Series Connections

When exercising LVS for either library cells or a block-level netlist, a flow option is available to accept a layout connection to a logically equivalent device or black box cell pin as a match to the netlist (see Figure 18.5).

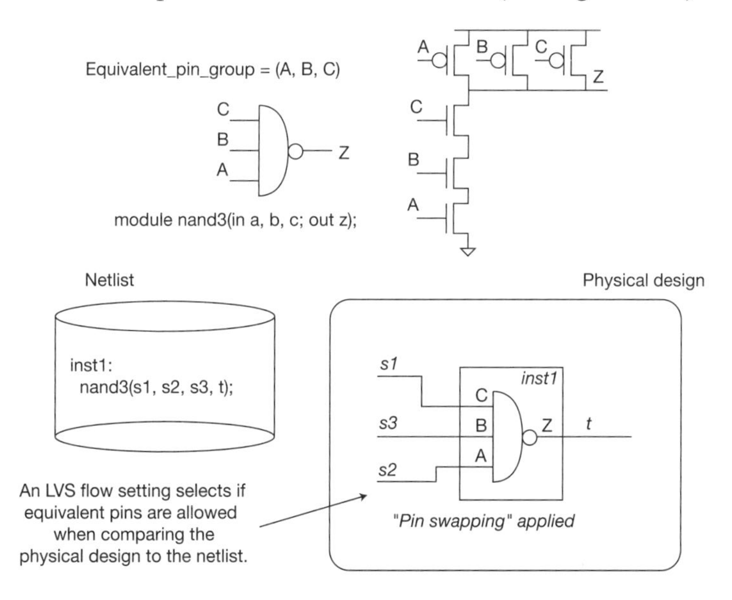


Note: Complex equivalent pin groups are possible -- e.g., a 2-2 And-Or: ((A1, A2), (B1, B2))
A1 and A2 are equivalent, and (A1,A2) as a group is equivalent to (B1, B2).

**Figure 18.5** An LVS flow setting determines whether exact or logically equivalent layout connections to cell pins (or devices) will be checked for matching connectivity to the netlist.

For example, during block-level routing, a pin-swap optimization algorithm may be invoked to reduce the routing congestion and/or improve critical path timing. The LVS tool would then require the cell library data to include information on the "pin groups" within which equivalence is valid. The figure highlights that multi-level pin groups may be defined for complex cells.

### 18.2.3 Application of "Global" Power Connections

When running the LVS flow on library cells, the power rails may be absent, having been allocated to the block construction flow. The layout may simply have pin shapes present for the power rails. These global connections will be treated as equivalent to the schematic connections to VDD and GND.

A more intricate case occurs when exercising the LVS flow on a routed block or the full chip. The (synthesized) cell netlist may not include explicit connections to the supply rails, as illustrated in Figure 18.6. In addition to verifying that the netlist signal connectivity is correct in the layout, the LVS flow confirms that the power and ground pins in the cell layout abstracts are correctly covered by the corresponding power and ground grids, as specified in the "global signals" input to the LVS flow.

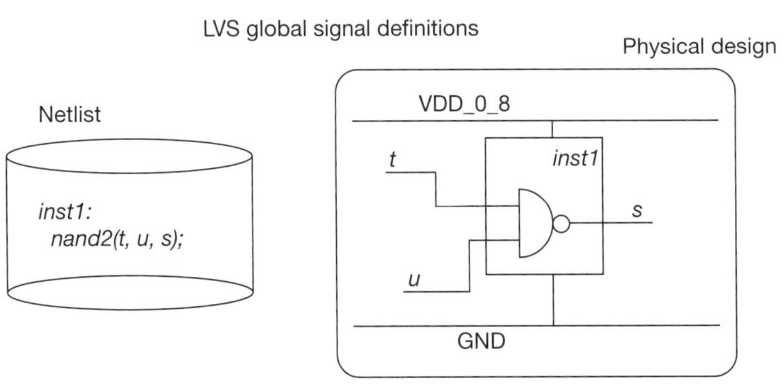


**Figure 18.6** Application of a global signal list in the LVS flow. The cell layout power pin shapes are to be connected to the corresponding global signal.

A special case is introduced for the physical implementation of a block with power gating. The SoC methodology, CAD, and library teams need to address the representation of the internal gated rail, the sleepFET cells, and the pwr_enable control inputs. Any state retention cells in the power gated block require connection to both the gated and ungated rails. A similar motivation to include explicit supply rail connections in the block netlist is offered by using level shifter cells at block inputs to resolve VDD supply differences between blocks in different voltage domains. Another example arises from cells that require unique power connections to the (triple) well for devices where the "substrate" connection is not directly tied to a global power or ground rail. The most straightforward approach would be to provide a cell netlist with explicit connections to all power rails. This *PG netlist* format is depicted in Figure 18.7.

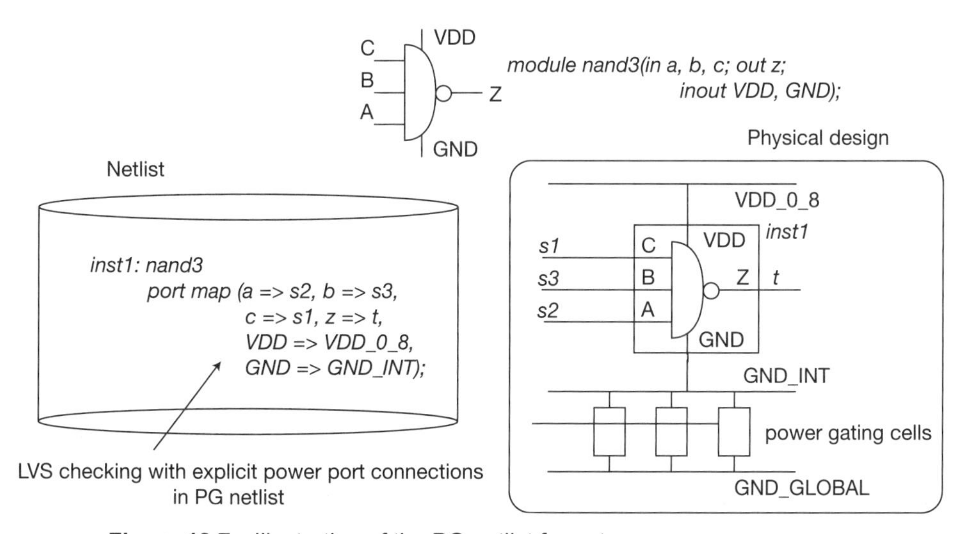


**Figure 18.7** Illustration of the PG netlist format.

### 18.2.4 Miscellaneous Layout Cells

The traversal of physical signal and power connectivity by the LVS flow encounters layout cell instances that are unrelated to the netlist graph comparison, such as spare cells, explicit decoupling capacitance cells, and end cap cells (which provide a more uniform layout pattern to the cells placed at the row ends). The LVS flow also includes an "ignore list" of layout cell names to exclude from the layout graph for efficiency.

EDA vendors have focused on maintaining LVS tool performance in advanced process nodes with similar approaches as described for DRC (i.e., multi-threaded execution, hierarchical verification, and promotion of child cell instances). In addition, LVS tools utilize matching algorithms for signal name properties in the two models to expedite throughput. Perhaps the most difficult example that stresses these algorithms is LVS verification of a large SRAM array macro. The schematic designer may slice the array hierarchy in a manner that is most conducive to (high-sigma) pre-layout circuit simulation, while the layout designer may seek to assemble the banks of the array from a different instance hierarchy. The LVS tool must overcome the lack of hierarchy and signal correspondence between the two models.

### 18.2.5 LVS Debug

The user interface provided by the EDA vendor for LVS error debugging is more complex than for depicting DRC errors. The correct correspondence between a layout and netlist (or schematic) signal is relatively straightforward to display. However, missing or incorrectly wired nets without correspondence in the two models are more difficult to illustrate. A particularly vexing case to debug is a "VDD-to-GND short" connection, as a significant percentage of the layout data are highlighted in such a case. The EDA vendor may incorporate additional analysis features in the LVS tool to provide guidance on the layout changes necessary to address the connectivity errors—for example, to allow the designer to temporarily alter the layout connectivity graph by re-assigning net names to polygons in the debug platform and re-evaluating the schematic/netlist (golden model) comparison.

## 18.3 Electrical Rule Checking (ERC)

The ERC flow applies runset commands and rule checks against the schematic and layout models for both cell detail and block-level verification. The purpose of the ERC flow is to identify potential electrical issues (schematic) and yield-detracting structures (layout) that are not identified by the DRC or LVS flows. Examples of typical ERC checks include the following:

- Identifying any extraneous "floating" devices, wells, or interconnect segments (layout, schematic)
- Identifying any high-voltage (VDD_IO) nets connected to regular (thin gate oxide) VDD-compatible device nodes (schematic)

- Exercising design rule checks specifically related to the maximum voltage difference between shapes (e.g., increased spacing between two metal segments at VDD_IO and GND)
- Identifying incorrect layout topologies and/or electrical resistances outside PDK limits for the ESD protect cells merged with chip I/O circuits (see Section 16.3)

The ERC commands expand on the LVS net tracing and measurement features to include the capability to calculate electrical parameters for layout net topologies, as required for the resistance calculations in the ESD cell connectivity to an I/O circuit. In addition, ERC commands include features to trace "through" device topologies to determine what voltages may be present on metal segments for the voltage-dependent metal spacing layout design rules and for the thin/thick oxide device gate voltage checks.

The ERC flow applies checks to (cell-based or schematic-based) netlists and layouts that are a combination of foundry rules and any additional design requirements established by the SoC methodology team. The CAD team is typically tasked with coding the ERC runset for additional verification checks to merge with the ERC runset in the foundry PDK. Examples of project-specific topologies that might be flagged include the following:

- Limits on (acceptable) pFET and nFET device dimensions and complementary pullup/pulldown device stack beta ratios, $\beta$ = Weff_p / Weff_n
- No "dotted" cell outputs (Designers may attempt to wire multiple cells together at the block level in an attempt to increase signal drive strength with greater switching crossover current.)
- Antenna checks to confirm the addition of reverse-biased diodes by the router to interconnect wires, as described in Section 9.4 (This ERC limit leverages foundry PDK yield optimization guidelines.)

It is likely evident but worth mentioning that the physical verification tool software license investment (almost) always selects DRC, LVS, and ERC tools from the same EDA vendor. The CAD team needs to be knowledgeable about the runset commands. The SoC layout design team needs to be as productive as possible working with the debug platform. Selecting these tools from a single EDA vendor means consistency in these characteristics, which is a productivity advantage.

## 18.4 Lithography Process Checking (LPC)

The scaling of advanced process node design rules has progressed faster than the reduction in lithographic exposure wavelength available for high-volume wafer fabrication. As a result, two key characteristics have evolved:

- **The mask data differs (rather drastically) from the tapeout physical database.** The mask data adjustments have evolved throughout process node generations. Initially, a set of *optical proximity corrections* (*OPC*) were applied, mostly to improve the exposure fidelity of the corners of drawn shapes. In subsequent process nodes, a set of *sub-resolution assist features* (*SRAFs*) were added to the mask data, separate from the drawn layout, to provide constructive and destructive interference illumination intensity when exposing the photoresist-coated wafer. Presently, an extremely complex set of algorithms, known as *source-mask optimization* (*SMO*), are applied to derive both the mask data and the (non-uniform) exposing illumination pattern.
- **Layout design rules became much more complex.** The layout data are now subject to an increasing set of design rules that arise from lithographic exposure resolution requirements or fabrication uniformity constraints:
  - Forbidden pitches
  - Additional dummy shapes required adjacent to active circuits to improve local exposure uniformity (e.g., unconnected active and gate fill shapes data)
  - Limited ranges for allowed dimensions of devices, contacts, vias, and wires for both improved exposure and etch uniformity (either a narrow range of continuous dimensions or, more commonly, a set of allowed discrete values)

Increasingly, the layout data for devices and local metals resembles an optical grating, with restrictions on segments in the non-preferred direction.

These two factors introduce a non-trivial risk that a tapeout layout database may be DRC clean yet present difficulties during mask data preparation to find a suitable SMO solution. For advanced process nodes, the foundries have introduced a new physical verification methodology flow requirement to perform LPC.

The foundry and EDA vendor collaborate closely to define a set of high-risk layout topologies to use with the LPC tool pattern matching features. In addition, the foundry may provide SMO-like algorithms to the EDA vendor to embed into the LPC software to further reduce the risk of an incomplete solution during subsequent mask data preparation. This combination of pattern ("rule") checking and algorithm ("model") analysis is computationally expensive. Further, LPC is really applicable just prior to tapeout. The assumption is that the design rules will indeed be sufficient, and the LPC flow will not uncover any SMO-related errors on the tapeout database. As a result, it would be unnecessary (and expensive) to exercise LPC on each successive version release of physical data during the design phase of the SoC schedule. Further, the set of LPC patterns and SMO models from the foundry evolves as the new process node matures; it is necessary to exercise LPC on the latest foundry PDK runset during the SoC schedule tapeout phase.

Because the demand for LPC software licenses is sporadic and the (full chip) computational resources are substantial, EDA vendors commonly provide LPC physical verification as a separate services business rather than lease on-premises software licenses. Thus, the EDA vendor chosen for stand-alone LPC verification could be separate from the provider of DRC, LVS, and ERC tools. The selection would likely be based on contracted services cost, availability of computational resources (that align with the SoC tapeout schedule), and verification turnaround time.

With the increasingly restrictive design rules for new process nodes, the SMO data preparation risk is likely to be adequately addressed by the rule-based, pattern-matching method, obviating the need for the model-based analysis. The layout regularity associated with FinFET fins, device gates, and wiring track multipatterning color pre-assignment reduces the custom layout variability, as depicted in Figure 18.8.

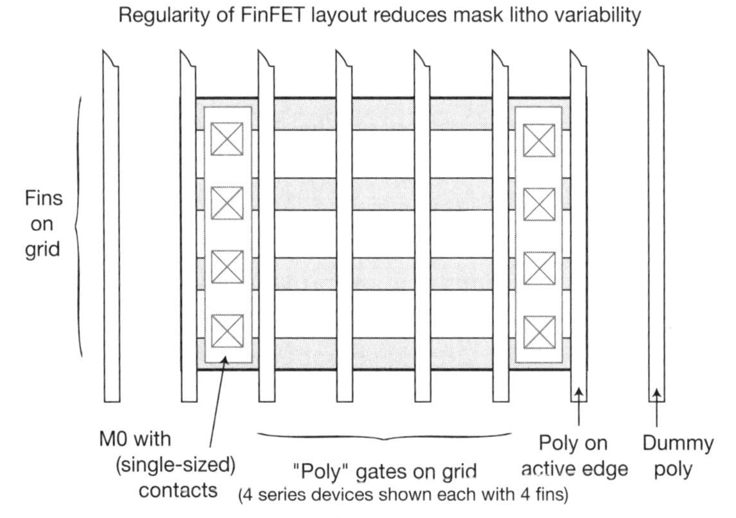


**Figure 18.8** Example of a FinFET technology layout. The reduced variability in active, gate, and local metallization topologies reduces the complexity of LPC verification.

LPC flow execution as a tapeout prerequisite may be achievable solely by applying pattern-matching features, using (existing) DRC software licenses.

## 18.5 DRC Waivers

A fabrication process evolves over time. Preproduction design rules from the PDK v0.x release versions are used by "early adopters" of the technology (e.g., IP developers, leading-edge customers) who want to have designs ready for volume ramp when the PDK v1.0 release is available. During this evolution, the fab is working on establishing a *process window* that results in suitable wafer-level yield across the manufacturing variations in deposition, lithography, and etch steps. Design rule updates in PDK v0.x releases may result.

After the PDK v1.0 release, the process continues to evolve as the foundry engineers pursue *continuous process improvement* (*CPI*) efforts to maximize yield and reduce costs. Again, design rule changes may be present in PDK releases v1.1, v2.0, and so on. As a result, designs verified to an earlier PDK release may have DRC violations when integrated into a subsequent SoC tapeout. An SoC project manager and foundry product engineering team need to review whether a *DRC waiver* will be applied. Such a waiver represents an assessment by the foundry that the wafer yield will not be adversely impacted.

For the SoC design team, the waiver request would be made only if the resource required to address the DRC error (and recharacterize the design) would have an impact on the tapeout schedule. The SoC project manager may commit to removing the waiver request in a subsequent tapeout if the prototype tapeout is allowed to proceed on its current schedule.

There are two project methodology aspects to addressing potential DRC waivers:

- **Early identification prior to tapeout**—To avoid schedule delays at tapeout while the foundry reviews a DRC waiver request, the project schedule should include full-chip DRC flow evaluation during preliminary physical integration steps. Some project managers include a *mock physical tapeout* milestone. Although this activity is resource intensive, the early identification of potential DRC waiver requests is critical to minimizing tapeout review delays at the foundry.

  During the SoC design phase, if a new PDK release is provided by the foundry, existing IP is re-validated. Any new DRC violations need to be reviewed to assess whether a layout update (and re-characterization) or a DRC waiver request should be pursued. This resource and schedule impact assessment is complicated by the use of design IP from external suppliers; their resource availability to update and rerelease a new IP version may or may not coincide with the SoC tapeout schedule. (This dependency on external vendor IP rerelease is especially problematic if new IP silicon testsite characterization is warranted.)
- **Project database management of DRC waivers**—The SoC project database maintains information about the PDK version used for verification of all library IP. In addition, the database includes all existing waivers that have been granted by the foundry. When the SoC design data and physical verification flow results are being compiled for tapeout submission, DRC errors need to be cross-referenced to the waivers in the project database and included in the supplementary tapeout materials. The prototype fabrication TAT from tapeout submission is used to establish project schedule dates for silicon bring-up tasks and resources. Any additional delays due to unsuccessful review of the tapeout materials at the foundry have a direct impact on that schedule. Appropriate project management of (requested and granted) waivers is critical.

## 18.6 Summary

The most computationally demanding sign-off methodology flows pertain to the four physical design verification requirements: DRC, LVS, ERC, and LPC. Project management attention to planning for this computational workload is required. Often, specific high core count, large memory compute servers are allocated to these tasks, or external resources may need to be pursued. The resources applied to a mock tapeout exercise are considerable, but the insights gained allow the SoC project manager to best prepare for the sign-off PDV steps. The project manager also needs to ensure the accuracy of all PDV flow results documentation for foundry review.

The PDV flow results criteria applied to the release of a block-level version update for global integration will certainly mitigate the risks of errors arising during the sign-off project phase. The key is to execute a PDV methodology plan that provides a predictable path to tapeout

## Further Research

### SoC Project Management and Physical Design Verification (PDV) Job Execution

This chapter highlights the significant resources required to complete full-chip physical design verification jobs. As a project approaches final tapeout preparations, the IT workload and software license demand to complete these sign-off flows is considerable. EDA vendors recognize the unique "burst" demand and high throughput requirements of the sign-off jobs and offer a few options to SoC project managers:

- Lease temporary software license keys, especially for multiprocessor distributed execution
- Provide a remix of software license quantities among the existing pool of products to support a shift in project emphasis toward PDV jobs as tapeout approaches
- Offer a private cloud at the EDA vendor facility, with SoC project access to both software licenses and (large memory) compute servers

An SoC project manager needs to address whether these options may be necessary to augment the existing software licenses and IT resources.

Describe the information that the SoC project manager needs to gather to prepare the sign-off execution plan, including:

- Full-chip PDV job resource estimates (e.g., runtime, tool licenses, compute servers)
- Internal IT resource availability (and reservation policies)
- Costs of EDA vendor "burst" throughput options

If an EDA vendor private cloud is indeed appropriate, there are additional considerations.

Describe the project status milestone criteria and design quality level to justify release of a full-chip database to the EDA vendor cloud. Describe the criteria used to estimate the duration for accessing the private cloud. And, significantly, describe the data security policies that need to be established with the EDA vendor, including:

- Encrypted full-chip database transmission to the cloud
- Job results data returned to the SoC project (or, potentially, a cloud-based interactive debugging environment across a secure link)
- Access controls for both SoC engineers accessing the cloud data and EDA vendor IT and tool support personnel
- EDA vendor private cloud firewall features
- SoC data scrubbing from cloud storage at project completion

CHAPTER 19

# Design for Testability Analysis

## 19.1 Stuck-at Fault Models and Automated Test Pattern Generation (ATPG)

This chapter briefly reviews some of the techniques developed to assist SoC designers prepare an efficient, yet thorough, set of production test patterns. The cost of tester time is a significant contribution to overall product cost and is thus a major consideration when the SoC design and methodology teams are defining the design for testability (DFT) architecture to be incorporated. The tester cost is weighed against the subsequent impact of discovering failures at final product testing or, worse, end customer failures. This DFT assessment helps establish production test coverage targets (and related *test escape* estimates). The DFT architecture is established to enhance the sensitization and detection of internal circuit faults—also known as *controllability and observability*—through the application of test patterns and test measures at the pins of the SoC.

The faults to detect are due to wafer fabrication (and, subsequently, package assembly) defects; the incorrect part behavior is immediately detectable. This differs from the part qualification and lifetime reliability assessment performed using accelerated stress techniques on a sample population over many hours, although production test patterns are also used as part of the qualification flow. The decision to apply *burn-in stress* and subsequent re-application of production test patterns to all parts isolates infant failures and is also a contributor to the overall test cost.

### 19.1.1 Stuck-at Logic Pin Faults

The test coverage targets (or, equivalently, the test escape rates) established during SoC project planning directly correspond to the internal circuit fault models that are assumed to be present after fabrication and the impact these faults have on the observable circuit behavior. As process nodes have evolved, the complexity of fault models commonly applied has also increased.

For digital logic circuits, the traditional fault model has assumed a *stuck-at logic value* failure at the input and output pins of each logic gate, as illustrated in Figure 19.1.

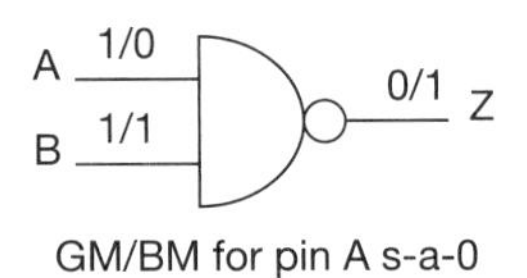


**Figure 19.1** Illustration of the stuck-at pin fault model for logic gates. The "good machine" and "bad machine" representation is highlighted.

The figure uses unique notation to represent the difference between the "good machine" and "bad machine" response—that is, GM/BM—in the presence of the assumed fault. The goal of test pattern development is to apply a value from sourcing logic that differs from the assumed pin fault and apply values to all other "side pins" in a selected logic path that propagates the GM/BM response to an observable point. Representative algorithms for *backward inference* and *forward propagation* of combinational logic test pattern values are described in References 19.1 and 19.2.

The stuck-at model assumption has several characteristics of note:

- A single fault is assumed to be present, and it affects a single logic gate pin. (Other fault models, discussed later in this chapter, add *bridging* between logic signals.)
- There are *equivalent faults* in the overall fault population to the *injected fault*.

For example, in Figure 19.1, input pin A s-a-0 and output pin Z s-a-1 of the NAND2 gate are equivalent. After a test pattern is defined for an assumed fault, the pattern is applied to the logic network by a *fault simulation* tool to evaluate other faults that are also detected, and they are removed from the remaining fault selection list. The fault simulator is similar to an event-driven functional simulation tool, with the expanded evaluation of both the GM/BM network response for the propagation of each pin fault under consideration.

Note that the GM/BM logic fault simulation of a test pattern assumes a zero-delay netlist model, synchronized only to the clocks applied during testing. Any timing races present in the design that may have satisfied timing verification are issues during fault simulation. A race is due to a captured state value that is dependent on the delays of two clock paths, two data paths, or a clock and data path dependency, as depicted in Figure 19.2; this is an issue for automated test pattern generation (ATPG). Multi-cycle paths defined in the constraint file input to static timing analysis also present unique ATPG considerations.

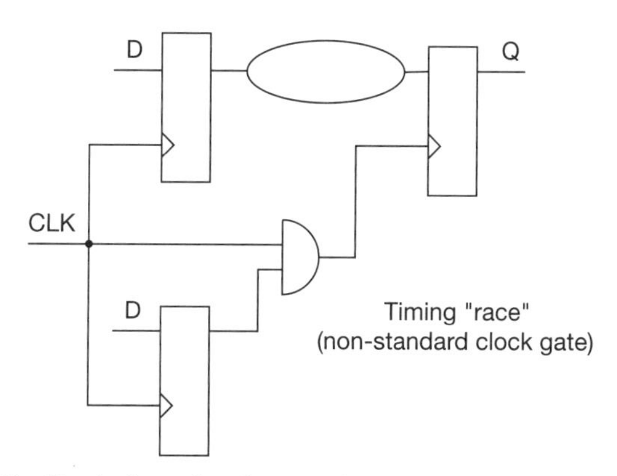


**Figure 19.2** Illustration of path races in a logic network, which are not supported in ATPG.

Section 19.2 describes a set of ATPG design checks that identify potential test pattern generation and application issues.

Internal faults of devices, contacts, and interconnects within the logic gate are abstracted to a stuck-at pin response. This abstraction assumption is extremely important. It reflects an engineering trade-off between the costs of test pattern complexity and test escapes. The early DFT research on CMOS fabrication in the 1970s evaluated a fault model where a single device of a complementary pair was defective (i.e., open or shorted), thus expanding the logic pin stuck-at fault model to the device level, as shown in Figure 19.3. As highlighted in the figure, this necessitated a sequential test pattern to stimulate an observable logic transition through the specific device with the assumed fault (e.g., the B1 input pFET in the figure). If the B1 pFET device were indeed open, no output transition current would result upon application of the second pattern, and the CMOS gate logic output value would remain unchanged (long enough for the GM/BM value difference to be detected, assuming that leakage currents were small—also known as a *gross delay* fault response). At the time, this sequential pattern methodology, where the GM value requires an output transition, proved to be quite expensive due to the growth in the size of the pattern set. The device-level CMOS logic fault model was not widely adopted at that time. The wafer test escape risk was suitably small with the cell pin-level stuck-at model.

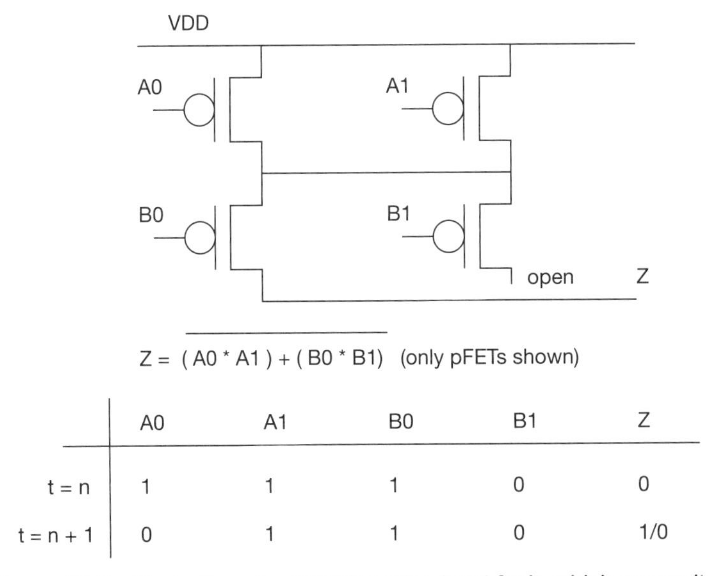


| | A0 | A1 | B0 | B1 | Z |
|---|---|---|---|---|---|
| t = n | 1 | 1 | 1 | 0 | 0 |
| t = n + 1 | 0 | 1 | 1 | 0 | 1/0 |

**Figure 19.3** Illustration of a CMOS device-level fault, which necessitates a sequential test pattern sequence.

However, as CMOS process scaling evolved, new classes of fabrication defects emerged; correspondingly, new fault models were introduced (as discussed in Sections 19.5 and 19.6). For example, the increasing contribution of interconnect delay for a signal resulted in greater sensitivity to via defects and/or wire RC parametric values that exceeded specifications; the need to simulate an output transition to observe a delay defect became a test requirement. The contributions of intra-gate device opens/shorts and local RC parametric excursions are once again being introduced to the overall fault population.[3] As originally pursued decades earlier, additional patterns are generated for a transition through a specific device and circuit path to produce the desired cell output transition.

Methods to mitigate the cost of increased pattern counts are also being pursued. Sections 19.3 and 19.4 describe the addition of circuitry to the SoC to generate pattern sequences and capture/compact the functional responses internally and at high clock speeds. These built-in self-test (BIST) macros add to the die area/cost, but their efficiency relative to the application of patterns externally using automated test equipment (ATE) is often a net benefit.

### 19.1.2 Parametric Faults and Tests

Prior to the application of production test patterns to each wafer die site, a set of parametric tests needs to be applied. These tests not only confirm specific circuit behavior but also serve as a screen to quickly discard dies with major fabrication defects. These tests include the following:

- **Static IDD current between VDD and GND**—This test detects shorts between P/G distribution grids and/or anomalous device leakage currents.
- **I/O pad driver/receiver voltage-level response**—The static DC interface voltage levels for all I/O pad circuits (i.e., VIH and VIL, VOH and VOL (at the output load specification)) need to be confirmed before patterns are applied.
- **On-die ring oscillator frequency**—The SoC die may include *performance-sensing ring oscillator* (*PSRO*) IP macros, which are used to (dynamically) detect where the die is operating over its performance range and adaptively adjust circuit response (e.g., modify the impedance of off-chip drivers). The response of each PSRO serves as a good test screen for a die operating outside the process window.
- **(Internally generated) clock response**—Production testing relies on valid clock behavior within the die. Often, a high-speed clock is generated internally, from a phase-locked loop (PLL) that is a synchronized multiple of an input reference clock. Whether a direct input from the ATE or a PLL-generated clock is used to source the test clock distribution,

specific DFT design steps are taken to propagate internal clocks to an observable point. Separate tests are initially evaluated to confirm that the internal clocks are being propagated successfully.

- **Isolation and testing of analog mixed-signal IP**—An SoC that integrates mixed-signal IP needs a DFT architecture that allows stimuli to be directly applied and IP responses to be observed. Data converters, sensors, and SerDes interfaces have specific test specifications that need to be part of the overall SoC test plan. In some cases, separate test equipment may be required from the ATE used for digital testing, requiring resource coordination with the test services provider.

There are several SoC design options to consider for parametric testing:

- **Probing of die test pads not connected to package pins may be pursued.** The wafer-level test fixture may be designed to probe pads for (parametric) tests that are not part of the package definition. The SoC DFT engineering team may choose to screen certain tests at the wafer level only, assuming no need to repeat the screen due to the low risk of a parametric failure being introduced during the packaging step.
- ***Blind build* parts may be packaged after (a subset of) parametric testing.** Blind build parts are unique: Prior to production testing, dies that pass selected wafer-level parametric tests may be forwarded directly to packaging. The SoC project manager might want to expedite the availability of a small quantity of packaged parts for structural/mechanical evaluation. Or the project manager might be willing to put minimally tested parts on (socketed) boards to get a head start on product evaluation, assuming that some percentage will be functional. The SoC test engineering team determines what parametric tests are to be applied for blind build parts and coordinates with the foundry and OSAT test service groups.
- **Parametric test results may be used for performance binning.** The PSRO data captured during parametric testing would be indicative of the overall die performance across the spectrum of fabrication variation (if all other parametric tests pass). If binning of functional yield by performance is a product revenue option, the PSRO measurements would be used to direct specific fully functional dies to the corresponding packaged part number. Multiple PSRO macros across various die locations may be appropriate to ensure that on-chip variations (OCVs) are observed.

One final comment on parametric testing and test probe fixtures: When screening for static IDD current fails, the probes should ideally contact all power pads. Flip-chip bump packaging enables a potentially very large number of VDD and GND bumps internal to the die perimeter to accommodate larger supply currents for high-performance designs and to readily enable the design of multiple SoC voltage domains. The design of the test probe needs to address the trade-offs between parametric test coverage and probe cost.

### 19.1.3 DFT and Automated Test Pattern Generation (ATPG)

Based on the traditional stuck-at logic pin fault model, algorithms to generate test patterns for combinational logic cones were developed, as illustrated in Figure 19.4. The logic cone inputs are chosen to provide the GM value for the (single) stuck-at pin fault as well as to propagate the GM/BM value to a cone endpoint. If a pattern can be found, the fault is *detectable* and added to the overall stuck-at coverage percentage; otherwise, the fault is reported as *undetectable*.

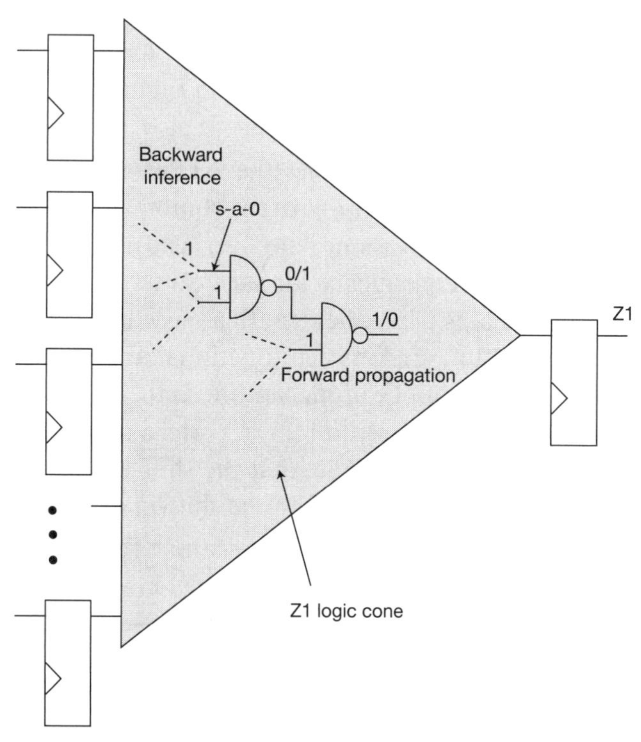


**Figure 19.4** Combinational test patterns are generated as inputs to logic cones to detect faults. The sequential registers are assumed to provide general pattern stimulus and response capture.

A fundamental assumption of the ATPG algorithm is that all combinational logic cone inputs can be independently assigned, and at least one of the potentially many paths from an assumed logic fault to any endpoint serves to capture the GM/BM observation response.

As highlighted earlier, a single test pattern detects many stuck-at faults in the logic network, in addition to the originally injected fault. Indeed, ATPG tools from EDA vendors may incorporate a procedure to apply and fault simulate a short sequence of simple (or *pseudo-random*) patterns to reduce the remaining fault list to the subset requiring more detailed logic value inference and propagation analysis.

To address the ATPG assumption that cone inputs can be assigned independent test pattern values and that any cone output is an observation point, sequential flops and registers need to be capable of loading arbitrary pattern values, then clocked to capture the GM/BM results, and then cone endpoint values observed without overwriting. The universally adopted DFT architecture to provide these features is to connect all sequential elements in a *serial shift-register scan chain* in addition to their functional connections. Two of the possible scan implementations are illustrated in Figure 19.5; a pattern application and response capture clocking sequence is also depicted.

The scan chain approach satisfies the ATPG algorithm assumptions—that is, application of an arbitrary pattern, combinational network response capture, and response observation. The overhead is the tester time for the serial shift-in and shift-out sequence for each pattern. To minimize this overhead, multiple scan chains of smaller length are used on an SoC design, where SoC pads may be multiplexed between functional and scan-shift modes. (For on-die logic BIST pattern application, capture, and compaction, many smaller scan chains are used.) To truly optimize test pattern time, balancing the length of scan chains is often pursued. Note that the shift-out of a captured pattern response is performed concurrently with the shift-in of the next test pattern.

There is a circuit PPA overhead to the scan-shift implementation, as well. The input mux in a mux_scan design or the dual-clocked latch in the clocked scan (or level-sensitive scan design, LSSD) example add to circuit area and impose a performance penalty on the functional timing path. The additional scan_out-to-scan_in connections to implement the shift-register chain add to physical routing congestion. The additional load on the flop output for the next scan_in fanout adds to signal delay and switching power.

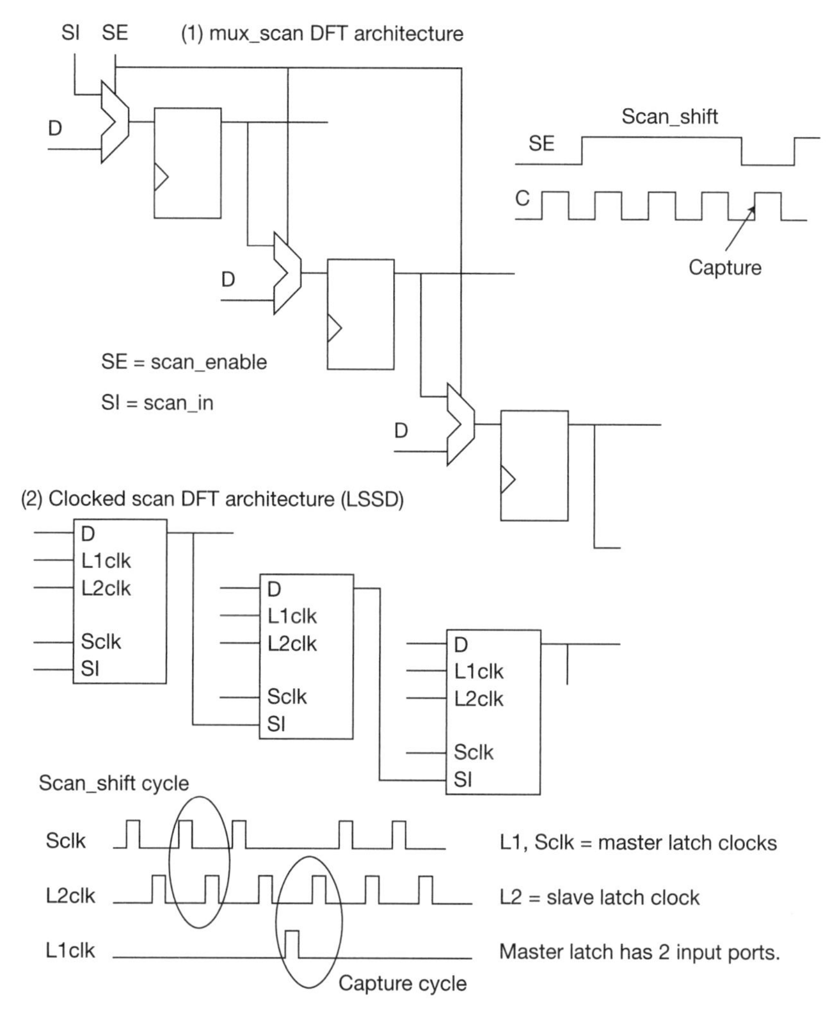


**Figure 19.5** Illustration of potential serial shift-register scan chain topologies, with the related clocking sequences.

During initial cell placement, a low priority is given to scan connections in the netlist between flops; the focus of timing-driven placement is on critical timing paths. Placement tools include a subsequent optimization to modify the scan chain order of the flops from the original netlist; this optimization involves selecting a sequence of connections to minimize the (post-placed) scan route lengths. In addition, physical design tools may insert a buffer (or delay) cell in the chain to minimize the interconnect loading on the timing-critical flop output. Nevertheless, the switching activity of scan connections in functional operation mode adds to the SoC power dissipation. The post-PD output netlist with the final scan order is provided to the ATPG algorithm.

Static timing analysis is exercised in functional mode, with scan paths disabled (but with the capacitive loading of the scan connection present in the parasitic model). Static timing analysis also needs to include a mode specific to production testing that focuses on scan clocking and scan chain paths. Figure 19.6 illustrates timing paths of interest in production test mode for the mux_scan DFT architecture. Hold time verification ensures that clock arrival skews are compared to the scan chain path delay, which has a short logic depth. Setup time verification confirms that the scan_enable arrival time is sufficient for the transitions between scan and capture operations. (A set of delay-fault patterns may also be added to the production test set. Section 19.5 describes the unique timing paths of interest in this test mode.)

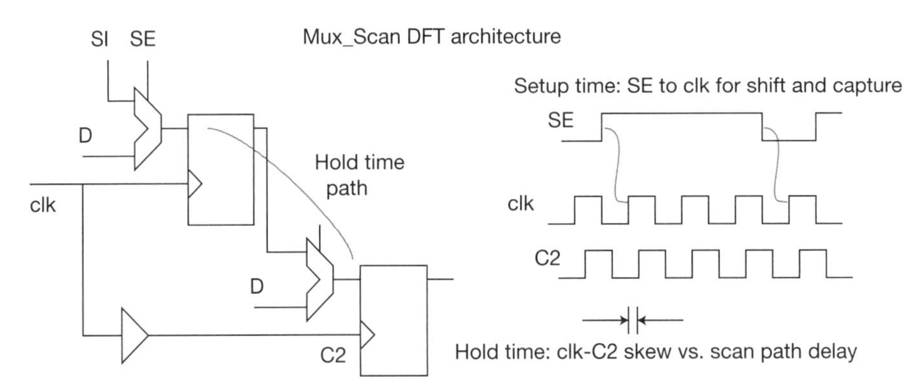


**Figure 19.6** Illustration of hold time and setup time path analysis in production test mode.

Despite this PPA overhead, a scan-based DFT architecture has been pervasively adopted to enable high fault coverage with an efficient pattern set. The alternative would be to eschew building scan chains and apply a reset to the SoC, followed by a "functional pattern set" sequence, assuming that a faulty machine response would be observed at functional SoC outputs. Although functional patterns may be included as part of the production test set to exercise specific (non-scan) behavior, a serial-shift DFT architecture offers the capability for improved *fault diagnosis*, when a systematic wafer test error is observed, as discussed in Section 19.7.

A similar DFT serial-shift architecture is commonly adopted for the I/O pad circuits on the SoC, where a scan flop connected to the pad driver and receiver circuits can be loaded with a pattern value, multiplexed into the functional I/O path, and clocked to capture a response. This offers several potential design advantages:

- Driver ($V_{out}$) and receiver ($V_{in}$) parametric testing are greatly simplified.
- Reduced probe count test fixtures could be used, where the scan flop in the I/O cell provides the internal stimulus/response measurements, rather than requiring probes in contact with all signal I/O pads.
- Product testing of board-level connectivity is simplified, using the I/O scan chains of different parts as the pattern stimulus and response endpoints between modules.

Because multiple parts (potentially from different vendors) would need to adopt a serial-shift DFT architecture to enable this board-level connectivity testing, the IEEE has standardized an I/O DFT architecture, known informally as *boundary scan* (IEEE JTAG 1149). Subsequently, the IEEE has also offered a similar standard for IP providers of complex cores to embed a serial-shift chain around the pins of the IP core, and in the case of black-box hard IP, include the production test patterns for the core. This standard is informally known as the IP *wrap test* DFT architecture, or IEEE P1500.

For the SoC methodology team, several key decisions are required in the project definition phase:

- Overall DFT architecture (with global clocking and test_enable multiplexing, cell library support, and scan access from SoC pins)
- I/O boundary scan and embedded IP wrap test architecture (and access from SoC pins)
- Stuck-at fault coverage targets
- SoC pin specifications need to be developed for scan chains, multiplexing controls, and clocking; the ATPG flow inputs a command file to the EDA vendor tool with these definitions
- SoC memory blocks that incorporate MBIST macros
- SoC functional diagnosability features to be added as part of the product specification

The fault coverage targets are applied at full chip, but may also be evaluated as part of a block-level version release method. A block with poor fault coverage may be required to investigate the cause and make micro-architectural changes. Undetected faults may require the addition of an observability endpoint flop within combinational logic cones in the block. ATPG analysis at the block level assumes that all PIs are controllable and all POs are observable, maximizing the fault coverage; untested faults within the block are not testable with the full chip integration netlist.

The last methodology decision above refers to the opportunity to use the scan-based DFT architecture functionally. Although added to enhance test quality and improve production test efficiency, the DFT architecture could also be incorporated into field testing and failure diagnosis. When an internal operating error is detected (e.g., an illegal instruction op code), the SoC error recovery procedure may assume control of clocks and test enables to scan out the current machine state to a log file for further analysis. Also, MBIST sequences could be initiated and results scanned out as part of the product feature set. This would be useful to determine if a "hard" field failure has arisen (as compared to a memory "soft error") upon the detection of a memory storage error by parity checking or ECC logic.

The test engineering team needs to collaborate with the SoC methodology and project management teams and apply these project architecture decisions to prepare block-level and SoC-level tool configuration inputs and to provide production test cost estimates. The CAD team needs to provide compute resource estimates and ensure that the ATPG flow results are appropriately captured in the SoC project status management database. The CAD team should also ensure that the physical design and equivalency flows provide the correct (scan chain reordered) netlist. The library team should provide the fault models for each cell, expanding each complex cell into its Boolean gate-level equivalent for the ATPG algorithm.

## 19.2 DFT Design Rule Checking

To enable scan-based ATPG support, the internal design needs to be toggled between the shift and capture operations directly from SoC pins. This DFT architecture implies several network topology restrictions:

- No combinational logic loops are allowed.
- All sequential (non-memory array) elements are incorporated into a scan chain. The question of whether register files are exercised as

embedded memory arrays or have all bits connected in a scan chain (with corresponding scan_enable and scan clocking) requires several DFT trade-off assessments.

- Test and capture clocks need to be directly controllable. Gated or logically combined clocks need to be configured to support scanning. Figure 19.7 illustrates a clock gating cell modified to support a scan testing methodology.

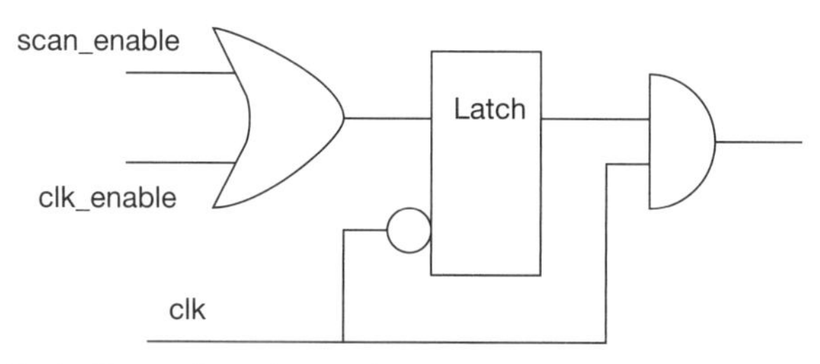


**Figure 19.7** Illustration of a clock gating cell with scan test mode propagation.

- All state elements should retain defined values when the scan and capture clock(s) are off.
- Any timing race condition results in an undefined capture value.
- Flops with direct set/reset input pins require glitch-free signaling (in a zero-delay ATPG simulation).

A precursor to running ATPG flows would be to exercise a *DFT design rule check* on the netlist and pin configuration input files.[4] To some degree, the DFT DRC algorithms are similar to the methods used in RTL synthesis and physical design that select the correct library flops, insert scan connections, and complete the DFT clocking and scan_enable signal distribution. EDA vendors commonly un-bundle the DFT DRC checking tool from their ATPG product and offer a reduced license cost option to iterate on DFT design preparation. The DFT rule checks are therefore consistent with the ATPG toolset, which may be from a different EDA vendor than the selected EDA supplier of synthesis and physical design tools. In addition, the DFT rule checker allows for the examination of embedded hard IP that will be integrated outside the synthesis and PD flows.

Note that logic inversions in the scan chain are typically allowed during DFT checking. Indeed, logic inversions could be an integral part of the scan chain. SoC designs have occasionally adopted a functional initialization sequence that applies a static value at scan inputs and exercises a number of scan-shift cycles. Logic inverter cells in the chain connectivity are used to initialize different subsets of the scan flops to 0 and 1 values.

Topological errors identified by the DFT rules checker need to be thoroughly reviewed by the SoC methodology and test engineering teams. As mentioned in the previous section, this flow would be applied initially at the block level, ideally as soon as a netlist with test logic insertion is available. Design issues reported by the DFT rule checking flow may require revisions at the HDL level; addressing DFT topology errors as early as possible can minimize the impact to the overall project schedule. Some DFT check issues with the netlist may still allow ATPG to proceed, as would be the case for an "undefined" network value propagating to a scan flop for a specific test pattern. During production wafer testing, as the captured values are shifted out and compared to the predicted values from ATPG simulation, the undefined capture flop positions in the chain are masked from comparison to prevent the die site from being designated as a failure. Other, non-ATPG pattern sets and clocking sequences could be added to focus on exercising and observing these logic networks.

## 19.3 Memory Built-in Self-Test (MBIST)

The fault models used for embedded memory IP are typically much more complex than the stuck-at pin assumption used for logic gates. Faults in the address decoder or word line driver circuitry introduce unique considerations:

- A bitcell may never be accessed.
- A bitcell may be accessed by multiple applied addresses.
- For an applied address, multiple bitcells in different words may be accessed concurrently.

In addition, faults present in the bit lines, sense amplifiers, or column decoders may have far-reaching impacts. The manifestation of memory array faults spans a range of anomalous behavior, from a single bitcell error to entire row or column values that do not match after write/read operations. Memory pattern sequences need to be applied to focus on detecting these conditions.

The nature of array physical design and the accumulated history of fabrication defects has led to the introduction of neighboring storage location pattern sensitivity fault models. The correct bit operation at each array site is evaluated using various write/read transitions at neighboring physical locations to confirm that an upset error has not occurred. Examples of complex array write/read sequences include "marching" 0s and 1s through incrementing and decrementing addresses, "walking 0s and 1s," checkerboard, GALPAT, and sliding diagonal. Given the very large number of test patterns that need to be applied and read access responses measured for the embedded array, an efficient method of pattern generation and application is required.

The initial approach that was commonly adopted for embedded memory testing was to design a test mode that multiplexed SoC I/O pads to directly connect to the IP pins, as illustrated in Figure 19.8.

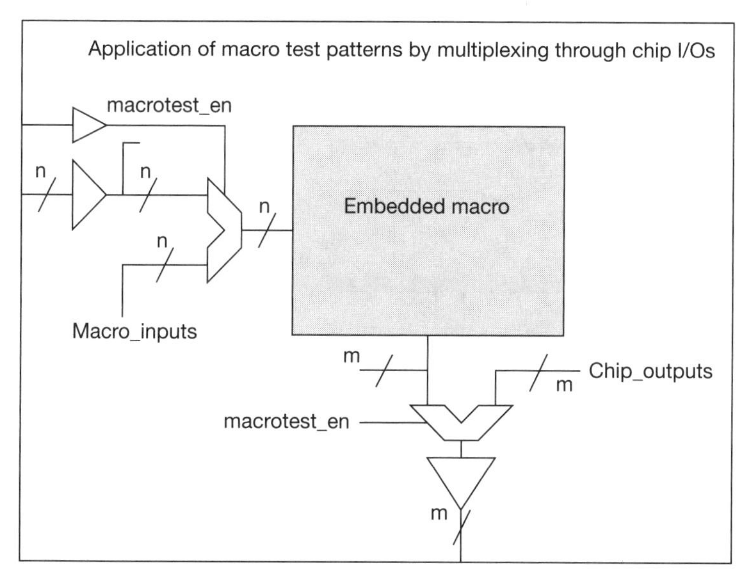


**Figure 19.8** Illustration of embedded memory array test architecture using "macro isolation" to control and observe array pins directly from I/O pads.

The patterns would be applied and responses would be measured by using ATE equipment, with the multiplexing logic isolating the embedded macro in test mode. However, this embedded macro test method does not scale well. The integration of multiple large arrays adds to the difficulty in developing a direct access design (without adversely impacting routing resources and performance) and requires significant additional tester time cost. Increasingly, SoC designs are allocating the die area to integrate a built-in self-test macro.

The BIST macro controller would connect to (one or more) embedded arrays to provide programmable pattern sequences. Figure 19.9 depicts a BIST architecture example, connected to the embedded array. Note that an embedded ROM array, cache memory array, and/or non-volatile memory array could also be exercised by the BIST engine.

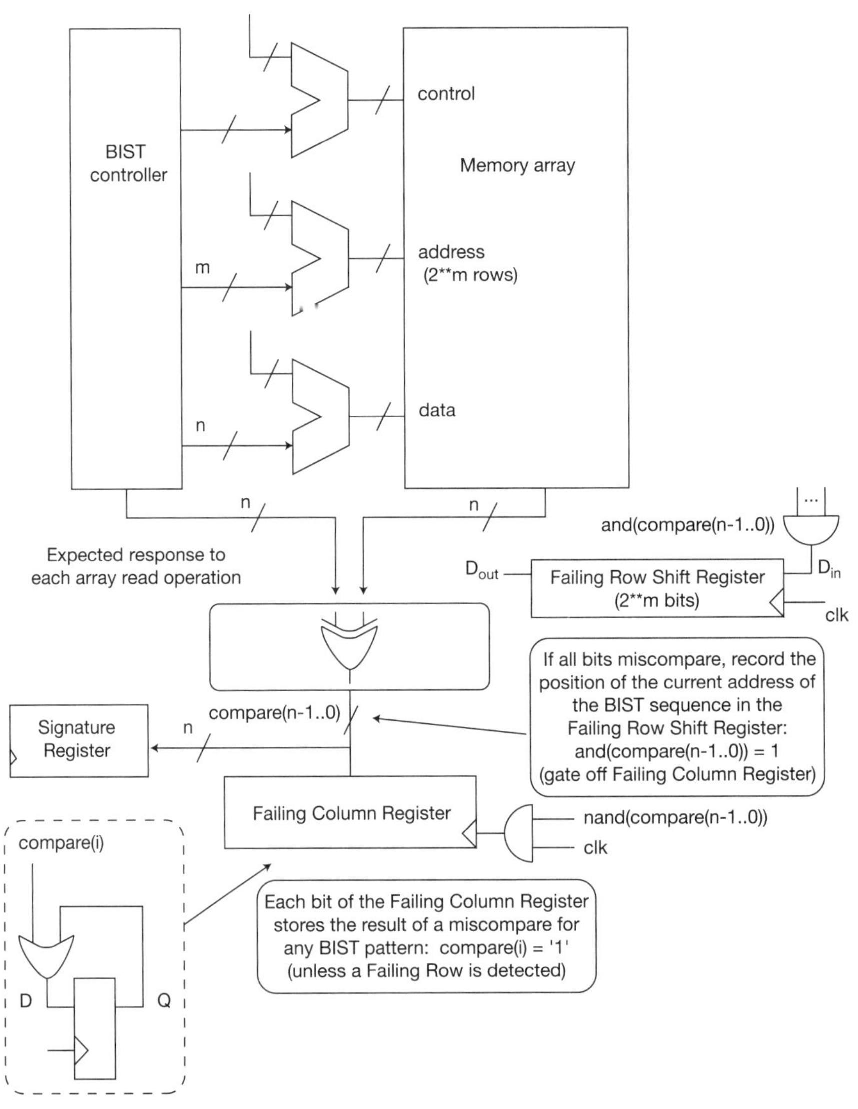


**Figure 19.9** Illustration of an MBIST architecture for embedded memory array testing, including "built-in self-diagnosis" (BISD) logic.[5]

The functional inputs to the array would be multiplexed with the BIST controller, with patterns coming from the controller in test mode. The BIST macro also includes features to evaluate the array responses on a read operation:

- The expected values written in an earlier pattern are calculated by the controller and compared to the array outputs (using a wide vector of XOR2 gates).
- A set of registers may be added to capture errors, supporting the identification of individual bitcell, full column, and full row miscomparisons.
- A *compactor* register is also typically included to record a *signature* of the result of the compare vector for the entire BIST pattern sequence. (A compactor design based on a linear-feedback shift register topology is discussed shortly.)

The BIST controller can apply patterns at a faster clock rate than the patterns originating from the ATE. The controller functionality could be expanded by programming additional (microcoded) routines to explore new fault mechanisms. The compaction algorithm reduces the full comparison output sequence to a single vector with a low risk of an *alias* (where the fault-free and faulty signatures would be the same). The additional registers in Figure 19.9 are provided to help identify the cause of a failing signature, a method denoted as built-in self-diagnosis (BISD). After a BIST pattern set is applied, a scan shift out of the signature register provides the overall pass/fail result. In case additional failure diagnosis is required, the other register values offer additional insights. Figure 19.10 depicts the values in the BISD registers for the cases of a failing array bitcell, row, or column.

EDA vendors have addressed the increased demand for on-chip test support by introducing software products to generate (synthesizable) MBIST RTL models. These MBIST tools are provided with a detailed specification file input, describing the array configuration, timing interface, and desired test pattern set programmability. The specification file would also support multiple array descriptions if a single BIST controller would be shared across arrays. The SoC physical design team would need to assess the area, routing congestion, and performance trade-off of distributed versus centralized BIST control.

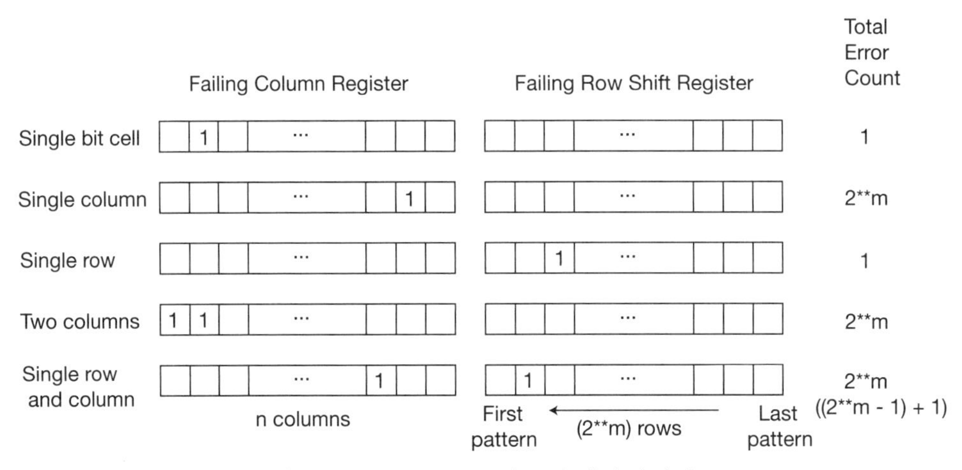


**Figure 19.10** Illustration of the BISD register values for different memory array response errors. An error counter is included for additional diagnostic information (not shown in Figure 19.9). The register values are reset between each set of patterns.[5]

These EDA products may also include support for *built-in self-repair* (*BISR*). The results of the BISD registers would be scanned out to identify a specific array row/column that, if replaced with a redundant row/column, would result in an error-free array. Memory array IP may include additional row and/or column resources. Additional array inputs insert the redundant resources into the array circuits to bypass the failing row/column. Figure 19.11 illustrates how the redundant resource is incorporated into the array.

To support array repair, the SoC would include an additional register (to be loaded with the specific redundancy pattern at power-on) or include an embedded (laser or electrical) fuse array that could be permanently programmed. An e-fuse array has an advantage in that programming can (typically) be done directly at the tester, whereas laser programming of fuses requires processing of the wafer lot after production testing with separate equipment. For a fuse array, during power-on initialization, the SoC reset controller transfers the fuse data through a shift register chain to the BISR redundancy register of the array. The BIST controller is invoked after shifting the repair values to confirm the correct array behavior, both during production die retesting after fuse programming and as part of the SoC functional reset sequence. Note that a similar repair value shift and BIST confirmation procedure is required if the array will be transitioning from a sleep mode power state during SoC operation.

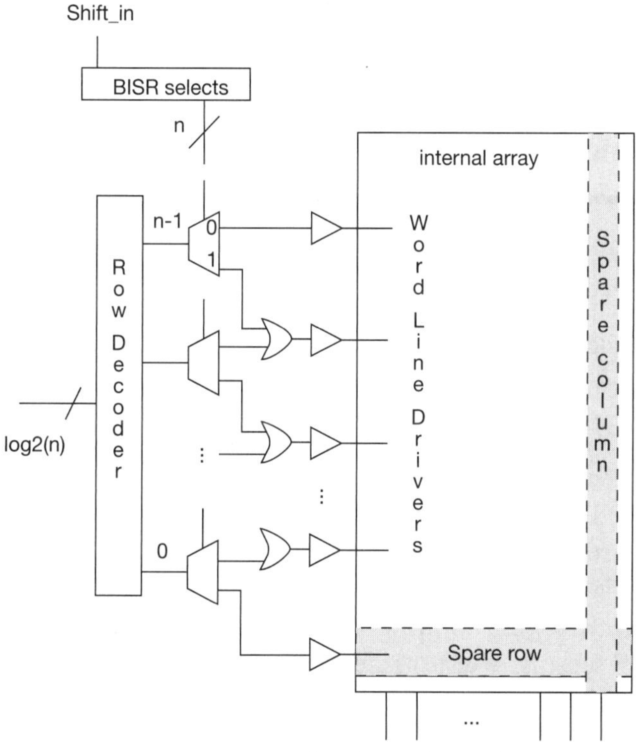


**Figure 19.11** Illustration of the insertion of a redundant resource into the operational memory array. A single spare row is shown (with simple decoding). The demultiplexer select inputs are shifted into redundancy register during power-on reset.

An SoC project manager needs to assess several trade-offs when deciding which arrays, if any, warrant the addition of BIST macros. There are area and performance impacts to adding BIST to evaluate against the test time cost reduction and increased fault coverage. The inserted BIST RTL logic also becomes an integral part of the functional validation plan, with testbenches required to ensure the correct array connectivity. The (synthesized) gate-level BIST implementation becomes an integral part of timing analysis, as well, to confirm that the array interfaces are correct in both functional and BIST pattern application timing modes. A project manager must also evaluate the trade-offs associated with incorporating redundancy with specific arrays and the means to program (and re-verify) the array decoders to reroute around failing bit locations. The integration of on-die fuses results in additional fabrication layers and wafer costs, as well.

The CAD team needs to ensure that the flows are in place to provide BIST logic generation and integration of the resulting RTL model. The BIST tool specification files are added to the design management database. Foundry PDK support for the functional modeling and physical layout design of fuse arrays is also needed if this redundancy programming approach is adopted.

The test engineering team has a broad set of responsibilities:

- Review the test pattern sensitivities recommended by the embedded array IP provider.
- Develop the BIST program code to generate the pattern sets and the scan-shift strategies during testing to initialize the BIST controller registers and microcode program storage.
- Ensure that the logical-to-physical mapping configuration is correct so that pattern sensitivity tests are indeed exercising physically neighboring cells.
- Review the BISD features of the controller—specifically how faulty rows and columns are identified for each array.
- Collaborate with the foundry engineering support team to understand what programmable fuse structures are available.
- Collaborate with the global SoC physical floorplanning team to integrate the fuse array.
- Develop a BISR strategy with the wafer-level test provider for embedded arrays with redundant rows/columns.

During the qualification of a new fabrication process node, the foundry R&D team commonly uses a large array with MBIST as the bring-up testsite. The array bitcell is likely to use aggressive design rules that are not generally available for logic and analog IP designs to maximize the areal bit density. The small process window associated with the aggressive lithography, combined with the high sensitivity to the fabrication defect density, makes a large array an ideal bring-up design. Specifically, the foundry may use multiple arrays on the testsite, incorporating multiple bit-cell types, such as high density (smallest area) or high drive strength (fastest read access performance). The foundry seeks to achieve suitable and consistent yields on the array testsite as a process qualification milestone. The measures are typically quoted as *natural yield* (with no repair) and *yield with repair*. A distinct benefit of the large array(s) as the foundry bring-up vehicle is the early identification of new MBIST test pattern sensitivities and array repair requirements to be shared with SoC designs licensing the foundry IP.

## 19.4 Logic Built-in Self-Test (LBIST)

The motivations described in the previous section for incorporating MBIST circuitry to exercise on-chip arrays extend to testing logic networks, as well. The additional connectivity between state elements to form serial scan-shift registers vastly improves the controllability and observability of combinational logic networks. However, the relative ratio between the available external ATE scan pattern channels and the volume of internal SoC logic faults is increasingly problematic. The test pattern set size combined with the tester time to exercise the pattern shift-in, response capture, and shift-out contributes to a high wafer test cost per die. Methods for applying and verifying logic test sequences on-chip, while minimizing the required connection bandwidth to the ATE, are being widely adopted in advanced process SoC designs. The following sections briefly describe several logic built-in self-test (LBIST) approaches in use.

### 19.4.1 LBIST Pattern Generation

A general LBIST architecture is depicted in Figure 19.12. A logic network is identified for internal testing (the *circuit under test* [*CUT*]), and additional logic is synthesized and added to the netlist to provide a sequence of generated patterns and to capture the CUT output response. The communication with the tester is minimized, providing only the instructions to the LBIST controller for the initialization, execute, and compare sequences. The signals from the LBIST circuitry to the CUT are multiplexed into the logic design during testing mode and are bypassed during CUT functional operation.

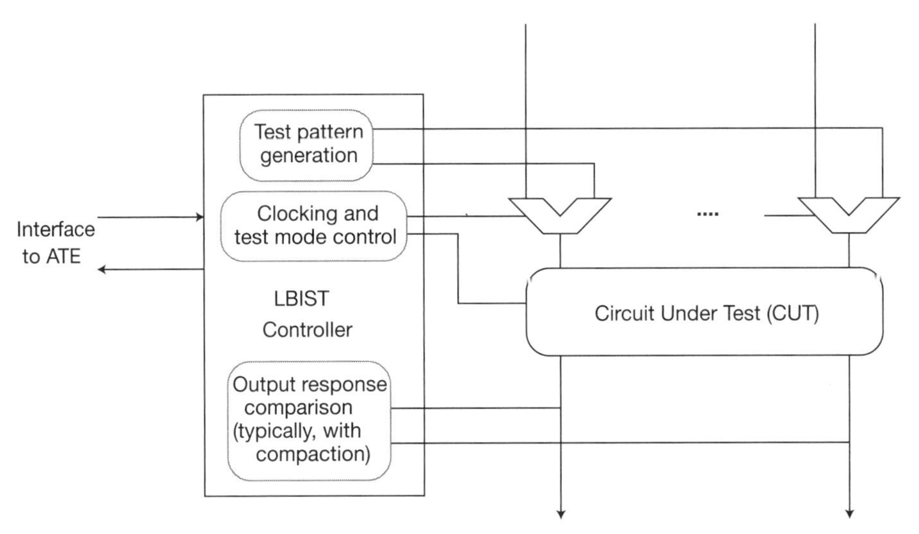


**Figure 19.12** LBIST architecture.

For a CUT with only combinational logic and n inputs, the test pattern generator could be as simple as an n-bit counter, which would provide exhaustive testing over 2**n internal clock cycles. Alternatively, a *pseudo-random pattern generator* (*PRPG*) could be used to provide a subset of the 2**n stimulus patterns yet still provide sufficient fault coverage. Extensive research has been undertaken on the design of a *linear feedback shift register* (*LFSR*) to provide a pseudo-random pattern sequence. In this case, *pseudo-random* implies that the sequence will repeat after m cycles, where m ≤ ((2**n) − 1) for an n-stage LFSR register. Figure 19.13 illustrates two general LFSR architectures, consisting of flops and XOR gates.

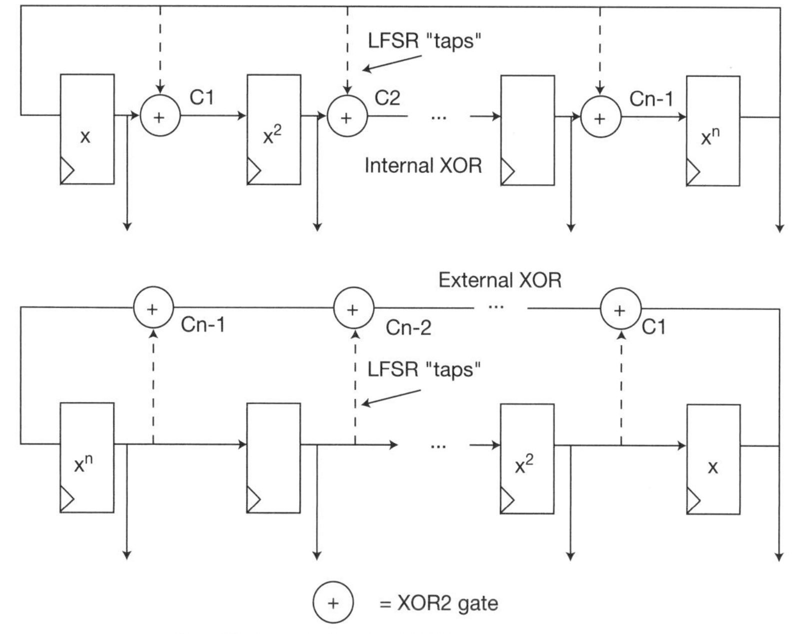


**Figure 19.13** Examples of linear feedback shift register (LFSR) architectures used for BIST pattern generation.

In the figure, each coefficient $C_i$ represents a node where an XOR gate and *feedback tap* may be included or omitted. The specific gates and feedback connections define the personalization of the LFSR. A convenient notation for a personalized LFSR is a polynomial, where the terms represent the included connections ($C_i$ = '1'), and the omitted gates are absent ($C_i$ = '0'). The polynomial representation enables an efficient mathematical solution to the LFSR output vector x(t+1) from x(t) and readily supports calculation of the unique length of the LFSR pattern sequence. Two small three-stage LFSR examples are shown in Figure 19.14: one of them has a personalization that supports a maximal pattern set ((2**3) − 1), and one has a shorter sequence.

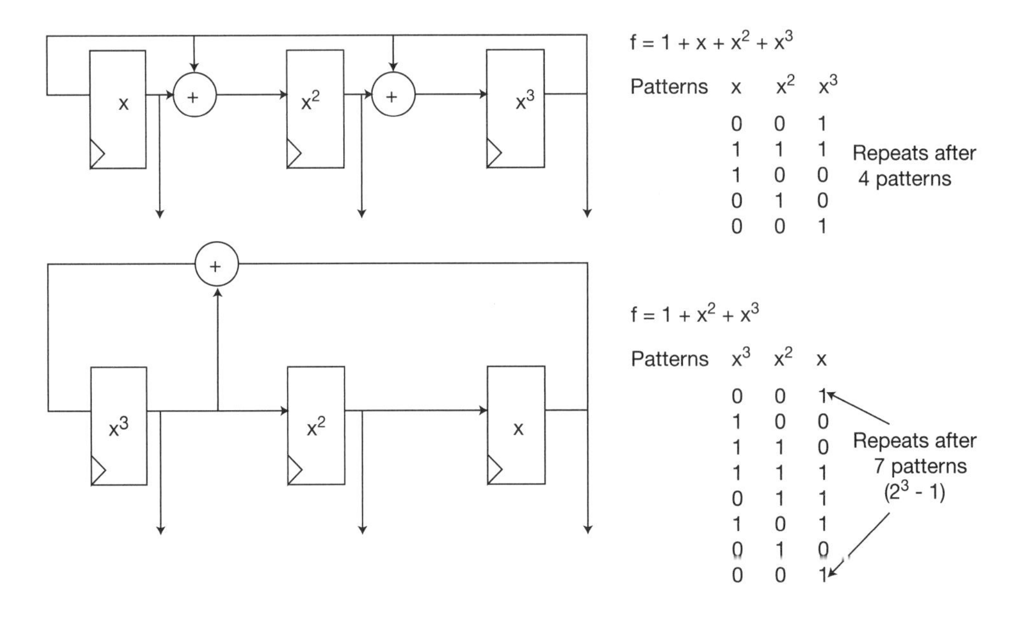


**Figure 19.14** Illustration of the (pseudo-random, repeating) output sequence generated by an LFSR. Two three-stage examples are depicted, with both external and internal XOR taps.

The sequence of patterns from the LFSR generated each cycle would be applied to a fault simulator to determine the cumulative coverage. For a CUT with sequential registers within the logic, the evaluation of the LFSR coverage involves more extensive fault simulation over the sequential depth of the network. Note that the LFSR outputs in Figure 19.13 can also be *biased* with the addition of logic gates between the LFSR and the CUT to generate specific patterns to improve fault coverage. This would be considered if the CUT contained *random pattern-resistant* faults (e.g., a high fan-in NAND gate output s-a-1 fault). Note also that the LFSR (as PRPG) in Figure 19.13 is *autonomous*: After initialization (e.g., using a short scan sequence into the LFSR flops), the pattern generation sequence is defined solely by the LFSR personalization. Note that the "all zeros" pattern is not supported in the autonomous LFSR; thus, the maximal sequence is (2**n – 1). It is also possible to inject external values through additional XOR gates into the LFSR during the pattern generation sequence; this architecture is discussed shortly, as part of a test pattern *compression* method.

### 19.4.2 LBIST Compaction

The response of the CUT to a sequence of applied LBIST patterns is multiplexed to logic that compacts the CUT outputs over the pattern duration into a single vector. The result is denoted as the *signature* of the overall pattern sequence applied to the CUT. The most common implementation of the compactor is a *multiple-input signature register* (*MISR*). The general architecture for an MISR is depicted in Figure 19.15. As with the PRPG, it consists of flops and XOR gates, with specific feedback taps. The MISR flops are clocked synchronously with the PRPG. Compaction could also consist of reducing the output vector width ("space compaction") in addition to the sequential circuit MISR ("time compaction").

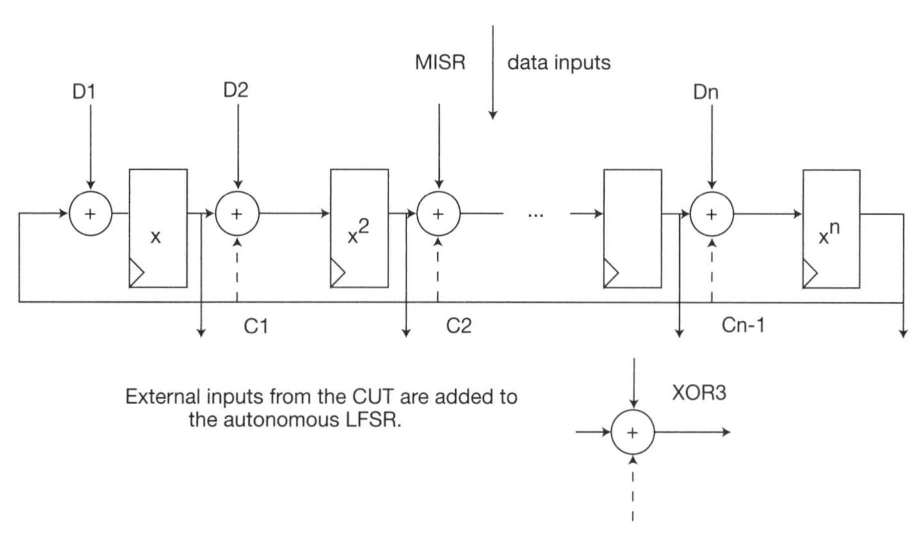


**Figure 19.15** Illustration of a multiple-input signature register (MISR), used to compact the responses from a BIST network into a signature.

The compactor is lossy in that faulty responses from the CUT during the pattern sequence could be overwritten, and an error could be masked. A faulty CUT could provide a result that is an identical alias of the correct signature and could thus be undetected. For example, the XOR of two combinational signals both in error in one cycle or the propagation of a captured error through MISR feedback could potentially cancel an error in subsequent cycles. The goal of the MISR design is to minimize the alias probability.

### 19.4.3 X-Values in Fault Simulation

Of particular note is the issue of handling unknown or uninitialized values in an LBIST implementation, typically denoted as X or U in simulation. In a general scan-based design, if the fault simulator encounters an X-value or a U-value propagating to a capture flop, the tester can be programmed to disregard that bit position in the scan chain during the scan_out shift operation. The scan_out value is not checked for good machine/bad machine differences for the X or U scan shift position, with a minor loss in fault coverage. However, an X-value or a U-value presented to the sequential MISR compactor quickly results in a very lossy signature. The CUT must be designed to be thoroughly initialized at the start of the pattern sequence and exclude logic that might introduce an X-value. Alternatively, additional test logic could be added in front of the MISR to provide a deterministic value and block a potential X or U response from being captured.

A general LBIST implementation minimizes the die test cost overhead by reducing the tester communication and by applying the internally generated SoC clocks to the PRPG and MISR at frequencies higher than the ATE channel could provide. However, it is very difficult to define suitable logic partitions that are sufficiently independent and have a manageable number of CUT input and output signals to make a general LBIST implementation viable. Features of the general LBIST architecture have been applied to alternative approaches that are more widely adopted; these BIST approaches also leverage the extensive experience with scan-based design and ATPG algorithms.

### 19.4.4 Scan-Based Pseudo-Random Pattern Generation

An early application of LBIST features to a scan-based SoC design multiplexed a PRPG to parallel scan chain inputs, as depicted in Figure 19.16.

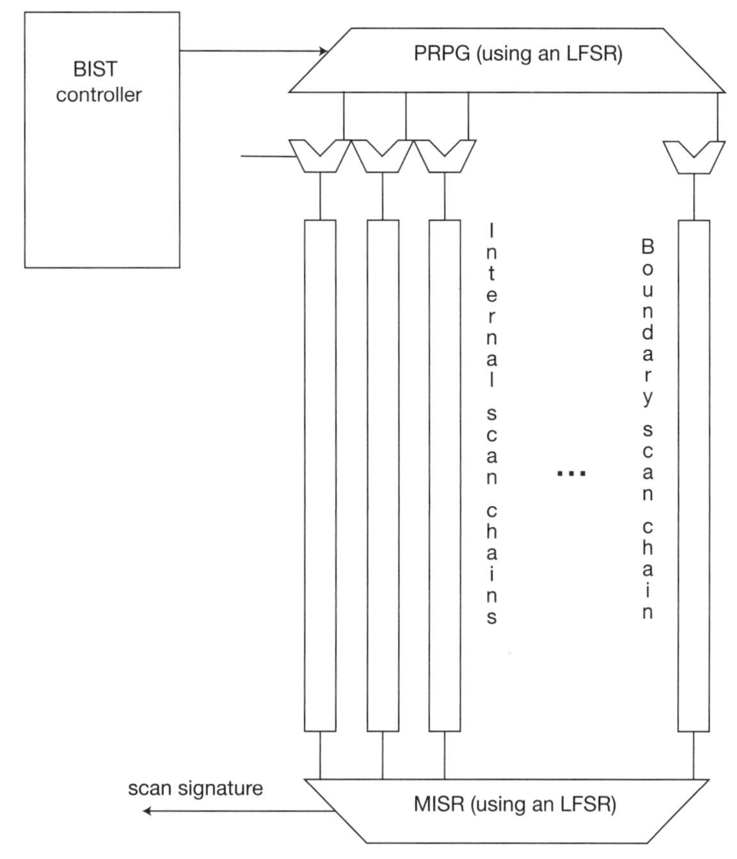


**Figure 19.16** Illustration of a PRPG to generate patterns in a scan-based DFT architecture.

The values from the PRPG are shifted into the chains, followed by a capture cycle, and then the scan chain output data is shifted into the MISR to provide the response signature (while shifting in a new PRPG pattern). There are several advantages to this method:

- Existing scan-shift, capture, scan-shift DFT architectures are utilized.
- The required width of the PRPG and MISR is associated with the number of scan chains, a more manageable resource than the number of signal inputs and outputs for a logic CUT partition.
- Pseudo-random patterns can be very effective in detecting a high percentage of stuck-at faults.
- Communication with the ATE is minimal.

The faults that are detected by the PRPG patterns can be removed from the fault list. The ATPG tool subsequently only needs to analyze the remaining faults to generate scan-based test patterns. (The ATPG step after PRPG fault simulation is often referred to as the "top off" method.) The production die test sequence would initially exercise the PRPG/MISR logic, followed by application of the ATPG patterns. The specific test mode would select whether LBIST or tester channels are multiplexed to the scan chains. Although the APTG patterns require extensive tester channel bandwidth, the number of patterns is reduced significantly due to the coverage provided by the scan-based LBIST.

Building on the scan-based LBIST implementation above, a new approach has emerged for current SoC designs. The number of internal scan chains is increased dramatically—much higher than the number of available ATE data channels—and ATPG-generated data are delivered to the die in an encoded pattern. This method is generally referred to as *test compression.*

### 19.4.5 Test Compression

The growth of the available logic (and sequential) gates in advanced processes has exacerbated the constraints of the number of ATE tester channels and pattern data rates. The longer scan chains and increased pattern counts required by current SoC designs extends the overall test time and cost. To address this issue, current DFT architectures are adopting a test compression strategy, as shown in Figure 19.17.

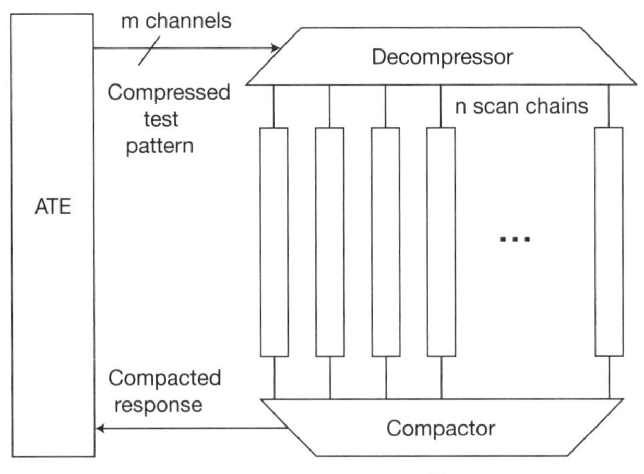


**Figure 19.17** Test compression architecture.[6]

The SoC design is divided into a very large number of (relatively short) scan chains. The combinational ATPG algorithms for fault sensitization and propagation to a scannable register are applied as usual. This is in contrast to the general and scan-based LBIST approaches, in which a somewhat iterative approach is needed to design, fault simulate, and reoptimize the PRPG stimulus pattern set.

The key innovation in test compression is the integration of a pattern *decompression logic* network. Typically, ATPG assigns values to a small subset of any scan chain length to detect an injected fault. In a traditional ATPG approach, the remaining scan chain positions are assigned 1 and 0 values to complete the full pattern to shift-in from the ATE. The full pattern would be fault simulated to identify additional detected faults and would reduce the list for selecting the next assumed fault. In the test compression approach, the focus is solely on the assigned stimulus values in the test pattern, with the remaining scan register positions left as "don't_care" values. The goal is for a minimal amount of external input data to be required for each pattern to "seed" the decompressor so that the test vector shifted into the internal scan chains contains the assigned values prior to the capture cycle.

Referring again to Figure 19.17, the number of ATE channels providing inputs to the decompressor, *m*, is much less than the number of internal scan chains, *n*. The decompressor consists of an LFSR and additional XOR logic; the width of the LFSR is less than *n*, and the XOR gates provide the fanout to the scan chain inputs. With the pattern seed from the ATE, the *n* signal values from the decompressor each cycle are shifted down the chains.

As illustrated in Figure 19.14, the LFSR personalization results in a *deterministic* sequence of output values. (Test compression is also referred to as an *embedded deterministic test* [EDT].) For the case of an autonomous LFSR, the output equations for each cycle are either a shifted bit value or the XOR of bit values from the preceding cycle, based on the personalization. The equations for any cycle can be fully expanded back to the LFSR initial state. In other words, the initial values are inputs to a set of equations that fully define the autonomous LFSR outputs at each cycle.

For a decompressor, the same approach to deriving the output equations is applicable. The difference is that there are additional XOR terms in the equations from the external input seed values provided each cycle. The inputs to the decompressor equations are the initial LFSR values plus the cycle-specific external pattern data. The assigned values in the ATPG test pattern are the "required" outputs from the decompressor at the end of the shift through the $n$ scan chains of length $L$, as depicted in Figure 19.18.

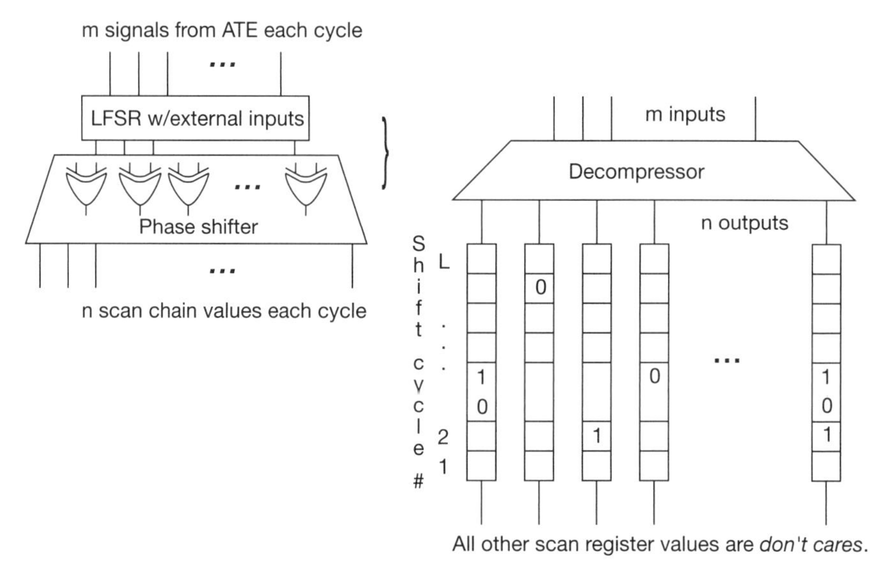


**Figure 19.18** Illustration of the decompressor values over shift cycles 1 through $L$ to solve for the target test pattern. The decompressor is composed of an LFSR with external pattern inputs and additional XOR bias gates.

The ATPG test compression algorithm is tasked with finding the external pattern inputs over cycles 1 through $L$ that "solve" the decompressor equations for cycles 1 through $L$ to provide the assigned pattern outputs among the $n$ scan chains. The number of equations to solve is vastly reduced, as all equations for don't_care values in the pattern can be discarded.

To summarize, given an LFSR personalization, the configuration of XOR bias gates, the XOR gates where external data is applied each cycle, and the number of shift cycles ($L$), a set of decompressor equations is defined. A solver is invoked to determine a compressed external sequence of length $L$, such that the given ATPG assigned pattern values will be present in the corresponding scan chain positions at the end of the $L$ shift cycles. Reference 19.6 provides more details on the solver algorithm, with examples.

Note that the set of XOR bias gates at the LFSR outputs of the decompressor is also known as a *phase shifter*. A phase shifter could be added in Figure 19.16 to bias the PRPG data, as well. And although the LFSR architecture has been depicted, other finite state machine implementations are also used.

Whereas a traditional DFT scan architecture effectively provides unconstrained shift register patterns, the test compression architecture depends on the decompressor equations. If a solution is not found for the sequence of external ATE inputs to realize the assigned values in an ATPG pattern, the assumed (combinational) fault is returned to the fault list to investigate whether different assigned pattern values can be identified for which a decompressor solution is found. To improve the success rate, an alternative method allows a sequence of multiple external seeds to be applied to the decompressor, only shifting the LFSR outputs into the scan chain by one position after the last seed is applied. For this method, the decompressor equations are more complex and the test time is extended, but the likelihood of finding a solution is improved.

Compared to a traditional DFT scan architecture, the test vector compression factor achievable with this method would be the ratio of the number of scan chains, $n$, divided by the width of the external channel, $m$. That is, a sequence of $m$ external values provided over $L$ cycles results in $( n * L )$ scan register bits:

$$C = ( n * L ) / ( m * L ) = ( n/m ) \qquad \text{(Eqn. 19.1)}$$

(This equation neglects the time to initialize the decompressor between patterns and assumes a single external seed per shift cycle.) Whereas traditional ATPG would fill in unassigned values to each scan vector present in the ATE buffer memory, for test compression, the ATE stores only the seed patterns. The values in unassigned bit positions in the scan chains for the pattern are defined by the decompressor equations not used by the solver.

Additional techniques have been researched further improve the compression ratio:

- **Add more assigned scan values for other faults.** As the percentage of assigned scan register values is typically small, the ATPG test compression tool may iteratively select additional faults to evaluate, as long as no conflicting seed pattern assignments from the ATE are generated. After an additional fault is injected, the solver would be invoked with the expanded set of decompressor equations. If a solution for the $i$th injected fault is not found, the $(i - 1)$th external pattern sequence is selected.

- **Attempt to solve for the assigned values using a "static" external seed.** The general test compression method assumes a new set of $m$ external values from the tester each cycle. If a single static external seed pattern could be identified (i.e., inputs $a_1$ through $a_m$ constant for all $L$ cycles), the tester pattern buffer memory requirements would be reduced.
- **Attempt to "broadcast" the same (random or assigned) values down additional sub-chain segments.** The pattern compression ratio increases with the number of internal scan chains. One method to increase the compression is illustrated in Figure 19.19. For some patterns, it may be feasible to divide the $n$ chains into additional segments (of length $< L$) and broadcast the same decompressor outputs to each.

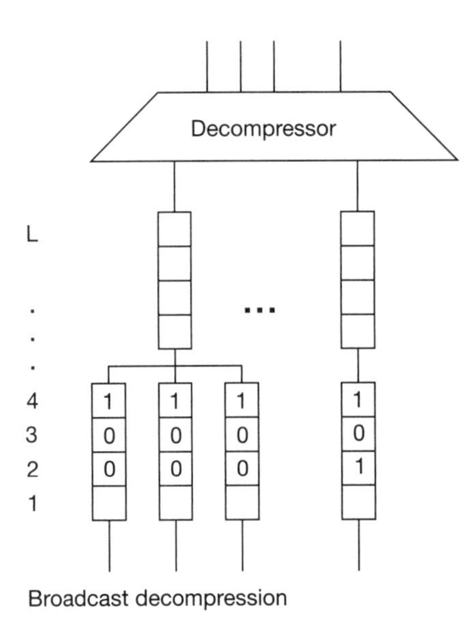


**Figure 19.19** Illustration of the broadcast method within the test compression architecture.

The scan-based LBIST method initially provided pseudo-random patterns to all chains to detect a high percentage of faults before ATPG top-off patterns are applied. A similar approach could be implemented with the decompressor, broadcasting identical patterns to the $n$ chains. Technically, only the ATPG patterns that necessitate shifting specific data into more than one of the $n$ chains would need a non-broadcast solution.

- **Add test points to improve fault coverage.** The technique of (judiciously) adding an observation point fanout in a combinational network to a scannable flop and/or adding a controlling fan-in point to a network during test mode is applicable to any DFT architecture. It may also be beneficial to improving the test compression ratio.

The SoC project manager and DFT architecture team are faced with a myriad of implementation alternatives when deciding on the suitability of LBIST methods to integrate. The principal advantage of scan-based and test-compression LBIST is the extension to existing ATPG technology for fault detection and error diagnosis. Further, new fault mechanisms identified for ATPG are directly supported by the scan-based and test-compression LBIST algorithms (discussed later in this chapter). However, these LBIST methods impose a die area overhead for the additional circuitry and introduce greater demand on routing beyond the already challenging problem of (optimal) scan chain ordering and scan connectivity between flops. Scan chains may also be used functionally, such as to "unload" internal machine state for system diagnostics upon detection of a runtime error. The scan chain configuration for LBIST may differ from a functional purpose, requiring additional multiplexing logic and test versus functional mode controls. This is especially true for the test compression LBIST approach, where the optimal number of internal scan chains is large.

The key decision factor in integrating LBIST is test cost, as mentioned at the beginning of this section. The SoC test engineering team needs to evaluate several design and production test metrics to estimate the tester time and cost:

- Number of SoC scannable flops
- Test pattern set size, which depends upon:
  - Fault mechanisms
  - Fault coverage targets
  - The ratio of the number of SoC logic gates to flops
- Design suitability to LBIST compaction, including:

    - The ability to eliminate uninitialized U and unknown X values during testing
    - MISR alias probability (due to space and time compaction)
    - Random pattern-resistant logic
- ATE equipment features available at the foundry and outsourced assembly and test (OSAT) services provider, including:
    - The number of scan pattern channels
    - Pattern memory buffer size
    - Pattern application rates and memory buffer load time

These calculations need to include appropriate wafer-level, package-level, and post-burn-in test environments. The cost of mixed-signal IP testing and array MBIST (with repair) testing also needs to be included.

The CAD team needs to provide corresponding flow support for LBIST, including:

- Scan chain definition to synthesis flows for DFT insertion (including chain length balancing and multiplexing logic for functional versus test modes)
- LBIST logic insertion
- Scan chain reordering in physical implementation flows (e.g., disregarding scan chain connectivity during initial placement, defining an optimal scan order at the end of placement or during routing)
- DFT rules checking (e.g., scan chain identification, scan control signal definitions, initialization, X-generation checks)
- ATPG (with test compression)
- Fault simulation

Any issues that arise during application of these flows are likely to necessitate design iterations. To accelerate getting the design into the DFT-related flows, a low-effort synthesis netlist from initial RTL versions is often used (as described in Section 7.11). As with most SoC flows, the methodology management tracking tool records the DFT flow status for the netlist version release.

### 19.4.6 Hierarchical Testing

The discussion in this chapter has not yet mentioned the methodology option for hierarchical testing. As an SoC will likely integrate many different types of IP, the methodology team may need to approach the development of the production test pattern set as a combination of test methods specific to each IP block. Hard IP designs from external vendors may not include internal fault models for ATPG flows but may instead provide a localized logic wrapper and test pattern set. The SoC may integrate multiple instances of a core IP design, which offers the opportunity to implement a replicated DFT implementation for each core with a simpler global test controller. Indeed, the SoC DFT architecture team has numerous implementation alternatives available for cost optimization, given the SoC design hierarchy.

## 19.5 Delay Faults

With the increasing contribution of interconnect R*C parasitics to overall timing path delay, a new transient fault mechanism was introduced in addition to the traditional node s-a-0 and s-a-1 static fault model. As illustrated in Figure 19.20, a "delay defect" implies that a timing path may demonstrate correct logical behavior if captured at a low frequency (i.e., it is not a stuck-at fault), but the path would not behave correctly at functional frequencies. Rather than the s-a-0 or s-a-1 pin fault model, a "slow-to-rise" and/or "slow-to-fall" signal transition fault is defined.

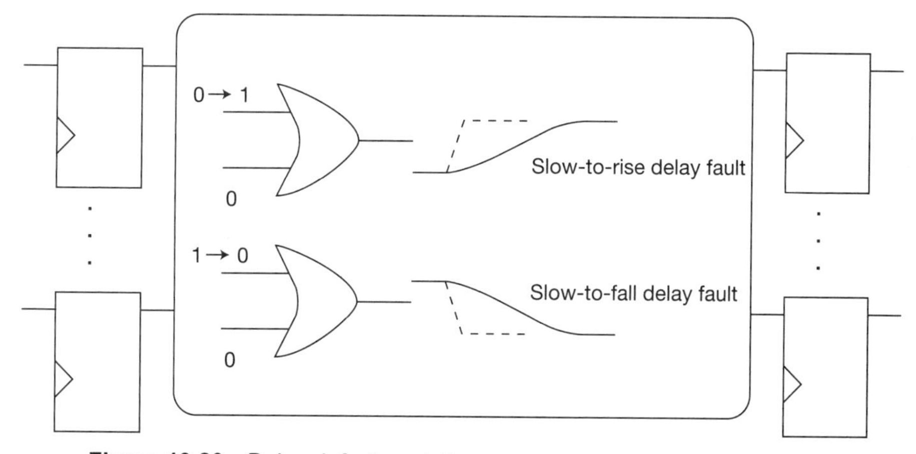


**Figure 19.20** Delay defect modeling.

The combinational ATPG fault sensitization and propagation algorithms used for stuck-at models are extended to provide a sensitized transition that propagates to a scannable capture flop. Initiating the transition requires a sequence of two applied patterns. As depicted in Figure 19.21, the preceding ATPG scan shift pattern establishes the starting signal logic values, and the next scan shift initiates the transition. The functional capture clock timing interval after the last pattern shift establishes the transient fault delay threshold. The last pattern shift value in the capture scan flop must differ from the new output transition value in order for the fault to be detected.

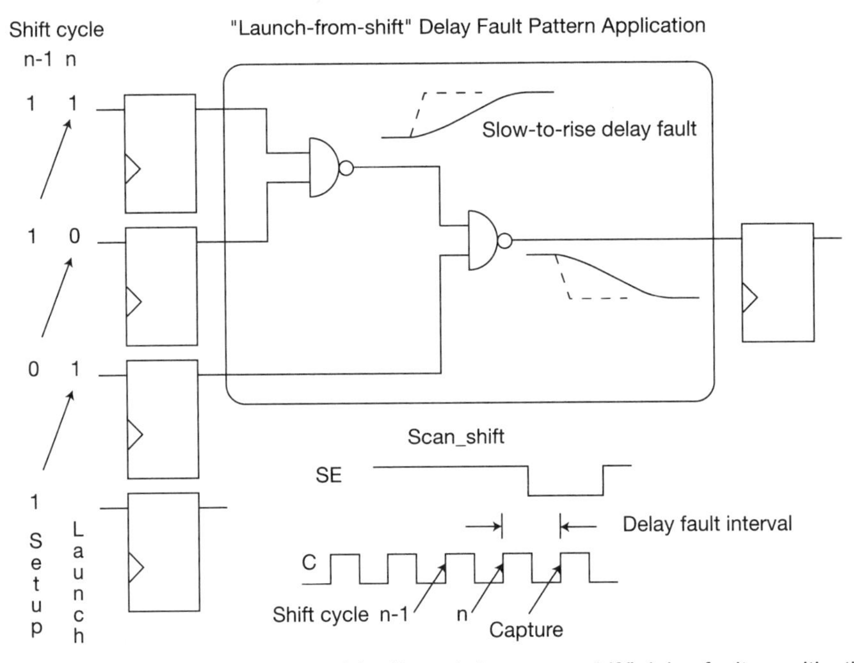


**Figure 19.21** Illustration of the "launch from scan shift" delay fault sensitization method.

This DFT method for delay fault testing is referred to as "launch-from-shift." This approach imposes constraints on the ATPG algorithm, as the setup and launch pattern vectors in the scan chain are now tightly coupled. As a result, the achievable delay fault coverage is limited.

### 19.5.1 Enhanced Scan with "Hold Latches"

To alleviate the constraints on the assigned values for the two patterns in the launch-from-shift method, it would be necessary to fully shift in independent setup and launch vectors before presenting the values to the combinational network. Figure 19.22 illustrates a scannable flop design with an additional output "hold latch." This enhanced scan design allows a pattern to be shifted into the chain without affecting the outputs (hold = '0'). After the shift is completed, the hold signal is raised, and the outputs provide the new "launch" values. (During functional operation, hold = '1'.)

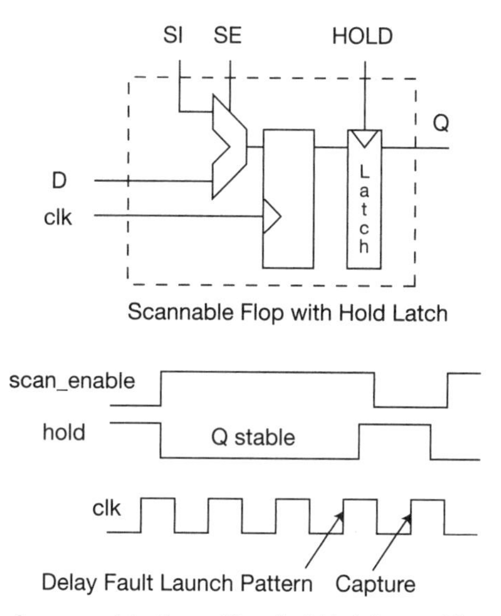


**Figure 19.22** A scannable flop with a hold latch provides an alternative method for launching the delay fault pattern.

This enables the ATPG algorithm to improve delay fault coverage, at the expense of increased flop circuit area and the additional clock-to-Q functional delay due to the embedded hold latch. Note that the hold latch design also reduces the power dissipation during a scan shift sequence, as the combinational network is no longer actively switching each shift cycle.

### 19.5.2 "Gross" and Distributed Delay Fault Models

In advanced lithography nodes, the prevalent fabrication process defects include:

- Highly resistive contacts/vias
- Interconnect resistance above spec
- Composite interlayer dielectrics with an effective dielectric constant outside spec (primarily due to thickness variations)

These are not "hard" defects, but they result in electrical parasitics outside the characterization and extraction corners. Although chemical-mechanical polishing (CMP) has greatly improved the surface topography for interconnects, the polishing step may result in local thickness fluctuation (although metal fill data reduces this variation). The *end point detect* method used during fabrication to define the appropriate etch time for contacts/vias may result in an incomplete opening. The use of a limited set of contact/via sizes in layout improves etch rate uniformity across the wafer, and having redundant contacts/vias improves yield. However, there may be individual openings that are more resistive. The impact of these fabrication outliers may result in either a single gross delay defect at a circuit node or a distributed small delay increase for multiple signals.

For the ATPG algorithm specifically, the delay fault model may be:

- **A node transition fault**—The gross delay defect is present at a single network node such that *any* path involving a transition on this node will exceed the launch-to-capture interval. This delay fault can be detected at any fanout endpoint, regardless of the logic path length to the capture flop.
- **A path delay fault**—Multiple delay defects that are smaller in magnitude may be distributed along a path. The ATPG pattern sensitization focuses on generating logical transitions along specific paths, such as the critical paths reported by the static timing analysis flow.

Given the delay test architecture adopted by the SoC (e.g., launch-from-shift or enhanced-scan) and a set of critical paths of interest, the ATPG tool generates node transition and path fault test patterns and reports the corresponding coverage. These ATPG delay fault patterns are consistent with the LBIST test compression method described in the previous section. The decompressor equations would reflect the two ATPG delay fault patterns to solve. Although pseudo-random patterns are quite effective for stuck-at fault coverage, these patterns are typically quite poor for delay fault coverage due to the specific sequence of setup and launch patterns required.

### 19.5.3 Internal Cell Delay Faults

This discussion of delay defects has focused on the slow-to-rise and slow-to-fall behavior of signals between netlist cells. The ATPG algorithm utilizes the (fault-free) cell logic models to assign values in the setup and launch patterns to provide the signal transitions. A new fault model introduces delay faults within library cells, where specific logic circuit pullup and pulldown paths are to be exercised by additional delay fault patterns. Figure 19.23 illustrates a specific circuit pullup transition of interest and a corresponding cell input pattern sequence. The support for internal cell delay faults is an extension to the existing ATPG tool algorithm for determining sensitization pattern sequences for the nets between cells. In this case, the library model for the cell would include the specific input patterns for ATPG, as shown in the figure.

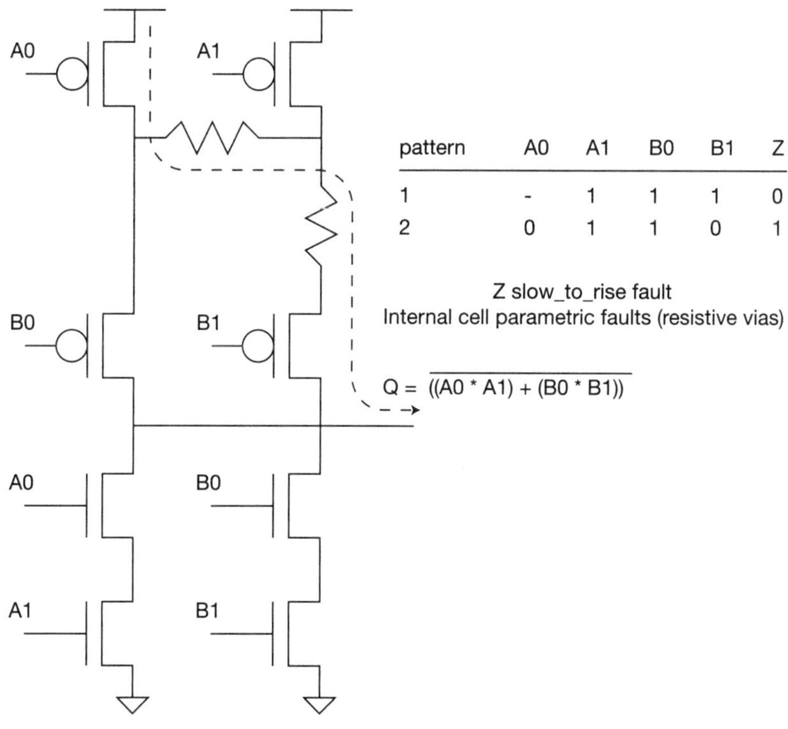


**Figure 19.23** Illustration of an internal cell delay fault.

The SoC project manager and test engineering team need to evaluate the extent to which (gross and distributed) delay defect testing will be required of the ATPG tool and the target transition fault and path coverage measures. Different setup and launch pattern methods impact the die area and functional path timing. The SoC test team also needs to work to establish the optimum order in which various ATPG pattern sets are applied. Typically, delay tests are prioritized (right after clock and scan-chain integrity tests), as these tests also provide significant stuck-at fault coverage; the ATPG static fault top-off list is reduced. The integration of external vendor hard IP with a wrap test architecture requires evaluation of the delay fault coverage specification from the vendor, as well as the contribution to the production test pattern count. In addition, the cell library IP team needs to develop and release internal cell delay defect patterns.

The methodology and CAD teams need to extend the ATPG flow for path delay fault testing that imports the critical timing paths of interest from the static timing analysis flow. For high-performance SoC designs, the number of STA paths with timing slack near zero could be very large. A judicious sampling of that timing distribution may be required to manage ATPG tool runtime and, ultimately, the production test pattern set size and tester time.

## 19.6 Bridging Faults

Another defect mechanism of interest in advanced process nodes is due to a *bridging fault*, in which a logical dependency is introduced between two separate signals, as depicted in Figure 19.24.

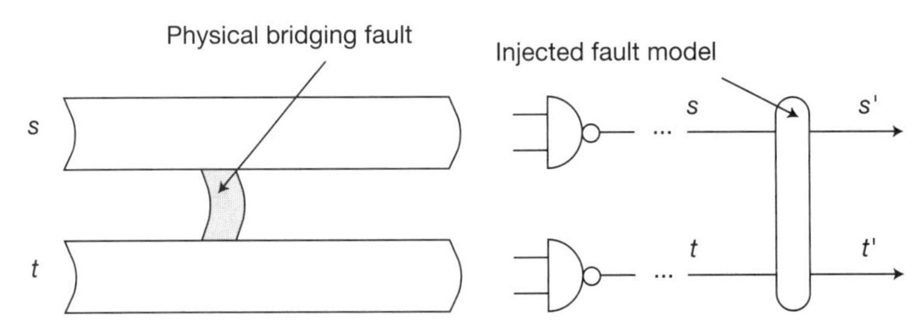


**Figure 19.24** Illustration of a physical bridging fault and the injected fault model combining the two signals.

A physical fabrication defect that retains a metal "sliver" between interconnects may result in interdependent behavior between the two nets. The fault model injected into the network could be quite varied in terms of the logic expressions for s'(s, t) and t'(s, t) in the figure. Common fault expression examples are a "dominant signal" (e.g., s' = s, t' = s ) or a "pseudo-logical gate":

$$s' = \text{AND}(s, t);\; t' = \text{AND}(s, t) \qquad \text{(Eqn. 19.2)}$$

The drive strengths of the signal sources would likely influence the resulting voltage levels on the two nets and the fault model selected. The bridging fault is a static fault. Unlike a delay fault, the ATPG tool only needs to generate a single pattern to distinguish the GM/BM value, given the injected fault expression.

Due to the unique defect source, bridging faults are only considered between physically adjacent interconnect segments on a metal layer. The SoC test engineering team needs to establish the criteria for a bridging fault candidate (e.g., a parallel run length for two segments on metal layer Mn exceeding some threshold). The CAD team needs to develop the flow to detect those segments in the physical layout and reference the layout-versus-schematic (LVS) correspondence flow results to determine the two signals (and the sourcing gates) in the netlist. The ATPG flow is then provided with these signal pairs and the fault expression for bridging fault pattern generation.

Note that there is an additional requirement to the CAD flow that identifies and submits bridging fault signal pairs to ATPG. If a logical path exists in the netlist between signals s and t, the injection of the bridging fault expression may introduce feedback and thus oscillatory or sequential behavior in a combinational network, invalidating a fundamental ATPG assumption (see Figure 19.25).

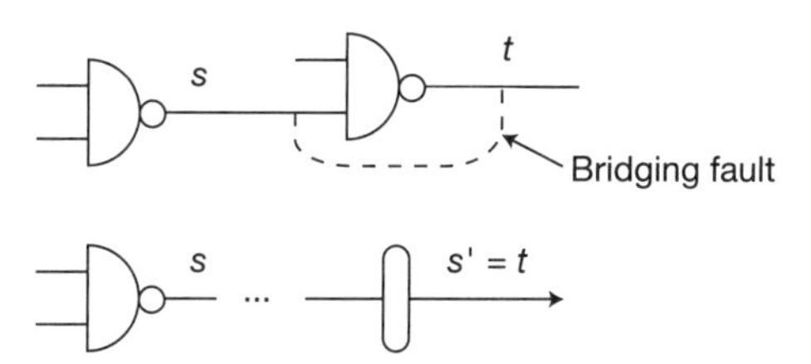


**Figure 19.25** Example of a bridging fault candidate signal pair that would introduce a feedback loop in a combinational network and is thus discarded.

As a result, candidate physical bridging fault signal pairs that are part of a transitive logic fanout path are discarded.

## 19.7 Pattern Diagnostics

The SoC test and product engineering support teams closely monitor the wafer-level and package-level yields throughout SoC prototype bring-up and, subsequently, production release. The foundry offers a target (maximum) defect density that is used in the expected yield calculation. Typically, failing parts originate from random fabrication defects; the patterns that identify the defect(s) are also randomly distributed within the test pattern set. However, if the yield is below expectations and, especially, if failing patterns are repetitive, a systematic defect may be present. The potential sources of non-random defects are many:

- A mask defect (introduced after mask inspection)
- Process equipment drift/failure
- Fabrication material and/or chemical impurities
- A circuit design sensitivity to process corners

When submitting a prototype design tapeout, "split lots" are often requested of the foundry, and specific fabrication module variation settings are introduced in key process steps. The purpose is to uncover circuit sensitivities to process variability. Nevertheless, *circuit-limited yield* is an ongoing product engineering concern.

It is crucial to determine the root cause of systematic yield loss. The test pattern responses are likely the only information initially available. Fault diagnosis is the task of correlating failing test pattern responses to potential physical die locations and/or circuits that could be the source of the failure. Diagnosis could proceed using either of the following:

- Analysis of the (combinational and sequential) networks, given the observed error response to a single pattern
- Analysis of the faults that correlate to the complete set of failing patterns

The choice of which analysis method would be more productive depends on whether a single-fault or multiple-fault assumption is made, as discussed next.

### 19.7.1 Diagnostic Network Analysis

The initial patterns in a test set typically demonstrate that internal clock distribution is functional, followed by a scan-shift test. A sequence of binary values is shifted through scan chains to evaluate for transition faults specifically in the scan flops. As illustrated in Figure 19.26, a sequential bit position *i* in a chain with a gross delay fault is readily diagnosed based on the value miscomparison at the ith shift cycle, using a scan test pattern with '0'/'1' transitions.

A (single) stuck-at fault in the scan chain is readily detected but more difficult to diagnose with only scan-shift patterns. In this case, functional capture of an ATPG pattern to initialize the chain is required.

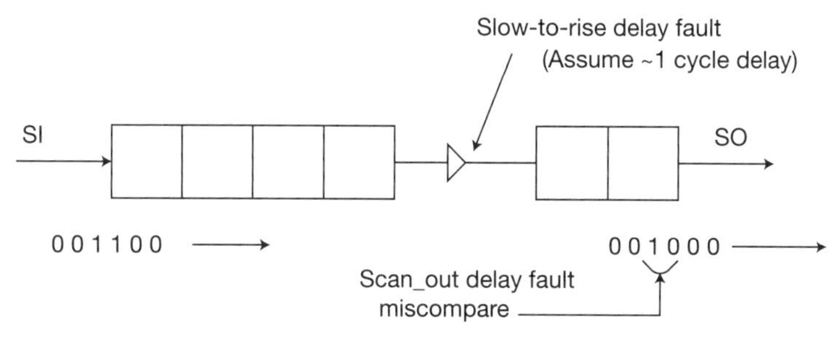


**Figure 19.26** Example of scan chain delay fault diagnosis.

Diagnostic analysis of a combinational network fault after scan-shift testing utilizes an observation of the scan chain capture positions that miscompare to the fault-free network model. Figure 19.27 depicts a case where multiple scan values are in error after functional capture and scan-out, with the overlapping logic cones that fan out to these bit positions. The gates and signals in this fanout intersection are "high likelihood" candidates for the source of the fault. Note that this fault candidate list can be pruned further by the logic cones that provide fault-free responses captured in other scan register positions.

This diagnostic method is useful where there is limited test data available, such as when the tester is programmed to stop on first fail. It is also applicable when a single-fault assumption is made, and the intersection of logic cones with endpoint errors is calculated. If multiple faults were assumed to be present, the union of endpoint error logic cones would be required, rapidly resulting in an intractable set of physical die locations to investigate.

Fault diagnosis requires a logic node-to-physical location correspondence; the physical location of a fault could potentially span a large number of layout shapes in each net. To accelerate the fault diagnosis investigation, it is critical to reduce the candidate fault list as much as possible, even if additional ATPG tool and tester time investment is required. As a result, fault correlation to the complete set of test patterns is more commonly pursued than the network analysis method.

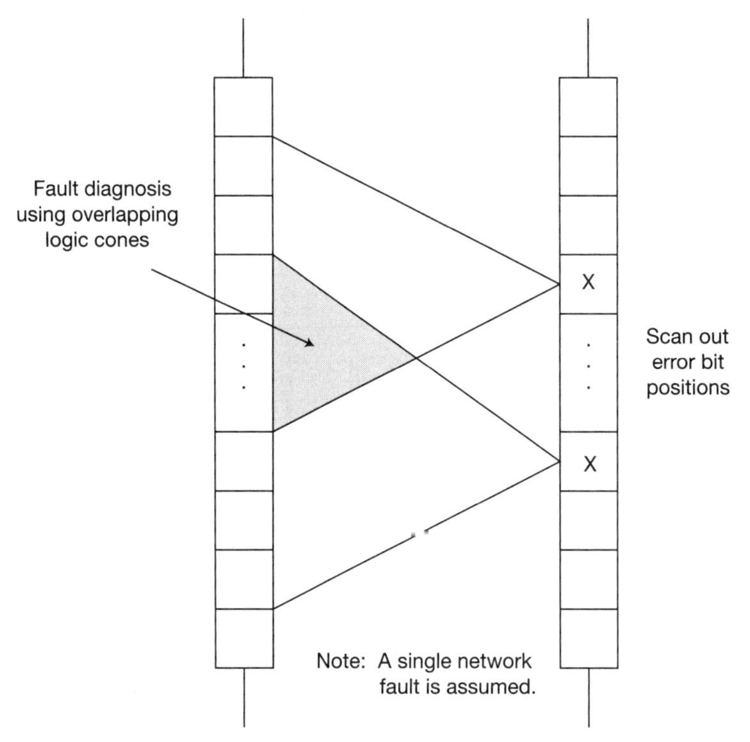


**Figure 19.27** Illustration of a diagnostic method for localizing a potential fault from multiple capture miscomparisons, using overlapping combinational logic cones.

### 19.7.2 Diagnostic Pattern-to-Fault Correlation

Recall that the ATPG tool assumes a single model fault, attempts to sensitize and propagate a GM/BM difference, and then invokes a fault simulation algorithm with the chosen pattern to see if additional, as yet undetected, faults may be removed from the remaining fault list. Additional faults may be detected either by another iteration of the pattern generation algorithm consistent with existing assigned values (e.g., using test compression) or by random filling of don't_care values in the pattern (using general scan-shift pattern application). In either case, fault simulation of the pattern data is used to further reduce the undetected fault list. For diagnostic purposes, however, fault simulation needs to be applied to the entire fault population to provide a comprehensive list of faults detected by each pattern. This database of faults per pattern is commonly referred to as the *fault dictionary*. Given a collection of failing test patterns applied to the die, the SoC test engineering team seeks to identify the candidate faults that are common to that collection.

This method is more amenable to the case of multiple faults present on the tested die, yet it does require the pass/fail results from exercising many patterns rather than stop on first fail. EDA fault diagnosis tool development has focused on optimizing the representation of the fault dictionary. Figure 19.28 depicts a simple table format and a binary tree representation of the dictionary.

Fault Dictionary Table

Analysis of passing and failing tests (for all patterns) used to identify candidate faults

(e.g., Tests 2 and 3 fail, all others pass, Fault 2 is a candidate)

| | Test 1 | Test 2 | Test 3 | ··· | Test m |
|---|---|---|---|---|---|
| Fault 1 | 1 | 0 | 0 | | 1 |
| Fault 2 | 0 | 1 | 1 | 0 | 0 ← Test syndrome |
| Fault 3 | 1 | 0 | 1 | | 0 |
| ⋮ | | | | | |
| Fault n | 0 | 1 | 0 | | 1 |

'1' = the test detects the fault, '0' = the test does not detect the fault

Note that duplicate test syndromes are not distinguishable in fault diagnosis.

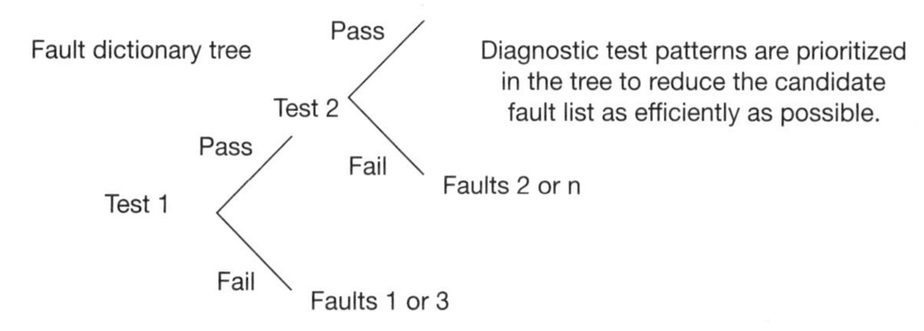


**Figure 19.28** Illustration of potential data structures used for representation of the fault dictionary.

If the results for all test patterns are available, analysis of the passing and failing column entries in the table provides a candidate fault list. Each row of the fault dictionary table data structure in the figure is known as the *test syndrome* for the fault. If the tester stopped on first pattern fail and the die is being retested specifically for diagnostics, the tree representation of the fault dictionary guides the priority order for pattern application.

The ATPG flow is focused on finding maximal fault coverage with the fewest patterns. The fault dictionary table would have a minimal number of columns, with a high density of entries in each. As a result, the candidate fault list for (a small number of) failing patterns may still be quite large. To assist with fault diagnosis, an additional collection of diagnostic test patterns may be generated, focused on differentiating faults rather than maximizing the fault coverage per pattern. These diagnostic patterns effectively extend the fault dictionary tree and reduce duplicate test syndromes.

Ideally, a diagnostic test set is used only during design prototype bring-up. In production, it would be much more difficult to disrupt the flow of material to inject additional diagnosis testing. Only in the case of a catastrophic "yield bust" would work-in-progress be halted.

### 19.7.3 Diagnosis with Test Response Compaction

Section 19.4 describes methods for test stimulus ("lossless") compression and test response ("lossy") compaction. The scan chain responses of a test compression architecture are provided to a multiple-input signature register (MISR), which offers time-based (and potentially space-based) compaction. An error present in the response sequence would result in an incorrect signature, indicating a failing part. Fault diagnosis is significantly complicated by the signature data compaction; it is extremely difficult to develop a candidate fault list from an erroneous signature. To enable separate diagnostic testing for an SoC with MISR compaction, additional test modes may be adopted into the SoC DFT design. The simplest approach would be to forgo the time-based signature compaction and scan-shift every MISR register bit from each scan chain every cycle. Another method would be to include additional logic between the scan chains and the MISR inputs to enable direct control over individual input values, as shown in Figure 19.29. If the correct signature is observed with an overridden MISR input pattern, the corresponding failing scan chain bits can be isolated for diagnosis, with little additional circuit overhead.

The SoC test engineering team is tasked with fault diagnosis analysis during SoC bring-up if systematic yield detractors are observed. The project schedule available for diagnosis is typically very compressed (no pun intended). In preparation for the wafer-level test results, the test engineering team should run extensive fault simulation and, potentially, diagnostic pattern generation. Usually, this occurs in the interval between prototype tapeout and wafers out from the fab. Given the constraints on allocating tester time at the foundry, the SoC project manager may need to investigate using separate engineering test (and failure analysis) service providers if extensive diagnosis of prototype yield issues is warranted. This constraint also definitely applies once the design is in production. The product engineering support team may need to consider engaging additional resources outside the foundry and OSAT to assist with yield-improvement diagnosis.

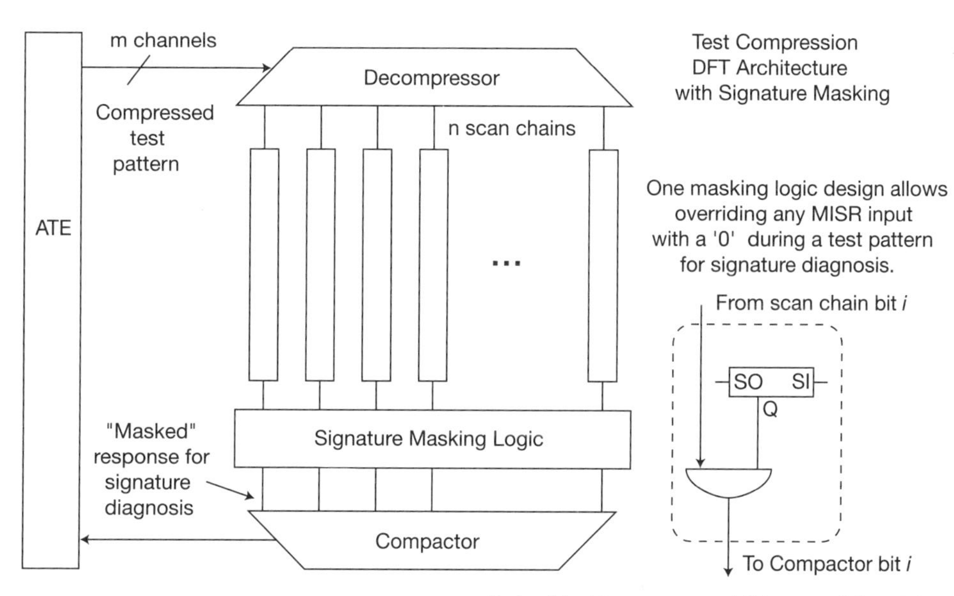


**Figure 19.29** Illustration of an MISR with overriding input values in an attempt to distinguish a passing signature from a failing signature.

The ATPG EDA tool vendor needs to provide additional support for the following:

- Fault dictionary generation (with extensive fault simulation)
- Additional diagnostic pattern generation (to reduce the number of common fault syndromes)
- Correspondence between pin and interconnect faults with physical layout data (incorporating LVS flow output results for reference)

Fault diagnosis requires both extensive data about the defects detected by the test pattern set and the collective insights and experience of test engineers, foundry process control engineering, and circuit designers to effectively isolate and determine the root causes of a fabrication yield below expectations. Once identified, corrective design measures (and, potentially, process control monitoring changes) can be pursued to improve yields.

### 19.7.4 Technologies to Rework Failing Dies

For prototype dies, the project schedule expects a suitable quantity of functional parts to be available from first-pass tapeout for product development. There are several non-production fabrication options available to cut interconnections and/or deposit additional overlay metal wires at specific locations on a small quantity of dies (e.g., laser cuts and *focused ion beam* [*FIB*] deposition). (Product bring-up may be able to proceed with some degree of reduced SoC functionality.) Although these technology options are beyond the scope of this book, their value has been proven repeatedly; these corrective steps have enabled many SoC development projects to avoid major schedule impacts. An SoC project management team should be prepared to address the value of sourcing these additional fabrication steps once fault diagnosis conclusions are available for a low-yield wafer lot. These corrective steps are also to be considered for a prototype design with a sufficient quantity of parts but for which a major functional bug has been detected early in product bring-up that could not be easily deferred until a subsequent tapeout revision.

## 19.8 Summary

This chapter briefly describes some of the design-for-test architectural approaches commonly incorporated to improve internal signal controllability and observability. The complexity of current SoC designs is driving the growing trend of on-chip integration of additional logic to exercise test patterns with minimal interface bandwidth to the tester, through test compression and built-in self-test methods.

The tester time associated with each die site is an increasing factor in SoC production costs. The need for product burn-in stress test screening, followed by retesting, adds to these costs. There is also an indirect cost associated with production test escapes, namely packaged parts with an undetected manufacturing defect that proceed to system build.

The SoC project manager, the methodology team, and the test engineering team are faced with several critical trade-offs to select the appropriate DFT architecture, fault models, fault coverage targets, and pattern generation plus response observation approaches that are appropriate for the product goals. The key to achieving these goals is to conduct this trade-off assessment as a fundamental part of the initial project planning phase. Although the test pattern set deliverable is typically not required until after tapeout, the ultimate success in realizing the shipped product quality and cost targets requires adhering to DFT guidelines throughout the SoC development schedule.

## References


[1] Roth, J.P., et al., "Programmed Algorithms to Compute Tests to Detect and Distinguish Between Failures in Logic Circuits," *IEEE Transactions on Electronic Computers*, October 1967, pp. 567–580.

[2] Goel, Prabhakar, "An Implicit Enumeration Algorithm to Generate Tests for Combinational Logic Circuits," *IEEE Transactions on Computers*, March 1981, pp. 215–222.

[3] Hapke, F., et al., "Cell-Aware Test," *IEEE Transactions on Computer-Aided Design of Integrated Circuits and Systems*, Volume 33, Issue 9, September 2014, pp. 1396–1409.

[4] Tsai, K.H., and Sheng, S., "Design Rule Check on the Clock Gating Logic for Testability and Beyond," *IEEE International Test Conference*, 2013, pp. 1–8.

[5] Mukherjee, N., et al., "High Volume Diagnosis in Memory BIST Based on Compressed Failure Data," *IEEE Transactions on Computer-Aided Design of Integrated Circuits and Systems*, Volume 29, Issue 3, March 2010, pp. 441–453.

[6] Rajski, J., et al., "Embedded Deterministic Test," *IEEE Transactions on Computer-Aided Design of Integrated Circuits and Systems*, Volume 23, Issue 5, May 2004, pp. 776–792.

## Further Research

### Backward Inference and Forward Propagation

The key facet of combinational ATPG is the ability to efficiently find logic cone inputs that sensitize the assumed fault and propagate the GM/BM signal to an observable endpoint. The algorithms for backward and forward traversal through the combinational network to assign, sensitize, and propagate logic values for the fault-free model attempt to avoid an unrealizable condition that necessitates "backtracking."

Research and describe the algorithms for combinational ATPG, specifically the approach that involves minimizing the logic assignments that subsequently require backtracking. (References 19.1 and 19.2 are a good starting point.)

### Fault Simulation

The discussion in this chapter highlights the application of fault simulation to reduce the fault list during ATPG and to derive the test syndrome for diagnostic patterns. The computational workload for fault simulation is considerable, and it necessitates algorithms focused on throughput.

Research and describe the differentiating characteristics of concurrent fault simulation, deductive fault simulation, and parallel fault simulation. (Although a fault simulation algorithm would support a general logic design, assume a scan-based serial shift register DFT architecture.)

### Memory Array Fault Models

Memory bitcells utilize the most aggressive process lithography design rules due to the high PPA impact on a typical SoC design with large embedded arrays. As a result, however, unique fault models are considered for arrays, as process variation may result in "weak bits." Array fault models include the sensitivity to read/write operations at neighboring cells, which might disturb the stored value. The fault models also involve demonstrating a bitcell read/write operation with specific patterns of neighboring cell values. Unique pattern sequences have been developed to stress these sensitivities at production testing; the programmability of the MBIST controller needs to support generation of these pattern sets.

Research and describe the MBIST write/read pattern sequences commonly used and the sensitivity that the pattern set is attempting to isolate—for example, *marching* (or *galloping*) patterns, *checkerboard* patterns, etc.

### Physical "Repair" Technologies for Prototype Dies (Advanced)

In the unlikely case of a "yield bust" from the prototype wafer lot fabrication, diagnostic activity may identify a root cause systematic failure in the design. The SoC design team may be tasked with identifying an incremental change to the physical implementation that would enable (reduced) SoC functionality to allow (a subset of) silicon bring-up activities to proceed (see Chapter 21, "Post-Silicon Debug and Characterization ('Bring-up') and Product Qualification").

The semiconductor equipment manufacturers have enabled such incremental updates to existing silicon dies for a very limited number of parts. (Concurrently, an SoC project manager is faced with the decision on whether to accelerate a tapeout revision release to the foundry with the implementation change—a cost-versus-bring-up schedule trade-off.)

Research and describe the technical capabilities available for implementing laser cuts and focused ion beam (FIB) metal deposition applied to a fabricated design. Describe the wafer preparation required, the SoC design data provided, the resolution available (for cuts and additions), etc.

CHAPTER 20

# Preparation for Tapeout

## 20.1 Introduction to Tapeout Preparation

This brief chapter summarizes the SoC project manager and methodology team tasks required in preparation for tapeout data release to the foundry. Many of these tasks have been introduced in previous sections; this chapter consolidates those earlier discussions.

It is common for an SoC project manager to be responsible for capturing design status information into a *tapeout checklist*. To assist with this task, the SoC methodology and CAD teams will have developed and maintained a methodology manager application—often referred to as the project *scoreboard*—which records design checking and analysis flow results for the SoC model hierarchy and successive model version releases. The tapeout checklist builds on the scoreboard status of the design version targeted for tapeout to include specific foundry data requirements.

In addition to the financial commitment associated with tapeout, the SoC project bring-up schedule is based on the expected duration from tapeout release to parts received. Any discrepancies or deficiencies in the tapeout database subsequently detected by the foundry (or OSAT package assembly provider) may result in additional costs and schedule delays. In short, tapeout is a major SoC project milestone, and it is crucial to ensure that the release data and documentation are correct and complete.

## 20.2 Foundry Interface Release Tapeout Options

Fabrication alternatives including the following will have been reviewed with the foundry support team long before tapeout:

- Metallization stack
- Additional process options (e.g., device threshold voltages, thick oxide devices, MIM capacitors, triple well devices, precision resistors, fuses)
- Die attach metallurgy (e.g., solder bumps, Cu pillars, bond wires)
- Fabrication lot priority (e.g., regular, an expedited "hot lot," a "rocket" lot)

In addition, there are options for the scope of the release data to be confirmed with the foundry and IP providers, including the following:

- Cell instances released as black boxes in the model hierarchy to be inserted by the foundry during mask data preparation (e.g., "protected" IP cores and macros, on-die process control monitors)
- Fill data

Many layers in advanced process nodes require a degree of uniformity in layout density for lithographic patterning and material etch rate consistency (e.g., active areas, device gates, metal segments). The fill data added to achieve the desired uniformity could be included in the tapeout release or managed by the foundry. Due to the impact on the parasitic electrical model at advanced process nodes, metal fill data is now routinely required in the design model. However, the foundry may wish to retain the fill algorithm for non-electrical data to allow for ongoing process improvement engineering.

### 20.2.1 Multipatterning Decomposition Color Assignment

In advanced process nodes, the layout data on specific mask layers is decomposed into multiple subsets for patterning; iterative steps of "litho-etch" are pursued to realize the full layer pattern. For example, a minimum (width plus space) pitch dimension of ~80nm for a mask layer using a 193nm immersion exposure system requires an LELE process sequence, with the decomposition of layer data into two subsets, A and B, often denoted using two colors. As described in Section 18.1, the layout design rule checking runset from the foundry invokes EDA tool DRC algorithms to confirm that no cyclic dependencies are present and that decomposition during mask data preparation will be successful. As a result, the specific task of layer color assignment could be deferred to the foundry. However, the foundry may indicate that the electrical characteristics of the mask layer subsets will differ. The nominal *bias* between drawn and fabricated wafer-level dimensions and/or the resulting statistical variation in the fabricated layer parameters may differ between the successive litho-etch steps. In this case, the highest accuracy settings for parasitic extraction require a color assignment to individual shapes on the layer. The tapeout release to the foundry may therefore also be precolored with separate layer designations assigned to each subset, as shown in Figure 20.1.

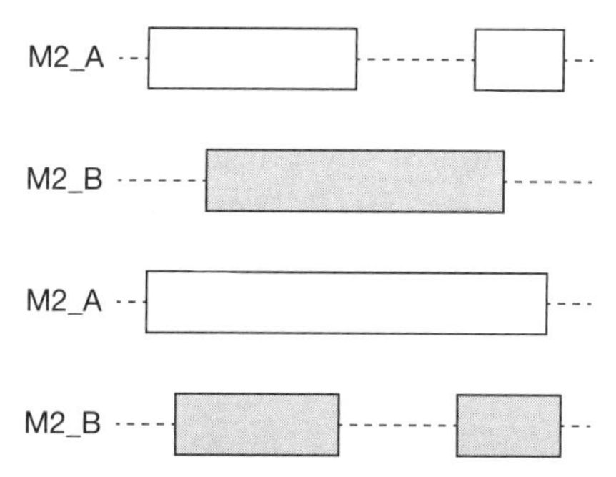


**Figure 20.1** Illustration of a pre-assigned multipatterning decomposition to layout shapes in the tapeout database. A separate mask layer designation is typically used to differentiate the color assignment.

### 20.2.2 FEOL and BEOL Layer Data

The mask layout layers are commonly divided into front-end-of-line (FEOL) and back-end-of-line (BEOL) subsets to distinguish between device fabrication (FEOL) and the addition of metal/via connections (BEOL). (The addition of local metallization—also known as layer M0—to connect device active and gate nodes without separate contact openings blurs the FEOL and BEOL distinction. The term *middle-end-of-line* [*MEOL*] is often used to describe this technology option.) It may be feasible for the foundry to accept the tapeout data in two stages: a FEOL release followed by a BEOL release. Mask preparation and prototype wafer lot fabrication would commence with FEOL data. BEOL layer data would subsequently be released on an agreed date with the foundry (typically, a few weeks later). This would give the SoC design team the opportunity to address (a small number of) final analysis flow issues or yield optimization steps, as long as the design updates were confined to the BEOL layers.

This tapeout release strategy is also fundamental to a metal-only design revision, in which case a BEOL release reflects new signal connections between existing cells and IP. Spare cells are typically added to the FEOL data to enable expedited logic ECOs between releases, as described in Section 17.1. An inventory of prototype wafers is held after FEOL processing to enable expedited fabrication turnaround time for a BEOL-only tapeout.

### 20.2.3 Tapeout Data Volume

The SoC design team and foundry benefit from initiatives to keep the size of the tapeout database in check. The runtime of the full chip design checking flows is accelerated if the physical model leverages cell and block instance hierarchy to the greatest extent possible. EDA tool vendors have optimized their checking algorithms to leverage the physical database hierarchy. Several methods, including the following, are available to minimize the dataset size:

- Common patterns of individual shapes are merged into a layout cell, with cell instances rather than "flat" shapes data added to the full chip design (e.g., power/ground grid patterns, via arrays in wide metal grids, fill data).
- Device gate bias data is maintained in an overlay cell.

They foundry may support a small drawn gate length extension lithography option (see Figure 20.2). An optimization flow exercised prior to tapeout may extend a device gate length to reduce leakage current power dissipation if the signal arrival time has sufficient timing slack. These gate bias shapes would be defined in an overlay cell, and a data union operation would be performed during mask prep; in this case, the "uniquification" of library cells, and thus the tapeout dataset size, is minimized.

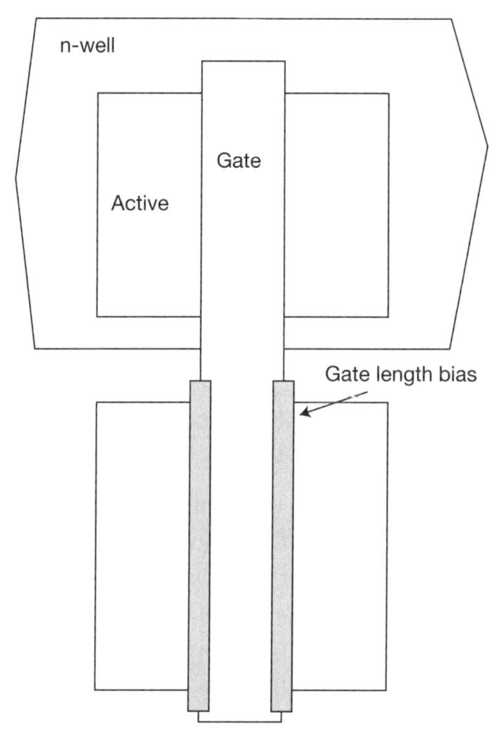


**Figure 20.2** Illustration of a "gate length bias" foundry option, added to devices as a leakage power optimization late in the SoC schedule. The bias shapes data is released as part of an overlay cell.

### 20.2.4 FinFET Data

The implementation of an FET device as a number of parallel fins is described in Section 1.4. The integral number of fins for any device, with a common pitch throughout the design, allows a single active shape to accurately represent the device dimension without having individual fins drawn, as depicted in Figure 20.3.

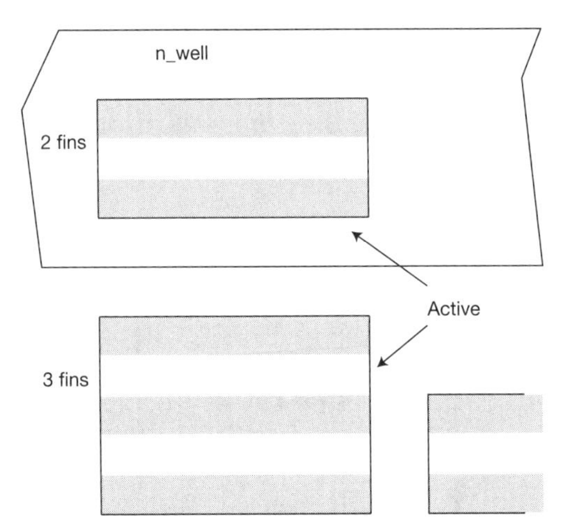


**Figure 20.3** The "fins" of a FinFET device are typically not explicitly drawn to reduce the dataset size. An active area shape of appropriate width and (aligned) origin defines the fins to be fabricated.

### 20.2.5 Metal Lines and "Cut Metal" Shapes

In advanced process nodes, for the densest routing layers, the traditional method of fabricating metal line segments is being replaced by a more complex "pattern and cut" process, as illustrated in Figure 20.4. Whereas the drawn metal endpoint spacing in mature process nodes could be directly resolved and patterned using a single lithographic step, the scaling in advanced process nodes necessitates a separate *cut metal* mask and an "artificial" alignment of metal segment endpoints (for both design routes and metal fill data).

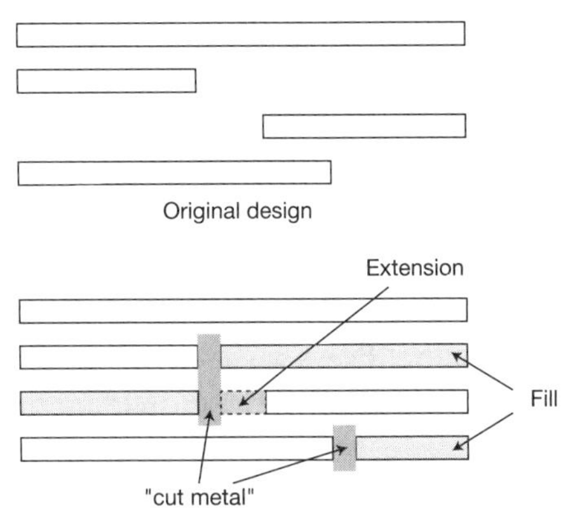


**Figure 20.4** Illustration of a "cut metal" shape across multiple metal wires, defining the metal line ends.

Optimization of the tapeout data representation with regular metal tracks (with pre-assigned multipatterning colors) and the cut metal personalization is an ongoing area of development.

The layout data is exported (or *streamed*) from the physical design tool platform, using a utility provided by the EDA tool vendor. This utility has numerous options and filters available, including:

- Expanding the line segment data with the "width plus endpoint extension" property to two-dimensional shapes
- Mapping the drawn layer designations in the physical design platform to the corresponding foundry layer values while merging layer data, as appropriate
- Removing information unrelated to fabrication (e.g., pin shapes and pin text in library cells, extraneous text labels)
- Invoking an algorithm for calculating data consistency (cyclic redundancy check, or CRC) values to append to the tapeout data for verification at the foundry of correct file transmission

Separate drawn layers may have been used for clarity during SoC physical design, and they may need to be merged into a single foundry mask layer; for example, M6_VDD and M6_GND would be merged into the foundry layer number for M6 shapes data. (Note that similar mapping operations would be used when streaming out physical layout data for analysis and verification flows using the foundry PDK runset.)

The mapping file used by the stream utility is developed and maintained by the CAD team, working with the SoC project manager and foundry customer support. This mapping file is critically important to the integrity of the tapeout data and needs to undergo extensive qualification testing. One option would be to stream the exported data back into the physical design tool platform and XOR the original and stream-in physical data to ensure that no design information has been corrupted.

The EDA industry is also focused on establishing more efficient data formats for tapeout release. An industry standard—Open Artwork System Interchange Standard (OASIS)—has recently been released to improve the dataset size from the de facto standard graphic design system representation (GDS-II) that has traditionally been used.

In addition to the physical data, related documentation is included in the tapeout release. The beginning of this section highlights the need to provide the foundry with the specific process options implemented on the design. The next section describes other key information that the foundry needs to review as part of the tapeout acceptance procedure.

## 20.3 Tapeout Checklist Review

The SoC project manager needs to conduct a final tapeout checklist review before data release to the foundry. The methodology tracking tool provides the following project status information:

- All analysis and checking flows have been exercised on the SoC release version designated for tapeout.
  - Flow dependencies have been correctly managed (i.e., flows have been run in the correct order).
  - All full-chip, tapeout-level flows used this SoC design release version.
  - Flows that may combine block-level analysis then full chip analysis with abstracted block models have used the correct hierarchical design configuration.
- All flow steps used an approved EDA software release version.
  - Potentially, more than one version is acceptable, as long as all have been reviewed/approved (e.g., IP blocks may have been analyzed using an earlier tool version than was available for the full chip design schedule phase).
- All analysis and verification flow steps used an approved release of foundry PDK techfiles, runsets, and simulation models.

The SoC project manager needs to ascertain from the foundry which PDK releases are acceptable. If the process is already in volume production, the foundry may indicate something like "PDK v1.0 or higher." Subsequent PDK v1.x releases are typically associated with process improvements, such as tighter statistical model variations as a result of ongoing process learning improvements with (hopefully) few design rule updates. Nevertheless, the foundry may require that a specific PDK release be used, potentially necessitating the rerunning of flows for existing IP blocks to a more recent PDK version. If the tapeout is aggressively targeting a preproduction process, there are likely to be significant techfile, runset, and model changes in the PDK v0.1, v0.5, and v1.0 release sequence. An IP block or cell library developed using an older PDK release may or may not be appropriate for use as part of a tapeout design if a more recent PDK has been issued. The SoC project manager needs to resolve any conflicts between the PDK version recorded for the various flows and the foundry expectations.

### 20.3.1 Internal Project and Foundry Tapeout Design Waivers

In addition to the tool and PDK version information, the methodology status tracking tool will have parsed and recorded pertinent log file results for each analysis and verification flow for the SoC hierarchy. The design closure on targets for path timing, signal noise margins, power dissipation (and signal slews), power distribution voltage drop margins, and FIT reliability estimates are closely evaluated, using the recorded log file data. In addition, the functional validation team provides the status of any open design bugs. If some targets are not fully achieved, or if open bugs remain, a critical decision is whether to proceed to tapeout, with consideration given to several criteria:

- **Prototype bring-up versus production-directed release**—If the tapeout is for (first-pass) prototype bring-up activity, some diminished functionality and/or performance-limiting paths may be acceptable. Product development and software testing may be able to utilize parts with known deficiencies. If the tapeout is anticipated for a production ramp, the design closure expectations are more stringent.
- **Severity of the functional bugs**—If the features associated with open hardware functional bugs have software workarounds with only moderate performance impact, a (low-priority) bug could be waived. Or perhaps a new feature could be deferred from the overall product goals if there are significant validation issues remaining. Given the extensive resources required to resolve a functional bug, spanning HDL design through physical implementation and analysis, the review of open functional issues should occur frequently throughout the SoC project schedule. Well before the project tapeout phase, an assessment of which open issues to address and which to defer should for the most part have already been made.
- **Path timing closure—both setup and hold timing checks—on all corners**—For a prototype tapeout, it may be acceptable for the SoC

path timing status to have setup time failures (of a hopefully small magnitude), assuming that product bring-up could proceed with a (slightly) reduced clock frequency. Conversely, hold time check failures need to be resolved for any tapeout, as they indicate clock skew versus data path issues, regardless of clock frequency.

The evaluation of path timing issues for the tapeout checklist is complicated by the increasing number of process corners and voltage/temperature operating modes. It is common to excise a circuit netlist for failing paths from the STA flow and exercise a statistical circuit simulation to determine a confidence level for the timing error to be present in production silicon. Sections 11.1 and 11.3 discuss some of the sources of design margin that may be present in cell characterization models, delay calculation, and path timing propagation. The results of this circuit-level margin analysis would also be considered during the tapeout review.

- **FIT reliability issues (specifically powerEM and sigEM flow results)**—For prototype parts, reliability estimates are a less critical tapeout checklist consideration. Although (instantaneous $j_{PEAK}$) peak current density electromigration flow errors need to be addressed, the contribution to the FIT calculation from average current EM analysis is less of a concern.
- **Test coverage**—If the fault coverage percentage is below targets, the tapeout checklist review team needs to make a difficult assessment. The test engineering team may have additional time between tapeout release and wafers out to enhance the test pattern set to improve coverage without requiring any logic design changes. Alternatively, it may be decided that additional controllability and observability logic additions are required. As with the case of functional bugs, the impact of test logic updates can be significant. Fault coverage assessments should be an ongoing project evaluation activity so that DFT logic-related decisions will have already been addressed well before tapeout approaches.

The decision to proceed to tapeout with open analysis flow issues involves assigning a waiver to the issue in the tapeout checklist. The following documentation accompanies the waiver:

- A list of the members of the tapeout review team (usually the lead engineers from all project-wide teams)

- The list of those who need to be informed of the waiver status
- The resource and target date for fixing the issue in a subsequent tapeout (or a permanent waiver)

For any deferred or disabled functional features related to a tapeout waiver, it is also important to ensure that the product marketing team is aware. Preliminary marketing materials based on original product specifications may need to be updated.

Analysis flow issues and functional validation bugs covered by tapeout waivers are typically managed within the SoC project team and do not require review by foundry customer support engineering. However, if there are any design rule checking errors for the tapeout database, those errors *must* be reviewed with the foundry. Ideally, no DRC errors would be present, using the foundry-recommended PDK runset version. Nevertheless, errors may arise—such as from older IP layouts that do not meet current DRC checks or from layer fill density and density gradients that may fail in a specific checking window. These DRC errors may be difficult to fix without affecting the tapeout schedule. The SoC project manager may opt to contact foundry customer support and request a tapeout "DRC waiver." An assessment will be made by the foundry on the impact of the existing layout on lithographic resolution; mask overlay sensitivity; deposition/etch process steps and resulting pattern fidelity; post-CMP surface topography and lithographic depth-of-focus issues; and, ultimately, the yield impact.

In addition, results from the design-for-yield (DFY) layout optimization algorithms may also need to be reviewed with the foundry (e.g., redundant via insertion percentages, wire spacing "critical layout area" versus defect size sensitivity calculations).

The foundry may or may not concur with an SoC tapeout request for DRC waivers. If such a request is granted, the waiver information is added to the methodology tracking tool and becomes an integral part of the tapeout release documentation.

The tapeout preparation phase of an SoC project schedule is typically a hectic time. Final full chip methodology flows are exercised on the design version targeted for tapeout. The CAD team is required to address any flow issues that arise with highest priority; ideally, flows will have been previously tested with full chip-sized datasets (perhaps as part of a mock tapeout exercise). The

IT support team typically needs to allocate specific (parallel and/or distributed) compute server CPUs with sufficient memory to ensure that these jobs execute efficiently. The checklist review team is assembled to scrutinize flow results and delve into open functional validation bugs. The decision to fix an issue rather than defer it sets in motion a high-priority task by members of the SoC architecture, logic, test, and physical implementation teams. The fix then requires iteration(s) on flows to confirm the validity of the update to the original tapeout dataset. The sooner the "fix versus defer" issue decision can be made, the less the potential impact on the tapeout schedule.

After the flurry of activity subsides once the tapeout data and documentation have been transmitted to the foundry, the focus of the SoC design team shifts to two (still rather urgent) activities:

- Preparation for silicon bring-up (discussed in Chapter 21, "Post Silicon Debug and Characterization ('Bring-up') and Product Qualification")
- Execution of the solution to design issues that were waived for tapeout (adhering to the resource constraints and schedule targets recorded with the waiver)

In addition, the test engineering team ramps up its activities for preparation of the full test pattern set, negotiating (digital and mixed-signal) tester resources for the date when wafers are due to exit fabrication. Similarly, the SoC package engineering team focuses on preparing all the assembly documentation for the OSAT service provider, as well as managing the logistics of ordering package substrates. During fabrication, the foundry customer support team provides ongoing status updates on the progress of the wafer lot(s), with the expected "wafers out" date. The SoC project manager coordinates these post-tapeout tasks to be aligned in schedule and duration for uninterrupted material flow from fabrication to wafer test to assembly (including burn-in and final test).

### 20.3.2 Post-tapeout ECOs

The tapeout database release is "frozen". This exact dataset needs to be available for reference during silicon bring-up to diagnose any errors found. A new configuration management (CM) version line is created after tapeout to allow engineering work to commence on deferred issues. Section 17.1 describes specific flows used for implementing engineering change orders (ECOs) throughout the methodology. (Although the term *ECO* originated as a means to manage change requests to design specifications and features—i.e., “orders”—it is generally applied to any proposed update to the tapeout baseline SoC dataset.) The proposed ECO update to apply to the design data is tagged in the CM repository with the specific deferred issue that is being addressed. The SoC tapeout checklist review team reviews and approves/rejects any proposed ECO before the update is merged into the latest CM revision to ensure consistency with requirements for the next pending tapeout (e.g., the number of mask layers modified for a layout update, any potential conflicts with other proposed ECOs). As with the information recorded for the original waiver request, the review approval details for the ECO would be saved as part of closing the waiver issue. Any existing tapeout DRC waivers that have been removed in preparation for a subsequent tapeout would be communicated to the foundry customer support team to allow the fab to prepare updates for any process control monitoring that had been put in place for the waiver.

## 20.4 Project Tapeout Planning

The ideal SoC design project supports volume production ramp from the initial tapeout release. However, first-pass success is not always achieved, for several reasons:

- Schedule pressures to get hardware to product bring-up before all open functional bugs and electrical analysis issues are resolved
- Bugs and issues uncovered during bring-up
- Problems arising during prototype fabrication (a specific concern for a preproduction early IP testsite or full chip tapeouts, using pre-v1.0 PDK data)

Ideally, any lithography exposure resolution and/or mask overlay issue arising during prototype fabrication would not require a design update. The foundry would take corrective steps and accelerate a backup wafer lot. The optical source illumination pattern for exposure could be updated, or (reluctantly) new masks could be generated with improved source-mask optimization (SMO) algorithms. The worst-case scenario would be a PDK design rule revision necessitating an SoC design ECO.

As an aside, to reduce the risk associated with a fabrication problem, the SoC project manager might choose to request multiple wafer lots, staggered in time. (As mentioned in Section 1.3, such a request might also include split lot process biasing to identify design sensitivities to specific process corners.) Multiple wafer lots may also be needed to satisfy a large volume of prototype parts requested by product bring-up and software development teams. Although multiple lots incur additional project development cost, the risk mitigation provides considerable justification. For advanced process nodes, the majority of the *non-recurring expense* (*NRE*) is associated with mask manufacture rather than subsequent wafer lot processing. Ordering multiple prototype lots is typically a worthwhile investment.

A conservative development schedule and project cost forecast that does not assume first-pass success is usually appropriate. The estimate for the number (and fab throughput priority) of all-layer and BEOL-only layer tapeout revisions required to achieve volume production ramp is the most important SoC project management responsibility. All schedule milestones, engineering team resource allocations, and NRE budgets are derived from this assumption. The tapeout iteration estimates are based on several project characteristics, such as the scope of new architectural features relative to previous product generations, the amount of new physical library IP, the target performance, and the foundry process maturity. The degree to which new methodology requirements are introduced—and, correspondingly, new EDA tools and flows are needed—is also a consideration in project tapeout planning. An increased methodology focus on flows such as functional validation coverage measures, formal model property checking, statistical analysis techniques, and so on may provide significant return-on-investment (ROI) if it provides sufficient confidence to reduce the tapeout iteration forecast.

If the design is to integrate new IP blocks, the project management team also must address the value of preparing a preliminary testsite tapeout to qualify the IP for the target SoC application. The value of silicon qualification on testsite dies needs to be assessed relative to the resource plus NRE applied to preparing the testsite for tapeout and (especially) the resource pulled from the design team for testsite bring-up. If the IP circuits of interest relate to external high-speed chip interfaces, the testsite development project may also require corresponding package design, fabrication, and OSAT assembly; the testsite package would need similar electrical characteristics as the final SoC package for accurate interface timing measurements.

Typical development expense and (cumulative) profit/loss curves over a product lifetime are depicted in Figure 20.5. If project development expense estimates are exceeded, the time to overall profitability is increased. More significantly, if production ramp is delayed, competitive market share is adversely impacted, and the initial and cumulative revenue are severely diminished.

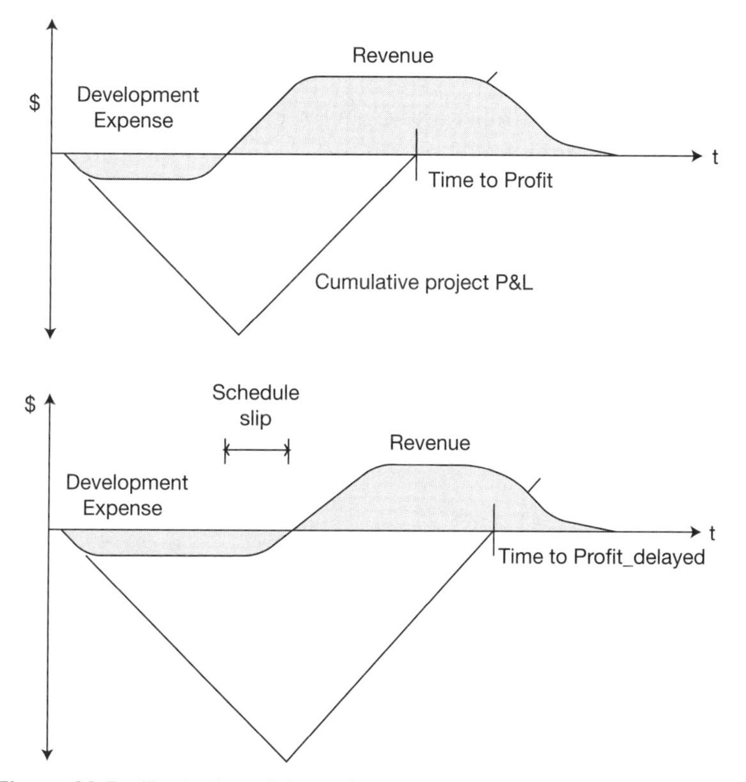


**Figure 20.5** Illustration of the typical product development expense, revenue, and time-to-profitability curves versus time. Schedule delays have a non-linear impact on the time to profit due to the likelihood of reduced competitive market share.

Project tapeout planning and (hopefully) flawless tapeout execution are crucial.

## Further Research

### Tapeout Checklist

Develop a tapeout checlist for use by the review team, incorporating the results of physical design verification flows, electrical analysis flows, and the functional validation bug tracking database.

### Outsourced Assembly and Test (OSAT) and Package Substrate Data Release

This chapter focuses on the release of SoC design data to the foundry. Another critical project milestone is the release of package design and assembly data to the package substrate supplier and OSAT service provider. This milestone may occur before or after the SoC tapeout, depending on the lead time for package substrates (e.g., multilayer ceramic or organic carriers).

Research and describe the data release information required by the package substrate supplier and OSAT service provider.

CHAPTER 21

# Post-Silicon Debug and Characterization ("Bring-up") and Product Qualification

## 21.1 Systematic Test Fails

During wafer testing, the set of failing dies may indicate that a systematic defect is present. As described in Section 19.7, test diagnostic procedures are pursued in an attempt to localize and determine the root cause of a defect. The ATPG tool flow is exercised in diagnostic mode, using the failing pattern syndrome to identify candidate faults. Physical locations on the die (and masks) are cross-referenced to the candidate faults for further investigation.

After wafer testing, good dies will be packaged and (likely) subjected to burn-in stress to screen infant fails. After (static or dynamic) burn-in, parts are retested. If there is significant population fallout at package testing, diagnostics are again pursued; in this case, however, a greater number of potential defect sources needs to be considered. The source of stuck-at, transition, or bridging fault identification could be due to defects accelerated by the burn-in stress conditions, including the following:

- "Weak" device gate oxide breakdown
- Device parameter drift (e.g., due to the presence of material contaminants, whose diffusivity is increased at elevated temperature and voltage)
- Material interface *delamination*, with structural cracks and/or permeability to humidity exposure (most often occurring at the interface between the die and the *underfill* material that seals the die to the package substrate)
- Die attach metallurgy fails (e.g., solder bump cracks or lifted bond wires)

For the root causes of these defect mechanisms, unique diagnostic techniques are applied. Subsequent sections of this chapter discuss methods of monitoring electrical parameters. For investigation of material defects, the package is removed, and high-resolution electron microscopy of the surface and/or cross-section of the die is performed.

The final stage of SoC bring-up testing occurs during end product evaluation. Fails that arise at this level are not typically related to die fabrication or assembly defects—hopefully burn-in screening has identified those issues—but rather to functional validation escapes or test escapes in the SoC methodology.

A validation escape implies that a feature of the SoC functional specification is incompletely or incorrectly designed, and SoC validation testbenches were insufficient. System software compiled to that specification likely uncovered the error. Product bring-up may be able to proceed if a software workaround is available or if it is possible to disable a feature; indeed, micro-architects may explicitly add disable signals to configuration registers in the SoC design for new features that are especially high risk. (These disable signals are commonly referred to as *chicken bits*.) The SoC bring-up team is tasked with assessing the impact of product failure on the logic design, validation testbench development, physical power/performance/area resources, and schedule. The product marketing group needs to evaluate this assessment against the value of the associated feature. Assuming that the validation escape has been detected as part of first-pass product bring-up and that a subsequent tapeout has been included in the SoC project plan, the overall impact should be manageable. However, if the SoC specification has experienced *feature creep* during development—that is, if new requirements have been added to the original specification late in the project design schedule—the risk of a validation escape on the last planned tapeout is higher. Assessing the trade-offs between the cost of another tapeout and the value of a differentiating feature requires challenging product management decisions.

Whereas a validation escape is pervasively evident at product bring-up (i.e., an SoC feature does not behave as expected), a test escape relates to a small population of bring-up parts with a fabrication or assembly defect that was not detected by the wafer/package test pattern set. The impact to the overall bring-up activity is usually small. (Ideally, the SoC on the bring-up board is socketed, and another prototype part is substituted for the defective part with the test escape.) The SoC test engineering team is responsible for fault diagnosis—not using the failing syndromes at the tester but with product runtime information on how the part with the test defect behaves differently than the population of good parts. The capability to breakpoint system functional operation and enter an SoC scan-shift test mode to observe runtime register data is extremely valuable in this regard; this is commonly referred to as a functional *scan dump* of the SoC internal state. The SoC test team is tasked with augmenting the existing test patterns to demonstrate that the test escape is covered; this may involve adding functional patterns to the patterns provided by the ATPG tool for the SoC DFT architecture.

## 21.2 "Shmoo" of Performance Dropout Versus Frequency

In addition to functional product testing, another bring-up task is evaluation of the performance of the prototype part distribution. The SoC design will have been subjected to static timing analysis at several "slow" process corners related to performance-limiting device and interconnect variations during fabrication. Path timing will have been closed at those corners to a target sigma of the process statistical distribution.

A sweep of the SoC clock frequency is applied to the parts during bring-up to evaluate correct functionality versus frequency. The cumulative distribution of pass/fail versus frequency in silicon is informative, as it should confirm the statistical variation design models. This distribution also offers insights into the percentage of parts that may be available in various performance bins, operating at a frequency higher than the baseline spec, as illustrated in Figure 21.1.

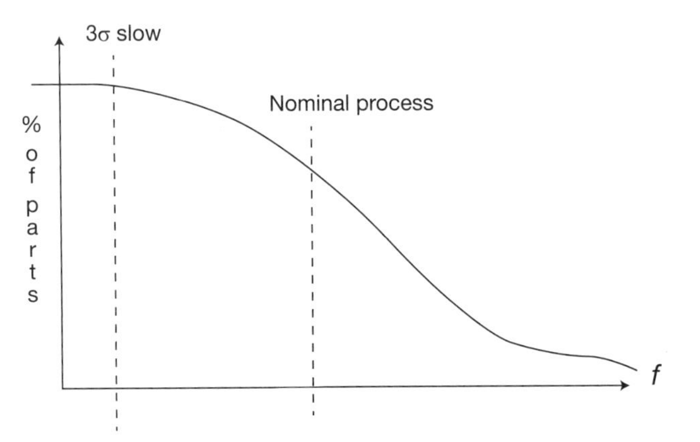


**Figure 21.1** Illustration of the performance of pass/fail parts as a function of operating clock frequency.

Bring-up may also involve a two-dimensional sweep to evaluate the functionality of a part population over ranges of applied voltage and frequency. An example of a two-dimensional pass/fail graph is provided in Figure 21.2; this plot format is familiarly known as a *shmoo plot*. Note that a shmoo plot is applicable to testing a single part, although it is most commonly used to represent a larger sample of the part population. It is also common to normalize the scale of the voltage and frequency axes, where 1.0 is the nominal value. Note that the voltage range on the shmoo extends beyond the published (VDD + n%) tolerance range specification.

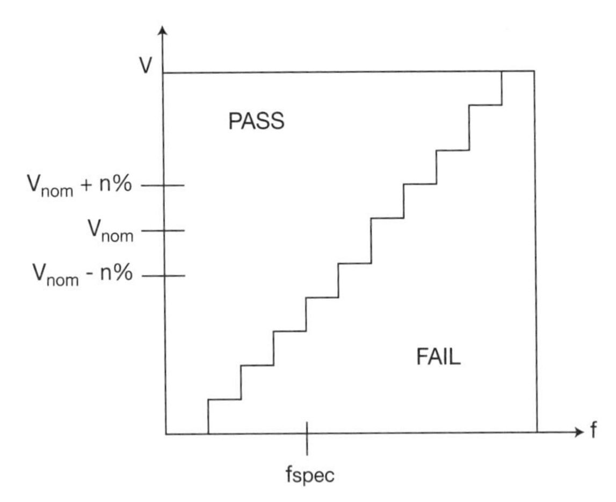


**Figure 21.2** Illustration of a two-dimensional shmoo of a sample of the part population versus applied voltage and operating clock frequency.

The bring-up data may be used as part of a dynamic voltage and frequency scaling (DVFS) product strategy, in which the SoC may support performance boost or (power-saving) throttle operating modes, with VDD changes coming from a voltage regulator.

There are some limitations with the traditional bring-up performance characterization methods. A statistically significant volume of parts needs to be measured to provide confidence in the binning and DVFS product strategies. The number of distinct power and clock domains on current SoC designs is large (and growing). For example, bring-up evaluation of large SRAM arrays may focus on a specific operating VDD_min for the arrays, using a distinct voltage domain from other IP. As a result, quickly and concisely capturing shmoo data for multiple domains during bring-up is becoming more difficult.

The specific circuit methods used to measure shmoo data also present an engineering challenge. Traditionally, the statistical fabrication parameters for an individual bring-up part would be adequately represented by a single, distinct measurement circuit added to the SoC design. A performance-sense ring oscillator (PSRO) circuit would be integrated, and the oscillator frequency would be measured as part of bring-up testing. (Reference 21.1 provides a description of methods to efficiently implement a PSRO frequency counter/decoder.) The *tracking* of fabrication variation across the die implied that the oscillator frequency would be representative of the overall SoC performance. For advanced process nodes, the tracking of circuit and interconnect parameter variations is more localized, and the correlation of PSRO data to overall performance diminishes. Designs began to incorporate multiple PSRO macros placed throughout the die. Still, these stand-alone circuit measurements are difficult to interpret for bring-up performance characterization. Instead, specific test programs are developed by the product bring-up team to exercise critical timing paths for performance shmoo data generation.

The bring-up team may be asked to delve further into the diagnosis of the shmoo data to determine specific paths that are the root causes of performance failures; hopefully, there is a strong correlation between the measured performance paths and the static timing analysis path reports. The SoC project plan may allocate design resources to focus on optimizing these performance-limiting paths in subsequent tapeouts. This optimization phase may be prior to initial production ramp, or it may be scheduled for a subsequent product release. Another tapeout release after initial volume sales may be appropriate to sustain the revenue and market share; this performance specification improvement is commonly known as a *mid-life kicker*.

The bring-up results from the initial tapeout identify functional errors and electrical issues to be addressed as part of preparation for the next tapeout iteration. The tapeout review team determines which proposed fixes should be accepted, and it merges the design change requirements with the previously deferred items. A key consideration is the ability to identify whether the necessary design ECO updates can be limited to BEOL mask layer revisions and existing spare cells or whether an all-layer tapeout release is required (along with the corresponding cost and schedule impacts).

## 21.3 Product Qualification

### 21.3.1 High-Temperature Operating Life (HTOL) Stress Testing

A separate activity that occurs concurrently to the prototype bring-up effort is initiating product qualification testing. A statistically significant sample of good parts from multiple wafer lots is subjected to high-temperature operating life (HTOL) stress testing; essentially, HTOL qualification is an extension to burn-in infant defect screening. The goal is to measure data related to the failure rate of the part. To accelerate the lifetime aging and parameter drift mechanisms, an applied voltage greater than the specification for $VDD_{max}$ (= $VDD_{nom}$ + n% tolerance) may also be part of the HTOL qualification procedure. In addition, the qualification test fixture could be developed in support of:

- **Static inputs**—Fixed chip input voltage signals
- **Dynamic inputs**—Patterns applied during the qualification stress testing, where the pattern set provides high internal net switching activity
- **Dynamic inputs with functional monitoring**—Patterns applied throughout the stress hours, where the chip output response is measured and compared to expected values

The qualification engineering cost is certainly higher for dynamic stress testing than for application of static inputs, because more sophisticated test chambers are required. For dynamic stress testing, the pattern application rate is greatly reduced compared to the functional part specification. Circuit timing paths are slower at the elevated temperatures.

There are specific HTOL qualification targets for different applications markets, such as consumer, medical, automotive (engine compartment and interior), and aerospace. Several industry standards organizations participate in establishing the qualification test criteria for integrated circuits, by application area, including:

- Joint Electron Device Engineering Council (JEDEC), specifically JESD47I
- Automotive Electronics Council (AEC), specifically AEC-Q100
- United States Department of Defense, specifically MIL-STD-883

These standards include definitions of the test sample size, the total stress duration, and the intermediate intervals within the total at which parts are withdrawn from the chamber and retested. The acceptable number of failing parts for each intermediate and final test is also specified. As might be expected, that number is explicitly set to zero for any high-reliability application.

The HTOL qualification test involves two temperatures: the ambient temperature of the chamber and the local (maximum) device junction temperature. The industry standard specifications refer to the operating junction temperature, necessitating a calculation of the appropriate external environment ($T_{ambient}$, $VDD_{qual}$) to achieve the target junction temperature, $T_j$, with estimates for the internal power dissipation for the qualification patterns. This calculation is more straightforward for static inputs. For dynamic patterns, the potentially wide variation in switching activity for different functional blocks may result in a significant thermal gradient across the die and some variation in $T_{junction}$. Ideally, the pattern set will have high switching activity across all SoC blocks to provide a comprehensive stress test.

Although less frequently employed than HTOL, a low-temperature operating life (LTOL) qualification stress test is also defined.

In addition to HTOL stress testing on devices and interconnects, product qualification includes tests focused on investigating other packaged part failure mechanisms.

### 21.3.2 Thermal Cycling

Parts may demonstrate structural reliability issues related to the mismatches in the *coefficient of thermal expansion* (*CTE*) between the die, die attach, and encapsulation materials. The chamber environment is cycled over time between high and cold temperatures, potentially with an applied voltage to the parts (e.g., multiple thermal excursions between extremes of −55°C and 125°C at ~10°C per minute; JEDEC JESD22-A104/A105; MIL-STD833, Methods 1007/1010). In additional to electrical retesting of the parts after *thermal shock* stress, a detailed visual inspection of the packages is required to search for external cracks.

A fatigue issue arising from CTE differences may result in internal cracking of a die-to-package pin connection and a corresponding I/O (logic or parametric) test failure. In addition, the CTE stress during cycling may propagate away from the die/package material interfaces, resulting in *delamination* of the top metal interconnect and (low-κ, structurally weak) dielectric layers below the die surface. Electrical retesting may then present pervasive failures that would be difficult to diagnose without detailed electron microscopy analysis.

### 21.3.3 Highly Accelerated Temperature/Humidity Stress Test (HAST)

To investigate the susceptibility to corrosion-related failures, parts are placed in a chamber with both elevated temperature and high relative humidity (e.g., $T_{ambient}$ = 105°C − 145°C, RH = 85%; JEDEC JESD22-A110/A118). To further accelerate the water vapor permeability, the chamber would typically be at elevated air pressure as well (>> 1 atm). HAST is also known as the *pressure cooker test* (*PCT*). The parts may or may not receive an applied voltage. HAST is unique in that the die power dissipation would typically reduce the moisture influx rate. However, an applied voltage that is cycled could provide additional acceleration of corrosion fails, allowing moisture condensation during periods when die power dissipation is zero.

### 21.3.4 Part Sampling for Qualification

The qualification effort includes electrostatic discharge (ESD) and latchup integrity testing of I/O circuits assembled in the specific package, as described in Sections 16.2 and 16.3.

Recall from earlier in this section that qualification uses "good" parts; that is, parts that have passed the production test pattern set are sampled, stressed, and retested. However, the SoC project manager may be faced with a dilemma. Qualification testing requires hundreds of hours and considerable costs. Ideally, if a prototype tapeout version is planned for volume production ramp, qualification is done concurrently with bring-up tasks. However, a critical decision point is reached if bring-up identifies a bug that necessitates a tapeout respin, with several key questions to address:

- Should the qualification activity be halted to wait for the updated revision prototype hardware? The costs incurred to that point would be lost, and additional budgeting (and time) for a new full qualification would be required.
- Should the qualification activity continue to completion? Would the qualification results be a gate to the subsequent tapeout revision release?
- Most significantly, if the full qualification passes on the initial version, what results are still applicable to a tapeout revision? Is a "partial" qualification investment (in time and cost) suitable for the silicon update, or is a full requalification required?

The SoC project manager may present a justification to the sustaining product engineering (PE) team that a logic ECO implemented in a tapeout revision does not invalidate the qualification results from the existing silicon die and package technology. Conservatively, the PE team (and end customers) may require full qualification on the final production version.

The SoC project manager and PE team must also address the question of qualification for silicon dies sourced from multiple fabs. Typically, the foundry uses different fabrication lines for (high-throughput, low-volume) prototype wafer lots than for production volume. Although the prototype qualification involves selecting a sample of parts from multiple wafer lots to incorporate lot-to-lot variation, the wafers are likely to be from a single fab. Also, once in production, the foundry may wish to move/re-balance orders between fabs supporting the same process node. The foundry provides data demonstrating that the fabs are all equivalently qualified to the foundry's criteria; the PE team needs to review that data to decide what (full or partial) requalification is required for multiple fab sourcing.

## 21.4 Summary

The bring-up and qualification phase of an SoC design project is of critical priority. The fact that these activities are pursued concurrently with the set of planned tapeout revision tasks amplifies its importance. Any issues identified during bring-up or qualification need to be urgently diagnosed, decisions need to be made promptly about the best corrective actions, and ECO updates need to be developed quickly to avoid impacting the (final) tapeout schedule.

Note that qualification failures are typically much more far-reaching and disruptive than functional validation escapes or test escapes. Any issues with package substrates, encapsulation materials, die-to-package pin connection metallurgy, and so on require engineering decisions involving the entire SoC product management organization, working with foundry and OSAT customer support. Referring again to Figure 20.5, qualification issues could result in significant product release delays and financial impact.

Although not described in this text, mechanical CAD (MCAD) software tool vendors provide algorithms to support analysis of material stress and elastic flow for complex three-dimensional geometries when subjected to mechanical forces and thermal gradients. The input to these algorithms provides the geometric model description, along with material strength and surface properties. These tools are indispensable for early investigation of potential structural issues between die, underfill, and package substrate. The MCAD structural analysis, the foundry's and OSAT's manufacturing qualification efforts, and the verification of ESD and latchup layouts by the SoC team will hopefully result in zero qualification failures, and thus no additional product delays, and no lost revenue opportunities.

## Reference

[1] Zick, K., and Hayes, J., "Low-Cost Sensing with Ring Oscillator Arrays for Healthier Reconfigurable Systems," *ACM Transactions on Reconfigurable Technology and Systems (TRETS)*, Volume 5, Issue 1, March 2012, pp. 1–26.

## Further Research

### Burn-in

Describe the capabilities (and capacities) of commercial burn-in systems.

### HTOL Qualification

Describe and contrast the procedures for HTOL qualification for the different standards described in Section 21.3.1 (JEDEC, AEC, and MIL-STD).

### SEM and TEM (Advanced)

Research and describe the features of scanning electron microscopy (SEM) and transmission electron microscopy (TEM), as applied to semiconductor failure analysis. Describe the characteristics of sample preparation.

# EPILOGUE

This book focused on SoC design methodologies and project management criteria for the advanced semiconductor fabrication processes available at this time. The design enablement provided by the foundry PDK is fundamental to supporting the design flows that comprise the overall methodology. The integration of intellectual property (IP) available from external vendors accelerates the SoC design and verification schedule and helps design teams focus on product differentiation features.

The continuing emphasis on additional product features, while concurrently addressing aggressive power, performance, and area (PPA) targets will no doubt result in major advances in the years to come. Some of these technology enhancements are already transitioning from fundamental research to design implementation. Alas, such emerging technologies are beyond the scope of this text. However, they may (quickly) become required expertise for microelectronics engineers and students. Briefly, the following are some of the areas deserving additional study:

- **Advanced packaging technologies**—A number of *2.5D and 3D* package offerings are available for integrating multiple dies into a single package; specifically, SoCs with (high-density) memory combinations provide attractive system performance gains over traditional single-die implementations located across a printed circuit board. The technologies for providing dense interconnects between multiple dies are rapidly evolving, especially where a vertical (3D) die stacking topology is used. Research topics of interest include HBM packaging, routing (and modeling) of interconnects through interposers, *wafer-level fanout* packaging, and the data required for an advanced package assembly design kit (ADK).

- **Analog/mixed-signal IP modeling**—The validation of AMS IP integration into an SoC design has traditionally involved (judicious) checking of the interfaces to SoC logic and memory blocks but with limited modeling capabilities for the analog circuit-level behavior within the overall SoC design. Functional validation of the full chip model may "stub out" the AMS block or utilize (manually written) simplified models. The microelectronics industry is focusing on defining behavioral modeling semantics for AMS IP to enable a more comprehensive validation methodology—especially for AMS IP connected to SoC I/O pins for external (SerDes or parallel bus) interfaces. Research topics of interest include the IBIS-AMI modeling standard and continuous-time modeling semantic features added to hardware description language standards (i.e., Verilog-AMS, VHDL-AMS).
- **Emerging memory IP technologies**—This book primarily discusses the integration of CMOS SRAM IP arrays on the SoC. For several process node generations, foundries have also offered non-volatile RAM (NVRAM) IP as an additional process option, usually as either one-time programmable (OTP) or electrically erasable programmable read-only memory (EEPROM). An emerging process option with more straightforward design integration and high density is the availability of magnetoresistive RAM (MRAM). A research topic of interest is spin-transfer torque magnetoresistive RAM (STT-MRAM).
- **New device materials for interconnects and dielectrics**—In the increasing effort to improve PPA, technology researchers are seeking improved fabrication material and process steps for devices, wires, vias, and dielectrics. Research topics of interest include the following:
    - **Dielectrics**—High-κ (composite) dielectric material layers for the device gate, providing high gate electric fields and robust dielectric breakdown reliability; also, air-gap dielectric regions between adjacent interconnects

  - **Interconnects/contacts/vias**—Cladding material layers for Cu interconnects, new metals for contacts/vias (Co, Ru), and lead-free die pad-to-package pin attach metallurgy at an aggressive layout pitch (e.g., small-diameter LF bumps, Cu pillars)
  - **Devices**—New stress/strain materials for carrier mobility enhancement, Ge and SiGe source/drain node composition, and nanowires and gate-all-around (GAA) device topologies (an evolution of FinFET device fabrication)
- **Integrated Si photonics structures**—A unique product opportunity is emerging based on the integration of silicon photonics structures. Optical fiber interconnects for long-distance data communication is a well-established technology. With the growing demand for higher bandwidth in short-range signaling, significant silicon process development is focused on the fabrication of on-die optical waveguides, optical structures for mixers and modulators, optoelectronic conversion circuits, etc. Research topics of interest in this field include silicon photonics waveguide cross-sections, photodiodes for optoelectronic conversion, (continuous wave) laser source technology integrated on-die, package assembly of optical fiber to the die, photonic structure simulation models, and, especially, EDA tool support required for the physical design, DRC, LVS, and model extraction of photonic cell layouts. (The modulation of the waveguide response by the application of surrounding electric fields requires specific attention to model accuracy.)
- **New behavioral modeling languages (and synthesis support)**—Section 4.1 briefly describes the use of digital modeling languages at a behavioral level, employing a higher degree of functional abstraction compared to the register-transfer level (RTL) coding style commonly used. Current hardware description language (HDL) standards (i.e., VHDL, SystemVerilog) include behavioral semantics as part of the language definition. However, new languages (and the related validation and synthesis tools) are being deployed to enable improved design representation efficiency and greater validation throughput. Specific IP algorithmic functionality and transaction-based behavior are well suited for high-level representation and validation. Subsequently, synthesis to an RTL model enables integration with other SoC IP and existing flows. Research topics of interest include SystemC modeling, MATLAB modeling, and EDA tool synthesis of these representations.

- **New EDA tools, flows, and methodologies**—As fabrication process nodes have scaled, new physical design, layout verification, electrical modeling and analysis, and test methods have been required. EDA vendors have collaborated closely with foundries to provide the corresponding tools. For example, the pace of technology scaling has far surpassed the lithographic wavelength used for mask exposure, necessitating mask layout data decomposition and multipatterning. These requirements have had a pervasive impact on the entire set of physical design and electrical analysis flows. The transition from 193nm (with immersion) wavelength to an extreme ultraviolet (EUV) wavelength exposure will no doubt bring additional requirements.

One trend will certainly persist: The magnitude of process variation across a die will continue to warrant growing investment in statistical-based models and analysis flows. The markets for advanced SoC designs should expect a high-sigma confidence level in electrical analysis closure. An SoC project management team must seek to improve the accuracy of circuit-limited yield projections as the development and NRE costs for advanced technology designs increase. The resolution and subsequent etch variation on the mask layers using EUV exposure will be improved over 193i lithography. Conversely, the transition to quad patterning from double patterning decomposition will increase variation.

The markets for advanced SoC designs will also insist upon a very high confidence level in fault defect detection, with "no test escapes" at product release. EDA tools for DFT architecture insertion and test pattern generation will be required to address this aggressive goal.

EDA tools will also certainly evolve to provide necessary productivity gains, as the volume of logic gates, memory, I/Os, and (especially) extracted parasitic detail grows with ongoing process scaling. Currently, for efficiency, RTL logic synthesis uses physical design estimates for optimization. The accuracy of those PD estimates, combined with the timing, noise, and power model algorithms in the synthesis tool, is critical to achieving design closure through the subsequent tapeout analysis flows. EDA vendors must strive to improve the correspondence between synthesis, physical implementation, and electrical analysis to expedite closure.

A trend that will accompany the continued growth in dataset size is the increased adoption of parallel and distributed computation by EDA tools. Already a staple for large-model parasitic extraction and physical layout design verification tools, multiprocessing algorithms will be introduced across a wider set of EDA tools in the SoC methodology. The CAD flow development and IT resource allocation teams must work closely with the SoC methodology and project management teams to ensure that these improved computational efficiencies are enabled.

## Summary

There are many challenges to the development of a comprehensive SoC design methodology that satisfies a diverse set of project goals and resource constraints. The preparation and testing of methodology flows is a crucial facet of the planning phase for a new design project. The SoC methodology team needs to ensure that design enablement deliverables are coordinated with many different engineering groups, including the following:

- **SoC architects**—HDL design guidelines; DFT architecture
- **SoC designers**—Development and management of block design constraints for synthesis and analysis flows; block PPA targets and budgets
- **Functional validation**—Simulation platforms; testbench development; coverage methods and measures; opportunities for formal model property checking
- **Test pattern development**—Fault models; coverage measures; DFT and BIST logic insertion
- **Physical implementation**—Floorplanning goals; hierarchical design management; preplacement and prerouting data; timing, noise, and power constraints; implementation of ECOs
- **Foundry interface team**—PDK releases; PDK and techfile model content; defect density targets and yield estimates
- **(External) IP vendors**—Models for hard IP and library cells; PVT corner support; IP qualification status
- **Packaging and assembly interface team**—OSAT data release requirements; package design rule checking; package electrical models
- **Bring-up and qualification team**—Bring-up test programs; shmoo testing plan; qualification and burn-in stress plan; FIT targets
- **CAD and IT teams**—Flow development; flow results reporting; methodology and flow dependency status tracking; configuration and version management tools/policies; foundry PDK installation and testing; EDA vendor tool technical evaluation, installation, and license management; project bug tracking tools/policies; IT resources, including job queuing and job dispatch priority policies, "large server" resources, and reservation policies

Although the vast majority of the SoC methodology coordination occurs during the initial project planning phase, the methodology team must closely monitor all aspects of the project execution phase, as well. Issues arising with tools and their corresponding flows need to be reviewed, testcases need to be developed, and resolution priorities (with "quick fix" tool software release schedules) need to be established with the EDA vendor. Neither the SoC project manager nor the EDA tool vendor is likely to be too keen on sending a full chip model to the EDA vendor for debugging; the methodology and CAD teams must work to provide more focused testcases that still demonstrate the tool issue. Flows that are not aligning with runtime and/or resource estimates need to be reviewed to determine whether the throughput can be improved or project resource plans should be adjusted accordingly. Flows may be required to include performance profiling support to identify bottlenecks (which may also be communicated to the EDA tool vendor or foundry PDK customer support).

During the SoC project execution schedule, new tool features may be released by the EDA vendor (i.e., a major software release, as opposed to an interim release to fix bugs). The methodology and CAD teams need to assess whether to move the SoC project to this new release as the required version for tapeout flows. This assessment requires the following steps:

1. Executing a test plan on the new release
2. Measuring the change in the *quality of results* (*QoR*) for synthesis and physical implementation tool updates (e.g., gate count/area, path timing results, routability)
3. Investigating the impact on current project data (especially focused on any inconsistencies in analysis flow results compared to current data)
4. Determining the relative value of the new features compared to the project impact

To be sure, a new major tool release occurring at or near the project tapeout phase would almost certainly be deferred to minimize the disruption to the tapeout preparation steps.

Being part of an SoC design methodology team is definitely challenging, but it also provides a unique experience, with exposure to all facets of chip design and project management. Hopefully, this book has provided insights into the scope and responsibilities of the methodology team and has inspired some of the readers to consider this role in the evolution of their engineering career.

# INDEX

## B

## C

## F

## I

## J



## K



## L

## N

## O



## P

## Q



## R

## S

## T

## U



## V



## W



## X-Y-Z